# Springer INdAM Series

Volume 13

More information about this series at http://www.springer.com/series/10283

Giovanna Citti • Maria Manfredini •
Daniele Morbidelli • Sergio Polidoro •
Francesco Uguzzoni
Editors

# Geometric Methods in PDE's

Springer

*Editors*
Giovanna Citti
Dipartimento di Matematica
Università di Bologna
Bologna, Italy

Maria Manfredini
Dipartimento di Matematica
Università di Bologna
Bologna, Italy

Daniele Morbidelli
Dipartimento di Matematica
Università di Bologna
Bologna, Italy

Sergio Polidoro
Dipartimento FIM
Università di Modena e Reggio Emilia
Modena, Italy

Francesco Uguzzoni
Dipartimento di Matematica
Università di Bologna
Bologna, Italy

ISSN 2281-518X ISSN 2281-5198 (electronic)
Springer INdAM Series
ISBN 978-3-319-02665-7 ISBN 978-3-319-02666-4 (eBook)
DOI 10.1007/978-3-319-02666-4

Library of Congress Control Number: 2015953212

Springer Cham Heidelberg New York Dordrecht London


Printed on acid-free paper

Springer International Publishing AG Switzerland is part of Springer Science+Business Media (www.springer.com)

*To our mentor Ermanno Lanconelli,*
*with profound gratitude*

# Preface

This volume celebrates the seventieth birthday of Professor Ermanno Lanconelli, whose scientific activity has strongly influenced the recent research on partial differential equations (PDEs) with non-negative characteristic form, and on the related potential theory. Beyond his distinguished scientific contributions, Ermanno Lanconelli has also been responsible for forming a prestigious school of mathematicians who share with him his infectious love for Mathematics.

The first notable contribution in Ermanno's scientific activity was a representation formula on the level sets of the fundamental solution of the heat equation, inspired by the work of Bruno Pini. This technique, until then used only for harmonic functions, has since been extended to increasingly large classes of equations, including the totally degenerate ones.

The scientific thought of Ermanno Lanconelli reached its full maturity with his works on totally degenerate equations. In the early 1980s, he introduced an original geometric approach for study of solutions to Grushin-type equations. Thereafter, he investigated the link between the geometric properties of the vector fields and the fundamental solutions of the associated second order operators in a long series of papers, culminating in a monograph, which is now considered one of the foundational references for potential theory in this setting.

He also addressed problems related to non-linear PDEs and proved a fundamental result for curvature-type equations, which opened up a new direction in the study of differential equations with non-linearity in the vector fields.

As a mathematician, Lanconelli has been constantly motivated by a strong desire to develop unifying techniques in the analysis of problems related to differential equations that classically were approached with separate and independent methods.

This volume contains 18 contributions that cover a wide range of topics that characterize Ermanno's scientific production. It brings together a selection of invited contributions from the main speakers at the conference "Geometric methods in PDEs: Indam Meeting on the occasion of the 70th birthday of Ermanno Lanconelli" and presents a wide cross-section of the most recent contributions on linear and non-linear differential equations and also on geometric problems that give rise to differential equations.

The first group of contributions in the volume deals with various kinds of functional inequalities: Friedrichs-type commutator lemmas, sharp inequalities of Hardy and Moser–Trudinger types, and Lusin theorems for *BV* functions.

Several contributions focus on the regularity theory of linear PDEs. They touch on Harnack-type estimates for equations associated with harmonic maps, subelliptic Fefferman–Phong type inequalities, estimates for parabolic equations involving Ornstein–Uhlenbeck terms, and the problem of existence and regularity of a fundamental solution for sum of squares of vector fields.

A third group of contributions deals with non-linear PDEs. Existence and multiplicity results for non-local eigenvalue problems are established; uniqueness problems for subelliptic semilinear and quasilinear equations are studied; existence and non-existence results for differential inequalities in Carnot groups are given; and gradient estimates with rigidity results for parabolic Modica-type PDEs are proven.

Some other contributions in the volume are concerned with fully non-linear PDEs of elliptic type, focusing on local and global gradient estimates for non-negative solutions and $C^{1,\gamma}$ regularity estimates for equations with sublinear first-order terms.

Also included are contributions concerning, first, the existence of solutions for a model to design reflectors and, secondly, some div-curl inequalities in Carnot groups.

Finally, the volume includes two surveys, the first of which is on free boundary problems. The second has been written by Ermanno Lanconelli's former students as a tribute to his career and to thank him for the guidance that he has provided throughout their research activities.

Beyond the scientific contents mentioned above, Andrea Bonfiglioli, Giovanna Citti, Giovanni Cupini, Maria Manfredini, Annamaria Montanari, Daniele Morbidelli, Andrea Pascucci, Sergio Polidoro, and Francesco Uguzzoni conceived this volume in order to offer researchers who have enjoyed collaborating with Ermanno the opportunity to share their scientific experiences.

The Editors

# Contents

# Contributors

**Annalisa Baldi** Dipartimento di Matematica, Università di Bologna, Bologna, Italy

**Agnid Banerjee** University of California, Irvine, CA, USA

**Gerassimos Barbatis** Department of Mathematics, University of Athens, Athens, Greece

**Isabeau Birindelli** Dipartimento di Matematica G. Castelnuovo, Sapienza Università di Roma, Roma, Italy

**Andrea Bonfiglioli** Dipartimento di Matematica, Università di Bologna, Bologna, Italy

**Marco Bramanti** Dipartimento di Matematica, Politecnico di Milano, Milano, Italy

**Luca Brandolini** Dipartimento di Ingegneria, Università degli studi di Bergamo, Dalmine (BG), Italy

**Italo Capuzzo Dolcetta** Dipartimento di Matematica, Sapienza Università di Roma, Roma, Italy

Istituto Nazionale di Alta Matematica "F. Severi", Roma, Italy

**Giovanna Citti** Dipartimento di Matematica, Università di Bologna, Bologna, Italy

**Giovanni Cupini** Dipartimento di Matematica, Università di Bologna, Bologna, Italy

**Francoise Demengel** Laboratoire d'Analyse et Géométrie, Université de Cergy Pontoise, Paris, France

**Daniela De Silva** Department of Mathematics, Barnard College, Columbia University, New York, NY, USA

**Lorenzo D'Ambrosio** Dipartimento di Matematica, Università degli Studi di Bari, Bari, Italy

**Sorin Dragomir** Dipartimento di Matematica, Informatica ed Economia, Università degli Studi della Basilicata, Potenza, Italy

**Giuseppe Di Fazio** Dipartimento di Matematica e Informatica, Università di Catania, Catania, Italy

**Maria Stella Fanciullo** Dipartimento di Matematica e Informatica, Università di Catania, Catania, Italy

**Fausto Ferrari** Dipartimento di Matematica, Università di Bologna, Bologna, Italy

**Bruno Franchi** Dipartimento di Matematica, Università di Bologna, Bologna, Italy

**Nicola Garofalo** Università di Padova, Padova, Italy

**Cristian E. Gutiérrez** Department of Mathematics, Temple University, Philadelphia, PA, USA

**Jorge Hounie** Departamento de Matemática, Universidade Federal de São Carlos, São Carlos, Brasil

**Nguyen Lam** Department of Mathematics, Wayne State University, Detroit, MI, USA

**Guozhen Lu** Department of Mathematics, Wayne State University, Detroit, MI, USA

**Marco Magliaro** Dipartimento di Matematica Informatica ed Economia, Università degli Studi della Basilicata, Potenza, Italy

**Maria Manfredini** Dipartimento di Matematica, Università di Bologna, Bologna, Italy

**Annamaria Montanari** Dipartimento di Matematica, Università di Bologna, Bologna, Italy

**Daniele Morbidelli** Dipartimento di Matematica, Università di Bologna, Bologna, Italy

**Enzo Mitidieri** Dipartimento di Matematica e Geoscienze, Università degli Studi di Trieste, Trieste, Italy

**Andrea Pascucci** Dipartimento di Matematica, Università di Bologna, Bologna, Italy

**Sergio Polidoro** Dipartimento FIM, Università di Modena e Reggio Emilia, Modena, Italy

**Enrico Priola** Dipartimento di Matematica "Giuseppe Peano", Università di Torino, Torino, Italy

**Patrizia Pucci** Dipartimento di Matematica e Informatica, Università degli Studi di Perugia, Perugia, Italy

**Ahmad Sabra** Department of Mathematics, Temple University, Philadelphia, PA, USA

**Sara Saldi** Dipartimento di Matematica e Informatica "U. Dini", Università degli Studi di Firenze, Firenze, Italy

Dipartimento di Matematica e Informatica, Università degli Studi di Perugia, Perugia, Italy

**Sandro Salsa** Dipartimento di Matematica, Politecnico di Milano, Milano, Italy

**András Telcs** Department of Quantitative Methods, Faculty of Economics, University of Pannonia, Veszprém, Hungary

Department of Computer Science and Information Theory, Budapest University of Technology and Economic, Budapest, Hungary

**Achilles Tertikas** Department of Mathematics and Applied Mathematics, University of Crete, Heraklion, Greece

Institute of Applied and Computational Mathematics, FORTH, Heraklion, Greece

**Francesca Tripaldi** Department of Mathematics, King's College London, Strand, London, UK

**Francesco Uguzzoni** Dipartimento di Matematica, Università di Bologna, Bologna, Italy

**Vincenzo Vespri** Dipartimento di Matematica ed Informatica Ulisse Dini, Universitá degli Studi di Firenze, Firenze, Italy

**Pietro Zamboni** Dipartimento di Matematica e Informatica, Università di Catania, Catania, Italy

# On Friedrichs Commutators Lemma for Hardy Spaces and Applications

**Jorge Hounie**

*Dedicated to Ermanno Lanconelli on the occasion of his 70th birthday*

**Abstract** We extend the classical Friedrichs commutator lemma—known for $L^p$ norms—to the case of local Hardy spaces $h^p(\mathbb{R}^N)$, $N/(N+1) < p \leq 1$, and apply the result to the study of the regularity in Sobolev-Hardy spaces of solutions of elliptic systems of vector fields with non smooth coefficients.

**Keywords** Elliptic systems • Friedrichs lemma • Hardy spaces

**AMS Classification:** 35J46, 46E35

## 1 Introduction

Let $h^p(\mathbb{R}^N)$ be the local Hardy space [2] for some $0 < p < \infty$ and consider a distribution $f(x) \in h^p(\mathbb{R}^N)$. Given a test function $\phi \in C_c^\infty(\mathbb{R}^N)$ with $\int \phi\, dx = 1$, let us denote the Friedrichs approximation of the identity by

$$J_\varepsilon f(x) = \phi_\varepsilon * f(x), \qquad \phi_\varepsilon(x) = \frac{1}{\varepsilon^N}\phi(x/\varepsilon), \quad \varepsilon > 0.$$

It is known that

$$\|J_\varepsilon f\|_{h^p} \leq C\,\|f\|_{h^p}, \qquad \lim_{\varepsilon\to 0} \|J_\varepsilon f - f\|_{h^p} = 0,$$

with $C > 0$ independent of $f \in h^p(\mathbb{R}^N)$.

For $p > 1$ we have $h^p(\mathbb{R}^N) = L^p(\mathbb{R}^N)$ and the classical Friedrichs lemma states that [4, p. 9] if $b(x)$ is a Lipschitz function and $f \in L^p(\mathbb{R}^N)$, $1 \leq p < \infty$ and $j = 1, \ldots, N$, then the commutators $[J_\varepsilon, b\partial_j]f(x) = \phi_\varepsilon * (b\partial_j f)(x) - b(x)\partial_j(\phi_\varepsilon * f)(x)$,

---

J. Hounie (✉)
Departamento de Matemática, Universidade Federal de São Carlos, São Carlos, Brasil
e-mail: hounie@dm.ufscar.br



G. Citti et al. (eds.), *Geometric Methods in PDE's*, Springer INdAM Series 13,
DOI 10.1007/978-3-319-02666-4_1

$\varepsilon > 0$, satisfy

$$\|[J_\varepsilon, b\partial_j]f\|_{L^p} \leq C\|b\|_{Lip}\|f\|_{L^p}, \qquad \lim_{\varepsilon\to 0}\|[J_\varepsilon, b\partial_j]f\|_{L^p} = 0, \tag{FL}$$

with $C > 0$ independent of $f \in L^p(\mathbb{R}^N)$, $b \in Lip(\mathbb{R}^N)$.

The main point about (FL) is that although the left hand side contains a derivative of $f$, the right hand side does not so one derivative is gained in the commutation. From the point o view of the calculus of pseudo-differential operators, $J_\varepsilon$ can be thought of as a pseudo-differential operator of order zero (uniformly with respect to $\varepsilon$) with symbol $\hat{\phi}(\varepsilon\xi)$ and, assuming $b(x)$ is smooth and bounded with bounded derivatives, $b\partial_j$ is a pseudo-differential of order zero. Hence, this fact is in agreement with the calculus of pseudo-differential operators, according to which the commutator of an operator of order zero with an operator of order one yields an operator of order zero. However, this calculus requires that $b(x)$ be smooth (or, at least that it possesses a large number of derivatives that increases with the dimension $N$ in order to grant that $[J_\varepsilon, b\partial_j]$ is bounded in $L^p(\mathbb{R}^N)$, furthermore, pseudo-differential operators of order zero are not bounded in $L^1(\mathbb{R}^N)$ in general). Friedrichs lemma is, in its original form, a frequent, useful and standard tool in the study of regularity properties for solutions of partial differential equations. For an extension valid for Hölder spaces see [6].

In this work we deal with the extension of (FL) below the threshold $p = 1$ within the framework of local Hardy spaces. In fact, we prove that (FL) holds for $N/(N+1) < p \leq 1$ as soon as we replace $L^p$ norms by $h^p$ "norms" and take the coefficient $b(x)$ in the Hölder space $\Lambda^{1+r}$ for any $r > N(p^{-1} - 1)$. In other words, we prove

$$\|[J_\varepsilon, b\partial_j]f\|_{h^p} \leq C\|b\|_{1+r}\|f\|_{h^p}, \qquad \lim_{\varepsilon\to 0}\|[J_\varepsilon, b\partial_j]f\|_{h^p} = 0, \tag{FL\#}$$

with $C > 0$ independent of $f \in h^p(\mathbb{R}^N)$, $b \in \Lambda^{1+r}(\mathbb{R}^N)$. This fact can be used as a tool to study the regularity of solutions of systems of vector fields in terms of Hardy-Sobolev "norms".

The paper is organized as follows. In Sect. 2 we present various basic facts about local Hardy spaces. In Sect. 3 we give sufficient regularity conditions for a continuous function $b(x)$ to be a multiplier in $h^p(\mathbb{R}^N)$ (i.e., $f(x) \in h^p(\mathbb{R}^N) \implies b(x)f(x) \in h^p(\mathbb{R}^N)$). In Sect. 4 we prove our version of Friedrichs lemma in local Hardy spaces and in Sect. 5 we present applications to the regularity of solutions of elliptic systems of complex vector fields with non smooth coefficients.

## 2 Some Background on Local Hardy Spaces

We recall how the localizable Hardy spaces $h^p(\mathbb{R}^N)$, introduced by Goldberg in [2], are defined. Fix, once for all, a radial nonnegative function $\varphi \in C_c^\infty(\mathbb{R}^N)$ supported in the unit ball with integral equal to 1. For $u \in \mathscr{S}'(\mathbb{R}^N)$ we define the *small maximal function* $m_\varphi u$ by

$$m_\varphi u(x) = \sup_{0<t<1} |(u * \varphi_t)(x)|$$

where $\varphi_t(x) = t^{-N}\varphi(x/t)$.

**Definition 2.1** Let $0 < p < \infty$. A tempered distribution $u \in \mathscr{S}'(\mathbb{R}^N)$ belongs to $h^p(\mathbb{R}^N)$ if and only if $m_\varphi u \in L^p(\mathbb{R}^N)$, i.e.,

$$\|u\|_{h^p} \doteq \|m_\varphi u\|_{L^p} < \infty.$$

For $p = \infty$, we set $h^\infty(\mathbb{R}^N) = L^\infty(\mathbb{R}^N)$.

The spaces $h^p(\mathbb{R}^N)$ are independent of the choice of the function $\varphi \in \mathscr{S}(\mathbb{R}^N)$ that is used to define $m_\varphi$ provided $\int_{\mathbb{R}^N} \varphi(x)\,dx \neq 0$. For $0 < p \leq 1$, the space $h^p(\mathbb{R}^N)$ is a complete metric space with the distance

$$d(u, v) = \|u - v\|_{h^p}^p, \quad u, v \in h^p(\mathbb{R}^N).$$

For $p = 1$, $\|u\|_{h^1}$ is a norm and $h^1(\mathbb{R}^N)$ is a normed space densely contained in $L^1(\mathbb{R}^N)$. For $p > 1$, $h^p(\mathbb{R}^N) = L^p(\mathbb{R}^N)$ and $\|u\|_{h^p}$ is a norm equivalent to the usual $L^p$ norm. Although $h^p(\mathbb{R}^N)$ is not locally convex for $0 < p < 1$ and $\|u\|_{h^p}$ is not truly a norm (it is a quasi-norm [9]), we will still refer to $\|u\|_{h^p}$ as the "norm" of $u$, as it is customary.

**Definition 2.2** Let $0 < r < 1$. A continuous function $f$ belongs to the homogeneous Hölder space $\dot{\Lambda}^r(\mathbb{R}^N)$ if there exist $c > 0$ such that

$$|f(x + h) - f(x)| \leq c|h|^r,$$

for every $x, h \in \mathbb{R}^N$. For $r = 1, f \in \dot{\Lambda}^1(\mathbb{R}^N)$ if there exist $c > 0$ such that

$$|f(x + h) + f(x - h) - 2f(x)| \leq c|h|^r,$$

and if $r = k + s$, $k = 1, 2, \ldots, 0 \leq s < 1, f \in \dot{\Lambda}^r(\mathbb{R}^N)$ if all derivatives $D^\alpha f \in \dot{\Lambda}^s(\mathbb{R}^N)$ for $|\alpha| \leq k$.

This is a locally convex topological vector space with the seminorm

$$|f|_{k+s} \doteq \sum_{|\alpha|\leq k} \sup_{\substack{x,h\in\mathbb{R}^N \\ h\neq 0}} \frac{|D^\alpha f(x + h) - D^\alpha f(x)|}{|h|^s}$$

modulo the subspace of those functions such that $|f|_r = 0$ which are the polynomials of degree $\leq m$ if $m$ is an integer such that $m - 1 < r \leq m$.

When $0 < p < 1$ the dual space of $h^p(\mathbb{R}^N)$ may be identified with the nonhomogeneous Hölder space $\Lambda^r(\mathbb{R}^N) \doteq \dot{\Lambda}^r(\mathbb{R}^N) \cap L^\infty(\mathbb{R}^N)$ for $r = N\left(\frac{1}{p} - 1\right)$ equipped with the norm $\|f\|_r = |f|_r + \|f\|_{L^\infty}$. Note that $N/(N + 1) < p < 1$ if and only if $0 < N\left(\frac{1}{p} - 1\right) < 1$. The dual of $h^1(\mathbb{R}^N)$ can be identified with the space $\mathrm{bmo}(\mathbb{R}^N)$ defined as the space of locally integrable functions $f$ which satisfy

$$\|f\|_{\mathrm{bmo}} \doteq \sup_{|Q|<1} \frac{1}{|Q|}\int_Q |f - f_Q| + \sup_{|Q|\geq 1} \frac{1}{|Q|}\int_Q |f| < \infty.$$

Here $Q$ is a cube in $\mathbb{R}^N$ with sides parallel to the axes and

$$f_Q \doteq \frac{1}{|Q|}\int_Q f(x)dx$$

where $|Q|$ is the Lebesgue measure of Q.

We now describe the atomic decomposition of $h^p(\mathbb{R}^N)$ [2, 7]. An $h^p(\mathbb{R}^N)$ atom is a bounded, compactly supported function $a(z)$ satisfying the following properties: there exists a cube $Q$ with sides parallel to the coordinate axes containing the support of $a$ such that

(1) $|a(z)| \leq |Q|^{-1/p}$, a.e., with $|Q|$ denoting the Lebesgue measure of $Q$.
(2) If $\|a\|_{L^\infty} > 1$, we further require that $\int z^\alpha a(z)\,dz = 0$, $\alpha \in \mathbb{N}^n$, $|\alpha| \leq N(p^{-1} - 1)$.

Any $f \in h^p$ can be written as an infinite linear combination of $h^p$-atoms, more precisely, there exist scalars $\lambda_j$ and $h^p$-atoms $a_j$ such that $\sum_j |\lambda_j|^p < \infty$ and the series $\sum_j \lambda_j a_j$ converges to $f$ both in $h^p$ and in $\mathscr{S}'$. Furthermore, $\|f\|_{h^p}^p \sim \inf \sum_j |\lambda_j|^p$, where the infimum is taken over all atomic representations. Another useful fact is that the atoms may be assumed to be smooth functions, in particular the inclusions $C_c^\infty(\mathbb{R}^N) \subset \mathscr{S}(\mathbb{R}^N) \subset h^p(\mathbb{R}^N)$ are dense. The atomic decomposition of $h^p(\mathbb{R}^N)$ is thus quite similar to the atomic decomposition of $H^p(\mathbb{R}^N)$ in terms of $H^p$-atoms [7], the difference being that the notion of $h^p$-atom is less restrictive than that of $H^p$-atom, as an $H^p$-atom $a$ must satisfy (1) and a stronger form of (2): moments are required to vanish regardless of the size of $\|a\|_{L^\infty}$.

Let $T : C_c^\infty(\mathbb{R}^N) \longrightarrow \mathscr{D}'(\mathbb{R}^N)$ be a linear weakly continuous operator in the sense that $\langle \phi_j, \psi\rangle \to \langle \phi_j, \psi\rangle \ \forall\, \psi \in C_c^\infty(\mathbb{R}^N) \implies \langle T\phi_j, \psi\rangle \to \langle T\phi_j, \psi\rangle \ \forall\, \psi \in C_c^\infty(\mathbb{R}^N)$).

**Proposition 2.1** *Given $0 < p \leq 1$, assume that for any smooth $h^p$-atom $a(x)$ we have $\|Ta\|_{h^p} \leq C$ for some fixed constant C. Then T can be extended as a bounded operator from $h^p(\mathbb{R}^N) \longrightarrow h^p(\mathbb{R}^N)$.*

*Proof* Indeed, if $f$ and $\psi \in C_c^\infty(\mathbb{R}^N)$ and $f = \sum_j \lambda_j a_j$ is a smooth atomic decomposition with $\|f\|_{h^p}^p \simeq \sum |\lambda_j|^p$, we have

$$\langle Tf, \psi\rangle = \sum_j \lambda_j \langle Ta_j, \psi\rangle$$

which for $\psi(x') = \varphi_t(x - x')$ gives $\langle Tf, \psi\rangle = Tf * \varphi_t(x)$ and we easily get

$$m_\varphi Tf(x) \leq \sum_j |\lambda_j|\, m_\varphi Ta_j(x).$$

Then

$$(m_\varphi Tf)^p(x) \leq \sum_j |\lambda_j|^p\, (m_\varphi Ta_j)^p(x).$$

Integrating in $x$ we obtain $\|Tf\|_{h^p} \leq C\|f\|_{h^p}$, $f \in C_c^\infty(\mathbb{R}^N)$, which allows the extension of $T$ to $h^p(\mathbb{R}^N)$ by density. □

## 3 Multipliers in $h^p(\mathbb{R}^N)$

Let $b(x)$ be a bounded measurable function on $\mathbb{R}^N$ and consider the multiplication operator $M_b f(x) = b(x)f(x)$, $f \in C_c^\infty(\mathbb{R}^N)$. It is clear that $M_b : C_c^\infty(\mathbb{R}^N) \longrightarrow \mathscr{D}'(\mathbb{R}^N)$ is a weakly continuous linear operator in the sense described in the previous section.

**Definition 3.1** We say that $b(x)$ is a multiplier in $h^p(\mathbb{R}^N)$ if the operator $M_b$ can be extended as a continuous linear operator $M_b : h^p(\mathbb{R}^N) \longrightarrow h^p(\mathbb{R}^N)$, i.e.,

$$\|M_b f\|_{h^p} \leq C\|f\|_{h^p}, \quad f \in C_c^\infty(\mathbb{R}^N), \tag{$*$}$$

for some fixed $C > 0$.

*Example 3.1* If $b(x) \in C^\infty(\mathbb{R}^N)$ and $D^\alpha b(x)$ is bounded for all $\alpha \in \mathbb{Z}_+^N$, then $M_b$ may be regarded as a pseudo-differential operator of order zero with symbol $b(x)$ in the Hörmander class $S^0_{1,0}$. Hence (see [2, 5] for the continuity o pseudo-differential on local Hardy spaces), $M_b$ is bounded in $h^p(\mathbb{R}^N)$ for all $p > 0$ and therefore $b(x)$ is a multiplier in $h^p(\mathbb{R}^N)$ for all $p > 0$.

Of course, for a fixed value of $p$, a function $b(x)$ does not need to possess an infinite number of bounded derivatives in order to grant that $M_b$ is a multiplier in $h^p(\mathbb{R}^N)$. The simplest situation occurs for $p > 1$ when $h^p(\mathbb{R}^N) = L^p(\mathbb{R}^N)$ so any measurable bounded function $b(x)$ is a multiplier in $h^p(\mathbb{R}^N)$, $p > 1$, a fact that is no longer true for $p \leq 1$. For instance, the function $b(x)$ given by $b(x) = 1$, $x \geq 0$, $b(x) = -1$, $x < 0$, is not a multiplier in $h^1(\mathbb{R})$. To see this, consider the interval $I_\rho = [-\rho, \rho]$, $0 < \rho < 1$, its characteristic function $\chi_\rho(x)$ and the function $f_\rho(x) = b(x)\chi_\rho(x)$. Then $f_\rho/2$ is an atom in $h^1(\mathbb{R})$ and in particular

$$\|f_\rho\|_{h^p} \leq C, \quad 0 < \rho < 1.$$

On the other hand, $b(x)f_\rho(x) = \chi_\rho(x) = |f_\rho(x)|$ and a simple computation shows that

$$m_\varphi \chi_\rho(x) \geq c_1 \min\left(\rho^{-1}, |x|^{-1}\right), \quad |x| \leq c_2,$$

for some constants $c_1, c_2 > 0$ independent of $\rho$. Integrating this inequality on $[-c_2, c_2]$ we see that for $0 < \rho < c_2$

$$\|M_b f_\rho\|_{h^1} \geq \int_{-c_2}^{c_2} m_\varphi \chi_\rho(x)\, dx \geq c_1(1 + |\log(c_2\rho)| \to \infty, \quad \rho \to 0.$$

By taking regularizations of the functions $f_\rho$ it is now easy to violate $(*)$ and show that $b(x)$ is not a multiplier in $h^1(\mathbb{R})$. Note that $b(x)f_\rho(x) = |f_\rho(x)|$ is in $L^1(\mathbb{R})$ uniformly in $0 < \rho < 1$ but does not belong uniformly to the class $L \log L$, a fact related to the $L \log L$ theorem of Stein [8].

In the last example $b(x)$ is not continuous and has a jump of size 2 at the origin, however, a refined construction in which the jump is avoided by modifying the graph of $b(x)$ close to the origin thanks to the introduction of a very steep straight segment so $b(x)$ passes rapidly although in a continuous way from the value $-1$ to the value 1, makes it possible to find a continuous $b(x)$ bounded by 1 such that $M_b$ has an arbitrary large operator norm in $h^1(\mathbb{R})$. On the other hand, the closed graph theorem implies that, should $b \mapsto M_b$ map $C(\mathbb{R}) \cap L^\infty(\mathbb{R})$ into $\mathscr{L}(h^p(\mathbb{R}))$ (the space of bounded linear operators on $h^p(\mathbb{R})$) then $b \mapsto M_b$ would be a bounded map from $C(\mathbb{R}) \cap L^\infty(\mathbb{R})$ to $\mathscr{L}(h^p(\mathbb{R}))$. Since the refined construction shows that this is not true, we conclude that there exists $b(x) \in C(\mathbb{R}) \cap L^\infty(\mathbb{R})$ such that $M_b$ is not bounded in $h^1(\mathbb{R})$.

Within the class of continuous functions, there is a standard way to describe regularity by introducing the concept of modulus of continuity. We recall that $\omega$ is a modulus of continuity if $\omega : [0, \infty) \longrightarrow \mathbb{R}^+$ is continuous, increasing, $\omega(0) = 0$ and $\omega(2t) \leq C\omega(t)$, $0 < t < 1$. A modulus of continuity determines the Banach space $C_\omega(\mathbb{R}^N)$ of bounded continuous functions $f : \mathbb{R}^N \longrightarrow \mathbb{C}$ such that

$$|f|_{C_\omega} \doteq \sup_{x \neq y} \frac{|f(y) - f(x)|}{\omega(|x - y|)} < \infty,$$

equipped with the norm $\|f\|_{C_\omega} = \|f\|_{L^\infty} + |f|_{C_\omega}$. Note that $C_\omega$ is only determined by the behavior of $\omega(t)$ for values of $t$ close to 0. If, for $0 < r < 1$, we set $\omega(t) = t^r$ the corresponding space $C_\omega(\mathbb{R}^N)$ is precisely the Hölder space $\Lambda^r(\mathbb{R}^N)$.

Continuous bounded functions which are regular enough in the sense of its modulus of continuity yield multipliers in $h^1(\mathbb{R}^N)$. Indeed, consider a modulus of continuity $\omega(t)$ that satisfies

$$\frac{1}{h^N} \int_0^h \omega(t) t^{N-1}\, dt \leq K \left(1 + \ln \frac{1}{h}\right)^{-1}, \quad 0 < h < 1, \tag{1}$$

and the corresponding space $C_\omega(\mathbb{R}^N)$. A proof of the following result can be found in [1, p. 374].

**Proposition 3.1** *Let $b(x) \in C_\omega(\mathbb{R}^N)$ and $f(x) \in h^1(\mathbb{R}^N)$. Then $b(x)f(x) \in h^1(\mathbb{R}^N)$ and there exists $C > 0$ such that*

$$\|bf\|_{h^1} \leq C\|b\|_{C_\omega}\|f\|_{h^1}, \quad b \in C_\omega(\mathbb{R}^N),\ f \in h^1(\mathbb{R}^N).$$

The next theorem shows that Hölder functions $b(x) \in \Lambda^r$, $r > 0$, are multipliers in $h^p(\mathbb{R}^N)$, $0 < p \leq 1$, provided $r > N(p^{-1} - 1)$. Note however that for $p = 1$ Proposition 3.1 gives a sharper result, as $\Lambda^r(\mathbb{R}^N)$ is strictly contained in $C_\omega(\mathbb{R}^N)$ for any $r > 0$.

**Theorem 3.1** *Let $0 < p \leq 1$ and $r > N(p^{-1} - 1)$. If $b(x) \in \Lambda^r(\mathbb{R}^N)$ then $M_b$ is a multiplier in $h^p(\mathbb{R}^N)$ and there exists $C > 0$ such that*

$$\|bf\|_{h^p} \leq C\|b\|_r\|f\|_{h^p}, \quad b \in \Lambda^r(\mathbb{R}^N),\ f \in h^p(\mathbb{R}^N).$$

*Proof* We are going to give the proof only for $N/(N+1) < p \leq 1$ which is the case we need for the applications we give in the next section. For these values of $p$, $N(p^{-1} - 1) < 1$ so it is enough to prove the theorem assuming that $N(p^{-1} - 1) < r < 1$ since $\Lambda^s(\mathbb{R}^N) \subset \Lambda^r(\mathbb{R}^N)$ when $s > r$.

Let $b(x) \in \Lambda^r(\mathbb{R}^N)$. It is enough to check that $\|ba\|_{h^p} \leq C\|b\|_r$ for every smooth $h^p$-atom $a$ with $C$ an absolute constant. This fact is obvious for atoms supported in balls $B$ with radius $\rho \geq 1$ without moment condition because $b$ is bounded so $ba/\|b\|_{L^\infty}$ is again an atom without moment condition. If $B = B(x_0, \rho)$, $\rho < 1$, we may write $a(x)b(x) = b(x_0)a(x) + (b(x) - b(x_0))a(x) = \beta_1(x) + \beta_2(x)$. Then $\beta_1(x)/\|b\|_{L^\infty}$ is again an atom while $\beta_2(x)$ is supported in $B$ and satisfies

$$\|\beta_2\|_{L^\infty} \leq C\rho^r\|a\|_{L^\infty} \leq \frac{C'\rho^r}{\rho^{N/p}} \tag{2}$$

$$\|\beta_2\|_{L^1} \leq C\|a\|_{L^\infty}\int_B |x - x_0|^r\,dx \leq C'\rho^{r+N(1-p^{-1})}. \tag{3}$$

We wish to conclude that $\|m_\varphi\beta_2\|_{L^p} < \infty$. Let $B^* = B(x_0, 2\rho)$. Since $m_\varphi\beta_2(x) \leq \|\beta_2\|_{L^\infty}$, we have

$$J_1 = \int_{B^*}(m_\varphi\beta_2)^p(x)\,dx \leq C|B^*|\frac{\rho^{rp}}{\rho^N} \leq C'.$$

It remains to estimate

$$J_2 = \int_{\mathbb{R}^N\setminus B^*}(m_\varphi\beta_2)^p(x)\,dx = \int_{2\rho\leq|x-x_0|\leq 2}(m_\varphi\beta_2)^p(x)\,dx \tag{4}$$

(observe that $m_\varphi \beta_2$ is supported in $B(x_0, 2)$ because $\operatorname{supp} \varphi \subset B(0,1)$). If $0 < \varepsilon < 1$ and $\varphi_\varepsilon * \beta_2(x) \neq 0$ for some $|x - x_0| \geq 2\rho$ we conclude that $\varepsilon \geq |x - x_0|/2$, which implies

$$|\varphi_\varepsilon * \beta_2(x)| \leq \left| \int \varphi_\varepsilon(y) \beta_2(x-y)\, dy \right| \leq \frac{C\|\beta_2\|_{L^1}}{\varepsilon^n} \leq C'|x - x_0|^{-N} \rho^{r+N(1-p^{-1})}$$

so

$$|m_\varphi \beta_2(x)|^p \leq \frac{C' \rho^{rp+N(p-1)}}{|x - x_0|^{Np}} \qquad \text{for} \quad |x - x_0| \geq 2\rho. \tag{5}$$

It follows from (4) and (5) that

$$J_2 \leq \int_{2\rho \leq |x-x_0| \leq 2} \frac{C' \rho^{rp+N(p-1)}}{|x - x_0|^{Np}}\, dx \leq C''$$

which leads to

$$\|ba\|_{h^p}^p \leq \|\beta_1\|_{h^p}^p + \|\beta_2\|_{h^p}^p \leq C_1 + J_1 + J_2 \leq C_2.$$

Tracking the estimates in the proof one sees that $C_2$ may be majorized by $C\|b\|_r^p$. Therefore, for every $h^p$-atom

$$\|ba\|_{h^p} \leq C\|b\|_r$$

which implies that $M_b$ may be extended as a bounded linear operator in $h^p(\mathbb{R}^N)$ and $\|M_b f\|_{h^p} \leq C\,\|b\|_r\,\|f\|_{h^p}$. □

## 4 Friedrichs Lemma in $h^p(\mathbb{R}^N)$

Let $f(x) \in h^p(\mathbb{R}^N)$, $0 < p < \infty$. Given a test function $\phi \in C_c^\infty(\mathbb{R}^N)$ with $\int \phi\, dx = 1$, let us denote the Friedrichs approximation of the identity by

$$J_\varepsilon f(x) = \phi_\varepsilon * f(x), \qquad \phi_\varepsilon(x) = \frac{1}{\varepsilon^N} \phi(x/\varepsilon), \quad \varepsilon > 0.$$

It is known that

$$\|J_\varepsilon f\|_{h^p} \leq C\|f\|_{h^p}, \qquad \lim_{\varepsilon \to 0} \|J_\varepsilon f - f\|_{h^p} = 0 \tag{6}$$

with $C > 0$ independent of $f \in h^p(\mathbb{R}^N)$.

For $p > 1$ we have $h^p(\mathbb{R}^N) = L^p(\mathbb{R}^N)$ and the classical Friedrichs lemma states that if $b(x)$ is a Lipschitz function and $f \in L^p(\mathbb{R}^N)$, $1 \leq p < \infty$ and $j = 1, \dots, N$, then the commutators $[J_\varepsilon, b\partial_j]f(x) = \phi_\varepsilon * (b\partial_j f)(x) - b(x)\partial_j(\phi_\varepsilon * f)(x)$, $\varepsilon > 0$, satisfy

$$\|[J_\varepsilon, b\partial_j]f\|_{L^p} \leq C\|b\|_{Lip}\|f\|_{L^p}, \qquad \lim_{\varepsilon\to 0} \|[J_\varepsilon, b\partial_j]f\|_{L^p} = 0,$$

with $C > 0$ independent of $f \in L^p(\mathbb{R}^N)$, $b \in Lip(\mathbb{R}^N)$. The following extension to local Hardy spaces holds.

**Lemma 4.1 (Friedrichs Lemma)** *Let $f(x) \in h^p(\mathbb{R}^N)$, $N/(N+1) < p \leq 1$, $b(x) \in \Lambda^{1+r}(\mathbb{R}^N)$, $r > N(p^{-1} - 1)$. Then*

$$\|[J_\varepsilon, b\partial_j]f\|_{h^p} \leq C\|b\|_{1+r}\|f\|_{h^p}, \qquad \lim_{\varepsilon\to 0} \|[J_\varepsilon, b\partial_j]f\|_{h^p} = 0, \tag{7}$$

*with $C > 0$ independent of $f \in h^p(\mathbb{R}^N)$, $b \in \Lambda^{1+r}(\mathbb{R}^N)$.*

*Proof* It is enough to prove just the estimate in (7) since $\lim_{\varepsilon\to 0} \|[J_\varepsilon, b\partial_j]f\|_{h^p} = 0$ clearly holds when $f \in C_c^\infty(\mathbb{R}^N)$ and, once the estimate has been proved, the conclusion about the limit follows in general by a density argument.

We will assume without loss of generality that $r < 1$. We may write after an integration by parts

$$\begin{aligned}
[J_\varepsilon, b\partial_j]f(x) &= \int \phi_\varepsilon(x-y)\big(b(y) - b(x)\big)\, \partial_j f(y)\, dy \\
&= \int (\partial_j\phi)_\varepsilon(x-y)\frac{b(y)-b(x)}{\varepsilon} f(y)\, dy - \int \phi_\varepsilon(x-y)(\partial_j b(y)) f(y)\, dy \\
&\doteq I_\varepsilon f(x) - J_\varepsilon(f\partial_j b)(x).
\end{aligned} \tag{8}$$

Since $\partial_j b(x) \in \Lambda^r(\mathbb{R}^N)$, Theorem 3.1 implies that $f\partial_j b \in h^p(\mathbb{R}^N)$ and the second term on the right hand side of (8), given by $J_\varepsilon(f\partial_j b)$, can be handled invoking (6). To deal with the first term on the right hand side of (8) we may use Taylor formula to write

$$\frac{b(y)-b(x)}{\varepsilon} = \sum_{k=1}^{N} \frac{y_k - x_k}{\varepsilon}\alpha_k(x,y)$$

where $x \mapsto \alpha_k(x,y) \in \Lambda^r(\mathbb{R}^N)$ uniformly in $y$ and $y \mapsto \alpha_k(x,y) \in \Lambda^r(\mathbb{R}^N)$ uniformly in $x$. Setting $\psi_k(x) \doteq -x_k\partial_j\phi(x)$, we have

$$I_\varepsilon f(x) = \sum_{k=1}^{N} \int (\psi_k)_\varepsilon(x-y)\alpha_k(x,y)f(y)\, dy \doteq \sum_{k=1}^{N} I_{\varepsilon,k}(\alpha_k f).$$

and we must prove

$$\|I_{\varepsilon,k}\alpha_k f\|_{h^p} \le C\|\alpha_k f\|_{h^p} \le C'\|f\|_{h^p}. \tag{9}$$

To prove (9) it is enough to prove it for atoms. To simplify the notation, we drop indexes and consider the limit $\lim_{\varepsilon\to 0} T_\varepsilon a(x)$ where

$$T_\varepsilon a(x) \doteq \int \psi_\varepsilon(x-y)\,\alpha(x,y)\,a(y)\,dy$$

while $a(x)$ is a smooth atom supported in $B(x_0,\rho)$, $\alpha(x,y)$ belongs to $\Lambda^r(\mathbb{R}^N)$ separately in each variable uniformly with respect to the other variable and $\psi \in C_c^\infty(\mathbb{R}^N)$. Since $y \mapsto \alpha(x,y)a(y) \in C_c(B(x_0,\rho))$ it is clear that $T_\varepsilon a(x)$ is supported in $B(x_0,\rho+1)$. We want to show that

$$\|T_\varepsilon a\|_{h^p} \le C \tag{10}$$

with $C$ independent of the particular atom. For $\rho \ge 1$, (10) follows easily from $\|T_\varepsilon a\|_{L^\infty} \le \|\psi\|_{L^1} \sup|\alpha|\,\|a\|_{L^\infty} \le C$. Assume that $\rho < 1$ and write

$$\begin{aligned} T_\varepsilon a(x) &= \alpha(x,x_0)\int \psi_\varepsilon(x-y)\,a(y)\,dy + \int \psi_\varepsilon(x-y)\,(\alpha(x,y)-\alpha(x,x_0))\,a(y)\,dy \\ &= \alpha(x,x_0)(\psi_\varepsilon * a(x)) + \beta_\varepsilon(x). \end{aligned}$$

Keeping in mind Theorem 3.1, the $h^p$ norm of the first term is majorized by

$$\|\alpha(\cdot,x_0)\|_{\Lambda^r}\|\psi_\varepsilon * a\|_{h^p} \le C\|\alpha(\cdot,x_0)\|_{\Lambda^r}\|a\|_{h^p} \le C'.$$

Note that

$$\|\beta_2\|_{L^\infty} \le C\rho^r\|\psi\|_{L^1}\|a\|_{L^\infty} \le \frac{C'\rho^r}{\rho^{N/p}}$$

and since $|\beta_2(x)| \le C\left(|\psi_\varepsilon| * A\right)(x)$ with $A(y) = |y-x_0|^r|a(y)|$,

$$\|\beta_2\|_{L^1} \le C\|A\|_{L^1} \le C_1\|a\|_{L^\infty}\int_B |x-x_0|^r\,dx \le C_2\rho^{r+N(1-p^{-1})}.$$

Hence, $\beta_2$ is supported in the ball $B(x_0,\rho+1)$ and $\|\beta_2\|_{L^\infty}$ as well as $\|\beta_2\|_{L^1}$ satisfy estimates analogous to the estimates (2) and (3) that were used in the proof of Theorem 3.1. A similar computation can then be carried out to show that

$$\int_{\mathbb{R}^N} (m_\varphi\beta_2)^p(x)\,dx \le C,$$

concluding the proof of (10). Thus, (9) holds and the lemma is proved. □

## 5 Elliptic Regularity in Hardy-Sobolev Spaces

Suppose that $\mathscr{L} = \{L_1, \ldots, L_n\}$ is a system of linearly independent vector fields with continuous complex coefficients defined on an open set $\Omega \subset \mathbb{R}^N$. We may consider the operator $\nabla_{\mathscr{L}}\, u \doteq (L_1 u, \ldots, L_n u)$ for $u \in C^\infty(\Omega)$ which corresponds to the operator $\nabla$ when $n = N$ and $L_j = \partial_{x_j}$. We are interested in the regularity of the equation

$$\nabla_{\mathscr{L}}\, u = f \tag{11}$$

in Hardy spaces. More precisely, if the coefficients of the vector fields $L_j$, $j = 1.\ldots, n$, are of class $\Lambda^{1+r}(\Omega)$ with $r = N(p^{-1} - 1)$ and $u \in h^p(\mathbb{R}^N)$, then $\nabla_{\mathscr{L}}\, u$ is a well defined distribution in $\mathscr{D}'(\Omega)$ due to the duality between $h^p(\mathbb{R}^N)$ and $\Lambda^r(\mathbb{R}^N)$. Roughly speaking, we wish to conclude that if the right hand side of (11) is in $h^p$ on a neighborhood of a point $x_0 \in \Omega$ then the derivatives $\partial_j u$, $j = 1, \ldots, N$, are also in $h^p$ on some neighborhood of $x_0$. To make things precise we introduce a definition.

**Definition 5.1** Let $p > 0$ and consider on $C_c^\infty(\mathbb{R}^N)$ the "norm"

$$\|f\|_{h^{1,p}(\mathbb{R}^N)} \doteq \sum_{j=1}^{N} \|\partial_j f\|_{h^p(\mathbb{R}^N)} = \|\nabla f\|_{h^p}.$$

The completion of $C_c^\infty(\mathbb{R}^N)$ for the norm $\|f\|_{h^{1,p}(\mathbb{R}^N)}$ may identified with a subspace of $\mathscr{S}'(\mathbb{R}^N)$ denoted by $h^{1,p}(\mathbb{R}^N)$ and called the Hardy-Sobolev space of order 1. If $f \in h^{1,p}(\mathbb{R}^N)$ then $\nabla f \in h^p(\mathbb{R}^N)$ and

$$\|f\|_{h^{1,p}(\mathbb{R}^N)} = \|\nabla f\|_{h^p}.$$

It is known that if $f \in h^{1,p}(\mathbb{R}^N)$ for some $p > N/(N+1)$ then $f$ is locally integrable. If $\Omega \subset \mathbb{R}^N$ is a domain the notation $h_c^p(\Omega)$ and $h_c^{1,p}(\Omega)$ will stand for $h^p(\mathbb{R}^N) \cap \mathscr{E}'(\Omega)$ and $h^{1,p}(\mathbb{R}^N) \cap \mathscr{E}'(\Omega)$ respectively, where $\mathscr{E}'(\Omega)$ denotes the distributions with compact support contained in $\Omega$.

The localizable spaces $h^p(\mathbb{R}^N)$ give rise to the local Hardy-Sobolev spaces $h^p_{\mathrm{loc}}(\Omega)$ which is the space of distributions $u \in \mathrm{d}'(\Omega)$ such that $\psi \in h^p(\mathbb{R}^N)$ for all $\psi \in C_c^\infty(\Omega)$. A sequence $u_j \in h^p_{\mathrm{loc}}(\Omega)$ converges to $u \in h^p_{\mathrm{loc}}(\Omega)$ if $\psi u_j \to \psi u$ in $h^p(\mathbb{R}^N)$ for all $\psi \in C_c^\infty(\Omega)$. It is enough to check the convergence for $\psi = \psi_k$, $k \in \mathbb{N}$, where $(\psi_k)$ is a partition of unity in $\Omega$ so the topology induced by this notion of convergence is metrizable. Similar remarks are valid for the localizable spaces $h^{1,p}(\mathbb{R}^N)$, i.e., the spaces $h^{1,p}(\mathbb{R}^N)$ give rise to the local Hardy-Sobolev spaces $h^{1,p}_{\mathrm{loc}}(\Omega)$.

We will always assume that

(i) $L_1, \ldots, L_n$ are everywhere linearly independent.
(ii) The system $\{L_1, \ldots, L_n\}$ is elliptic.

The later means that for any *real* 1-form $\omega$ (i.e., any section of $T^*(\Omega)$) such that $\langle \omega, L_j \rangle = 0$ implies $\omega = 0$. Consequently, the number $n$ of vector fields must satisfy

$$\frac{N}{2} \le n \le N.$$

Alternatively, if the coefficients of the vector fields are of class $C^2$, (ii) is equivalent to saying that the second order operator

$$\Delta_L \doteq L_1^* L_1 + \cdots + L_n^* L_n \tag{12}$$

is elliptic. Here, $L_j^* = \overline{L}_j^t, j = 1, \ldots, n$, where $\overline{L}_j$ denotes the vector field obtained from $L_j$ by conjugating its coefficients and $L_j^t$ is the formal transpose of $L_j$. If we write $\mathrm{div}_{\mathscr{L}} f = \mathrm{div}_{\mathscr{L}} (f_1, \ldots, f_n) = L_1^* f_1 + \cdots + L_n^* f_n$ we have

$$\Delta_L = \mathrm{div}_{\mathscr{L}} \nabla_{\mathscr{L}}.$$

**Theorem 5.1** *Assume* $N/(N+1) < p \le 1$ *and* $r > N(p^{-1} - 1)$. *If the coefficients of* $L_1, \ldots, L_n$ *are or class* $\Lambda^{1+r}(\Omega)$ *and* $u \in h^p_{loc}(\Omega)$ *satisfies* (11) *with* $f \in h^p_{loc}(\Omega)$ *then* $u \in h^{1,p}_{loc}(\Omega)$. *In particular, the distribution u is a locally integrable function.*

*Proof* Since we are dealing with a local question, it is enough to prove that given a point $x_0 \in \Omega$ there exists $\eta \in C_c^\infty(\Omega)$ such that $\eta(x_0) \neq 0$ and $\eta u \in h^{1,p}(\mathbb{R}^N)$ or, equivalently, that $\eta \nabla u \in h^p(\mathbb{R}^N)$. Thus, there is no loss of generality in assuming that $x_0 = 0$ and $\Omega$ is an open ball centered at the origin.

Assume first that the coefficients of $L_1, \ldots, L_n$ are smooth. By a standard result in the theory of pseudo-differential operators, we may find scalar operators $q(x, D)$ and $r(x, D)$ with symbols $q(x, \xi)$ of order $-2$ and $r(x, \xi)$ of order $-\infty$, such that

$$v = q(x, D)\Delta_L v + r(x, D)v, \quad v \in \mathscr{E}'(\Omega). \tag{13}$$

Pick $\psi \in C_c^\infty(\Omega)$ equal to 1 in a neighborhood of the origin and note that, by Leibniz's rule,

$$\Delta_L(\psi u) = \psi \Delta_L u + w = \psi \mathrm{div}_{\mathscr{L}} f + w$$

with $w \equiv 0$ on a neighborhood of the origin. Write (13) for $v = \psi u$ and take the gradient of the resulting identity to obtain

$$\nabla \psi u = \nabla q(x, D)(\psi \mathrm{div}_{\mathscr{L}} f) + w_2 = f_1 + w_2 \tag{14}$$

where $w_2$ is smooth on a neighborhood of the origin and $f_1 \in h^p(\mathbb{R}^N)$. We have used here that $f_1 = \nabla q(x, D)(\psi \mathrm{div}_{\mathscr{L}} f)$ may be written as a sum of pseudo-differential operators of order zero acting on $\psi f \in h^p(\mathbb{R}^N)$ or on $(\partial_j \psi) f \in h^p(\mathbb{R}^N)$,

$j = 1, \dots, N$, and pseudo-differential operators of order zero are bounded in $h^p(\mathbb{R}^N)$. Hence, if $\eta \in C_c^\infty(\Omega)$ is supported on the neighborhood where $\psi \equiv 1$ and $\eta(0) \neq 0$, multiplying (14) by $\eta$ we conclude that $\eta \nabla u \in h^p(\mathbb{R}^N)$ which is what we wanted to show.

When the coefficients are not smooth we will use a roundabout argument that involves a priori estimates. Still in the case of smooth coefficients, we make use of (13) with $v \in C_c^\infty(\mathbb{R}^N)$ and take the gradient on both sides to get

$$\nabla v = p_1(x, D)\nabla_{\mathscr{L}} v + p_2(x, D)v$$

where $p_1(x, D)$ and $p_2(x, D)$ are pseudo-differential operators of order zero, thus bounded in $h^p(\mathbb{R}^N)$. This implies the a priori estimate

$$\|\nabla v\|_{h^p} \leq C_1 \|\nabla_{\mathscr{L}} v\|_{h^p} + C_2 \|v\|_{h^p}, \quad v \in C_c^\infty(\Omega).$$

Given $\epsilon > 0$, we may shrink $\Omega$ to grant the estimate

$$\|v\|_{h^p} \leq \epsilon \|\nabla v\|_{h^p}, \quad v \in C_c^\infty(\Omega),$$

(*cf.* [3, Prop. 3.1]) which for $\epsilon > 0$ sufficiently small makes it possible to absorb the zero order term in the previous estimate yielding

$$\|\nabla v\|_{h^p} \leq C \|\nabla_{\mathscr{L}} v\|_{h^p}, \quad v \in C_c^\infty(\Omega). \tag{15}$$

Let us return to the case in which the coefficients of $\mathscr{L}$ are of class $\Lambda^{1+r}$. Call $\mathscr{L}_0$ the system with constant coefficients obtained by freezing the coefficients of $\mathscr{L}$ at the origin. Applying (15) to $\mathscr{L}_0$ we get

$$\|\nabla v\|_{h^p} \leq C \|\nabla_{\mathscr{L}_0} v\|_{h^p}, \quad v \in C_c^\infty(\Omega),$$

that, considering $\mathscr{L}$ as a perturbation of $\mathscr{L}_0$, implies

$$\|\nabla v\|_{h^p} \leq C \|\nabla_{\mathscr{L}} v\|_{h^p} + C \|\nabla_{(\mathscr{L} - \mathscr{L}_0)} v\|_{h^p}, \quad v \in C_c^\infty(\Omega).$$

The coefficients of the perturbation $\mathscr{L} - \mathscr{L}_0$ are of class $\Lambda^{1+r}$ and vanish at the origin, so they may be assumed to have arbitrarily small norm in $\Lambda^r$ provided we shrink $\Omega$. Taking advantage of Theorem 3.1 we obtain

$$\|\nabla v\|_{h^p} \leq C \|\nabla_{\mathscr{L}} v\|_{h^p} + O(\mathrm{diam}(\Omega)) \|\nabla v\|_{h^p}, \quad v \in C_c^\infty(\Omega),$$

and absorbing the second term on the right hand side we get the a priori estimate

$$\|\nabla v\|_{h^p} \leq C \|\nabla_{\mathscr{L}} v\|_{h^p}, \quad v \in C_c^\infty(\Omega). \tag{16}$$

Consider now $u \in h^p_{loc}(\Omega)$ and suppose that it satisfies (11) with $f \in h^p_{loc}(\Omega)$. We will now use Friedrichs mollifiers. Pick $\eta \in C^\infty_c(\Omega)$ satisfying $\eta(x_0) \neq 0$ and apply (16) to $v = J_\varepsilon(\eta u)$ for $\varepsilon > 0$ small to get

$$\begin{aligned}\|\nabla J_\varepsilon(\eta u)\|_{h^p} &\leq C\|\nabla_{\mathscr{L}} J_\varepsilon(\eta u)\|_{h^p}\\ &\leq C\|J_\varepsilon \nabla_{\mathscr{L}}(\eta u)\|_{h^p} + C\|[\nabla_{\mathscr{L}}, J_\varepsilon](\eta u)\|_{h^p}.\end{aligned}$$

Note that $g \doteq \nabla_{\mathscr{L}}(\eta u) = u\nabla_{\mathscr{L}}\eta + \eta f \in h^p(\mathbb{R}^N)$ and that $\|[\nabla_{\mathscr{L}}, J_\varepsilon](\eta u)\|_{h^p} \to 0$ as $\varepsilon \to 0$ by Friedrichs lemma (Lemma 4.1). Similarly, if we choose a sequence $\varepsilon_k \searrow 0$, set $w_k = J_{\varepsilon_k}\nabla(\eta u)$, $g_k = J_{\varepsilon_k} g$ and apply (16) to $v = J_{\varepsilon_k}(\eta u) - J_{\varepsilon_j}(\eta u)$ we get

$$\|w_k - w_j\|_{h^p} \leq C\|g_k - g_j\|_{h^p} + \rho(k,j)$$

where $\rho(k,j) \to 0$ as $k, j \to \infty$, showing that $w_k$ is a Cauchy sequence in $h^p(\mathbb{R}^N)$. This implies that $\nabla(\eta u) \in h^p(\mathbb{R}^N)$ so $\eta\nabla u \in h^p(\mathbb{R}^N)$ as we wished to prove. □

## References


1. Berhanu, S., Cordaro P., Hounie, J.: An Introduction to Involutive Structure. Cambridge University Press, Cambridge (2008)
2. Goldberg D.: A local version of real Hardy spaces. Duke Math. J. **46**, 27–42 (1979)
3. Hoepfner G., Hounie, J., Picon, T.: Div-Curl Type estimates for elliptic systems of complex vector fields. J. Math. Anal. Appl. **429**(2), 774–799 (2015)
4. Hörmander, L: The Analysis of Linear Partial Differential Operators III. Springer, Berlin/New York (1985)
5. Hounie, J., Kapp, R.: Pseudodifferential operators on local Hardy spaces. J. Fourier Anal. Appl. **15**, 153–178 (2009)
6. Hounie, J., dos Santos Filho, J.R.: A fractional Friedrich's lemma for Hölder norms and applications. Nonlinear Anal. **28**, 1063–1077 (1997)
7. Stein, E.: Harmonic Analysis: Real-Variable Methods Orthogonality, and Oscillatory Integrals. Princeton University Press, Princeton (1993)
8. Stein, E.: Note on the class $L \log L$. Studia Math. **32**, 305–310 (1969)
9. Triebel, H.: Theory of Function Spaces. Monographs in Mathematics, vol. 78. Birkhauser, Basel (1983)

# On the Hardy Constant of Some Non-convex Planar Domains

**Gerassimos Barbatis and Achilles Tertikas**

*Dedicated to Ermanno Lanconelli on the occasion of his 70th birthday*

**Abstract** The Hardy constant of a simply connected domain $\Omega \subset \mathbb{R}^2$ is the best constant for the inequality

$$\int_\Omega |\nabla u|^2 dx \geq c \int_\Omega \frac{u^2}{\mathrm{dist}(x, \partial\Omega)^2} dx , \qquad u \in C_c^\infty(\Omega).$$

After the work of Ancona where the universal lower bound 1/16 was obtained, there has been a substantial interest on computing or estimating the Hardy constant of planar domains. In Barbatis and Tertikas (J Funct Anal 266:3701–3725, 2014) we have determined the Hardy constant of an arbitrary quadrilateral in the plane. In this work we continue our investigation and we compute the Hardy constant for other non-convex planar domains. In all cases the Hardy constant is related to that of a certain infinite sectorial region which has been studied by E.B. Davies.

**Keywords** Distance function • Hardy constant • Hardy inequality

**2010 Mathematics Subject Classification:** 35A23, 35J20, 35J75 (46E35, 26D10, 35P15)

---

G. Barbatis
Department of Mathematics, University of Athens, Athens, Greece
e-mail: gbarbatis@math.uoa.gr

A. Tertikas (✉)
Department of Mathematics and Applied Mathematics, University of Crete, Heraklion, Greece

Institute of Applied and Computational Mathematics, FORTH, Heraklion, Greece
e-mail: tertikas@math.uoc.gr



G. Citti et al. (eds.), *Geometric Methods in PDE's*, Springer INdAM Series 13,
DOI 10.1007/978-3-319-02666-4_2

# 1 Introduction

The well-known Hardy inequality for $\mathbb{R}^N_+ = \mathbb{R}^{N-1} \times (0, +\infty)$ reads

$$\int_{\mathbb{R}^N_+} |\nabla u|^2 dx \geq \frac{1}{4} \int_{\mathbb{R}^N_+} \frac{u^2}{x_N^2} dx\,, \quad \text{for all } u \in C_c^\infty(\mathbb{R}^N_+), \tag{1}$$

where the constant $1/4$ is the best possible and equality is not attained in the appropriate Sobolev space. The analogue of (1) for a domain $\Omega \subset \mathbb{R}^N$ is

$$\int_\Omega |\nabla u|^2 dx \geq \frac{1}{4} \int_\Omega \frac{u^2}{d^2} dx\,, \quad \text{for all } u \in C_c^\infty(\Omega), \tag{2}$$

where $d = d(x) = \operatorname{dist}(x, \partial\Omega)$. However, (2) is not true without geometric assumptions on $\Omega$. The typical assumption made for the validity of (2) is that $\Omega$ is convex. A weaker geometric assumption introduced in [5] is that $\Omega$ is weakly mean convex, that is

$$-\Delta d(x) \geq 0\,, \quad \text{in } \Omega, \tag{3}$$

where $\Delta d$ is to be understood in the distributional sense. Condition (3) is equivalent to convexity when $N = 2$ but strictly weaker than convexity when $N \geq 3$ [2]. Other geometric assumptions on the domain that guarantee that the best Hardy constant is 1/4 were recently obtain in [3, 9].

For a general domain $\Omega$ we may still have a Hardy inequality provided that the boundary $\partial\Omega$ has some regularity. In particular it is well known that for any bounded Lipschitz domain $\Omega \subset \mathbb{R}^N$ there exists $c > 0$ such that

$$\int_\Omega |\nabla u|^2 dx \geq c \int_\Omega \frac{u^2}{d^2} dx\,, \quad \text{for all } u \in C_c^\infty(\Omega). \tag{4}$$

The best constant $c$ of inequality (4) is called the Hardy constant of the domain $\Omega$.

In general the Hardy constant depends on the domain $\Omega$; see [6] for results that concern properties of this dependence. In dimension $N \geq 3$ Davies [8] has constructed Lipschitz domains with Hardy constant as small as one wishes. On the other hand for $N = 2$ Ancona [1] has proved that for a simply connected domain the Hardy constant is always at least $1/16$; see also [11] where further results in this directions where obtained.

Davies [8] computed the Hardy constant of an infinite planar sector $\Lambda_\beta$ of angle $\beta$,

$$\Lambda_\beta = \{\, 0 < r,\ \ 0 < \theta < \beta\}.$$

He used the symmetry of the domain to reduce the computation to the study of a certain ODE; see (9) below. In particular he established the following two results, which are also valid for the circular sector of angle $\beta$:

(a) The Hardy constant is $1/4$ for all angles $\beta \leq \beta_{cr}$, where $\beta_{cr} \cong 1.546\pi$.
(b) For $\beta_{cr} \leq \beta \leq 2\pi$ the Hardy constant of $\Lambda_\beta$ strictly decreases with $\beta$ and at the limiting case $\beta = 2\pi$ the Hardy constant is $\cong 0.2054$.

Our interest is to determine the Hardy constant of certain domains in two space dimensions; see [4, 10] for relevant questions. In this direction, in our recent work [7] we have established

**Theorem** *Let $\Omega$ be a non-convex quadrilateral with non-convex angle $\pi < \beta < 2\pi$. Then the Hardy constant of $\Omega$ depends only on $\beta$. The Hardy constant, which we denote from now on by $c_\beta$, is the unique solution of the equation*

$$\sqrt{c_\beta}\tan\big(\sqrt{c_\beta}(\frac{\beta-\pi}{2})\big) = 2\Bigg(\frac{\Gamma(\frac{3+\sqrt{1-4c_\beta}}{4})}{\Gamma(\frac{1+\sqrt{1-4c_\beta}}{4})}\Bigg)^2, \tag{5}$$

*when $\beta_{cr} \leq \beta < 2\pi$ and $c_\beta = 1/4$ when $\pi < \beta \leq \beta_{cr}$. The critical angle $\beta_{cr}$ is the unique solution in $(\pi, 2\pi)$ of the equation*

$$\tan\big(\frac{\beta_{cr}-\pi}{4}\big) = 4\Bigg(\frac{\Gamma(\frac{3}{4})}{\Gamma(\frac{1}{4})}\Bigg)^2. \tag{6}$$

Actually the constant $c_\beta$ coincides with the Hardy constant of the sector $\Lambda_\beta$, so Eq. (5) provides an analytic description of the Hardy constant computed numerically in [8].

In this work we continue our investigation and determine the Hardy constant for other families of non-convex planar domains. Our first result reads as follows; see Fig. 1.

**Theorem 1.1** *Let $\Omega = K \cap \Lambda_\beta$, $\beta \in (\pi, 2\pi]$, where $K$ is a bounded convex planar set and the vertex of $\Lambda_\beta$ is an interior point of $K$. Let $\gamma_+$ and $\gamma_-$ denote the interior angles of intersection of $K$ with $\Lambda_\beta$. There exists an angle $\gamma_\beta \in (\pi/2, \pi)$ such that if $\gamma_+, \gamma_- \leq \gamma_\beta$, then the Hardy constant of $\Omega$ is $c_\beta$, where $c_\beta$ is given by (5) and (6).*

Detailed information on the angle $\gamma_\beta$ is given in Lemma 27 and Theorem 3.1. We note that Theorem 1.1 can be extended to cover the case where $\Omega$ is unbounded and the boundary of the convex set $K$ does not intersect the boundary of the sector $\Lambda_\beta$; see Theorem 3.2.

We next study the Hardy constant for a family of domains $E_{\beta,\gamma}$ which may have two non-convex angles. The boundary $\partial E_{\beta,\gamma}$ of such a domain consists of the segment $OP$ and two half lines starting from $O$ and from $P$ with interior angles $\beta$

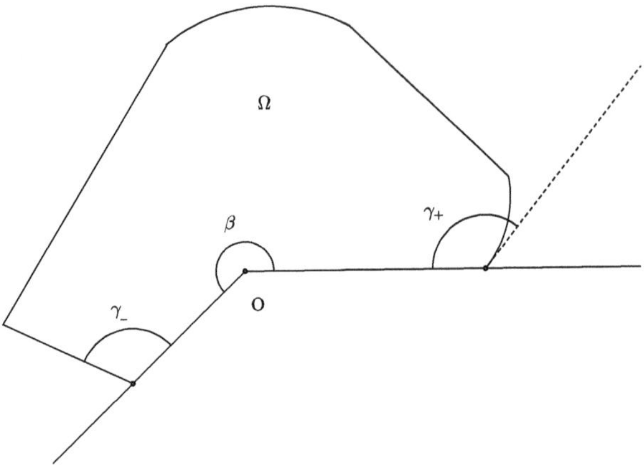


**Fig. 1** A typical domain $\Omega$ for Theorem 1.1

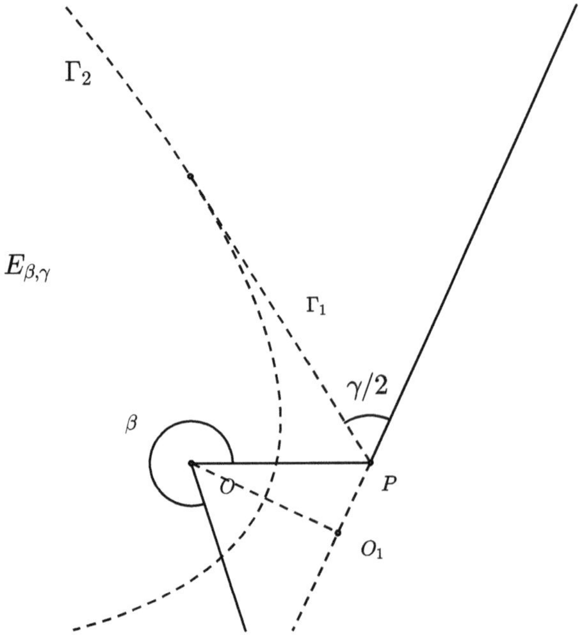


**Fig. 2** A typical domain $E_{\beta,\gamma}$, $\gamma < \pi < \beta$

and $\gamma$; hence $\beta + \gamma \le 3\pi$; see Fig. 2 in case $\gamma < \pi$ and Fig. 3 in case $\gamma > \pi$. We then have the following result

**Theorem 1.2**

*(i) If* $0 < \gamma \le \pi \le \beta \le 2\pi$ *then the Hardy constant of* $E_{\beta,\gamma}$ *is* $c_\beta$.
*(ii) If* $\pi \le \beta, \gamma \le 2\pi$ *then the Hardy constant of* $E_{\beta,\gamma}$ *is* $c_{\beta+\gamma-\pi}$,

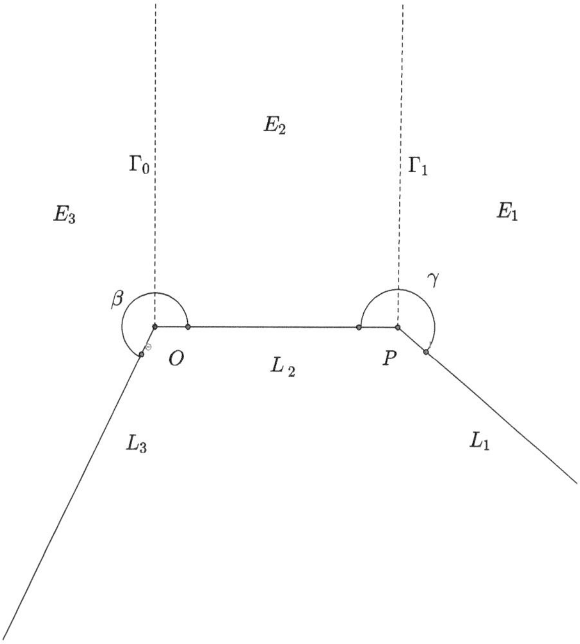


**Fig. 3** A typical domain $E_{\beta,\gamma}$, $\beta, \gamma > \pi$

*provided that*

$$|\beta - \gamma| \leq \frac{2}{c_{\beta+\gamma-\pi}} \arccos(2\sqrt{c_{\beta+\gamma-\pi}}). \tag{7}$$

It is interesting to notice that in case (i) where we have only one non-convex angle, the Hardy constant is related to the non-convex angle $\beta$, whereas in case (ii) where we have two non-convex angles, the Hardy constant is related to the angle $\beta+\gamma-\pi$ formed by the two halflines.

Our technique can actually be applied to establish best constant for Hardy inequality with mixed Dirichlet-Neumann boundary conditions. We consider a bounded domain $D_\beta$ whose boundary $\partial D_\beta$ consists of two parts, $\partial D_\beta = \Gamma_0 \cup \Gamma$. On $\Gamma_0$ we impose Dirichlet boundary conditions and it is from $\Gamma_0$ that we measure the distance from, $d(x) = \mathrm{dist}(x, \Gamma_0)$. On the remaining part $\Gamma$ we impose Neumann boundary conditions. The curve $\Gamma_0$ is the union of two line segments which have as a common endpoint the origin $O$ where they meet at an angle $\beta$, $\pi < \beta \leq 2\pi$. We assume that the curve $\Gamma$ is the graph in polar coordinates of a Lipschitz function $r(\theta)$,

$$\Gamma = \{(r(\theta), \theta) : 0 \leq \theta \leq \beta\};$$

see Fig. 4.

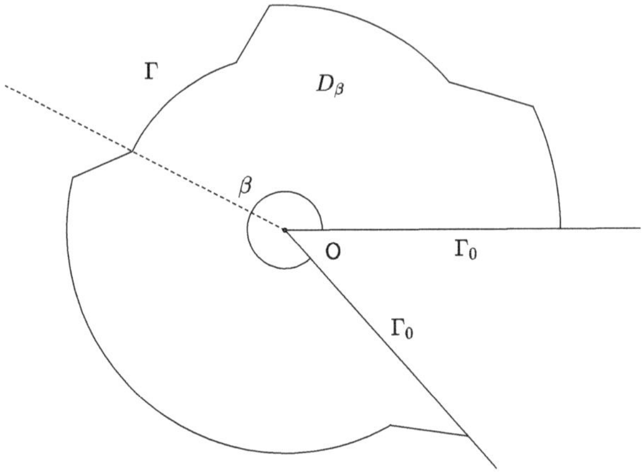


**Fig. 4** A typical domain $D_\beta$. Note that $\Gamma$ is not necessarily the boundary of a convex set

We then have

**Theorem 1.3** *Let $D_\beta$ be as above, $\pi < \beta \leq 2\pi$. If $\Gamma$ is such that*

$$r'(\theta) \leq 0, \quad 0 \leq \theta \leq \frac{\beta}{2},$$

$$r'(\theta) \geq 0, \quad \frac{\beta}{2} \leq \theta \leq \beta\,,$$

*then for all functions $u \in C^\infty(\overline{D_\beta})$ that vanish near $\Gamma_0$ there holds*

$$\int_{D_\beta} |\nabla u|^2 dx\,dy \geq c_\beta \int_{D_\beta} \frac{u^2}{d^2} dx\,dy\,.$$

*The constant $c_\beta$ is the best possible.*

The structure of the paper is simple: in Sect. 2 we prove various auxiliary results, while in Sects. 3–5 we prove the theorems.

## 2 Auxiliary Results

Let $\beta > \pi$ be fixed. We define the potential $V(\theta)$, $\theta \in (0, \beta)$,

$$V(\theta) = \begin{cases} \dfrac{1}{\sin^2\theta}, & 0 < \theta < \frac{\pi}{2}, \\ 1, & \frac{\pi}{2} < \theta < \beta - \frac{\pi}{2}, \\ \dfrac{1}{\sin^2(\beta - \theta)}, & \beta - \frac{\pi}{2} < \theta < \beta. \end{cases} \tag{8}$$

For $c > 0$ we then consider the following boundary-value problem:

$$\begin{cases} -\psi''(\theta) = cV(\theta)\psi(\theta), \ 0 \le \theta \le \beta, \\ \psi(0) = \psi(\beta) = 0. \end{cases} \tag{9}$$

It was proved in [8] that the Hardy constant of the sector $\Lambda_\beta$ coincides with the largest positive constant $c$ for which (9) has a positive solution. Due to the symmetry of the potential $V(\theta)$ this also coincides with the largest constant $c$ for which the following boundary value problem has a positive solution:

$$\begin{cases} -\psi''(\theta) = cV(\theta)\psi(\theta), \ 0 \le \theta \le \beta/2, \\ \psi(0) = \psi'(\beta/2) = 0\,. \end{cases} \tag{10}$$

The largest angle $\beta_{cr}$ for which the Hardy constant is $1/4$ for $\beta \in [\pi, \beta_{cr}]$ was computed numerically in [8] and analytically in [7, 12] where (6) was established; the approximate value is $\beta_{cr} \cong 1.546\pi$.

We define the hypergeometric function

$$F(a, b, c; z) := \frac{\Gamma(c)}{\Gamma(a)\Gamma(b)} \sum_{n=0}^{\infty} \frac{\Gamma(a+n)\Gamma(b+n)}{\Gamma(c+n)} \frac{z^n}{n!}.$$

The boundary value problem (10) was studied in [7] where the following lemma was proved:

**Lemma 2.1**

(i) *Let $\beta > \beta_{cr}$. The boundary value problem (10) has a positive solution if and only if $c = c_\beta$. In this case the solution is given by*

$$\psi(\theta) = \begin{cases} \dfrac{\sqrt{2}\cos\left(\sqrt{c}(\beta-\pi)/2\right)\sin^{\alpha}(\theta/2)\cos^{1-\alpha}(\theta/2)}{F(\frac{1}{2}, \frac{1}{2}, \alpha+\frac{1}{2}; \frac{1}{2})} F(\dfrac{1}{2}, \dfrac{1}{2}, \alpha+\dfrac{1}{2}; \sin^2(\dfrac{\theta}{2})), \\ \qquad\qquad\qquad\qquad\qquad\qquad\qquad\qquad \text{if } 0 < \theta \le \frac{\pi}{2}, \\ \cos\left(\sqrt{c}(\frac{\beta}{2} - \theta)\right), \qquad\qquad\qquad\qquad \text{if } \frac{\pi}{2} < \theta \le \frac{\beta}{2}, \end{cases}$$

*where $\alpha$ is the largest solution of $\alpha(1-\alpha) = c$.*

(ii) *Let $\pi < \beta \le \beta_{cr}$. The largest value of $c$ so that the boundary value problem (10) has a positive solution is $c = 1/4$. For $\beta = \beta_{cr}$ the solution is*

$$\psi(\theta) = \begin{cases} \dfrac{\cos\left(\frac{\beta_{cr}-\pi}{4}\right)\sin^{1/2}\theta}{F(\frac{1}{2}, \frac{1}{2}, 1; \frac{1}{2})} F(\dfrac{1}{2}, \dfrac{1}{2}, 1; \sin^2(\dfrac{\theta}{2})), \ 0 < \theta \le \frac{\pi}{2}, \\ \cos\left(\frac{1}{2}(\frac{\beta_{cr}}{2} - \theta)\right), \qquad\qquad \frac{\pi}{2} < \theta \le \frac{\beta_{cr}}{2}, \end{cases},$$

*while for $\beta_{cr} < \beta < 2\pi$ and $0 < \theta < \pi/2$ it has the form*

$$\psi(\theta) = c_1 \sin^{1/2}(\frac{\theta}{2}) \cos^{1/2}(\frac{\theta}{2}) F(\frac{1}{2}, \frac{1}{2}, 1; \sin^2(\frac{\theta}{2}))$$
$$+ c_2 \sin^{1/2}(\frac{\theta}{2}) \cos^{1/2}(\frac{\theta}{2}) F(\frac{1}{2}, \frac{1}{2}, 1; \sin^2(\frac{\theta}{2}))$$
$$\times \int_{\sin^2(\theta/2)}^{1/2} \frac{dt}{t(1-t)F^2(\frac{1}{2}, \frac{1}{2}, 1; t)}$$

*for suitable $c_1$, $c_2$.*

For our purposes it is useful to write the solution of (10) in case $\beta \geq \beta_{cr}$ as a power series

$$\psi(\theta) = \theta^\alpha \sum_{n=0}^{\infty} a_n \theta^n \ , \tag{11}$$

where $\alpha$ is the largest solution of the equation $\alpha(1-\alpha) = c_\beta$ in case $\beta > \beta_{cr}$ and $\alpha = 1/2$ when $\beta = \beta_{cr}$. We normalize the power series setting $a_0 = 1$; simple computations then give

$$a_1 = 0 \ , \qquad a_2 = -\frac{\alpha(1-\alpha)}{6(1+2\alpha)}. \tag{12}$$

We also define the auxiliary functions

$$f(\theta) = \frac{\psi'(\theta)}{\psi(\theta)} \ , \qquad \theta \in (0, \beta) \ , \tag{13}$$

and

$$g(\theta) = \frac{\psi'(\theta)}{\psi(\theta)} \sin\theta \ , \qquad \theta \in (0, \beta) \ , \tag{14}$$

where $\psi$ is the normalized solution of (9) described in Lemma 2.1. We note that these functions depend on $\beta$. Simple computations show that they respectively solve the differential equations

$$f'(\theta) + f^2(\theta) + c_\beta V(\theta) = 0 \ , \qquad 0 < \theta < \beta \tag{15}$$

and

$$g'(\theta) = -\frac{1}{\sin\theta}\Big[ g(\theta)^2 - \cos\theta \, g(\theta) + c_\beta \Big] \ , \quad 0 < \theta \leq \pi/2. \tag{16}$$

We shall also need the following

**Lemma 2.2** *Let $\pi \le \beta \le 2\pi$ and $\gamma \ge 0$ with $\beta + 2\gamma \le 3\pi$. Then*

$$f(\theta)\cos(\theta+\gamma) + \alpha[1+\sin(\theta+\gamma)] \ge 0\,, \quad \frac{\pi}{2} \le \theta \le \frac{3\pi}{2} - \gamma\,.$$

*Proof* We first note that

$$f(\theta) = \sqrt{c_\beta}\tan\Big(\sqrt{c_\beta}(\frac{\beta}{2}-\theta)\Big), \qquad \frac{\pi}{2} \le \theta \le \frac{3\pi}{2} - \gamma,$$

and

$$-\frac{\pi}{4} \le \sqrt{c_\beta}(\frac{\beta}{2}-\theta) \le \frac{\pi}{4}\,, \qquad \frac{\pi}{2} \le \theta \le \frac{3\pi}{2} - \gamma.$$

It follows that the required inequality is written equivalently,

$$\alpha(1+\sin(\gamma+\theta))\cos(\sqrt{c_\beta}(\frac{\beta}{2}-\theta)) \tag{17}$$

$$+\sqrt{c}\sin(\sqrt{c_\beta}(\frac{\beta}{2}-\theta))\cos(\gamma+\theta) \ge 0\,, \quad \frac{\pi}{2} \le \theta \le \frac{3\pi}{2} - \gamma. \tag{18}$$

But, since $\alpha \ge \sqrt{c_\beta}$,

$$\begin{aligned}
&\alpha\ (1+\sin(\theta+\gamma))\cos(\sqrt{c_\beta}(\frac{\beta}{2}-\theta)) + \sqrt{c_\beta}\sin(\sqrt{c_\beta}(\frac{\beta}{2}-\theta))\cos(\theta+\gamma)\\
&\ge \sqrt{c_\beta}\Big\{(1+\sin(\theta+\gamma))\cos(\sqrt{c_\beta}(\frac{\beta}{2}-\theta)) + \sin(\sqrt{c_\beta}(\frac{\beta}{2}-\theta))\cos(\theta+\gamma)\Big\}\\
&= 2\sqrt{c_\beta}\sin\Big[\sqrt{c_\beta}(\frac{\beta}{2}-\theta)+\frac{\pi}{4}+\frac{\theta}{2}+\frac{\gamma}{2}\Big]\sin(\frac{\pi}{4}+\frac{\theta}{2}+\frac{\gamma}{2}).
\end{aligned} \tag{19}$$

The second sine is clearly non-negative, so it only remains to prove that the first sine is also non-negative. For this we use the monotonicity of $\sqrt{c_\beta}(\frac{\beta}{2}-\theta)+\frac{\pi}{4}+\frac{\theta}{2}+\frac{\gamma}{2}$ with respect to $\theta$ to obtain

$$\begin{aligned}
\sqrt{c_\beta}(\frac{\beta}{2}-\theta)+\frac{\pi}{4}+\frac{\theta}{2}+\frac{\gamma}{2} &\le \sqrt{c_\beta}(\frac{\beta}{2}-(\frac{3\pi}{2}-\gamma)) + \frac{\pi}{4} + \frac{\frac{3\pi}{2}-\gamma}{2} + \frac{\gamma}{2}\\
&= \sqrt{c_\beta}\frac{\beta+2\gamma-3\pi}{2} + \pi \le \pi,
\end{aligned} \tag{20}$$

by our hypothesis $\beta + 2\gamma \le 3\pi$. This completes the proof. □

We shall need to consider the initial value problem (21) below. Although this is a strongly singular problem, we shall see that standard comparison arguments hold. In particular we shall establish existence, uniqueness and monotonicity with respect to a parameter.

**Lemma 2.3** *Consider the singular initial value problem*

$$\begin{cases} h'(\theta) = -\dfrac{1}{\sin\theta}\Big(\alpha h(\theta)^2 - \cos\theta h(\theta) + 1 - \alpha\Big), \; 0 < \theta \le \frac{\pi}{2}, \\ h(0) = 1. \end{cases} \tag{21}$$

*(i) If $\alpha \in (1/2, 1)$ then the problem has a classical solution which is unique. The solution $h(\alpha,\theta)$ depends monotonically on $\alpha$: if $\alpha_1 < \alpha_2$ then $h(\alpha_1,\theta) < h(\alpha_2,\theta)$ for all $\theta \in (0,\pi/2]$.*

*(ii) For $\alpha = 1/2$ we do not have uniqueness. Indeed we have a continuum of positive solutions.*

*(iii) Let $1/2 < \alpha < 1$ and in addition let $\overline{h} \in C[0,\pi/2] \cap C^1(0,\pi/2]$ be an upper solution of problem (21), that is*

$$\begin{cases} \overline{h}'(\theta) \ge -\dfrac{1}{\sin\theta}\Big(\alpha\overline{h}(\theta)^2 - \cos\theta\overline{h}(\theta) + 1 - \alpha\Big), \; 0 < \theta \le \frac{\pi}{2}, \\ \overline{h}(0) \ge 1. \end{cases} \tag{22}$$

*Then*

$$h(\alpha,\theta) \le \overline{h}(\theta), \qquad 0 \le \theta \le \frac{\pi}{2}.$$

*Proof*

(i) By Lemma 2.1 the function

$$\psi(\theta) = \sin^{\alpha}(\theta/2)\cos^{1-\alpha}(\theta/2)F(\frac{1}{2},\frac{1}{2},\alpha+\frac{1}{2};\sin^2(\frac{\theta}{2}))$$

solves the differential equation

$$\psi''(\theta) + \alpha(1-\alpha)\frac{\psi(\theta)}{\sin^2\theta} = 0\,, \quad 0 < \theta < \frac{\pi}{2}.$$

It is then easily verified that the function

$$h(\theta) = \frac{1}{\alpha}\frac{\psi'(\theta)}{\psi(\theta)}\sin\theta$$

is a solution of the initial-value problem (21).

We next establish the uniqueness of a solution. Let $h_1, h_2$ be two solutions of the initial value problem (21). Then the function $z = h_2 - h_1$ solves the singular linear initial value problem

$$\begin{cases} z'(\theta) = -\frac{1}{\sin\theta}\Big(\alpha(h_1+h_2) - \cos\theta\Big)z(\theta), \\ z(0) = 0. \end{cases}$$

Let us assume the $z$ is not identically zero. By the standard uniqueness theorem, $z$ cannot have any positive zeros, hence we may assume that $z(\theta) > 0$ for all $\theta \in (0, \pi/2)$. However we have $\alpha(h_1 + h_2) - \cos\theta > 0$ near $\theta = 0$, hence $z$ decreases near zero, which is a contradiction.

The monotonicity of the solution $h$ with respect to $\alpha$ will follow from the monotonicity of the nonlinearity with respect to $\alpha$. Let

$$V(\theta, h, \alpha) = -\frac{1}{\sin\theta}\Big(\alpha h^2 - \cos\theta h + 1 - \alpha\Big).$$

For $0 < h < 1$ and $0 < \theta < \pi/2$ we then have

$$\frac{\partial V}{\partial \alpha} = \frac{1-h^2}{\sin\theta} > 0. \tag{23}$$

Now, let $1/2 < \alpha_1 < \alpha_2 < 1$. By (23) we have $h(\alpha_2, \theta) > h(\alpha_1, \theta)$ near $\theta = 0$. Once we are away from $\theta = 0$ we can apply the standard comparison arguments to complete the proof.

(ii) By Lemma 2.1 the general solution of the equation

$$\psi''(\theta) + \frac{1}{4}\frac{\psi(\theta)}{\sin^2\theta} = 0\,, \quad 0 < \theta < \frac{\pi}{2},$$

is

$$\begin{aligned} \psi(\theta) &= c_1 \sin^{1/2}(\frac{\theta}{2})\cos^{1/2}(\frac{\theta}{2})F(\frac{1}{2}, \frac{1}{2}, 1; \sin^2(\frac{\theta}{2})) \\ &\quad + c_2 \sin^{1/2}(\frac{\theta}{2})\cos^{1/2}(\frac{\theta}{2})F(\frac{1}{2}, \frac{1}{2}, 1; \sin^2(\frac{\theta}{2})) \\ &\quad \times \int_{\sin^2(\theta/2)}^{1/2} \frac{dt}{t(1-t)F^2(\frac{1}{2}, \frac{1}{2}, 1; t)}. \end{aligned}$$

This is positive in $(0, \pi/2]$ when $c_1 > 0$ and $c_2 \geq 0$. For any such $\psi$ the function

$$h(\theta) = \frac{2\psi'(\theta)}{\psi(\theta)}\sin\theta$$

then satisfies

$$h'(\theta) = -\frac{1}{2\sin\theta}\Big(h(\theta)^2 - 2\cos\theta\, h(\theta) + 1\Big)\,, \quad h(0) = 1.$$

Actually after some computations we find that the function $h$ is given in this case by

$$h(\theta) = \cos\theta + \sin^2\theta \frac{F(\frac{3}{2},\frac{3}{2},2;\sin^2(\frac{\theta}{2}))}{4F(\frac{1}{2},\frac{1}{2},1;\sin^2(\frac{\theta}{2}))} - \frac{\lambda}{F^2(\frac{1}{2},\frac{1}{2},1;\sin^2(\frac{\theta}{2}))\Big(1+\lambda\int_{\sin^2(\theta/2)}^{1/2}\frac{dt}{t(1-t)F^2(\frac{1}{2},\frac{1}{2},1;t)}\Big)},$$

where $\lambda = c_2/c_1 \geq 0$.

(iii) When $\overline{h}(0) > 1$ the result follows immediately by combining continuity with standard comparison arguments. Assume now that $h(0) = 1$. The function $z = \overline{h} - h$ then satisfies

$$\begin{cases} z'(\theta) \geq -\frac{1}{\sin\theta}\Big(\alpha(\overline{h}+h) - \cos\theta\Big)z(\theta), \\ z(0) = 0. \end{cases} \tag{24}$$

The quantity $\alpha(\overline{h}+h) - \cos\theta$ is positive near $\theta = 0$, say in $(0,\theta_0)$. We shall establish that $z \geq 0$ in this interval; the result for $(0,\pi/2)$ will then follow immediately. Suppose on the contrary that there exists an interval $(\theta_1,\theta_2) \subset (0,\theta_0)$ such that $z < 0$ in $(\theta_1,\theta_2)$. By (24) we conclude that $z$ is actually strictly increasing in $(\theta_1,\theta_2)$. This contradicts the initial value $z(0) = 0$. □

From Lemma 2.3 it follows that the case $\alpha = 1/2$ is critical and needs a different approach. This will be done in the next lemma. In order to make explicit the dependence on $\beta$ we denote

$$g(\beta,\theta) = \frac{\psi_\theta(\beta,\theta)}{\psi(\beta,\theta)}\sin\theta\,,$$

where $\psi(\beta,\theta)$ is the solution of (9) and $\psi_\theta(\beta,\theta)$ is the derivative with respect to $\theta$.

**Lemma 2.4** *Suppose $\pi \leq \beta \leq \beta_{cr}$. Then $g(\beta,\theta)$, $0 < \theta \leq \pi/2$, is strictly increasing as a function of $\beta$, that is, if $\pi \leq \beta_1 < \beta_2 \leq \beta_{cr}$ then $g(\beta_1,\theta) < g(\beta_2,\theta)$ for all $\theta \in (0,\pi/2]$.*

*Proof* The function $g(\beta,\theta)$ solves the differential equation

$$\frac{\partial g}{\partial\theta} = -\frac{1}{\sin\theta}\Big(g^2 - g\cos\theta + \frac{1}{4}\Big). \tag{25}$$

Since

$$g(\beta, \frac{\pi}{2}) = \frac{1}{2} \tan(\frac{\beta - \pi}{4}),$$

which is strictly increasing with respect to $\beta$, the result follows from a standard comparison argument. □

Let us note here that for $\pi \leq \beta \leq \beta_{cr}$ we have $g(\beta, 0) = 1/2$. So the functions $g(\beta, \cdot)$, $\pi \leq \beta \leq \beta_{cr}$, all solve the same initial value problem.

**Lemma 2.5** *Let $\beta \in [\pi, 2\pi]$. There exists an angle $\gamma^*_\beta$ so that for all $0 < \gamma \leq \gamma^*_\beta$ we have*

$$g(\beta, \theta)\cos(\theta + \frac{\gamma}{2}) + \alpha \cos\frac{\gamma}{2} \geq 0\,, \qquad 0 \leq \theta \leq \frac{\pi}{2}. \tag{26}$$

*Moreover $\gamma^*_\beta$ is a strictly decreasing function of $\beta$ and in particular:*

$$\begin{aligned} &\textit{for } \pi \leq \beta \leq \beta_{cr} \quad \textit{we have} \quad 0.701\pi \approx \gamma^*_{\beta_{cr}} \leq \gamma^*_\beta \leq \gamma^*_\pi \approx 0.867\pi \\ &\textit{for } \beta_{cr} \leq \beta \leq 2\pi \quad \textit{we have} \quad 0.673\pi \approx \gamma^*_{2\pi} \leq \gamma^*_\beta \leq \gamma^*_{\beta_{cr}} \approx 0.701\pi. \end{aligned} \tag{27}$$

*Proof* Inequality (26) is written equivalently

$$\cot\frac{\gamma}{2} \geq \frac{\sin\theta}{\cos\theta + \frac{\alpha}{g(\beta,\theta)}}, \tag{28}$$

so what matters is the maximum of the function at the RHS of (28). For each $0 < \theta \leq \pi/2$ this function is strictly monotone as a function of $\beta$; this follows from Lemma 2.3 for $\beta_{cr} \leq \beta \leq 2\pi$ and from Lemma 2.4 for $\pi \leq \beta \leq \beta_{cr}$.

The angle $\gamma^*_\beta \in (0, \pi)$ defined by

$$\cot\frac{\gamma^*_\beta}{2} = \max_{[0,\pi/2]} \frac{\sin\theta}{\cos\theta + \frac{\alpha}{g(\beta,\theta)}}$$

is then a strictly increasing function of $\beta$. The approximate values in the statement have been obtained by numerical computations; see however Lemma 2.6. □

It would be nice to have good estimates on $\gamma^*_\beta$ without using a numerical solution of the differential equation (16) solved by $g(\theta)$. This will be done for $\beta_{cr} \leq \beta \leq 2\pi$ by obtaining very good upper estimates on $g(\beta, \theta)$. We define

$$\overline{g}(\beta, \theta) = a - \frac{a}{2(2a+1)}\theta^2 + \frac{a(4a^2 + 2a + 3)}{24(2a+1)(4a^2 + 8a + 3)}\theta^4, \quad 0 < \theta < \frac{\pi}{2},$$

where $a$ is the largest solution of $a(1-a) = c_\beta$. We define the auxiliary quantity $\gamma_\beta^{**} \in (0,\pi)$ by

$$\cot\frac{\gamma_\beta^{**}}{2} = \max_{[0,\pi/2]} \frac{\sin\theta}{\cos\theta + \frac{\alpha}{\overline{g}(\beta,\theta)}}.$$

**Lemma 2.6** *Let $\beta_{cr} \le \beta \le 2\pi$. Then we have*

$$(i)\ \ g(\beta,\theta) \le \overline{g}(\beta,\theta)\,, \qquad 0 < \theta < \frac{\pi}{2}.$$

$$(ii)\ \ \gamma_\beta^{**} \le \gamma_\beta^{*}.$$

*Actually we have [cf. (27)]*

$$\gamma_{\beta_{cr}}^{**} \approx 0.700\pi\,, \qquad \gamma_{2\pi}^{**} \approx 0.672\pi\,.$$

*Proof* We have $g(\beta,0) = \overline{g}(\beta,0) = \alpha$. Therefore, given that $g(\beta,\theta)$ satisfies

$$\frac{\partial g}{\partial\theta} = -\frac{1}{\sin\theta}\Big(g^2 - g\cos\theta + c_\beta\Big), \tag{29}$$

it is enough to show that

$$\frac{\partial\overline{g}}{\partial\theta} \ge -\frac{1}{\sin\theta}\Big(\overline{g}^2 - \overline{g}\cos\theta + c_\beta\Big). \tag{30}$$

The function $\overline{g}(\beta,\theta)$ is decreasing with respect to $\theta$, hence

$$\begin{aligned} &\sin\theta\frac{d\overline{g}}{d\theta} + \overline{g}^2 - (\cos\theta)\overline{g} + c_\beta \\ &\qquad \ge \Big(\theta - \frac{\theta^3}{6} + \frac{\theta^5}{120}\Big)\frac{d\overline{g}}{d\theta} + \overline{g}^2 - \Big(1 - \frac{\theta^2}{2} + \frac{\theta^4}{24}\Big)\overline{g} + c_\beta\,. \end{aligned} \tag{31}$$

Now, a direct computation shows that the RHS of (31) is equal to

$$\begin{aligned} &\frac{a(1-a)\theta^6[16(2a+3)(2a+1)(22a^2+2a+3) - (12a^2+2a+3)(4a^2+2a+3)\theta^2]}{2880(2a+1)^2(4a^2+8a+3)^2} \\ &\qquad \ge \frac{a(1-a)(12a^2+2a+3)(4a^2+2a+3)\theta^6(16-\theta^2)}{2880(2a+1)^2(4a^2+8a+3)^2} \\ &\qquad \ge 0. \end{aligned}$$

We note that in our argument we only used that $\alpha \in [1/2, 1)$.

We now establish (i) for $\beta_{cr} < \beta \le 2\pi$. The function

$$\overline{h}(\alpha, \theta) = \frac{\overline{g}(\beta, \theta)}{\alpha}$$

(where, as usual, $\alpha$ is the largest solution of $\alpha(1-\alpha) = c_\beta < 1/4$) is an upper solution to the initial value problem (21). Hence applying (iii) of Lemma 2.3 we obtain the comparison.

To obtain (i) for $\beta = \beta_{cr}$ we use the monotonicity with respect to $\alpha$ of $h(\alpha, \theta)$. Passing to the limit $\alpha \to 1/2+$ we conclude that

$$H(\theta) := \lim_{\alpha \to 1/2+} h(\alpha, \theta) \le \overline{h}(\frac{1}{2}, \theta) \le 2\overline{g}(\beta_{cr}, \theta) , \qquad 0 < \theta < \frac{\pi}{2}.$$

The function $H(\theta)$ is then the maximal solution of the singular initial value problem (21) and therefore coincides with the function $2g(\beta_{cr}, \theta)$. This completes the proof of (i). Part (ii) then follows immediately from (i). □

## 3 Proof of Theorem 1.1

In this section we give the proofs of our theorems. We start with a proposition that is fundamental in our argument and will be used repeatedly. We do not try to obtain the most general statement and for simplicity we restrict ourselves to assumptions that are sufficient for our purposes.

Let $U$ be a domain and assume that $\partial U = \Gamma \cup \Gamma_0$ where $\Gamma$ is Lipschitz continuous. We denote by $\mathbf{n}$ the exterior unit normal on $\Gamma$.

**Proposition 3.1** *Let $\phi \in H^1_{\mathrm{loc}}(U)$ be a positive function such that $\nabla\phi/\phi \in L^2(U)$ and $\nabla\phi/\phi$ has an $L^1$ trace on $\Gamma$ in the sense that $v\nabla\phi/\phi$ has an $L^1$ trace on $\partial U$ for every $v \in C^\infty(\overline{U})$ that vanishes near $\Gamma_0$. Then*

$$\int_U |\nabla u|^2 dx\,dy \ge -\int_U \frac{\Delta\phi}{\phi} u^2 dx\,dy + \int_\Gamma \frac{\nabla\phi}{\phi} \cdot \mathbf{n} u^2 dS \tag{32}$$

*for all smooth functions $u$ which vanish near $\Gamma_0$ and $\Delta\phi$ is understood in the weak sense. If in particular there exists $c \in \mathbb{R}$ such that*

$$-\Delta\phi \ge \frac{c}{d^2}\phi , \tag{33}$$

*in the weak sense in $U$, where $d = \mathrm{dist}(x, \Gamma_0)$, then*

$$\int_U |\nabla u|^2 dx\,dy \ge c\int_U \frac{u^2}{d^2} dx\,dy + \int_\Gamma u^2 \frac{\nabla\phi}{\phi} \cdot \mathbf{n} dS \tag{34}$$

*for all functions $u \in C^\infty(\overline{U})$ that vanish near $\Gamma_0$.*

*Proof* Let $u$ be a function in $C^\infty(\overline{U})$ that vanishes near $\Gamma_0$. We denote $\mathbf{T} = -\nabla\phi/\phi$. Then

$$\int_U u^2 \mathrm{div}\mathbf{T}\, dx\, dy = -2\int_U u\nabla u\cdot\mathbf{T}\, dx\, dy + \int_\Gamma u^2\mathbf{T}\cdot\mathbf{n}\, dS$$
$$\leq \int_U |\mathbf{T}|^2 u^2 dx\, dy + \int_U |\nabla u|^2 dx\, dy + \int_\Gamma u^2\mathbf{T}\cdot\mathbf{n}\, dS\,,$$

that is

$$\int_U |\nabla u|^2 dx\, dy \geq \int_U (\mathrm{div}\mathbf{T} - |\mathbf{T}|^2)u^2 dx\, dy - \int_\Gamma \mathbf{T}\cdot\mathbf{n}u^2 dS\,.$$

Using assumption (33) we obtain (34). □

For $\beta \in (\pi, 2\pi]$ we denote by $\Pi_\beta$ the class of all planar polygons which have precisely one non-convex vertex and the angle at that vertex is $\beta$. Given a polygon in $\Pi_\beta$ we denote by $\gamma_+$ and $\gamma_-$ the angles at the vertices next to the non-convex vertex.

**Theorem 3.1** *Let $\beta \in (\pi, 2\pi]$. Let $\Omega$ be a polygon in $\Pi_\beta$ with*

$$\gamma_+, \gamma_- \leq \min\{\gamma^*_\beta, \frac{3\pi - \beta}{2}\} \tag{35}$$

*where $\gamma^*_\beta \in (0,\pi)$ is defined by*

$$\cot\frac{\gamma^*_\beta}{2} = \max_{[0,\pi/2]} \frac{\sin\theta}{\cos\theta + \frac{\alpha}{g(\beta,\theta)}}.$$

*Then the Hardy constant of $\Omega$ is $c_\beta$.*

*Proof* We denote by $A_-$, $A_+$ the vertices next to the non-convex vertex $O$, so that $A_-$, $O$ and $A_+$ are consecutive vertices with respective angles $\gamma_-$, $\beta$ and $\gamma_+$. We may assume that $O$ is the origin and that $A_+$ lies on the positive $x$-semiaxis. We write the boundary $\partial\Omega$ as

$$\partial\Omega = S_1 \cup S_2$$

where $S_1 = OA_+ \cup OA_-$ and $S_2 = \partial\Omega \setminus S_1$. We then define the equidistance curve

$$\Gamma = \{x \in \partial\Omega : \mathrm{dist}(x, S_1) = \mathrm{dist}(x, S_2)\}.$$

Hence $\Gamma$ divides $\Omega$ into two sets $\Omega_1$ and $\Omega_2$, whose nearest boundary points belong in $S_1$ and $S_2$ respectively. It is clear that $\Gamma$ can be parametrized by the polar angle $\theta \in [0, \beta]$.

The curve $\Gamma$ consists of line segments and parabola segments. Starting from $\theta = 0$ we have line segments $L_1, \dots, L_k$; then from $\theta = \pi/2$ to $\theta = \beta - \pi/2$ we have parabola segments $P_1, \dots, P_m$; and from $\theta = \beta - \pi/2$ to $\theta = \beta$ we have again line segments $L'_1, \dots, L'_n$.

Let $u \in C_c^\infty(\Omega)$ be given. Let $\mathbf{n}$ denote the unit normal along $\Gamma$ which is outward with respect to $\Omega_1$. Applying Proposition 3.1 with $\phi(x, y) = \psi_\beta(\theta)$, where $\theta$ is the polar angle of the point $(x, y)$, we obtain

$$\int_{\Omega_1} |\nabla u|^2 dx\,dy \geq c_\beta \int_{\Omega_1} \frac{u^2}{d^2} dx\,dy + \int_\Gamma \frac{\nabla\phi}{\phi} \cdot \mathbf{n}\, u^2 dS. \tag{36}$$

We next apply Proposition 3.1 on $\Omega_2$ for the function $\phi_1(x, y) = d(x, y)^\alpha$ (we recall that $\alpha$ is the largest solution of $\alpha(1 - \alpha) = c_\beta$). In $\Omega_2$ the function $d(x, y)$ coincides with the distance from $S_2$ and this implies that

$$-\Delta d^\alpha \geq \alpha(1 - \alpha) \frac{d^\alpha}{d^2}, \qquad \text{on } \Omega_+ .$$

Applying Proposition 3.1 we obtain that

$$\int_{\Omega_+} |\nabla u|^2 dx\,dy \geq c \int_{\Omega_+} \frac{u^2}{d^2} dx\,dy - \int_\Gamma \frac{\alpha\nabla d}{d} \cdot \mathbf{n}\, u^2 dS\,. \tag{37}$$

Adding (36) and (37) we conclude that

$$\int_\Omega |\nabla u|^2 dx\,dy \geq c \int_\Omega \frac{u^2}{d^2} dx\,dy + \int_\Gamma \Big(\frac{\nabla\phi}{\phi} - \alpha\frac{\nabla d}{d}\Big) \cdot \mathbf{n}\, u^2 dS\,. \tag{38}$$

We emphasize that in the last integral the values of $\nabla\phi/\phi$ are obtained as limits from $\Omega_1$ and, more importantly, those of $\nabla d/d$ are obtained as limits from $\Omega_2$.

It remains to prove that the line integral in (38) is non-negative. For this we shall consider the different segments of $\Gamma$. Due to the symmetry of our assumptions with respect to $\theta = \beta/2$ it is enough to establish the result for $0 \leq \theta \leq \beta/2$.

(i) Let $L$ be one of the line segments $L_1, \dots, L_k$. The points on this segment $L$ are equidistant from the side $OA_+$ and some side $E$ of $\partial\Omega \setminus (OA_+ \cup OA_-)$. Let $\gamma$ be the angle formed by the line $E$ and the $x$-axis so that the outward normal vector along $E$ is $(\sin\gamma, \cos\gamma)$ and $E$ has equation $x\cos\gamma + y\sin\gamma + c = 0$ for some $c \in \mathbb{R}$. Elementary geometric considerations then give $\gamma \in (-\pi/2, \pi)$. Now, simple computations give

$$\Big(\frac{\nabla\phi}{\phi} - \alpha\frac{\nabla d}{d}\Big) \cdot \mathbf{n} = \frac{1}{d}\Big(g(\theta)\cos(\theta + \frac{\gamma}{2}) + \alpha\cos(\frac{\gamma}{2})\Big)\,, \qquad \text{on } L. \tag{39}$$

It remains to show that the RHS of (39) is non-negative for $0 \leq \theta \leq \pi/2$. In the case $0 < \gamma < \pi$ this is equivalent to showing that

$$\cot\frac{\gamma}{2} \geq \frac{\sin\theta}{\cos\theta + \frac{\alpha}{g(\theta)}}, \qquad 0 \leq \theta \leq \frac{\pi}{2}. \tag{40}$$

This is true since $\gamma \leq \gamma_+ \leq \gamma^*_\beta$.

In the case $-\pi/2 < \gamma \leq 0$ we have $\cos(\theta + \frac{\gamma}{2}) \geq 0$ for all $0 \leq \theta \leq \pi/2$ and the RHS is clearly non-negative.

(ii) Let $P$ be one of the parabola segments $P_1, \ldots, P_m$. The points on $P$ are equidistant from the origin $O$ and some side $E$ of $\partial\Omega \setminus (OA_+ \cup OA_-)$. As in (i) above, let $\gamma$ be the angle formed by the line $E$ and the $x$-axis so that the outward normal vector along $E$ is $(\sin\gamma, \cos\gamma)$ and $E$ has equation $x\cos\gamma + y\sin\gamma + c = 0$ for some $c \in \mathbb{R}$. Then $\gamma \in [\pi - \frac{\beta}{2}, \pi]$. We note that the axis of the parabola has an asymptote at angle $\theta = \frac{3\pi}{2} - \gamma$. Indeed we shall prove the required inequality for all $\theta \in [\frac{\pi}{2}, \frac{3\pi}{2} - \gamma] \supset [\frac{\pi}{2}, \frac{\beta}{2}]$.

Simple computations on $P$ give

$$\Big(\frac{\nabla\phi}{\phi} - \alpha\frac{\nabla d}{d}\Big)\cdot \mathbf{n} = \frac{1}{r\sqrt{2 + 2\sin(\theta+\gamma)}}\Big(f(\theta)\cos(\theta+\gamma) + \alpha[1 + \sin(\theta+\gamma)]\Big). \tag{41}$$

Hence, noting that $\gamma \leq \gamma_+$, the result follows from Lemma 2.2. This completes the proof. □

*Proof of Theorem 1.1* This follows easily by approximating the convex set $K$ by a sequence of convex polygons and using Theorem 3.1; see Fig. 1. □

**Remark** In case $\beta \leq \beta_{cr}$ we have $\gamma^*_\beta \leq \gamma^*_{\beta_{cr}} \approx 0.701\pi$ and therefore the condition $\gamma_+, \gamma_- \leq \min\{\gamma^*_\beta, \frac{3\pi - \beta}{2}\}$ of Theorems 1.1 and 3.1 takes the simpler form

$$\gamma_+, \gamma_- \leq \gamma^*_\beta.$$

If the convex set $K$ is unbounded and $\partial K$ does not intersect the boundary of $\Lambda_\beta$ then there is no need for any restriction. In particular

**Theorem 3.2** *Let $\Omega = K \cap \Lambda_\beta$, where $K$ is an unbounded convex set and $\Lambda_\beta$ is a sector of angle $\beta \in (\pi, 2\pi]$ whose vertex is inside $K$. Assume that the boundaries of $K$ and $\Lambda_\beta$ do not intersect. Then the Hardy constant of $\Omega$ is $c_\beta$, where $c_\beta$ is given by (5) and (6).*

*Proof* Let $u \in C^\infty_c(\Omega)$ be fixed. There exists a bounded convex set $K_1$ such that $\Omega_1 := K_1 \cap S_\beta$ satisfies all the assumptions of Theorem 1.1 and in addition

$$\mathrm{dist}(x, \partial\Omega) = \mathrm{dist}(x, \partial\Omega_1)\,, \qquad x \in \mathrm{supp}(u)\,;$$

of course, $K_1$ depends on $u$. Applying Theorem 1.1 to $\Omega_1$ we obtain the required Hardy inequality. □

**Remark** Of course, one could state an intermediate result where the intersection $\partial K \cap \partial \Lambda_\beta$ is exactly one point forming an angle $\gamma$; in this the assumption $\gamma \le \min\{\gamma^*_\beta, \frac{3\pi-\beta}{2}\}$ should hold.

## 4 Domains $E_{\beta,\gamma}$ with Two Non-convex Angles

We recall from the Introduction that given angles $\beta$ and $\gamma$, we denote by $E_{\beta,\gamma}$ the domain shown in Fig. 2 in case $\gamma < \pi$ and in Fig. 3 in case $\gamma > \pi$. Its boundary $\partial E_{\beta,\gamma}$ consists of three parts $L_1$, $L_2$ and $L_3$. $L_2$ is a line segment and meets the halflines $L_3$ and $L_1$ at the origin O and the point $P(1,0)$ respectively. We assume that $\beta + \gamma \le 3\pi$ so that the halflines $L_1$ and $L_3$ do not intersect. Without loss of generality we assume that $\beta \ge \gamma$ and since we are interested in the non-convex case, we assume that $\beta > \pi$.

*Proof of Theorem 1.2 Part (i)* We denote by $\Gamma$ the curve

$$\Gamma = \{(x,y) \in E_{\beta,\gamma} : \text{dist}((x,y), L_1) = \text{dist}((x,y), L_2 \cup L_3)\}.$$

The curve $\Gamma$ divides $E_{\beta,\gamma}$ in two sets $E_- = \{(x,y) \in E_{\beta,\gamma} : d(x,y) = \text{dist}((x,y), L_2 \cup L_3)\}$ and $E_+ = \{(x,y) \in E_{\beta,\gamma} : d(x,y) = \text{dist}((x,y), L_1)\}$. We denote by $\mathbf{n}$ the unit normal along $\Gamma$ which is outward with respect to $E_-$.

Once again we shall use Proposition 3.1. We distinguish two cases: Case A, where $0 \le \gamma \le \pi/2$ and Case B, where $\pi/2 \le \gamma \le \pi$.

*Case A* $(0 \le \gamma \le \pi/2)$ We distinguish two subcases.

*Subcase Aa* $\beta + \gamma < 2\pi$. In this case $\Gamma$ consists of three parts: a line segment $\Gamma_1$ which bisects the angle at $P$; a parabola segment $\Gamma_2$, whose points are equidistant from the origin and the line $L_1$; and a halfline $\Gamma_3$ whose points are equidistant from $L_1$ and $L_3$. We parametrize $\Gamma$ by the polar angle $\theta$, so that $\Gamma_1 = \{0 \le \theta \le \frac{\pi}{2}\}$, $\Gamma_2 = \{\frac{\pi}{2} \le \theta \le \beta - \frac{\pi}{2}\}$, and $\Gamma_3 = \{\beta - \frac{\pi}{2} \le \theta < \frac{\beta+\pi-\gamma}{2}\}$.

Let $u \in C_c^\infty(E_{\beta,\gamma})$. We apply Proposition 3.1 with $U = E_-$, $\Gamma_0 = L_2 \cup L_3$ and for the function $\phi(x,y) = \psi(\theta)$, where $\psi = \psi_\beta$ and $\theta$ is the polar angle of $(x,y)$. We obtain that

$$\int_{E_-} |\nabla u|^2 dx\,dy \ge c_\beta \int_{E_-} \frac{u^2}{d^2} dx\,dy + \int_\Gamma \frac{\nabla\phi}{\phi} \cdot \mathbf{n} u^2\, dS\,. \tag{42}$$

We next apply Proposition 3.1 to the domain $E_+$ and the function $\phi_1(x,y) = d(x,y)^\alpha$. We obtain that

$$\int_{E_+} |\nabla u|^2 dx\,dy \geq c_\beta \int_{E_+} \frac{u^2}{d^2} dx\,dy - \alpha \int_\Gamma \frac{\nabla d}{d} \cdot \mathbf{n} u^2 dS\,. \tag{43}$$

Adding (42) and (43) we conclude that

$$\int_{E_{\beta,\gamma}} |\nabla u|^2 dx\,dy \geq c_\beta \int_{E_{\beta,\gamma}} \frac{u^2}{d^2} dx\,dy + \int_\Gamma \Big(\frac{\nabla\phi}{\phi} - \alpha\frac{\nabla d}{d}\Big) \cdot \mathbf{n}\, u^2 dS\,. \tag{44}$$

We note that in the last integral the values of $\nabla\phi/\phi$ are obtained as limits from $E_-$ while those of $\nabla d/d$ are obtained as limits from $E_+$. It remains to prove that the last integral in (44) is non-negative. For this we shall consider the different parts of $\Gamma$.

(i) The segment $\Gamma_1$ $(0 \leq \theta \leq \pi/2)$. Simple computations give that

$$\frac{\nabla\phi}{\phi} - \alpha\frac{\nabla d}{d} = \frac{1}{d}\Big(g(\theta)\cos(\theta + \frac{\gamma}{2}) + \alpha\cos(\frac{\gamma}{2})\Big)\,, \qquad 0 < \theta \leq \frac{\pi}{2}\,;$$

this is non-negative by Lemma 27, since $\gamma^*_\beta > \pi/2$.

(ii) The segment $\Gamma_2$ $(\pi/2 \leq \theta \leq \beta - \pi/2)$. In this case we have

$$\Big(\frac{\nabla\phi}{\phi} - \alpha\frac{\nabla d}{d}\Big)\cdot\mathbf{n} = \frac{1}{r\sqrt{2 + 2\sin(\theta+\gamma)}}\Big(f(\theta)\cos(\theta+\gamma) + \alpha[1 + \sin(\theta+\gamma)]\Big),$$

this is non-negative by Lemma 2.2, since $\beta - \frac{\pi}{2} < \frac{3\pi}{2} - \gamma$.

(iii) The segment $\Gamma_3$ $(\beta - \frac{\pi}{2} \leq \theta < \frac{\beta+\pi-\gamma}{2})$. The line containing $\Gamma_3$ has equation

$$x\cos(\frac{\beta-\gamma}{2}) + y\sin(\frac{\beta-\gamma}{2}) = \frac{\sin\gamma}{2\sin(\frac{\beta+\gamma}{2})}\,,$$

hence the outer (with respect to $E_-$) unit normal along $\Gamma_3$ is $(\cos(\frac{\beta-\gamma}{2}), \sin(\frac{\beta-\gamma}{2}))$.

Using the fact that $d = r\sin(\beta - \theta)$ on $\Gamma_3$, we have along $\Gamma_3$,

$$\Big(\frac{\nabla\phi}{\phi} - \alpha\frac{\nabla d}{d}\Big)\cdot\mathbf{n} = [\frac{1}{r}\frac{\psi'(\theta)}{\psi(\theta)}(-\sin\theta, \cos\theta) + \alpha\frac{(\sin\gamma, \cos\gamma)}{d}] \cdot\ (\cos(\frac{\beta-\gamma}{2}), \sin(\frac{\beta-\gamma}{2}))$$

$$= \frac{1}{r}\Big[\frac{\psi'(\theta)}{\psi(\theta)}\sin(\frac{\beta-\gamma}{2}-\theta)+\alpha\frac{\sin(\frac{\beta+\gamma}{2})}{\sin(\beta-\theta)}\Big]$$
$$\geq 0,$$

since both terms in the last sum are non-negative (the first one, as the product of two non-positive terms).

*Subcase Ab* $\beta+\gamma \geq 2\pi$. In this case $\Gamma$ consists of only two parts $\Gamma_1$ and $\Gamma_2$, described exactly as in subcase Aa, the only difference being that the range of $\theta$ in $\Gamma_2$ is $\frac{\pi}{2} \leq \theta < \frac{3\pi}{2}-\gamma$. This means that the parabola segment goes all the way to infinity. As before we have

$$\Big(\frac{\nabla\phi}{\phi}-\alpha\frac{\nabla d}{d}\Big)\cdot\mathbf{n} = \frac{1}{r\sqrt{2+2\sin(\theta+\gamma)}}\Big(\frac{\psi'(\theta)}{\psi(\theta)}\cos(\theta+\gamma)+\alpha[1+\sin(\theta+\gamma)]\Big)$$

and the result follows again from Lemma 2.2. This completes the proof in the case $0<\gamma\leq\pi/2$.

*Case B* $(\pi/2 \leq \gamma \leq \pi)$. On $E_-$ we again consider the function $\phi(x,y) = \psi(\theta)$ and apply Proposition 3.1 as in the previous case. We fix a function $u \in C_c^\infty(E_{\beta,\gamma})$ and we obtain

$$\int_{E_-}|\nabla u|^2dx\,dy \geq c_\beta\int_{E_-}\frac{u^2}{d^2}dx\,dy+\int_\Gamma(\frac{\nabla\phi}{\phi}\cdot\mathbf{n})u^2dS\,. \tag{45}$$

In $E_+$ we consider a new orthonormal coordinate system with cartesian coordinates denoted by $(x_1,y_1)$ and polar coordinates denoted by $(r_1,\theta_1)$. The origin $O_1$ of this system is located on the line $L_1$ and is such that the line $OO_1$ is perpendicular to $L_1$. The positive $x_1$ axis is then chosen so as to contain $L_1$ (Figure 2) We note that this choice is such that

$$\text{the point on } \Gamma_1 \text{ for which } \theta=\tfrac{\pi}{2}-\tfrac{\gamma}{2} \text{ satisfies also } \theta_1=\tfrac{\pi}{2}-\tfrac{\gamma}{2}. \tag{46}$$

We apply Proposition 3.1 on $E_+$ with the function $\phi_1(x,y) = \psi(\theta_1)$. This function clearly satisfies $-\Delta\phi_1 = c\,d^{-2}\phi_1$, hence we obtain

$$\int_{E+}|\nabla u|^2dx\,dy \geq c\int_{E_+}\frac{u^2}{d^2}dx\,dy-\int_\Gamma(\frac{\nabla\phi_1}{\phi_1}\cdot\mathbf{n})u^2\,dS\,, \tag{47}$$

where, as before, $\mathbf{n}$ is the interior to $E_+$ unit normal along $\Gamma$.

Adding (45) and (47) we conclude that

$$\int_{E_{\beta,\gamma}}|\nabla u|^2dx\,dy \geq c_\beta\int_{E_{\beta,\gamma}}\frac{u^2}{d^2}dx\,dy+\int_\Gamma\Big(\frac{\nabla\phi}{\phi}-\frac{\nabla\phi_1}{\phi_1}\Big)\cdot\mathbf{n}\,u^2dS\,. \tag{48}$$

The rest of the proof is devoted to showing that the last integral in (48) is non-negative.

As in the case $0 < \gamma \leq \pi/2$, we need to distinguish two subcases: Subcase Ba, where $\beta + \gamma < 2\pi$, and Subcase Bb, where $\beta + \gamma \geq 2\pi$.

*Subcase Ba* $\beta + \gamma < 2\pi$. The curve $\Gamma$ consists of three parts: a line segment $\Gamma_1$ which bisects the angle at $P$; a (part of a) parabola $\Gamma_2$, whose points are equidistant from the origin and the line $L_1$; and a halfline $\Gamma_3$ whose points are equidistant from $L_1$ and $L_3$. As before, we consider separately each segment and we parametrize $\Gamma$ by the polar angle $\theta$ so that

$$\Gamma_1 = \{\theta \in \Gamma : 0 \leq \theta \leq \frac{\pi}{2}\}, \quad \Gamma_2 = \{\frac{\pi}{2} \leq \theta \leq \beta - \frac{\pi}{2}\},$$
$$\Gamma_3 = \{\beta - \frac{\pi}{2} \leq \theta < \frac{\beta + \pi - \gamma}{2}\}.$$

(i) The segment $\Gamma_1$ ($0 \leq \theta \leq \pi/2$). We have

$$\frac{\nabla\phi}{\phi} \cdot \mathbf{n} = \frac{\psi'(\theta)}{r\psi(\theta)} \cos(\theta + \frac{\gamma}{2}), \qquad \text{on } \Gamma_1.$$

and similarly

$$\frac{\nabla\phi_1}{\phi_1} \cdot \mathbf{n} = -\frac{\psi'(\theta_1)}{r_1\psi(\theta_1)} \cos(\theta_1 - \frac{\gamma}{2}), \qquad \text{on } \Gamma_1.$$

Since $r_1 \sin\theta_1 = r\sin\theta$ along $\Gamma_1$, it is enough to prove the inequality

$$g(\theta)\cos(\theta + \frac{\gamma}{2}) + g(\theta_1)\cos(\theta_1 - \frac{\gamma}{2}) \geq 0, \qquad 0 \leq \theta \leq \frac{\pi}{2}. \tag{49}$$

This has been proved in [7]; we include a proof here for the sake of completeness. Recalling (46) and applying the sine law we obtain that along $\Gamma_1$ the polar angles $\theta$ and $\theta_1$ are related by

$$\cot\theta_1 = -\cos\gamma\cot\theta + \sin\gamma\,. \tag{50}$$

**Claim** There holds

$$\theta_1 \geq \theta + \gamma - \pi\,, \qquad \text{on } \Gamma_1\,. \tag{51}$$

*Proof of Claim* We fix $\theta \in [0, \pi/2]$ and the corresponding $\theta_1 = \theta_1(\theta)$. If $\theta + \gamma - \pi \leq 0$, then (51) is obviously true, so we assume that $\theta + \gamma - \pi \geq 0$. Since $0 \leq \theta + \gamma - \pi \leq \pi/2$ and $0 \leq \theta_1 \leq \pi/2$, (51) is written equivalently $\cot\theta_1 \leq \cot(\theta + \gamma - \pi)$; thus, recalling (50), we conclude that to prove the claim it is enough to show that

$$-\cos\gamma\cot\theta+\sin\gamma\le\cot(\theta+\gamma)\,,\quad \pi-\gamma\le\theta\le\frac{\pi}{2},$$

or, equivalently (since $\pi\le\theta+\gamma\le 3\pi/2$),

$$-\cos\gamma\cot^2\theta+(-\cos\gamma\cot\gamma-\cot\gamma+\sin\gamma)\cot\theta+1+\cos\gamma\ge 0\,,\pi-\gamma\le\theta\le\frac{\pi}{2}. \tag{52}$$

The left-hand side of (52) is an increasing function of $\cot\theta$ and therefore takes its least value at $\cot\theta=0$. Hence the claim is proved. □

For $0\le\theta\le\pi/2-\gamma/2$ (49) is true since all terms in the left-hand side are non-negative. So let $\pi/2-\gamma/2\le\theta\le\pi/2$ and $\theta_1=\theta_1(\theta)$. From (50) we find that

$$\begin{aligned}\frac{d\theta_1}{d\theta}-1&=-\frac{\cos\gamma(1+\cot^2\theta)+1+\cot^2\theta_1}{1+\cot^2\theta_1}\\&=-\frac{1+\sin^2\gamma+\cos\gamma-2\sin\gamma\cos\gamma\cot\theta+\cos\gamma(1+\cos\gamma)\cot^2\theta}{1+\cot^2\theta_1}.\end{aligned}$$

The function

$$h(x):=1+\sin^2\gamma+\cos\gamma-2\sin\gamma\cos\gamma x+\cos\gamma(1+\cos\gamma)x^2$$

is a concave function of $x$. We will establish the positivity of $h(\cot\theta)$ for $\pi/2-\gamma/2\le\theta\le\pi/2$. For this it is enough to establish the positivity at the endpoints. At $\theta=\pi/2$ positivity is obvious, whereas

$$h(\tan(\frac{\gamma}{2}))=1+\sin^2\gamma+\cos\gamma-2\cos\gamma\sin^2\frac{\gamma}{2}\ge 0.$$

From (46) we conclude that $\theta_1\le\theta$ for $\pi/2-\gamma/2\le\theta\le\pi/2$. Now, it was proved in [7, Lemma 4] that the function $g$ is decreasing. Hence for $\pi/2-\gamma/2\le\theta\le\pi/2$ we have,

$$\begin{aligned}g(\theta)\cos(\theta+\frac{\gamma}{2})+g(\theta_1)\cos(\theta_1-\frac{\gamma}{2})&\ge g(\theta)[\cos(\theta+\frac{\gamma}{2})+\cos(\theta_1-\frac{\gamma}{2})]\\&=2g(\theta)\cos(\frac{\theta+\theta_1}{2})\cos(\frac{\theta-\theta_1+\gamma}{2})\\&\ge 0,\end{aligned}$$

where for the last inequality we made use of the claim. Hence (49) has been proved.

(ii) The segment $\Gamma_2$ ($\frac{\pi}{2} \le \theta \le \beta - \frac{\pi}{2}$). After some computations we obtain that

$$\Big(\frac{\nabla\phi}{\phi} - \frac{\nabla\phi_1}{\phi_1}\Big)\cdot \mathbf{n} = \frac{1}{r\sqrt{2+2\sin(\theta+\gamma)}} \Big\{ f(\theta)\cos(\theta+\gamma)$$
$$- f(\theta_1)\sin\theta_1[\sin(\theta_1-\theta-\gamma)-\cos\theta_1]\Big\},$$

where $\theta$ and $\theta_1$ are related by $\cot\theta_1 = -\cos(\theta+\gamma)$. The result then follows by applying [7, Lemma 6].

(iii) The segment $\Gamma_3$ ($\beta - \frac{\pi}{2} \le \theta < \frac{\beta+\pi-\gamma}{2}$). Simple computations yield that along $\Gamma_3$ we have

$$\Big(\frac{\nabla\phi}{\phi} - \frac{\nabla\phi_1}{\phi_1}\Big)\cdot \mathbf{n} = \frac{\psi'(\theta)}{r\psi(\theta)}\sin\Big(\frac{\beta-\gamma}{2}-\theta\Big) + \frac{\psi'(\theta_1)}{r_1\psi(\theta_1)}\sin\Big(\frac{\beta+\gamma}{2}-\theta_1\Big). \tag{53}$$

The first summand in the right-hand side of (53) is non-negative since $\psi'(\theta)$ and $\sin(\frac{\beta-\gamma}{2}-\theta)$ are non-positive in the given range of $\theta$. Moreover, two applications of the sine law yield that along $\Gamma_3$ the coordinates $(r,\theta)$ and $(r_1,\theta_1)$ are related by

$$r_1\sin\theta_1 = r\sin(\beta-\theta)\ , \qquad \tan\theta_1 = -\frac{\sin(\beta-\theta)}{\cos(\theta+\gamma)}.$$

It follows in particular that $0 \le \theta_1 \le \pi/2$, and hence $\pi/4 \le \frac{\beta+\gamma}{2} - \theta_1 \le \pi$. Hence the second summand in the right-hand side of (53) is also non-negative, completing the proof in this case.

*Subcase B2* $\beta+\gamma \ge 2\pi$. In this case $\Gamma$ consists only of two parts $\Gamma_1$ and $\Gamma_2$, described as in Case B1. The only difference is that the range of $\theta$ in $\Gamma_2$ now is $\frac{\pi}{2} \le \theta < \frac{3\pi}{2} - \gamma$; the result follows as before. This completes the proof of the theorem. □

*Proof of Theorem 1.2 Part (ii)* We set for simplicity $\psi = \psi_{\beta+\gamma-\pi}$. We divide $E_{\beta,\gamma}$ in three parts $E_1$, $E_2$ and $E_3$ as in Figure 3, and denote $L_i = (\partial E_i) \cap \partial E_{\beta,\gamma}$. We also set $\Gamma_i = \{(i,y) : y \ge 0\}$, $i = 0,1$, the halflines that are the common boundaries of the $E_j$'s. We first apply Proposition 3.1 to the domain $E_1$. For this we introduce polar coordinates $(r_1,\theta_1)$ centered at $P$, so that the positive $x_1$ axis coincides with the halfline $L_1$. Let $u \in C_c^\infty(E_{\beta,\gamma})$ be fixed. Applying Proposition 3.1 with $\phi(x,y) = \psi(\theta_1)$ we obtain

$$\int_{E_1} |\nabla u|^2 dx\,dy \ge c_{\beta+\gamma-\pi}\int_{E_1}\frac{u^2}{d^2}dx\,dy + \frac{\psi'(\gamma-\frac{\pi}{2})}{\psi(\gamma-\frac{\pi}{2})}\int_{\Gamma_1}\frac{u^2}{y}dy\,. \tag{54}$$

On $E_3$ we use the standard polar coordinates $(r,\theta)$ and the function $\phi(x,y) = \psi(\beta-\theta)$. We obtain

$$\int_{E_3} |\nabla u|^2 dx\,dy \geq c_{\beta+\gamma-\pi} \int_{E_3} \frac{u^2}{d^2} dx\,dy + \frac{\psi'(\beta-\frac{\pi}{2})}{\psi(\beta-\frac{\pi}{2})} \int_{\Gamma_0} \frac{u^2}{y} dy\,. \tag{55}$$

Without loss of generality we assume that $\beta \geq \gamma$ and we therefore have

$$\frac{\psi'(\gamma-\frac{\pi}{2})}{\psi(\gamma-\frac{\pi}{2})} = -\frac{\psi'(\beta-\frac{\pi}{2})}{\psi(\beta-\frac{\pi}{2})} \geq 0.$$

Now, we have $u(1,y)^2 - u(0,y)^2 = 2\int_0^1 uu_x dx$, hence, using also the one-dimensional Hardy inequality we have for any $\epsilon > 0$,

$$\begin{aligned}\int_{\Gamma_0} \frac{u^2}{y} dy - \int_{\Gamma_1} \frac{u^2}{y} dy &\leq \epsilon \int_{E_2} \frac{u^2}{y^2} dx\,dy + \frac{1}{\epsilon}\int_{E_2} u_x^2 dx\,dy \\ &\leq (\epsilon - \frac{1}{4\epsilon}) \int_{E_2} \frac{u^2}{y^2} dx\,dy + \frac{1}{\epsilon}\int_{E_2} u_y^2 dx\,dy + \frac{1}{\epsilon}\int_{E_2} u_x^2 dx\,dy\end{aligned}$$

and therefore

$$\int_{E_2} |\nabla u|^2 dx\,dy \geq \Big(\frac{1}{4} - \epsilon^2\Big) \int_{E_2} \frac{u^2}{y^2} dx\,dy + \epsilon \int_{\Gamma_0} \frac{u^2}{y} dy - \epsilon \int_{\Gamma_1} \frac{u^2}{y} dy\,. \tag{56}$$

This is also true for $\epsilon = 0$. We choose $\epsilon = \psi'(\gamma - \frac{\pi}{2})/\psi(\gamma-\frac{\pi}{2})$ and we note that by (7) we have

$$c_{\beta+\gamma-\pi} \leq \frac{1}{4} - c_{\beta+\gamma-\pi} \tan^2\Big(\sqrt{c_{\beta+\gamma-\pi}}\frac{\beta-\gamma}{2}\Big) = \frac{1}{4} - \Big(\frac{\psi'(\gamma-\frac{\pi}{2})}{\psi(\gamma-\frac{\pi}{2})}\Big)^2 = \frac{1}{4} - \epsilon^2\,.$$

Adding (54), (55) and (56) we obtain the inequalities in all cases.

We now prove the sharpness of the constant. Let $C$ denote the best Hardy constant for $E_{\beta,\gamma}$. We extend the halflines $L_1$ and $L_3$ until they meet at a point $A$, and we call $D_0$ the resulting infinite sector, whose angle is $\beta+\gamma-\pi$. We introduce a family of domains $D_\epsilon$ that are obtained from $E_{\beta,\gamma}$ by moving $L_2$ parallel to itself towards $A$ so that it is a distance $\epsilon$ from $A$. All these domains $D_\epsilon$ have the same Hardy constant as $E_{\beta,\gamma}$. Let $d_\epsilon(x) = \mathrm{dist}(x, \partial D_\epsilon)$ and $d_0(x) = \mathrm{dist}(x, \partial D_0)$. Then clearly $d_\epsilon(x) \to d_0(x)$ for all $x \in D_0$.

Let $u \in C_c^\infty(D_0)$ vanish near $\Gamma_0$. This can be used as a test function for the Hardy inequality in $D_\epsilon$, therefore we have

$$\int_{D_\epsilon} |\nabla u|^2 dx\,dy \geq C \int_{D_\epsilon} \frac{u^2}{d_\epsilon^2} dx\,dy\,,$$

which can be written equivalently

$$\int_{D_0} |\nabla u|^2 dx\,dy \geq C \int_{D_0} \frac{u^2}{d_\epsilon^2} dx\,dy\,.$$

Passing to the limit $\epsilon \to 0$ we therefore obtain

$$\int_{D_0} |\nabla u|^2 dx\,dy \geq C \int_{D_0} \frac{u^2}{d_0^2} dx\,dy\,.$$

Since the best Hardy constant of $D_0$ is $c_{\beta+\gamma-\pi}$, we conclude that $C \leq c_{\beta+\gamma-\pi}$, which establishes the sharpness. □

## 5 A Dirichlet–Neumann Hardy Inequality

We finally prove Theorem 1.3.

*Proof of Theorem 1.3* Let $u \in C^\infty(\overline{D_\beta})$. Applying Proposition 3.1 for $\phi(x,y) = \psi(\theta)$ we have

$$\begin{aligned}\int_{D_\beta} |\nabla u|^2 dx\,dy &\geq -\int_{D_\beta} \frac{\Delta\phi}{\phi} u^2 dx\,dy + \int_\Gamma \frac{\nabla\phi}{\phi} \cdot \mathbf{n} u^2 dS \\ &= c_\beta \int_{D_\beta} \frac{u^2}{d^2} dx\,dy + \int_\Gamma \frac{\nabla\phi}{\phi} \cdot \mathbf{n} u^2 dS.\end{aligned}$$

A direct computation gives that along $\Gamma$ we have

$$\frac{\nabla\phi}{\phi} \cdot \mathbf{n} = -\frac{r'(\theta)}{r(\theta)\sqrt{r(\theta)^2 + r'(\theta)^2}} \cdot \frac{\psi'(\theta)}{\psi(\theta)},$$

which establishes the inequality. The fact that $c_\beta$ is sharp follows by comparing with the corresponding Dirichlet problem. □

## References

1. Ancona, A.: On strong barriers and an inequality of Hardy for domains in $\mathbb{R}^n$. J. Lond. Math. Soc. **34**(2), 274–290 (1986)
2. Armitage, D.H., Kuran, U.: The convexity and the superharmonicity of the signed distance function. Proc. Am. Math. Soc. **93**(4), 598–600 (1985)
3. Avkhadiev, F.: Families of domains with best possible Hardy constant. Russ. Math. (Iz. VUZ) **57**, 49–52 (2013)

4. Banũelos, R.: Four unknown constants. Oberwolfach Report No. 06 (2009)
5. Barbatis, G., Filippas, S., Tertikas, A.: A unified approach to improved $L^p$ Hardy inequalities with best constants. Trans. Am. Math. Soc. **356**, 2169–2196 (2004)
6. Barbatis, G., Lamberti, P.D.: Shape sensitivity analysis of the Hardy constant. Nonlinear Anal. Theory Methods Appl. Ser. A Theory Methods **103**, 98–112 (2014)
7. Barbatis, G., Tertikas, A.: On the Hardy constant of non-convex planar domains: the case of the quadrilateral. J. Funct. Anal. **266**, 3701–3725 (2014)
8. Davies, E.B.: The Hardy constant. Q. J. Math. Oxford Ser. **184**(2), 417–431 (1995)
9. Gkikas, K.: Hardy-Sobolev inequalities in unbounded domains and heat kernel estimates. J. Funct. Anal. **264**, 837–893 (2013)
10. Laptev, A.: Lecture Notes, Warwick, 3–8 April 2005 (unpublished). http://www2.imperial.ac.uk/alaptev/Papers/ln.pdf
11. Laptev, A., Sobolev, A.: Hardy inequalities for simply connected planar domains. In: Suslina, T. (ed.) Spectral Theory of Differential Operators. American Mathematical Society Translation Seriers 2, vol. 225, pp. 133–140. American Mathematical Society, Providence, RI (2008)
12. Tidblom, J.: Improved $L^p$ Hardy inequalities. Ph.D. Thesis, Stockholm University (2005)

# Sharp Singular Trudinger-Moser-Adams Type Inequalities with Exact Growth

**Nguyen Lam and Guozhen Lu**

*Dedicated to Ermanno Lanconelli on the occasion of his 70th birthday, with friendship*

**Abstract** The main purpose of this paper is two fold. On the one hand, we review some recent progress on best constants for various sharp Moser-Trudinger and Adams inequalities in Euclidean spaces $\mathbb{R}^N$, hyperbolic spaces and other settings, and such sharp inequalities of Lions type. On the other hand, we present and prove some new results on sharp singular Moser-Trudinger and Adams type inequalities with exact growth condition and their affine analogues of such inequalities (Theorems 1.1, 1.2 and 1.3). We also establish a sharpened version of the classical Moser-Trudinger inequality on finite balls (Theorem 1.4).

**Keywords** Best constants • Sharp Adams inequalities • Sharp inequalities with exact growth condition • Sharp Moser-Trudinger inequalities

**Mathematics Subject Classification:** 26D10, 46E35

## 1 Introduction

The Trudinger-Moser-Adams inequalities are the replacements in the borderline case for the Sobolev embeddings. When $\Omega \subset \mathbb{R}^n$ is a bounded domain and $kp < n$, it is well-known that $W_0^{k,p}(\Omega) \subset L^q(\Omega)$ for all $1 \leq q \leq \frac{np}{n-kp}$. However, by counterexamples, $W_0^{k,\frac{n}{k}}(\Omega) \nsubseteq L^\infty(\Omega)$. In this situation, Yudovich [65], Pohozaev [58] and Trudinger [63] obtained independently that $W_0^{1,n}(\Omega) \subset L_{\varphi_n}(\Omega)$ where $L_{\varphi_n}(\Omega)$ is the Orlicz space associated with the Young function $\varphi_n(t) = \exp\left(\alpha |t|^{n/(n-1)}\right) - 1$ for some $\alpha > 0$. These results are refined in the 1971 paper [55] by J. Moser. In fact, we have the following Moser-Trudinger inequality:

N. Lam • G. Lu (✉)
Department of Mathematics, Wayne State University, Detroit, MI, USA
e-mail: nguyenlam@wayne.edu; gzlu@wayne.edu



G. Citti et al. (eds.), *Geometric Methods in PDE's*, Springer INdAM Series 13,
DOI 10.1007/978-3-319-02666-4_3

**Theorem (J. Moser [55])** *Let $\Omega$ be a domain with finite measure in Euclidean n-space $\mathbb{R}^n$, $n \geq 2$. Then there exists a sharp constant $\beta_n = n\omega_{n-1}^{\frac{1}{n-1}}$, where $\omega_{n-1}$ is the area of the surface of the unit n-ball, such that*

$$\frac{1}{|\Omega|}\int_{\Omega} \exp\left(\beta\,|u|^{\frac{n}{n-1}}\right)dx \leq c_0$$

*for any $\beta \leq \beta_n$, any $u \in W_0^{1,n}(\Omega)$ with $\int_{\Omega} |\nabla u|^n\, dx \leq 1$. This constant $\beta_n$ is sharp in the sense that if $\beta > \beta_n$, then the above inequality can no longer hold with some $c_0$ independent of u.*

This result has been studied and extended in many directions. For instance, we refer the reader to the sharp Moser inequality with mean value zero by Chang and Yang [12], Lu and Yang [51], Leckband [41], sharp Moser-Trudinger trace inequalities and sharp Moser-Trudinger inequalities without boundary conditions by Cianchi [14, 15], Moser-Trudinger inequality for Hessians by Tian and Wang [62], a singular version of the Moser-Trudinger inequality by Adimurthi and Sandeep in [4], etc. We also refer to the survey articles of Chang and Yang [13] and Lam and Lu [31] for descriptions of applications of such inequalities to nonlinear PDEs.

Recently, using the $L^p$ affine energy $\mathscr{E}_p(f)$ of $f$ instead of the standard $L^p$ energy of gradient $\|\nabla f\|_p$, where

$$\mathscr{E}_p(f) = c_{n,p}\left(\int\limits_{S^{n-1}} \|D_v f\|_p^{-n}\, dv\right)^{-1/n},$$

$$c_{n,p} = \left(\frac{n\omega_n\omega_{p-1}}{2\omega_{n+p-2}}\right)^{1/p}(n\omega_n)^{1/n},$$

$$\|D_v f\|_p = \left(\int\limits_{\mathbb{R}^n} |v\cdot\nabla f(x)|^p\, dx\right)^{1/p}.$$

The authors proved in [16] proved a sharp version of affine Moser-Trudinger inequality by replacing the constraint $||\nabla f||_n \leq 1$ by $\mathscr{E}_p(f) \leq 1$ in Moser's inequality. Namely,

**Theorem (Cianchi et al. [20])** *Let $\Omega$ be a domain with finite measure in Euclidean n-space $\mathbb{R}^n$, $n \geq 2$. Then there exists a constant $m_n > 0$ such that*

$$\frac{1}{|\Omega|}\int_{\Omega} \exp\left(\alpha\,|u|^{\frac{n}{n-1}}\right)dx \leq m_n$$

*for any* $\alpha \leq \alpha_n$, *any* $u \in W_0^{1,n}(\Omega)$ *with* $\mathscr{E}_n(u) \leq 1$. *The constant* $\alpha_n$ *is sharp in the sense that if* $\alpha > \alpha_n$, *then the above inequality can no longer hold with some* $m_n$ *independent of* $u$.

It is worthy to note that by the Holder inequality and Fubini's theorem, we have that

$$\mathscr{E}_p(f) \leq \|\nabla f\|_p$$

for every $f \in W^{1,p}(\mathbb{R}^n)$ and $p \geq 1$. Moreover, since the ratio $\frac{\|\nabla f\|_p}{\mathscr{E}_p(f)}$ is not uniformly bounded from above by any constant, these affine Moser-Trudinger inequalities are actually stronger than the standard Moser-Trudinger inequality. In [39], Lam, Lu and Tang used this $L^n$ affine energy $\mathscr{E}_n(f)$ to study several improved versions of the Moser-Trudinger type inequality in unbounded domains of Lions type.

In [26], Haberl, Schuster and Xiao used an asymmetric $L^p$ affine energy

$$\mathscr{E}_p^+(f) = 2^{1/p} c_{n,p} \left( \int_{S^{n-1}} \left\| D_v^+ f \right\|_p^{-n} dv \right)^{-1/n},$$

$$D_v^+ f(x) = \max \left\{ D_v f(x), 0 \right\},$$

to study an asymmetric affine version of the Moser-Trudinger inequality in the spirit of [16] by replacing the constraint $\mathscr{E}_n(u) \leq 1$ in the inequality of [16] by $\mathscr{E}_n^+(u) \leq 1$.

We note here that

$$\mathscr{E}_p^+(f) \leq \mathscr{E}_p(f) \leq \|\nabla f\|_p.$$

Hence, the theorem of [26] is an improvement of the classical Moser-Trudinger inequality.

Concerning Moser's type inequality with respect to high order derivatives, D.R. Adams, using a different approach, investigated in [2] the following Moser-Trudinger inequality in the high order case, which is now known as the Adams inequality:

**Theorem (D.R. Adams [2])** *Let* $\Omega$ *be an open and bounded set in* $\mathbb{R}^n$. *If* $m$ *is a positive integer less than* $n$, *then there exists a constant* $C_0 = C(n,m) > 0$ *such that for any* $u \in W_0^{m,\frac{n}{m}}(\Omega)$ *and* $\|\nabla^m u\|_{L^{\frac{n}{m}}(\Omega)} \leq 1$, *then*

$$\frac{1}{|\Omega|} \int_{\Omega} \exp(\beta |u(x)|^{\frac{n}{n-m}}) dx \leq C_0$$

*for all* $\beta \le \beta(n,m)$ *where*

$$\beta(n,\, m) \;=\; \begin{cases} \dfrac{n}{w_{n-1}}\left[\dfrac{\pi^{n/2}2^m\Gamma(\frac{m+1}{2})}{\Gamma(\frac{n-m+1}{2})}\right]^{\frac{n}{n-m}} & \text{when } m \text{ is odd} \\ \dfrac{n}{w_{n-1}}\left[\dfrac{\pi^{n/2}2^m\Gamma(\frac{m}{2})}{\Gamma(\frac{n-m}{2})}\right]^{\frac{n}{n-m}} & \text{when } m \text{ is even} \end{cases}.$$

*Furthermore, for any* $\beta > \beta(n,m)$*, the integral can be made as large as possible.*

The Adams inequality was extended recently by Tarsi [61]. More precisely, Tarsi used the Sobolev space with Navier boundary conditions $W_N^{m,\frac{n}{m}}(\Omega)$ which contains the Sobolev space $W_0^{m,\frac{n}{m}}(\Omega)$ as a closed subspace: It was shown that the sharp constants in this case are the same as in the classical Adams' inequities.

We stress here that the method of Adams was used successfully to establish the Moser-Trudinger-Adams inequalities in many settings, see on the Riemannian manifolds by Fontana [24], on the Heisenberg group in [17] and on the CR spheres in [19] by Cohn and Lu, sharp Moser-Onofri type inequalities on spheres by Beckner [7] and CR spheres by Branson et al. [8], and generalizations in other settings [6, 18, 20, 25, 38].

It can be noted that the Moser-Trudinger-Adams inequalities are senseless when the domains have infinite volume. Thus, it is interesting to investigate versions of the Moser-Trudinger-Adams inequalities in this setting. There are attempts to extend the Moser-Trudinger inequality to infinite volume domains by Cao [9] and Ogawa [56] in dimension two and by Do O in high dimension [22]. Moreover, Adachi and Tanaka established the best constants for all dimensions in [1]. Interestingly enough, by the sharp result of Adachi and Tanaka, the Moser-Trudinger type inequality can only be established for the subcritical case while they use the norm $\left(\int_{\mathbb{R}^n} |\nabla u|^n\, dx\right)^{1/n}$. Indeed, Adachi and Tanaka in [1] have proved that

$$\sup_{\int_{\mathbb{R}^n}|\nabla u|^n dx \le 1} \int_{\mathbb{R}^n} \phi\left(\beta\, |u|^{\frac{n}{n-1}}\right) dx < \infty$$

if $\beta < \beta_n$. Moreover, their results are actually sharp in the sense that the supremum is infinity when $\beta \ge \beta_n$. The result of Adachi and Tanaka was investigated in the affine case in [39] where the authors used the $L^p$ affine energy $\mathcal{E}_p(f)$ of $f$ instead of the standard $L^p$ energy of gradient $\|\nabla f\|_p$. In fact, it was proved in [39] that

**Theorem (Lam et al. [39])** *For any* $\beta \in (0,n)$ *and* $\alpha \in \left(0, \left(1-\frac{\beta}{n}\right)\alpha_n\right)$*, there exists a constant* $C_{\alpha,\beta} > 0$ *such that*

$$\int\limits_{\mathbb{R}^n} \frac{\phi_{n,1}\left(\alpha\, |u|^{\frac{n}{n-1}}\right)}{|x|^{\beta}} dx \le C_{\alpha,\beta}\, \|u\|_n^{n-\beta},$$

*for any* $u \in W^{1,n}(\mathbb{R}^n)$ *with* $\mathscr{E}_n(u) \leq 1$. *This inequality is false for* $\alpha \geq \left(1 - \frac{\beta}{n}\right)\alpha_n$ *in the sense that if* $\alpha \geq \left(1 - \frac{\beta}{n}\right)\alpha_n$, *then the above inequality can no longer hold with some* $C_{\alpha,\beta}$ *independent of* $u$. *Here*

$$\phi_{p,q}(t) = e^t - \sum_{j=0}^{j_{\frac{p}{q}}-2} \frac{t^j}{j!},\ j_{\frac{p}{q}} = \min\left\{j \in \mathbb{N} : j \geq \frac{p}{q}\right\} \geq \frac{p}{q}.$$

To achieve the critical case $\beta = \beta_n$, Ruf [59] and then Li and Ruf [45] need to use the full form of the norm in $W^{1,n}$, namely, $||u||_{1,n} = \left(\int_{\mathbb{R}^n} |u|^n\, dx\right)^{1/n} + \left(\int_{\mathbb{R}^n} |\nabla u|^n\, dx\right)^{1/n}$.

**Theorem (Ruf [59] and Li and Ruf [45])** *For* $\alpha \leq \alpha_n$, *there exists a constant* $C_\alpha > 0$ *such that*

$$\int_{\mathbb{R}^n} \phi_{n,1}\left(\alpha\, |u|^{\frac{n}{n-1}}\right) dx \leq C_\alpha,$$

*for any* $u \in W^{1,n}(\mathbb{R}^n)$ *with* $||u||_{1,n} \leq 1$. *This inequality is false for* $\alpha > \alpha_n$ *in the sense that if* $\alpha > \alpha_n$, *then the above inequality can no longer hold with some* $C_\alpha$ *independent of* $u$.

The singular version of Ruf and Li-Ruf inequalities was given in [5], and sharp Moser's type inequality on unbounded domains into Lorentz-Sobolev spaces was established by Cassani and Tarsi [11]. We recall that Lions is the first one who established an improvement of Moser's result by sharpening the constant $\beta_n$ [48]. The result of Lions was further improved by Adimurthi and Druet in [3]. Very recently, the authors in [39] used a rearrangement-free argument, initiated in [34, 35], to study a sharp version of the affine and improved Moser-Trudinger type inequality on domains of infinite volume in the spirit of Lions [48], namely,

**Theorem (Lam et al. [39])** *Let* $0 \leq \beta < n$ *and* $\tau > 0$. *Then there exists a constant* $C = C(n, \beta) > 0$ *such that for all* $u \in C_0^\infty(\mathbb{R}^n) \setminus \{0\}$, $\mathscr{E}_n(u) < 1$, *we have*

$$\int_{\mathbb{R}^n} \frac{\phi_{n,1}\left(\frac{2^{\frac{1}{n-1}}\left(1-\frac{\beta}{n}\right)\alpha_n}{\left(1+\mathscr{E}_n(u)^n\right)^{\frac{1}{n-1}}}\, |u|^{\frac{n}{n-1}}\right)}{|x|^\beta} dx \leq C(n,\beta)\, \frac{\|u\|_n^{n-\beta}}{\left|1 - \mathscr{E}_n(u)^n\right|^{1-\frac{\beta}{n}}}.$$

*As a consequence, we have that there exists a constant* $C = C(n, \beta, \tau) > 0$ *such that*

$$M_{4,\alpha} \leq M_{3,\alpha} \leq M_{1,\alpha} \leq C(n,\beta,\tau),$$
$$M_{4,\alpha} \leq M_{2,\alpha} \leq M_{1,\alpha} \leq C(n,\beta,\tau),$$

*for all* $0 \leq \alpha \leq \left(1 - \frac{\beta}{n}\right) \alpha_n$*, where*

$$M_{1,\alpha} = \sup_{u \in W^{1,n}(\mathbb{R}^n),\ \mathcal{E}_n(u)^n + \tau \|u\|_n^n \leq 1} \int_{\mathbb{R}^n} \frac{\phi_{n,1}\left(\frac{2^{\frac{1}{n-1}} \alpha}{(1+\mathcal{E}_n(u)^n)^{\frac{1}{n-1}}} |u|^{\frac{n}{n-1}}\right)}{|x|^\beta} dx$$

$$M_{2,\alpha} = \sup_{u \in W^{1,n}(\mathbb{R}^n),\ \mathcal{E}_n(u)^n + \tau \|u\|_n^n \leq 1} \int_{\mathbb{R}^n} \frac{\phi_{n,1}\left(\alpha |u|^{\frac{n}{n-1}}\right)}{|x|^\beta} dx$$

$$M_{3,\alpha} = \sup_{u \in W^{1,n}(\mathbb{R}^n),\ \|\nabla u\|_n^n + \tau \|u\|_n^n \leq 1} \int_{\mathbb{R}^n} \frac{\phi_{n,1}\left(\frac{2^{\frac{1}{n-1}} \alpha}{(1+\|\nabla u\|_n^n)^{\frac{1}{n-1}}} |u|^{\frac{n}{n-1}}\right)}{|x|^\beta} dx$$

$$M_{4,\alpha} = \sup_{u \in W^{1,n}(\mathbb{R}^n),\ \|\nabla u\|_n^n + \tau \|u\|_n^n \leq 1} \int_{\mathbb{R}^n} \frac{\phi_{n,1}\left(\alpha |u|^{\frac{n}{n-1}}\right)}{|x|^\beta} dx.$$

*Moreover, the constant* $\left(1 - \frac{\beta}{n}\right) \alpha_n$ *in the above inequality and supremum is sharp in the sense that when* $\alpha > \left(1 - \frac{\beta}{n}\right) \alpha_n$, $M_{1,\alpha} = M_{2,\alpha} = M_{3,\alpha} = M_{4,\alpha} = \infty$.

The Trudinger type inequalities for high order derivatives on domains of infinite volume were studied by Ozawa [57], Kozono et al. [30] with non-optimal constants. The sharp constants were recently established by Ruf and Sani [60] in the case of even derivatives and by Lam and Lu in all order of derivatives including fractional orders [32, 33, 35, 39]. The idea of Ruf and Sani [60] is to use the comparison principle for polyharmonic equations (thus only dealt with the case of even order of derivatives) and thus involves some difficult construction of auxiliary functions. The argument in [32, 35] uses the representation of the Bessel potentials and thus avoids dealing with such a comparison principle. Moreover, the argument in [35] does not use the symmetrization method and thus also works for the sub-Riemannian setting such as the Heisenberg groups [34, 40].

We state here the Adams inequality on unbounded domains in the most general form for fractional orders.

**Theorem (Lam and Lu [35])** *Let* $0 < \gamma < n$ *be an arbitrary real positive number,* $p = \frac{n}{\gamma}$ *and* $\tau > 0$*. There holds*

$$\sup_{u \in W^{\gamma,p}(\mathbb{R}^n),\ \left\|(\tau I - \Delta)^{\frac{\gamma}{2}} u\right\|_p \leq 1} \int_{\mathbb{R}^n} \phi\left(\beta_0(n,\gamma) |u|^{p'}\right) dx < \infty$$

*where*

$$\phi(t) = e^t - \sum_{j=0}^{j_p-2} \frac{t^j}{j!},$$

$$j_p = \min\{j \in \mathbb{N} : j \geq p\} \geq p.$$

*Here*

$$p' = \frac{p}{p-1},$$

$$\beta_0(n,\gamma) = \frac{n}{\omega_{n-1}} \left[ \frac{\pi^{n/2} 2^{\gamma} \Gamma(\gamma/2)}{\Gamma\left(\frac{n-\gamma}{2}\right)} \right]^{p'}.$$

*Furthermore this inequality is sharp, i.e., if $\beta_0(n,\gamma)$ is replaced by any $\beta > \beta_0(n,\gamma)$, then the supremum is infinite.*

The main idea in [35] is to write $u$ in the Bessel potential form to avoid the rearrangement of the high order derivatives. Therefore, we can first establish a version of the Adams type inequality on domains of finite volume. Using a delicate decomposition of $\mathbb{R}^n$ into level sets of the functions under consideration, we can get the desired result. This method does not apply symmetrization argument which is not available on high order Sobolev spaces and Heisenberg groups.

There is also an improved version of the Adams type inequality in the spirit of Lions [48] as follows:

**Theorem (Lam et al. [39])** *Let $0 \leq \beta < n$, $n \geq 3$, and $\tau > 0$. Then there exists a constant $C = C(n,\beta) > 0$ such that for all $u \in C_0^\infty(\mathbb{R}^n) \setminus \{0\}$, $\|\Delta u\|_{\frac{n}{2}} < 1$, $u \geq 0$, we have*

$$\int_{\mathbb{R}^n} \frac{\phi_{n,2}\left( \frac{2^{\frac{2}{n-2}}\left(1-\frac{\beta}{n}\right)\beta(n,2)}{\left(1+\|\Delta u\|_{\frac{n}{2}}^{\frac{n}{2}}\right)^{\frac{2}{n-2}}} |u|^{\frac{n}{n-2}} \right)}{|x|^{\beta}} dx \leq C(n,\beta) \frac{\|u\|_{\frac{n}{2}}^{\frac{n}{2}-\frac{\beta}{2}}}{\left|1-\|\Delta u\|_{\frac{n}{2}}^{\frac{n}{2}}\right|^{1-\frac{\beta}{n}}}.$$

*Consequently, we have that there exists a constant $C = C(n,\beta,\tau) > 0$ such that*

$$\sup_{u \in W^{2,\frac{n}{2}}(\mathbb{R}^n),\ \int_{\mathbb{R}^n} |\Delta u|^{\frac{n}{2}} + \tau |u|^{\frac{n}{2}} \leq 1} \int_{\mathbb{R}^n} \frac{\phi_{n,2}\left( \frac{2^{\frac{2}{n-2}}\alpha}{\left(1+\|\Delta u\|_{\frac{n}{2}}^{\frac{n}{2}}\right)^{\frac{2}{n-2}}} |u|^{\frac{n}{n-2}} \right)}{|x|^{\beta}} dx \leq C(n,\beta,\tau),$$

*for all* $0 \le \alpha \le \left(1-\frac{\beta}{n}\right)\beta(n,2)$. *When* $\alpha > \left(1-\frac{\beta}{n}\right)\beta(n,2)$, *the supremum is infinite.*

The next aim is to study the sharp subcritical Adams type inequalities in some special cases. More precisely, we have proved that

**Theorem (Lam et al. [39])** *For any* $\alpha \in (0, \beta(n,2))$, *there exists a constant* $C_\alpha > 0$ *such that*

$$\int_{\mathbb{R}^n} \phi_{n,2}\left(\alpha |u|^{\frac{n}{n-2}}\right) dx \le C_\alpha \|u\|_{\frac{n}{2}}^{\frac{n}{2}}, \ \forall u \in W^{2,\frac{n}{2}}(\mathbb{R}^n), \ \|\Delta u\|_{\frac{n}{2}} \le 1. \quad (1)$$

**Theorem (Lam et al. [39])** *For any* $\alpha \in (0, \beta(2m,m))$, *there exists a constant* $C_\alpha > 0$ *such that*

$$\int_{\mathbb{R}^{2m}} \phi_{2m,m}\left(\alpha |u|^2\right) dx \le C_\alpha \|u\|_2^2, \ \forall u \in W^{m,2}\left(\mathbb{R}^{2m}\right), \ \|\nabla^m u\|_2 \le 1. \quad (2)$$

It was proved in [30] that the inequality (1) does not hold when $\alpha > \beta(n,2)$, neither does inequality (2) when $\alpha > \beta(2m,m)$. Nevertheless, we still cannot verify the borderline case $\alpha = \beta(n,2)$ in the second order case and $\alpha = \beta(2m,m)$ in the high order case in the above two theorems.

Sharp Moser-Trudinger inequalities were also recently established on hyperbolic spaces Mancini and Sandeep [54] on conformal discs and by Lu and Tang in all dimensions [49, 50] including singular versions of Adachi-Tanaka type inequalities [1] and those of Ruf [59] and Li and Ruf type [45]. Sharp Moser-Trudinger inequalities on unbounded domains of the Heisenberg groups were also established by Lam, Lu and Tang [34, 37, 39]. We also mention that extremal functions for Moser-Trudinger inequalities on bounded domains were studied by Carleson and Chang [10], de Figueiredo et al. [21], Flucher [23], Lin [47], and on Riemannian manifolds by Li [42, 43], and on unbounded domains by Ruf [59], Li and Ruf [45], Ishiwata [28] and Ishiwata et al. [29]. In [36], Lam, Lu and Zhang studied some variants of the Moser-Trudinger type inequalities and their extremals. More precisely, it was proved in [36] that

**Theorem (Lam et al. [36])** *Let* $N \ge 2$ *and* $0 \le \beta < N$. *Then for all* $0 \le \alpha < \alpha_N\left(1-\frac{\beta}{N}\right)$, $q \ge 1$ *and* $p > q\left(1-\frac{\beta}{N}\right)$ $(p \ge q$ *if* $\beta = 0)$, *there exists a positive constant* $C_{p,N,\alpha,\beta} > 0$ *such that*

$$\int_{\mathbb{R}^N} \frac{\exp\left(\alpha |u|^{\frac{N}{N-1}}\right)|u|^p}{|x|^\beta} dx \le C_{N,p,q,\alpha,\beta} \|u\|_q^{q\left(1-\frac{\beta}{N}\right)},$$

$$\forall u \in D^{1,N}\left(\mathbb{R}^N\right) \cap L^q\left(\mathbb{R}^N\right), \ \|\nabla u\|_N \le 1.$$

*This constant* $\alpha_N\left(1-\frac{\beta}{N}\right)$ *is sharp. Moreover, the supremum can be achieved.*

Very little is known for existence of extremals for Adams inequalities. The only known cases are in the second order derivatives on compact Riemannian manifolds and bounded domains in dimension four by Li and Ndiaye [44] and Lu and Yang [52] respectively.

Very recently, while trying to study critical Moser-Trudinger-Adams type inequalities in the infinite volume domain cases when using the norm $\left(\int_{\mathbb{R}^n} |\nabla u|^n \, dx\right)^{1/n}$, Ibrahim et al. [27] and Masmoudi and Sani [54] set up the following Moser-Trudinger-Adams type inequalities with exact growth in dimension two and dimension four respectively:

**Theorem A (Ibrahim et al. [27])** *There exists some uniform constant $C > 0$ such that*

$$\sup_{u \in W^{1,2}(\mathbb{R}^2), ||\nabla u||_{L^2(\mathbb{R}^2)} \leq 1} \int_{\mathbb{R}^2} \frac{e^{4\pi |u|^2} - 1}{(1 + |u|^2)} dx \leq C||u||^2_{L^2(\mathbb{R}^2)}.$$

*Moreover, the power* 2 *in the denominator cannot be replaced with any* $p < 2$.

**Theorem B (Masmoudi and Sani [54])** *There exists some uniform constant $C > 0$ such that*

$$\sup_{u \in W^{2,2}(\mathbb{R}^4), ||\Delta u||_{L^2(\mathbb{R}^4)} \leq 1} \int_{\mathbb{R}4} \frac{e^{32\pi^2 |u|^2} - 1}{(1 + |u|^2)} dx \leq C||u||^2_{L^2(\mathbb{R}^4)}.$$

*Moreover, the power* 2 *in the denominator cannot be replaced with any* $p < 2$.

A different type of improvement of Moser's result involving a remainder term in the norms was given by Wang and Ye [64]. More recently, Lu and Tang [50] have established a singular version of sharp Moser-Trudinger inequalities on hyperbolic spaces with exact growth. Thus, the result of [50] extends those of [49] in the same spirit of [27, 54]. To describe the main result in [50], we need to introduce some notions.

Let $B^n = \{x \in R^n : |x| < 1\}$ denote the unit open ball in the Euclidean space $\mathbb{R}^n$. The space $B^n$ endowed with the Riemannian metric $g_{ij} = (\frac{1}{1-|x|^2})^2 \delta_{ij}$ is called the ball model of the hyperbolic space $\mathbb{H}^n$. Denote the associated hyperbolic volume by $dV = (\frac{2}{1-|x|^2})^n dx$. For any measurable set $E \subset \mathbb{H}^n$, set $|E| = \int_E dV$. Let $d(0,x)$ denote the hyperbolic distance between the origin and $x$. It is known that $d(0,x) = \ln \frac{1+|x|}{1-|x|}$ for $x \in \mathbb{H}^n$. The hyperbolic gradient $\nabla_g$ is given by $\nabla_g = (\frac{1-|x|^2}{2})^2 \nabla$.

Let $\Omega \subset \mathbb{H}^n$ be a bounded domain. Denote $\|f\|_{n,\Omega} = (\int_\Omega |f|^n dV)^{\frac{1}{n}}$. Then we have the following:

$$\|\nabla_g f\|_{n,\Omega} = (\int_\Omega < \nabla_g f, \nabla_g f >_g^{n/2} dV)^{\frac{1}{n}} = (\int_\Omega |\nabla f|^p dx)^{\frac{1}{n}}.$$

Let $\|f\|_n = (\int_{\mathbb{H}^n} |f|^n dV)^{\frac{1}{n}}$. Then we have

$$\|\nabla_g f\|_n = (\int_{\mathbb{H}^n} < \nabla_g f, \nabla_g f >_g^{n/2} dV)^{\frac{1}{n}} = (\int_{B^n} |\nabla f|^p dx)^{\frac{1}{n}}.$$

We use $W_0^{1,n}(\Omega)$ to express the completion of $C_0^\infty(\Omega)$ under the norm

$$\|u\|_{W_0^{1,n}(\Omega)} = (\int_\Omega |f|^n dV + \int_\Omega |\nabla f|^n dx)^{\frac{1}{n}}.$$

We will also use $W^{1,n}(\mathbb{H}^n)$ to express the completion of $C_0^\infty(\mathbb{H}^n)$ under the norm

$$\|u\|_{W_0^{1,n}(\mathbb{H}^n)} = (\int_{\mathbb{H}^n} |f|^n dV + \int_{\mathbb{H}^n} |\nabla f|^n dx)^{\frac{1}{n}}.$$

Then the authors have established the following in [50]:

**Theorem (Lu and Tang [50])** *Let $\alpha_n = n\omega_{n-1}^{\frac{1}{n-1}}$, then there exists a constant $C > 0$ such that for any $u \in W^{1,n}(\mathbb{H}^n)$ satisfying $\|\nabla_g u\|_n \leq 1$,*

$$\int_{\mathbb{H}^n} \frac{\Phi_n(\alpha_n |u|^{\frac{n}{n-1}})}{(1+|u|)^{\frac{n}{n-1}}} dV \leq C\|u\|_n^n,$$

*where $\Phi_n(t) = e^x - \sum_{j=0}^{n-2} \frac{t^j}{j!}$. The result is sharp in the sense that: if the power $\frac{n}{n-1}$ in the denominator is replaced by any $p < \frac{n}{n-1}$, there exists a sequence of function $\{u_k\}$ such that $\|\nabla_g u_k\|_n \leq 1$, but*

$$\frac{1}{\|u_k\|_n^n} \int_{\mathbb{H}^n} \frac{[\Phi_n(\alpha_n(|u_k|)^{\frac{n}{n-1}})]}{(1+|u|)^p} dV \to \infty.$$

More recently, motivated by the works of [27, 50, 54], H. Tang, M. Zhu and the second author of this paper have established in [53] the sharp second order Adams inequality with the exact growth in $\mathbb{R}^n$ in general dimension $n \geq 3$. Thus, we obtained the extension of the work of [54] to all dimension $n \geq 3$.

The results of [53] are as follows.

**Theorem (Lu et al. [53])** *There exists a constant $C > 0$ such that for all $f \in W^{2,\frac{n}{2}}(\mathbb{R}^n)$ $(n \geq 3)$ satisfying $\|\Delta f\|_{\frac{n}{2}} \leq 1$,*

$$\int_{\mathbb{R}^n} \frac{\Phi(\beta_n |f|^{\frac{n}{n-2}})}{(1+|f|)^{\frac{n}{n-2}}} dx \leq C\|f\|_{\frac{n}{2}}^{\frac{n}{2}}.$$

*Where* $\Phi(t) = \exp(t) - \sum_{j=0}^{j_{\frac{n}{2}}-2} \frac{t^j}{j!}$, $j_{\frac{n}{2}} = \min\{j \in \mathbb{N} : j \geq \frac{n}{2}\} \geq n/2$ *and* $\beta_n = \beta(n,2) = \frac{n}{\omega_{n-1}}[\frac{\pi^{\frac{n}{2}} 4}{\Gamma(n/2-1)}]^{\frac{n}{n-2}}$.

We remark that both the power $\frac{n}{n-2}$ and the constant $\beta$ are optimal. These can be justified by the following theorem.

**Theorem (Lu et al. [53])** *If the power* $\frac{n}{n-2}$ *in the denominator is replaced by any* $p < \frac{n}{n-2}$, *there exists a sequence of function* $\{f_k\}$ *such that* $\|\Delta u_k\|_{\frac{n}{2}} \leq 1$, *but*

$$\frac{1}{\|u_k\|_{\frac{n}{2}}^{\frac{n}{2}}} \int_{\mathbb{R}^n} \frac{\Phi(\beta_n(|f_k|)^{\frac{n}{n-2}})}{(1+|f_k|)^p} dx \to \infty.$$

*Moreover, if* $\alpha > \beta_n$, *there exists a sequence of function* $\{f_k\}$ *such that* $\|\Delta u_k\|_{\frac{n}{2}} \leq 1$, *but*

$$\frac{1}{\|f_k\|_{\frac{n}{2}}^{\frac{n}{2}}} \int_{\mathbb{R}^n} \frac{\Phi(\alpha(|f_k|)^{\frac{n}{n-1}})}{(1+|f_k|)^p} dx \to \infty,$$

*for any* $p \geq 0$.

The main purpose of the remaining part of this paper is to establish sharp singular Moser-Trudinger and Adams inequalities and then extend the above Theorems A and B to the singular versions.

**Theorem 1.1 (Sharp Singular Moser-Trudinger Inequality)** *Let* $0 \leq \beta < 2$ *and* $0 < \alpha \leq 4\pi\left(1 - \frac{\beta}{2}\right)$. *Then there exists a constant* $C = C(\alpha, \beta) > 0$ *such that for all* $u \in W^{1,2}(\mathbb{R}^2) : \mathscr{E}_2^+(u) \leq 1$, *there holds*

$$\int_{\mathbb{R}^2} \frac{e^{\alpha u^2} - 1}{\left(1 + |u|^{2-\beta}\right)|x|^\beta} dx \leq C \|u\|_2^{2-\beta}.$$

*Moreover, the power* $2-\beta$ *in the denominator cannot be replaced with any* $q < 2-\beta$.

**Theorem 1.2 (Sharp Singular Adams Inequality)** *Let* $0 \leq \beta < 4$ *and* $0 < \alpha \leq 32\pi^2\left(1 - \frac{\beta}{4}\right)$. *Then there exists a constant* $C = C(\alpha, \beta) > 0$ *such that*

$$\int_{\mathbb{R}^4} \frac{e^{\alpha u^2} - 1}{\left(1 + |u|^{2-\beta/2}\right)|x|^\beta} dx \leq C \|u\|_2^{2-\frac{\beta}{2}} \text{ for all } u \in W^{2,2}(\mathbb{R}^4) : \|\Delta u\|_2 \leq 1.$$

*Moreover, the power* $2 - \beta/2$*in the denominator cannot be replaced with any* $q < 2 - \beta/2$.

As the reader will see, the key ingredients in the proofs of Theorems 1.1 and 1.2 are symmetrization arguments and the Radial Sobolev inequalities: Theorems 2.1 and 2.2 established in Ibrahim et al. [27] and in Masmoudi and Sani [54] (see Sect. 2). It is worthy noting that such a Radial Sobolev inequality in all dimensions has been recently established by Lu and Tang [50]:

**Theorem (Lu and Tang [50])** *Let $N \geq 2$. There exists a constant $C > 0$ such that for any nonnegative nonincreasing radial function $u \in W_{rad}^{1,N}\left(\mathbb{R}^N\right)$ satisfying $u(R) > 1$ and*

$$\omega_{N-1}\int_R^\infty \left|u'(t)\right|^N t^{N-1}dt \leq K$$

*for some $R,\ K > 0$, then we have*

$$\frac{\exp\left(\frac{\alpha_N u^{\frac{N}{N-1}}(R)}{K^{\frac{1}{N-1}}}\right)}{u^{\frac{N}{N-1}}(R)}R^N \leq C\frac{\int_R^\infty |u(t)|^N t^{N-1}dt}{K^{\frac{N}{N-1}}}.$$

With this Radial Sobolev inequality, we can prove a sharp version of the singular affine Moser-Trudinger type inequality with exact growth for $N$-dimensional case that we will state here and omit the proof:

**Theorem 1.3 (Sharp Singular Moser-Trudinger Inequality on $\mathbb{R}^N$)** *Let $0 \leq \beta < N$ and $0 < \alpha \leq \alpha_N\left(1-\frac{\beta}{N}\right)$. Then there exists a constant $C = C(\alpha,\beta,N) > 0$ such that for all $u \in W^{1,N}\left(\mathbb{R}^N\right) : \mathscr{E}_N^+(u) \leq 1$, there holds*

$$\int_{\mathbb{R}^N}\frac{\phi_{N,1}\left(\alpha u^{\frac{N}{N-1}}\right)}{\left(1+|u|^{\frac{N}{N-1}\left(1-\frac{\beta}{N}\right)}\right)|x|^\beta}dx \leq C\,\|u\|_N^{N-\beta}.$$

*Moreover, the power $\frac{N}{N-1}\left(1-\frac{\beta}{N}\right)$ in the denominator cannot be replaced with any $q < \frac{N}{N-1}\left(1-\frac{\beta}{N}\right)$.*

Obviously, we can also obtain the singular version of the second order Adams inequality in $\mathbb{R}^n$ for all $n \geq 3$ by combining the techniques in [53] and the method here. We shall not present the proof here. We also remark that Theorem 1.2 is true for all $N \geq 3$ by incorporating the result in [53].

Our last main result is an improved version of the classical Moser-Trudinger inequality in $W_0^{1,n}(B)$ where $B$ is the unit ball in $\mathbb{R}^n$.

**Theorem 1.4** *Let $B$ be the unit ball in $\mathbb{R}^n$. Then there holds*

$$\sup_{u\in W_0^{1,n}(B),\ \mathcal{E}_n^+(u)\leq 1} \int_B \left(\ln\frac{1}{|x|^n}+1-\alpha_n|u|^{\frac{n}{n-1}}\right)e^{\alpha_n|u|^{\frac{n}{n-1}}}dx<\infty.$$

*This constant $\alpha_n$ is sharp in the sense that if $\alpha>\alpha_n$, then*

$$\sup_{u\in W_0^{1,n}(B),\ \mathcal{E}_n^+(u)\leq 1} \int_B \left(\ln\frac{1}{|x|^n}+1-\alpha_n|u|^{\frac{n}{n-1}}\right)e^{\alpha|u|^{\frac{n}{n-1}}}dx=\infty.$$

It is worthy noting that by symmetrization arguments, we have

$$\begin{aligned}&\sup_{u\in W_0^{1,n}(B),\ \|\nabla u\|_n\leq 1}\int_B \exp\left(\beta|u|^{\frac{n}{n-1}}\right)dx\\ &=\sup_{u\in W_0^{1,n}(B),\ \|\nabla u\|_n\leq 1,\ u\text{ is radially nonincreasing}}\int_B \exp\left(\beta|u|^{\frac{n}{n-1}}\right)dx.\end{aligned}$$

Also, if $u\in W_0^{1,n}(B)$, $\|\nabla u\|_n\leq 1$ and is radially nonincreasing, then we could show that

$$\alpha_n|u|^{\frac{n}{n-1}}\leq\ln\frac{1}{|x|^n}. \tag{3}$$

Combining these two things and the fact that $\mathcal{E}_p^+(f)\leq\|\nabla f\|_p$, we obtain

$$\begin{aligned}&\sup_{u\in W_0^{1,n}(B),\ \|\nabla u\|_n\leq 1}\int_B \exp\left(\alpha_n|u|^{\frac{n}{n-1}}\right)dx\\ &\leq\sup_{u\in W_0^{1,n}(B),\ \mathcal{E}_n^+(u)\leq 1}\int_B\left(\ln\frac{1}{|x|^n}+1-\alpha_n|u|^{\frac{n}{n-1}}\right)e^{\alpha_n|u|^{\frac{n}{n-1}}}dx.\end{aligned}$$

Hence our Theorem 1.4 is indeed an improvement of the classical Moser-Trudinger inequality [55] on $W_0^{1,n}(B)$.

The organization of the paper is as follows. In Sect. 2, we will recall some preliminaries on the symmetrization rearrangement and collect some known results that we will need to prove our sharp singular Moser-Trudinger-Adams inequalities. Section 3 will give the proof of the sharp singular Moser-Trudinger inequality in $\mathbb{R}^2$ (Theorem 1.1 and Sect. 4 contains the proof of the sharp singular Adams inequality in $\mathbb{R}^4$ (Theorem 1.2. Finally, an improvement of the Moser-Trudinger inequality will be studied in Sect. 5 (Theorem 1.4). Section 6 is devoted to the verification of sharpness of the constants in Theorems 1.1 and 1.2.

## 2 Some Useful Results

In this section, we introduce some useful results that will be used in our proofs.

### 2.1 *Rearrangements*

Let $\Omega \subset \mathbb{R}^N$, $N \geq 2$, be a measurable set. We denote by $\Omega^{\#}$ the open ball $B_R \subset \mathbb{R}^N$ centered at 0 of radius $R > 0$ such that $|B_R| = |\Omega|$.

Let $u : \Omega \to \mathbb{R}$ be a real-valued measurable function. The distribution function of $u$ is the function

$$\mu_u(t) = |\{x \in \Omega : |u(x)| > t\}|$$

and the decreasing rearrangement of $u$ is the right-continuous, nonincreasing function $u^*$ that is equimeasurable with $u$ :

$$u^*(s) = \sup\{t \geq 0 : \mu_u(t) > s\}.$$

It is clear that *suppu*$^* \subseteq [0, |\Omega|]$. We also define

$$u^{**}(s) = \frac{1}{s}\int_0^s u^*(t)dt \geq u^*(s).$$

Moreover, we define the spherically symmetric decreasing rearrangement of $u$ :

$$u^{\#} : \Omega^{\#} \to [0, \infty]$$
$$u^{\#}(x) = u^*\left(\sigma_N |x|^N\right).$$

Then we have the following important result that could be found in [16, 26, 46]:

**Lemma 2.1 (Pólya-Szegö Inequality)** *Let* $u \in W^{1,p}(\mathbb{R}^n)$, $p \geq 1$. *Then* $f^{\#} \in W^{1,p}(\mathbb{R}^n)$,

$$\mathscr{E}_p^+\left(f^{\#}\right) = \mathscr{E}_p\left(f^{\#}\right) = \left\|\nabla f^{\#}\right\|_p$$

*and*

$$\mathscr{E}_p^+\left(f^{\#}\right) \leq \mathscr{E}_p^+(f)\,;\ \mathscr{E}_p\left(f^{\#}\right) = \mathscr{E}_p(f)\,;\ \left\|\nabla f^{\#}\right\|_p = \|\nabla f\|_p.$$

We now recall two theorems from [27] and [54] respectively and we also refer to [50, 53] for high dimensional cases.

**Theorem 2.1** *There exists a constant $C > 0$ such that for any nonnegative radially decreasing function $u \in W^{1,2}(\mathbb{R}^2)$ satisfying*

$$u(R) > 1 \text{ and } \int\limits_{\mathbb{R}^2 \setminus B_R} |\nabla u|^2 \, dx \le K \text{ for some } K > 0,$$

*we have*

$$\frac{e^{\frac{4\pi}{K} u^2(R)}}{u^2(R)} R^2 \le \frac{C}{K^2} \int\limits_{\mathbb{R}^2 \setminus B_R} |u|^2 \, dx.$$

**Theorem 2.2** *Let $u \in W^{2,2}(\mathbb{R}^4)$ and let $R > 0$. If $u^{\#}(R) > 1$ and $f = -\Delta u$ in $\mathbb{R}^4$ satisfies*

$$\int\limits_{|B_R|}^{\infty} \left[f^{**}(s)\right]^2 ds \le 4K \text{ for some } K > 0,$$

*then there exists a universal constant $C > 0$ such that*

$$\frac{e^{\frac{32\pi^2}{K} \left[u^{\#}(R)\right]^2}}{\left[u^{\#}(R)\right]^2} R^4 \le \frac{C}{K^2} \int\limits_{\mathbb{R}^4 \setminus B_R} \left|u^{\#}\right|^2 dx.$$

We note here that in the Theorem 2.2 and also in the rest of this paper, when we write $u^{\#}(R)$, we mean that $u^{\#}(x)$ for $|x| = R$.

We also recall the following result in [33]:

**Lemma 2.2** *Let $0 < \gamma \le 1$, $1 < p < \infty$ and $a(s,t)$ be a non-negative measurable function on $(-\infty, \infty) \times [0, \infty)$ such that (a.e.)*

$$a(s,t) \le 1, \text{ when } 0 < s < t, \tag{4}$$

$$\sup_{t>0} \left( \int_{-\infty}^{0} + \int_{t}^{\infty} a(s,t)^{p'} ds \right)^{1/p'} = b < \infty. \tag{5}$$

*Then there is a constant $c_0 = c_0(p, b, \gamma)$ such that if for $\phi \ge 0$,*

$$\int_{-\infty}^{\infty} \phi(s)^p ds \le 1, \tag{6}$$

*then*

$$\int_0^\infty e^{-F_\gamma(t)}dt \le c_0 \tag{7}$$

*where*

$$F_\gamma(t) = \gamma t - \gamma\left(\int_{-\infty}^{\infty} a(s,t)\phi(s)ds\right)^{p'}. \tag{8}$$

We will need the following strengthened lemma which is needed in the proof of Theorem 1.4 and is also of independent interest.

**Lemma 2.3** *Let* $1 < p < \infty$, $k \in \mathbb{N}$. *There exists a constant* $c_0 > 0$ *such that for* $\varphi(s) \ge 0$, *and*

$$\int_0^\infty |\varphi(s)|^p\, ds \le 1,$$

*it follows that*

$$\int_0^\infty (F(t)+1)^k\, e^{-F(t)}dt \le c_0$$

*where*

$$F(t) = t - \left(\int_0^t \varphi(s)ds\right)^{p'} \ge 0.$$

*Proof* Set

$$E_\lambda = \{t \ge 0 : F(t) \le \lambda\}.$$

Then we can show that $E_\lambda$ is empty for sufficiently small $\lambda$, and that there exist constant $A_1, A_2$ such that

$$|E_\lambda| \le A_1\lambda + A_2.$$

Now, we prove by induction.

$k = 0$ : this is the Lemma 2.2 with $\gamma = 1$.

$k = 1:$

$$\begin{aligned}
\int_0^\infty (F(t)+1)\, e^{-F(t)} dt &= \int_0^\infty \int_{F(t)}^\infty e^{-\lambda} \lambda d\lambda dt \\
&= \int_0^\infty \int_{F(t)\le\lambda} e^{-\lambda} \lambda dt d\lambda \\
&= \int_0^\infty |E_\lambda|\, e^{-\lambda} \lambda d\lambda \\
&\le c_0.
\end{aligned}$$

Assume that the result is true with $k = 0, 1, \ldots, m$. Then

$$\begin{aligned}
\int_0^\infty (F(t))^{m+1} e^{-F(t)} dt + (m+1) \int_0^\infty \int_{F(t)}^\infty e^{-\lambda} \lambda^m d\lambda dt &= \int_0^\infty \int_{F(t)}^\infty e^{-\lambda} \lambda^{m+1} d\lambda dt \\
&= \int_0^\infty \int_{F(t)\le\lambda} e^{-\lambda} \lambda^{m+1} dt d\lambda \\
&= \int_0^\infty |E_\lambda|\, e^{-\lambda} \lambda^{m+1} d\lambda \\
&\le c_0.
\end{aligned}$$

The proof now is completed.

We now state the Hardy-Littlewood inequality that could be found in [46].

**Lemma 2.4** *Let $f \in L^p(\mathbb{R}^N)$ and $g \in L^q(\mathbb{R}^N)$ where $\frac{1}{p} + \frac{1}{q} = 1,\ 1 \le p, q \le \infty$. Then*

$$\int_{\mathbb{R}^N} f(x)\, g(x)\, dx \le \int_{\mathbb{R}^N} f^{\#}(x)\, g^{\#}(x)\, dx.$$

Using the above Hardy-Littlewood inequality, we will prove the following result that will be used later in our proof of Theorems 1.1 and 1.2. This result is also of independent interest and its proof is not immediately trivial at its first glance. Nevertheless, it is indeed true after a careful analysis of the monotonicity of the function $\frac{e^{\alpha u^2}-1}{1+|u|^{2-\beta}}$.

**Lemma 2.5** *Let $0 \leq \beta < 2$ and $\alpha > 0$. There holds:*

$$\int_{\mathbb{R}^2} \frac{e^{\alpha u^2} - 1}{\left(1 + |u|^{2-\beta}\right)|x|^{\beta}} dx \leq \int_{\mathbb{R}^2} \frac{e^{\alpha \left(u^{\#}\right)^2} - 1}{\left(1 + |u^{\#}|^{2-\beta}\right)|x|^{\beta}} dx$$

*for any $u \in W^{1,2}\left(\mathbb{R}^2\right)$.*

*Proof* *Let $u \in W^{1,2}\left(\mathbb{R}^2\right)$ and set*

$$f(x) = (F \circ |u|)(x) = \frac{e^{\alpha u^2(x)} - 1}{\left(1 + |u(x)|^{2-\beta}\right)} \text{ and } g(x) = \frac{1}{|x|^{\beta}}$$

*where*

$$F(t) = \frac{e^{\alpha t^2} - 1}{1 + t^{2-\beta}}.$$

*First, we note here that*

$$g^{\#}(x) = \frac{1}{|x|^{\beta}} = g(x).$$

*Now, we claim that $F(t)$ is nondecreasing on $\mathbb{R}^+$. Indeed, we have*

$$\begin{aligned}
F'(t) &= \frac{2\alpha t e^{\alpha t^2}\left(1 + t^{2-\beta}\right) - (2-\beta)\, t^{1-\beta}\left(e^{\alpha t^2} - 1\right)}{\left(1 + t^{2-\beta}\right)^2} \\
&= \frac{e^{\alpha t^2}\left[2\alpha t + 2\alpha t^{3-\beta} - (2-\beta)\, t^{1-\beta}\right] + (2-\beta)\, t^{1-\beta}}{\left(1 + t^{2-\beta}\right)^2} \\
&= t^{1-\beta}\left[\frac{e^{\alpha t^2}\left[2\alpha t^{\beta} + 2\alpha t^2 - (2-\beta)\right] + (2-\beta)}{\left(1 + t^{2-\beta}\right)^2}\right] \\
&= \frac{t^{1-\beta} h(t)}{\left(1 + t^{2-\beta}\right)^2}
\end{aligned}$$

*where*

$$h(t) = e^{\alpha t^2}\left[2\alpha t^{\beta} + 2\alpha t^2 - (2-\beta)\right] + (2-\beta).$$

*Noting that*

$$
\begin{aligned}
h'(t) &= 2\alpha t e^{\alpha t^2}\left[2\alpha t^{\beta} + 2\alpha t^2 - (2-\beta)\right] + e^{\alpha t^2}\left[2\alpha\beta t^{\beta-1} + 4\alpha t\right] \\
&= e^{\alpha t^2}\left[2\alpha t\left(2\alpha t^{\beta} + 2\alpha t^2 - (2-\beta)\right) + 2\alpha\beta t^{\beta-1} + 4\alpha t\right] \\
&= e^{\alpha t^2}\left[2\alpha t\left(2\alpha t^{\beta} + 2\alpha t^2\right) - 2\alpha t(2-\beta) + 2\alpha\beta t^{\beta-1} + 4\alpha t\right] \\
&= e^{\alpha t^2}\left[2\alpha t\left(2\alpha t^{\beta} + 2\alpha t^2\right) + 2\alpha\beta t + 2\alpha\beta t^{\beta-1}\right] \\
&\geq 0,
\end{aligned}
$$

*hence*

$$h(t) \geq h(0) = 0 \, for \, t \geq 0.$$

*Thus,* $F'(t) \geq 0$ *when* $t \geq 0$, *which means that* $F$ *is nondecreasing on* $\mathbb{R}^+$.

*Hence, by a property of rearrangement (see [46]), we have that*

$$f^{\#}(x) = (F \circ |u|)^{\#}(x) = \left(F \circ u^{\#}\right)(x).$$

*By Lemma 2.4, we get our desired result.*

## 3 Sharp Singular Truding-Moser Type Inequality with Exact Growth

In this section, we will prove a version of sharp singular Moser-Trudinger type inequality with exact growth, namely Theorem 1.1. We follow [27] closely. We have chosen to present all the details for its completeness. Its proof is essentially an adaptation of the proof of the non-singular version given in [27] and an application of our Lemma 2.5, together with a careful decomposition of the integral domains in terms of the weight function $\frac{1}{|x|^{\beta}}$ and the norms of $u$.

**Theorem 3.1** *Let* $0 \leq \beta < 2$ *and* $0 < \alpha \leq 4\pi\left(1 - \frac{\beta}{2}\right)$. *Then there exists a constant* $C = C(\alpha, \beta) > 0$ *such that for all* $u \in W^{1,2}\left(\mathbb{R}^2\right) : \mathscr{E}_2^+(u) \leq 1$, *there holds*

$$\int_{\mathbb{R}^2} \frac{e^{\alpha u^2} - 1}{\left(1 + |u|^{2-\beta}\right)|x|^{\beta}} dx \leq C \|u\|_2^{2-\beta}.$$

*Proof* By the symmetrization arguments: the Pólya-Szegö inequality, the Hardy-Littlewood inequality; Lemma 2.5 and the density arguments, we may assume that $u$ is a smooth, nonnegative and radially decreasing function. Let $R_1 = R_1(u)$ be such that

$$\int_{B_{R_1}} |\nabla u|^2 \, dx = 2\pi \int_0^{R_1} u_r^2.rdr \leq 1 - \varepsilon_0,$$

$$\int_{\mathbb{R}^2 \backslash B_{R_1}} |\nabla u|^2 \, dx = 2\pi \int_{R_1}^{\infty} u_r^2.rdr \leq \varepsilon_0.$$

Here $\varepsilon_0 \in (0, 1)$ is fixed and does not depend on $u$.

By the Holder's inequality, we have

$$\begin{aligned} u(r_1) - u(r_2) &\leq \int_{r_1}^{r_2} - u_r dr \qquad (9) \\ &\leq \left( \int_{r_1}^{r_2} u_r^2.rdr \right)^{1/2} \left( \ln \frac{r_2}{r_1} \right)^{1/2} \\ &\leq \left( \frac{1 - \varepsilon_0}{2\pi} \right)^{1/2} \left( \ln \frac{r_2}{r_1} \right)^{1/2} \quad \text{for } 0 < r_1 \leq r_2 \leq R_1, \end{aligned}$$

and

$$u(r_1) - u(r_2) \leq \left( \frac{\varepsilon_0}{2\pi} \right)^{1/2} \left( \ln \frac{r_2}{r_1} \right)^{1/2} \quad \text{for } R_1 \leq r_1 \leq r_2. \qquad (10)$$

We define $R_0 := \inf \{ r > 0 : u(r) \leq 1 \} \in [0, \infty)$. Hence $u(s) \leq 1$ when $s \geq R_0$. WLOG, we assume $R_0 > 0$.

Now, we split the integral as follows:

$$\begin{aligned} \int_{\mathbb{R}^2} \frac{e^{\alpha u^2} - 1}{\left(1 + u^{2-\beta}\right) |x|^{\beta}} dx &= \int_{B_{R_0}} \frac{e^{\alpha u^2} - 1}{\left(1 + u^{2-\beta}\right) |x|^{\beta}} dx + \int_{\mathbb{R}^2 \backslash B_{R_0}} \frac{e^{\alpha u^2} - 1}{\left(1 + u^{2-\beta}\right) |x|^{\beta}} dx \\ &= I + J. \end{aligned}$$

First, we will estimate $J$. Since $u \leq 1$ on $\mathbb{R}^2 \setminus B_{R_0}$, we have

$$J = \int\limits_{\mathbb{R}^2 \setminus B_{R_0}} \frac{e^{\alpha u^2} - 1}{\left(1 + u^{2-\beta}\right) |x|^{\beta}} dx \tag{11}$$

$$\leq C \int_{\{u \leq 1\}} \frac{u^2}{|x|^{\beta}} dx$$

$$= C \int_{\{u \leq 1; |x| > \|u\|_2\}} \frac{u^2}{|x|^{\beta}} dx + C \int_{\{u \leq 1; |x| \leq \|u\|_2\}} \frac{u^2}{|x|^{\beta}} dx$$

$$\leq C \|u\|_2^{2-\beta}.$$

Hence, now, we just need to deal with the integral $I$.

*Case 1* $0 < R_0 \leq R_1$.

In this case, using (9), we have for $0 < r \leq R_0$ :

$$u(r) \leq 1 + \left(\frac{1 - \varepsilon_0}{2\pi}\right)^{1/2} \left(\ln \frac{R_0}{r}\right)^{1/2}.$$

By using

$$(a + b)^2 \leq (1 + \varepsilon)a^2 + \left(1 + \frac{1}{\varepsilon}\right) b^2,$$

we get

$$u^2(r) \leq \frac{1 - \varepsilon_0^2}{2\pi} \ln \frac{R_0}{r} + \left(1 + \frac{1}{\varepsilon_0}\right).$$

Thus, we can estimate the integral $I$ as follows:

$$I = \int\limits_{B_{R_0}} \frac{e^{\alpha u^2} - 1}{\left(1 + u^{2-\beta}\right) |x|^{\beta}} dx \tag{12}$$

$$\leq \int\limits_{B_{R_0}} \frac{e^{\alpha \frac{1-\varepsilon_0^2}{2\pi} \ln \frac{R_0}{|x|} + \alpha\left(1 + \frac{1}{\varepsilon_0}\right)}}{|x|^{\beta}} dx$$

$$\leq C R_0^{\alpha \frac{1-\varepsilon_0^2}{2\pi}} \int\limits_0^{R_0} r^{1 - \alpha \frac{1-\varepsilon_0^2}{2\pi} - \beta} dr$$

$$\leq CR_0^{2-\beta}$$

$$\leq C\left(\int_{B_{R_0}} 1dx\right)^{1-\frac{\beta}{2}}$$

$$\leq C\left\|u\right\|_2^{2-\beta}.$$

*Case 2* $0 < R_1 < R_0$.

We have

$$\begin{aligned}
I &= \int_{B_{R_0}} \frac{e^{\alpha u^2}-1}{\left(1+u^{2-\beta}\right)|x|^{\beta}}dx \\
&= \int_{B_{R_1}} \frac{e^{\alpha u^2}-1}{\left(1+u^{2-\beta}\right)|x|^{\beta}}dx + \int_{B_{R_0}\setminus B_{R_1}} \frac{e^{\alpha u^2}-1}{\left(1+u^{2-\beta}\right)|x|^{\beta}}dx \\
&= I_1 + I_2.
\end{aligned}$$

Using (10), we get

$$u\left(r\right)-u\left(R_0\right) \leq \left(\frac{\varepsilon_0}{2\pi}\right)^{1/2}\left(\ln\frac{R_0}{r}\right)^{1/2} \quad \text{for } r \geq R_1.$$

Hence

$$u\left(r\right) \leq 1+\left(\frac{\varepsilon_0}{2\pi}\right)^{1/2}\left(\ln\frac{R_0}{r}\right)^{1/2}$$

and again, by using

$$(a+b)^2 \leq (1+\varepsilon)a^2+\left(1+\frac{1}{\varepsilon}\right)b^2$$

we have

$$u^2\left(r\right) \leq \left(1+\varepsilon\right)\frac{\varepsilon_0}{2\pi}\ln\frac{R_0}{r}+\left(1+\frac{1}{\varepsilon}\right), \ \forall\varepsilon > 0.$$

So

$$
\begin{aligned}
I_2 &= \int_{B_{R_0}\setminus B_{R_1}} \frac{e^{\alpha u^2}-1}{\left(1+u^{2-\beta}\right)|x|^{\beta}}dx \\
&\leq C\int_{R_1}^{R_0} e^{\alpha(1+\varepsilon)\frac{\varepsilon_0}{2\pi}\ln\frac{R_0}{r}+\alpha\left(1+\frac{1}{\varepsilon}\right)} r^{1-\beta}dr \\
&= Ce^{\alpha\left(1+\frac{1}{\varepsilon}\right)} R_0^{\alpha(1+\varepsilon)\frac{\varepsilon_0}{2\pi}} \frac{R_0^{2-\beta-\alpha(1+\varepsilon)\frac{\varepsilon_0}{2\pi}} - R_1^{2-\beta-\alpha(1+\varepsilon)\frac{\varepsilon_0}{2\pi}}}{2-\beta-\alpha(1+\varepsilon)\frac{\varepsilon_0}{2\pi}} \\
&\leq \frac{Ce^{\alpha\left(1+\frac{1}{\varepsilon}\right)}}{2-\beta-\alpha(1+\varepsilon)\frac{\varepsilon_0}{2\pi}}\left(R_0^{2-\beta}-R_1^{2-\beta}\right) \\
&\leq C\left(R_0^2-R_1^2\right)^{1-\frac{\beta}{2}} \\
&\leq C\left(\int_{B_{R_0}\setminus B_{R_1}} 1dx\right)^{1-\frac{\beta}{2}} \\
&\leq C\|u\|_2^{2-\beta},
\end{aligned}
$$

(since $\alpha \leq 4\pi - 2\beta$, we can choose $\varepsilon > 0$ such that $2-\beta-\alpha(1+\varepsilon)\frac{\varepsilon_0}{2\pi} > 0$).

So, we need to estimate $I_1 = \int_{B_{R_1}} \frac{e^{\alpha u^2}-1}{\left(1+u^{2-\beta}\right)|x|^{\beta}}dx$ with $u(R_1) > 1$.

First, we define

$$v(r) = u(r) - u(R_1) \text{ on } 0 \leq r \leq R_1.$$

It's clear that $v \in W_0^{1,2}(B_{R_1})$ and that $\int_{B_{R_1}} |\nabla v|^2 dx = \int_{B_{R_1}} |\nabla u|^2 dx \leq 1-\varepsilon_0$.

Moreover, for $0 \leq r \leq R_1$ :

$$
\begin{aligned}
u^2(r) &= [v(r) + u(R_1)]^2 \\
&\leq (1+\varepsilon)v^2(r) + \left(1+\frac{1}{\varepsilon}\right)u^2(R_1).
\end{aligned}
$$

Hence

$$I_1 = \int_{B_{R_1}} \frac{e^{\alpha u^2} - 1}{\left(1 + u^{2-\beta}\right) |x|^\beta} dx \tag{13}$$
$$\leq \frac{e^{\alpha\left(1+\frac{1}{\varepsilon}\right)u^2(R_1)}}{u^{2-\beta}(R_1)} \int_{B_{R_1}} \frac{e^{(1+\varepsilon)\alpha v^2(r)}}{|x|^\beta} dx$$
$$= \frac{e^{\alpha\left(1+\frac{1}{\varepsilon}\right)u^2(R_1)}}{u^{2-\beta}(R_1)} \int_{B_{R_1}} \frac{e^{\alpha w^2(r)}}{|x|^\beta} dx$$

where $w = \sqrt{1+\varepsilon}v$.

It's clear that $w \in W_0^{1,2}(B_{R_1})$ and $\int_{B_{R_1}} |\nabla w|^2 dx = (1+\varepsilon) \int_{B_{R_1}} |\nabla v|^2 dx \leq (1+\varepsilon)(1-\varepsilon_0) \leq 1$ if we choose $0 < \varepsilon \leq \frac{\varepsilon_0}{1-\varepsilon_0}$. Hence, using the singular Moser-Trudinger inequality, we have

$$\int_{B_{R_1}} \frac{e^{w^2(r)}}{|x|^\beta} dx \leq C |B_{R_1}|^{1-\frac{\beta}{2}} \leq C R_1^{2-\beta}. \tag{14}$$

Also, using Theorem 2.1, we have

$$\frac{e^{\alpha\left(1+\frac{1}{\varepsilon}\right)u^2(R_1)}}{u^{2-\beta}(R_1)} R_1^{2-\beta} \leq \left[\frac{e^{\frac{2}{2-\beta}\alpha\left(1+\frac{1}{\varepsilon}\right)u^2(R_1)}}{u^2(R_1)} R_1^2\right]^{1-\frac{\beta}{2}} \tag{15}$$
$$\leq \left(\frac{C}{\varepsilon_0^2} \int_{\mathbb{R}^2 \setminus B_R} |u|^2 dx\right)^{1-\frac{\beta}{2}}$$
$$\leq \left(\frac{C}{\varepsilon_0^2} \|u\|_2^2\right)^{1-\frac{\beta}{2}}$$

if we choose $\varepsilon$ such that

$$\frac{1}{\varepsilon_0} - 1 \leq \frac{1}{\varepsilon} \leq \frac{2\pi(2-\beta)}{\alpha} \frac{1}{\varepsilon_0} - 1.$$

By (13), (14) and (15), the proof is now completed.

## 4 Sharp Singular Adams Type Inequality with Exact Growth

We will prove Theorem 1.2 in this section. Again, we will adapt the argument of [54] together with an application of the symmetrization lemma similar to our Lemma 2.5. We mention that the following theorem holds for all $N \geq 3$ using the result in [53]. But we only state and present its proof in dimension $N = 4$.

**Theorem 4.1** *Let* $0 \leq \beta < 4$ *and* $0 < \alpha \leq 32\pi^2\left(1 - \frac{\beta}{4}\right)$. *Then there exists a constant* $C = C(\alpha, \beta) > 0$ *such that*

$$\int_{\mathbb{R}^4} \frac{e^{\alpha u^2} - 1}{\left(1 + |u|^{2-\beta/2}\right)|x|^\beta} dx \leq C \|u\|_2^{2-\frac{\beta}{2}} \text{ for all } u \in W^{2,2}\left(\mathbb{R}^4\right) : \|\Delta u\|_2 \leq 1.$$

*Proof* Fix $u \in C_0^\infty\left(\mathbb{R}^4\right)$ such that $\|\Delta u\|_2 \leq 1$ and define

$$R_0 = R_0(u) = \inf\left\{r > 0 : u^\#(r) \leq 1\right\} \in [0, \infty).$$

We may assume that $R_0 > 0$.

Similar to the proof of Theorem 1.1, we will need a symmetrization lemma. By the Hardy-Littlewood inequality and a similar proof to Lemma 2.5, we have

$$\begin{aligned}\int_{\mathbb{R}^4} \frac{e^{\alpha u^2} - 1}{\left(1 + u^{2-\beta/2}\right)|x|^\beta} dx &\leq \int_{\mathbb{R}^4} \frac{e^{\alpha (u^\#)^2} - 1}{\left(1 + (u^\#)^{2-\beta/2}\right)|x|^\beta} dx \qquad (16)\\ &= \int_{B_{R_0}} \frac{e^{\alpha (u^\#)^2} - 1}{\left(1 + (u^\#)^{2-\beta/2}\right)|x|^\beta} dx + \int_{\mathbb{R}^4 \backslash B_{R_0}} \frac{e^{\alpha (u^\#)^2} - 1}{\left(1 + (u^\#)^{2-\beta/2}\right)|x|^\beta} dx\\ &= I + J.\end{aligned}$$

We now estimate $J$. Indeed, since $u^\#(r) \leq 1$ on $\mathbb{R}^4 \setminus B_{R_0}$, we get

$$\begin{aligned}J &\leq C \int_{\{u^\# \leq 1\}} \frac{(u^\#)^2}{|x|^\beta} dx \qquad (17)\\ &\leq C \int_{\left\{u^\# \leq 1;\ |x| \geq \|u\|_2^{1/2}\right\}} \frac{(u^\#)^2}{|x|^\beta} dx + C \int_{\left\{u^\# \leq 1;\ |x| < \|u\|_2^{1/2}\right\}} \frac{(u^\#)^2}{|x|^\beta} dx\\ &\leq C \left[\int_{\mathbb{R}^4} (u^\#)^2 dx\right]^{1-\frac{\beta}{4}}\\ &\leq C \|u\|_2^{2-\frac{\beta}{2}}.\end{aligned}$$

So, the difficult part is the integral $I$.

Set

$$f = -\Delta u$$

$$\beta = \int_0^\infty \left[f^{**}(s)\right]^2 ds.$$

Then since $\|\Delta u\|_2 \leq 1$, we have $\beta \leq 4$.

We now fix $\varepsilon_0 \in (0,1)$ (we note that $\varepsilon_0$ is independent of $u$) and choose $R_1 = R_1(u) > 0$ such that

$$\int_0^{|B_{R_1}|} \left[f^{**}(s)\right]^2 ds = \beta(1-\varepsilon_0) \tag{18}$$

$$\int_{|B_{R_1}|}^\infty \left[f^{**}(s)\right]^2 ds = \beta\varepsilon_0.$$

We have the following result:

$$u^{\#}(r_1) - u^{\#}(r_2) \leq \frac{\sqrt{2}}{16\pi}\left(\int_{|B_{R_1}|}^{|B_{R_2}|} \left[f^{**}(s)\right]^2 ds\right)^{1/2} \left(\ln\frac{r_2^4}{r_1^4}\right)^{1/2} \quad \text{for } 0 < r_1 < r_2. \tag{19}$$

Hence, by (18) and (19), we get

$$u^{\#}(r_1) - u^{\#}(r_2) \leq \left(\frac{1-\varepsilon_0}{32\pi^2}\right)^{1/2}\left(\ln\frac{r_2^4}{r_1^4}\right)^{1/2} \quad \text{for } 0 < r_1 < r_2 \leq R_1, \tag{20}$$

$$u^{\#}(r_1) - u^{\#}(r_2) \leq \left(\frac{\varepsilon_0}{32\pi^2}\right)^{1/2}\left(\ln\frac{r_2^4}{r_1^4}\right)^{1/2} \quad \text{for } R_1 \leq r_1 < r_2.$$

We distinguish two cases:

*Case 1* $0 < R_0 \leq R_1$

In this case, by (20), we have for $0 < r \leq R_0\ (\leq R_1)$ :

$$u^{\#}(r) \leq 1 + \left(\frac{1-\varepsilon_0}{32\pi^2}\right)^{1/2}\left(\ln\frac{R_0^4}{r^4}\right)^{1/2}$$

$$\left[u^{\#}(r)\right]^2 \leq \frac{1-\varepsilon_0^2}{32\pi^2}\ln\frac{R_0^4}{r^4} + \left(1+\frac{1}{\varepsilon_0}\right).$$

Hence,

$$\begin{aligned}
I &= \int\limits_{B_{R_0}} \frac{e^{\alpha\left(u^{\#}\right)^2}-1}{\left(1+\left(u^{\#}\right)^{2-\beta/2}\right)|x|^{\beta}}dx \qquad (21)\\
&\leq C\int\limits_0^{R_0} e^{\alpha\frac{1-\varepsilon_0^2}{32\pi^2}\ln\frac{R_0^4}{r^4}+\alpha\left(1+\frac{1}{\varepsilon_0}\right)} r^{3-\beta}dx\\
&\leq CR_0^{4\alpha\frac{1-\varepsilon_0^2}{32\pi^2}}\int\limits_0^{R_0} r^{3-\beta-4\alpha\frac{1-\varepsilon_0^2}{32\pi^2}}dr\\
&\leq CR_0^{4-\beta}\\
&\leq C\left(\int\limits_{B_{R_0}} 1dx\right)^{1-\frac{\beta}{4}}\\
&\leq C\left\|u\right\|_2^{2-\frac{\beta}{2}}.
\end{aligned}$$

*Case 2* $0 < R_1 < R_0$.

$$\begin{aligned}
I &= \int\limits_{B_{R_0}} \frac{e^{\alpha\left(u^{\#}\right)^2}-1}{\left(1+\left(u^{\#}\right)^{2-\beta/2}\right)|x|^{\beta}}dx\\
&= \int\limits_{B_{R_1}} + \int\limits_{B_{R_0}\setminus B_{R_1}} \frac{e^{\alpha\left(u^{\#}\right)^2}-1}{\left(1+\left(u^{\#}\right)^{2-\beta/2}\right)|x|^{\beta}}dx\\
&= I_1 + I_2.
\end{aligned}$$

Again, by (20), we have for $R_1 \leq r \leq R_0$ :

$$\left[u^{\#}(r)\right]^2 \leq \frac{\varepsilon_0(1+\varepsilon)}{32\pi^2}\ln\frac{R_0^4}{r^4} + \left(1+\frac{1}{\varepsilon}\right),\ \forall\varepsilon > 0.$$

Hence if we choose $0 < \varepsilon < \frac{32\pi^2\left(1-\frac{\beta}{4}\right)}{\alpha}\frac{1}{\varepsilon_0} - 1$, we get

$$\begin{aligned}
I_2 &= \int_{B_{R_0}\setminus B_{R_1}} \frac{e^{\alpha\left(u^\#\right)^2}-1}{\left(1+\left(u^\#\right)^{2-\beta/2}\right)|x|^\beta}dx \qquad (22)\\
&\le C\int_{R_1}^{R_0} e^{\alpha\frac{\varepsilon_0(1+\varepsilon)}{32\pi^2}\ln\frac{R_0^4}{r^4}+\alpha\left(1+\frac{1}{\varepsilon}\right)} r^{3-\beta}dr\\
&\le CR_0^{4\alpha\frac{\varepsilon_0(1+\varepsilon)}{32\pi^2}}\left[R_0^{4-\beta-4\alpha\frac{\varepsilon_0(1+\varepsilon)}{32\pi^2}} - R_1^{4-\beta-4\alpha\frac{\varepsilon_0(1+\varepsilon)}{32\pi^2}}\right]\\
&\le C\left[R_0^{4-\beta} - R_0^{4\alpha\frac{\varepsilon_0(1+\varepsilon)}{32\pi^2}}R_1^{4-\beta-4\alpha\frac{\varepsilon_0(1+\varepsilon)}{32\pi^2}}\right]\\
&\le C\left[R_0^{4-\beta} - R_1^{4-\beta}\right]\\
&\le C\left[R_0^4 - R_1^4\right]^{1-\frac{\beta}{4}}\\
&\le C\left(\int_{B_{R_0}\setminus B_{R_1}} 1dx\right)^{1-\frac{\beta}{4}}\\
&\le C\|u\|_2^{2-\frac{\beta}{2}}.
\end{aligned}$$

To estimate $I_1$, we first note that by (20), we have

$$\left[u^\#(r)\right]^2 \le (1+\varepsilon)\left[u^\#(r)-u^\#(R_1)\right]^2 + \left(1+\frac{1}{\varepsilon}\right)\left[u^\#(R_1)\right]^2, \text{ for all } 0\le r\le R_1;\ \varepsilon>0.$$

As a consequence, we get

$$\begin{aligned}
I_1 &= \int_{B_{R_1}} \frac{e^{\alpha\left(u^\#\right)^2}-1}{\left(1+\left(u^\#\right)^{2-\beta/2}\right)|x|^\beta}dx \qquad (23)\\
&\le \frac{1}{\left[u^\#(R_1)\right]^{2-\beta/2}}\int_{B_{R_1}}\frac{e^{\alpha\left(u^\#\right)^2}}{|x|^\beta}dx\\
&\le C\frac{e^{\alpha\left(1+\frac{1}{\varepsilon}\right)\left[u^\#(R_1)\right]^2}}{\left[u^\#(R_1)\right]^{2-\beta/2}}\int_{B_{R_1}}\frac{e^{\alpha(1+\varepsilon)\left[u^\#(|x|)-u^\#(R_1)\right]^2}}{|x|^\beta}dx.
\end{aligned}$$

By Theorem 2.2, we have that there exists a universal constant $C > 0$ such that

$$\frac{e^{\alpha\left(1+\frac{1}{\varepsilon}\right)\left[u^{\#}(R_1)\right]^2}}{\left[u^{\#}(R_1)\right]^{2-\beta/2}} = \left(\frac{e^{\alpha\frac{4}{4-\beta}\left(1+\frac{1}{\varepsilon}\right)\left[u^{\#}(R_1)\right]^2}}{\left[u^{\#}(R_1)\right]^2}\right)^{1-\frac{\beta}{4}} \tag{24}$$

$$\leq \left(\frac{C}{R_1^4}\int_{\mathbb{R}^4\setminus B_{R_1}} \left|u^{\#}\right|^2 dx\right)^{1-\frac{\beta}{4}} \leq \frac{C}{R_1^{4-\beta}}\|u\|_2^{2-\frac{\beta}{2}}.$$

Here we choose $\varepsilon$ such that $1+\frac{1}{\varepsilon} \leq \frac{32\pi^2\left(1-\frac{\beta}{4}\right)}{\alpha}\frac{1}{\varepsilon_0}$.

Now, we claim that

$$\frac{1}{R_1^{4-\beta}}\int_{B_{R_1}} \frac{e^{\alpha(1+\varepsilon)\left[u^{\#}(|x|)-u^{\#}(R_1)\right]^2}}{|x|^{\beta}}dx \leq C. \tag{25}$$

Indeed, we have for $0 < r < |B_{R_1}|$ :

$$0 \leq u^*(r) - u^*\left(|B_{R_1}|\right) \leq \frac{\sqrt{2}}{16\pi}\int_r^{|B_{R_1}|}\frac{f^{**}(s)}{\sqrt{s}}ds.$$

Hence,

$$\int_{B_{R_1}} \frac{e^{\alpha(1+\varepsilon)\left[u^{\#}(|x|)-u^{\#}(R_1)\right]^2}}{|x|^{\beta}}dx \leq C\int_0^{|B_{R_1}|}\frac{\exp\left[\sqrt{1-\frac{\beta}{4}}\frac{\sqrt{1+\varepsilon}}{2}\int_r^{|B_{R_1}|}\frac{f^{**}(s)}{\sqrt{s}}ds\right]^2}{r^{\beta/4}}dr$$

$$\leq C\left|B_{R_1}\right|^{1-\beta/4}\int_0^{\infty}\exp\left[\sqrt{1-\frac{\beta}{4}}\frac{\sqrt{1+\varepsilon}}{2}\int_{|B_{R_1}|e^{-t}}^{|B_{R_1}|}\frac{f^{**}(z)}{\sqrt{z}}dz\right]^2 e^{-t(1-\beta/4)}dt$$

$$\leq C\left|B_{R_1}\right|^{1-\beta/4}\int_0^{\infty}\exp\left((1-\beta/4)\left[\frac{\sqrt{1+\varepsilon}}{2}\int_{|B_{R_1}|e^{-t}}^{|B_{R_1}|}\frac{f^{**}(z)}{\sqrt{z}}dz\right]^2 - t(1-\beta/4)\right)dt$$

$$\leq C\left|B_{R_1}\right|^{1-\beta/4}\int_0^\infty \exp\left((1-\beta/4)\left[\frac{\sqrt{1+\varepsilon}}{2}\int_0^t \frac{f^{**}\left(\left|B_{R_1}\right|e^{-s}\right)}{\sqrt{\left|B_{R_1}\right|e^{-s}}}\left|B_{R_1}\right|e^{-s}ds\right]^2 -t\left(1-\beta/4\right)\right)dt.$$

Here, we make change $r = \left|B_{R_1}\right|e^{-t}$ in the second inequality and $z = \left|B_{R_1}\right|e^{-s}$ in the last one.

Now, using Lemma 2.2 with

$$\gamma = 1-\beta/4;\ p = 2$$

$$a(s,t) = \begin{cases} 1 \text{ if } 0 < s < t \\ 0 \text{ otherwise} \end{cases}$$

$$\phi(s) = \sqrt{\left|B_{R_1}\right|}\frac{\sqrt{1+\varepsilon}}{2}f^{**}\left(\left|B_{R_1}\right|e^{-s}\right)e^{-\frac{s}{2}}ds;\ s \geq 0$$

we can conclude (25). Indeed, we have

$$\begin{aligned}\int_0^\infty \phi^2(s)ds &= \left|B_{R_1}\right|\frac{1+\varepsilon}{4}\int_0^\infty \left[f^{**}\left(\left|B_{R_1}\right|e^{-s}\right)\right]^2 e^{-s}ds \\ &= \frac{1+\varepsilon}{4}\int_0^{\left|B_{R_1}\right|}\left[f^{**}(r)\right]^2 dr \\ &\leq (1+\varepsilon)(1-\varepsilon_0) \leq 1\end{aligned}$$

if we choose $\varepsilon$ such that

$$\frac{1}{\varepsilon_0} \leq 1 + \frac{1}{\varepsilon} \leq \frac{32\pi^2\left(1-\frac{\beta}{4}\right)}{\alpha}\frac{1}{\varepsilon_0}.$$

The proof now is completed.

## 5 Proof of Theorem 1.4

Let $u^{\#}$ be the symmetric decreasing rearrangement of $u$. By rearrangement properties: the Pólya-Szegö inequality and the Hardy-Littlewood inequality, we get

$$\left\|\nabla u^{\#}\right\|_n = \mathscr{E}_n^+\left(u^{\#}\right) \le \mathscr{E}_n^+(u),$$
$$\int\limits_B \left(\ln\frac{1}{|x|^n}+1\right) e^{\alpha_n |u|^{\frac{n}{n-1}}} dx \le \int\limits_B \left(\ln\frac{1}{|x|^n}+1\right) e^{\alpha_n \left|u^{\#}\right|^{\frac{n}{n-1}}} dx,$$
$$\int\limits_B \alpha_n |u|^{\frac{n}{n-1}} e^{\alpha_n |u|^{\frac{n}{n-1}}} dx = \int\limits_B \alpha_n \left|u^{\#}\right|^{\frac{n}{n-1}} e^{\alpha_n \left|u^{\#}\right|^{\frac{n}{n-1}}} dx.$$

Hence, we may assume that $u$ is radially symmetric and nonincreasing.

By changing of variable

$$|x|^n = e^{-t},$$
$$w(t) = n^{\frac{n-1}{n}} \omega_{n-1}^{1/n} u(x)$$

we have $w(t)$ is a $C'$-function and $0 \le t < \infty$ satisfying

$$w(0) = 0,\ w' \ge 0,\ \int\limits_0^\infty \left|w'(t)\right|^n dt \le 1$$

and we need to prove that

$$\int\limits_0^\infty \left(t - w^{\frac{n}{n-1}}(t) + 1\right) e^{w^{\frac{n}{n-1}}(t)-t} dt \le MT.$$

Set

$$F(t) = t - w^{\frac{n}{n-1}}(t),$$

so we need to show

$$\int\limits_0^\infty (F(t)+1)\, e^{-F(t)} dt \le MT.$$

Using Lemma 2.3 with $\varphi = w'$, we get the result.

## 6 Sharpness of Constants in Theorems 1.1 and 1.2

The main purpose of this section is to prove the sharpness of the constants in Theorems 1.1 and 1.2.

First, we will show that the Theorem 1.1 is sharp in the sense that the inequality

$$\int_{\mathbb{R}^2} \frac{F(u)}{|x|^{\beta}} dx \leq C \|u\|_2^{2-\beta} \text{ for all } u \in W^{1,2}\left(\mathbb{R}^2\right) : \mathscr{E}_2^+(u) \leq 1$$

fails if the growth

$$F(u) = \frac{e^{4\pi\left(1-\frac{\beta}{2}\right)u^2} - 1}{\left(1 + |u|^{2-\beta}\right)}$$

is replaced with the higher order growth

$$F(u) = \frac{e^{\alpha u^2} - 1}{(1 + |u|^q)}$$

where either $\alpha > 4\pi\left(1 - \frac{\beta}{2}\right)$ and $q = 2 - \beta$ or $\alpha = 4\pi\left(1 - \frac{\beta}{2}\right)$ and $q < 2 - \beta$.

The former case is easy since we can find some constant $C = C(\beta)$ such that

$$\int_{\mathbb{R}^2} \frac{e^{\alpha u^2} - 1}{\left(1 + |u|^{2-\beta}\right)|x|^{\beta}} dx \geq C \int_{\mathbb{R}^2} \frac{e^{\frac{\left(\alpha + 4\pi\left(1-\frac{\beta}{2}\right)\right)}{2}u^2} - 1}{|x|^{\beta}} dx.$$

Now, fix $q < 2 - \beta$ and let

$$F(x) = \frac{e^{4\pi\left(1-\frac{\beta}{2}\right)x^2} - 1}{(1 + |x|^q)}.$$

First, we choose sequences

$$1 \lll z_n \uparrow \infty \text{ and } A_n \uparrow A = \frac{1}{2\pi}$$

such that

$$c_n = \exp\left(\frac{-2\left(1 - \frac{\beta}{2}\right)z_n^2}{A_n}\right) z_n^{2\left(1-\frac{\beta}{2}\right)} F(z_n) \to \infty.$$

We can do that since

$$\lim_{|x|\to\infty}\frac{|x|^{2-\beta}F(x)}{e^{2\left(1-\frac{\beta}{2}\right)|x|^2/K}}=\infty.$$

Let

$$u_n(r)=\begin{cases} z_n & \text{if } r<T_n \\ z_n\frac{|\log r|}{|\log T_n|} & \text{if } T_n\le |x|<1 \\ 0 & \text{if } |x|\ge 1 \end{cases}$$

where

$$T_n=\exp\left(-\frac{z_n^2}{A_n}\right).$$

Then, it is clear that

$$\|\nabla u_n\|_2^2=2\pi z_n^2\int_{T_n}^{1}\frac{dr}{r\,|\log T_n|^2}=2\pi A_n<1;$$

$$\|u_n\|_2\sim\frac{\sqrt{2\pi}A_n}{z_n};$$

$$G(u_n)=\int_{\mathbb{R}^2}\frac{F(u_n)}{|x|^\beta}dx\ge\frac{2\pi}{2-\beta}T_n^{2-\beta}F(z_n)=\frac{2\pi}{2-\beta}\frac{c_n}{z_n^{2-\beta}}.$$

Hence

$$\frac{G(u_n)}{\|u_n\|_2^{2-\beta}}\ge Cc_n\to\infty.$$

Similarly, in Theorem 1.2, we just need to prove that for fix $q<2-\frac{\beta}{2}$, then the inequality

$$\int_{\mathbb{R}^4}\frac{e^{32\pi^2\left(1-\frac{\beta}{4}\right)u^2}-1}{(1+|u|^q)\,|x|^\beta}dx\le C\,\|u\|_2^{2-\frac{\beta}{2}}\ \text{ for all } u\in W^{2,2}\left(\mathbb{R}^4\right):\|\Delta u\|_2\le 1 \tag{26}$$

does not hold. Indeed, consider the sequence introduced by Lu and Yang in [52]:

$$u_n(r) = \begin{cases} \sqrt{\frac{1}{32\pi^2}\log\frac{1}{R_n}} - \frac{r^2}{\sqrt{8\pi^2 R_n \log\frac{1}{R_n}}} + \frac{1}{\sqrt{8\pi^2\log\frac{1}{R_n}}} & \text{if } 0 \le r \le \sqrt[4]{R_n} \\ \frac{1}{\sqrt{2\pi^2\log\frac{1}{R_n}}}\log\frac{1}{r} & \text{if } \sqrt[4]{R_n} < r \le 1 \\ \eta_n & \text{if } r > 1 \end{cases}$$

where $R_n \downarrow 0$ and $\eta_n$ is smooth and is chosen such that for some $R > 1$ :

$$\eta_n|_{\partial B_1} = \eta_n|_{\partial B_R} = 0,$$
$$\frac{\partial \eta_n}{\partial \nu}|_{\partial B_1} = \frac{1}{\sqrt{2\pi^2\log\frac{1}{R_n}}}; \quad \frac{\partial \eta_n}{\partial \nu}|_{\partial B_R} = 0$$

and $\eta_n$, $\Delta\eta_n$ are all $O\left(\frac{1}{\sqrt{\log\frac{1}{R_n}}}\right)$. Then, we can verify that

$$\|u_n\|_2 = O\left(\frac{1}{\sqrt{\log\frac{1}{R_n}}}\right)$$
$$1 \le \|\Delta u_n\|_2^2 = 1 + O\left(\frac{1}{\log\frac{1}{R_n}}\right).$$

Now, assume (26) holds. Then let

$$v_n = \frac{u_n}{\|\Delta u_n\|_2},$$

we get

$$\int_{\mathbb{R}^4} \frac{e^{32\pi^2\left(1-\frac{\beta}{4}\right)v_n^2} - 1}{(1+|v_n|^q)\,|x|^\beta}dx \le C\,\|v_n\|_2^{2-\frac{\beta}{2}} \le C\left(\frac{1}{\sqrt{\log\frac{1}{R_n}}}\right)^{2-\frac{\beta}{2}}.$$

Hence

$$\lim_{n\to\infty}\left(\sqrt{\log\frac{1}{R_n}}\right)^{2-\frac{\beta}{2}} \int_{\mathbb{R}^4} \frac{e^{32\pi^2\left(1-\frac{\beta}{4}\right)v_n^2} - 1}{(1+|v_n|^q)\,|x|^\beta}dx < \infty. \tag{27}$$

Also, noting that $u_n \geq \sqrt{\frac{1}{32\pi^2}\log\frac{1}{R_n}}$ on $B_{\sqrt[4]{R_n}}$, we have that

$$\begin{aligned}
\int_{\mathbb{R}^4} \frac{e^{32\pi^2\left(1-\frac{\beta}{4}\right)v_n^2}-1}{\left(1+|v_n|^q\right)|x|^\beta}dx &\geq \int_{B_{\sqrt[4]{R_n}}} \frac{e^{32\pi^2\left(1-\frac{\beta}{4}\right)v_n^2}-1}{\left(1+|v_n|^q\right)|x|^\beta}dx \gtrsim \int_{B_{\sqrt[4]{R_n}}} \frac{e^{32\pi^2\left(1-\frac{\beta}{4}\right)v_n^2}}{|v_n|^q|x|^\beta}dx \\
&\gtrsim \frac{\exp\left(\frac{32\pi^2\left(1-\frac{\beta}{4}\right)\frac{1}{32\pi^2}\log\frac{1}{R_n}}{\|\Delta u_n\|_2^2}\right)}{\left(\sqrt{\frac{1}{32\pi^2}\log\frac{1}{R_n}}\right)^q} R_n^{1-\frac{\beta}{4}} \\
&\approx \exp\left(\left(1-\frac{\beta}{4}\right)\log\frac{1}{R_n}\left(\frac{1}{\|\Delta u_n\|_2^2}-1\right)\right)\left(\log\frac{1}{R_n}\right)^{-\frac{q}{2}}.
\end{aligned}$$

Hence

$$\lim_{n\to\infty}\left(\sqrt{\log\frac{1}{R_n}}\right)^{2-\frac{\beta}{2}} \int_{\mathbb{R}^4} \frac{e^{32\pi^2\left(1-\frac{\beta}{4}\right)v_n^2}-1}{\left(1+|v_n|^q\right)|x|^\beta}dx \gtrsim \lim_{n\to\infty}\left(\sqrt{\log\frac{1}{R_n}}\right)^{2-\frac{\beta}{2}-q} = \infty$$

which is a contradiction of (27).

Our last main task is to verify (3) since if (3) holds, then with $\alpha > \alpha_n$ :

$$\begin{aligned}
&\sup_{u\in W_{0,rad}^{1,n}(B),\ \mathscr{E}_n^+(u)\leq 1} \int_B \left(\ln\frac{1}{|x|^n}+1-\alpha_n|u|^{\frac{n}{n-1}}\right)e^{\alpha|u|^{\frac{n}{n-1}}}dx \\
\geq &\sup_{u\in W_{0,rad}^{1,n}(B),\ \mathscr{E}_n^+(u)\leq 1} \int_B e^{\alpha|u|^{\frac{n}{n-1}}}dx = \infty.
\end{aligned}$$

Indeed, by changing of variable

$$\begin{aligned}
|x|^n &= e^{-t}, \\
w(t) &= n^{\frac{n-1}{n}}\omega_{n-1}^{1/n}u(x)
\end{aligned}$$

we have $w(t)$ satisfies

$$w(0)=0,\ w'\geq 0,\ \int_0^\infty \left|w'(t)\right|^n dt \leq 1.$$

Hence

$$w(t) = \int_0^t w'(s)ds \leq \left( \int_0^t |w'(s)|^n \, ds \right)^{1/n} \left( \int_0^t 1ds \right)^{\frac{n-1}{n}}$$
$$\leq t^{\frac{n-1}{n}}.$$

But this is exactly (3).

**Acknowledgements** This research is partly supported by a US NSF grant DMS#1301595.

# References


1. Adachi, S., Tanaka, K.: Trudinger type inequalities in $\mathbb{R}^N$ and their best exponents. Proc. Am. Math. Soc. **128**, 2051–2057 (1999)
2. Adams, D.R.: A sharp inequality of J. Moser for higher order derivatives. Ann. Math. (2) **128**(2), 385–398 (1988)
3. Adimurthi, A., Druet, O.: Blow-up analysis in dimension 2 and a sharp form of Trudinger-Moser inequality. Commun. Partial Differ. Equ. **29**(1–2), 295–322 (2004)
4. Adimurthi, A., Sandeep, K.: A singular Moser-Trudinger embedding and its applications. NoDEA Nonlinear Differ. Equ. Appl. **13**(5–6), 585–603 (2007)
5. Adimurthi, A., Yang, Y.: An interpolation of Hardy inequality and Trudinger-Moser inequality in $\mathbb{R}^N$ and its applications. Int. Math. Res. Not. IMRN **13**, 2394–2426 (2010)
6. Balogh, Z., Manfredi, J., Tyson, J.: Fundamental solution for the Q-Laplacian and sharp Moser-Trudinger inequality in Carnot groups. J. Funct. Anal. **204**(1), 35–49 (2003)
7. Beckner, W.: Sharp Sobolev inequalities on the sphere and the Moser-Trudinger inequality. Ann. Math. (2) **138**(1), 213–242 (1993)
8. Branson, T., Fontana, L., Morpurgo, C.: Moser-Trudinger and Beckner-Onofri's inequalities on the CR sphere. Ann. Math. (2) **177**(1), 1–52 (2013)
9. Cao, D.: Nontrivial solution of semilinear elliptic equation with critical exponent in $R^2$. Commun. Partial Differ. Equ. **17**(3–4), 407–435 (1992)
10. Carleson, L., Chang, S.Y.A.: On the existence of an extremal function for an inequality. J. Moser. Bull. Sci. Math. (2) **110**(2), 113–127 (1986)
11. Cassani, D., Tarsi, C.: A Moser-type inequality in Lorentz-Sobolev spaces for unbounded domains in $\mathbb{R}^N$. Asymptot. Anal. **64**(1–2), 29–51 (2009)
12. Chang, S.Y.A., Yang, P.: Conformal deformation of metrics on $S^2$. J. Differ. Geom. **27**, 259–296 (1988)
13. Chang, S.Y.A., Yang, P.: The inequality of Moser and Trudinger and applications to conformal geometry. Dedicated to the memory of Jürgen K. Moser. Commun. Pure Appl. Math. **56**(8), 1135–1150 (2003)
14. Cianchi, A.: Moser-Trudinger inequalities without boundary conditions and isoperimetric problems. Indiana Univ. Math. J. **54**(3), 669–705 (2005)
15. Cianchi, A.: Moser-Trudinger trace inequalities. Adv. Math. **217**(5), 2005–2044 (2008)
16. Cianchi, A., Lutwak, E., Yang, D., Zhang, G.: Affine Moser-Trudinger and Morrey-Sobolev inequalities. Calc. Var. **36**(3), 419–436 (2009)
17. Cohn, W.S., Lu, G.: Best constants for Moser-Trudinger inequalities on the Heisenberg group. Indiana Univ. Math. J. **50**(4), 1567–1591 (2001)

18. Cohn, W.S., Lu, G.: Best constants for Moser-Trudinger inequalities, fundamental solutions and one-parameter representation formulas on groups of Heisenberg type. Acta Math. Sin. (Engl. Ser.) **18**(2), 375–390 (2002)
19. Cohn, W.S., Lu, G.: Sharp constants for Moser-Trudinger inequalities on spheres in complex space $C^n$. Commun. Pure Appl. Math. **57**(11), 1458–1493 (2004)
20. Cohn, W.S., Lam, N., Lu, G., Yang, Y.: The Moser-Trudinger inequality in unbounded domains of Heisenberg group and sub-elliptic equations. Nonlinear Anal. Theory Methods Appl. **75**(12), 4483–4495 (2012)
21. de Figueiredo, D.G., do Ó, J.M., Ruf, B.: On an inequality by N. Trudinger and J. Moser and related elliptic equations. Commun. Pure Appl. Math. **55**(2), 135–152 (2002)
22. do Ó, J.M.: N-Laplacian equations in $R^N$ with critical growth. Abstr. Appl. Anal. **2**(3–4), 301–315 (1997)
23. Flucher, M.: Extremal functions for the Trudinger-Moser inequality in 2 dimensions. Comment. Math. Helv. **67** (1992), no. 3, 471–497.
24. Fontana, L.: Sharp borderline Sobolev inequalities on compact Riemannian manifolds. Commun. Math. Helv. **68**, 415–454 (1993)
25. Fontana, L., Morpurgo, C.: Adams inequalities on measure spaces. Adv. Math. **226**(6), 5066–5119 (2011)
26. Haberl, C., Schuster, F.E., Xiao, J.: An asymmetric affine Pólya–Szegö principle. Math. Ann. **352**(3), 517–542
27. Ibrahim, S., Masmoudi, N., Nakanishi, K.: Trudinger-Moser inequality on the whole plane with the exact growth condition. arXiv:1110.1712v1
28. Ishiwata, M.: Existence and nonexistence of maximizers for variational problems associated with Trudinger-Moser type inequalities in $R^N$. Math. Ann. **351**(4), 781–804 (2011)
29. Ishiwata, M., Nakamura, M., Wadade, H.: On the sharp constant for the weighted Trudinger-Moser type inequality of the scaling invariant form. Ann. I. H. Poincaré-AN (2013). http://dx.doi.org/10.1016/j.anihpc.2013.03.004
30. Kozono, H., Sato, T., Wadade, H.: Upper bound of the best constant of a Trudinger-Moser inequality and its application to a Gagliardo-Nirenberg inequality. Indiana Univ. Math. J. **55**(6), 1951–1974 (2006)
31. Lam, N., Lu, G.: The Moser-Trudinger and Adams inequalities and elliptic and subelliptic equations with nonlinearity of exponential growth. In: Recent Developments in Geometry and Analysis. Advanced Lectures in Mathematics (ALM), vol. 23, pp. 179–251. International Press, Somerville (2012)
32. Lam, N., Lu, G.: Sharp Adams type inequalities in Sobolev spaces $W^{m,\frac{n}{m}}(\mathbb{R}^n)$ for arbitrary integer $m$. J. Differ. Equ. **253**, 1143–1171 (2012)
33. Lam, N., Lu, G.: Sharp singular Adams inequalities in high order Sobolev spaces. Methods Appl. Anal. **19**(3), 243–266 (2012)
34. Lam, N., Lu, G.: Sharp Moser-Trudinger inequality on the Heisenberg group at the critical case and applications. Adv. Math. **231**, 3259–3287 (2012)
35. Lam, N., Lu, G.: A new approach to sharp Moser-Trudinger and Adams type inequalities: a rearrangement-free argument. J. Differ. Equ. **255**(3), 298–325 (2013)
36. Lam, N., Lu, G., Zhang, L.: Sharp singular Trudinger-Moser inequalities and their extremals. Preprint (2013)
37. Lam, N., Tang, H.: Sharp constants for weighted Moser-Trudinger inequalities on groups of Heisenberg type. Nonlinear Anal. **89**, 95–109 (2013)
38. Lam, N., Lu, G., Tang, H.: On nonuniformly subelliptic equations of Q-sub-Laplacian type with critical growth in the Heisenberg group. Adv. Nonlinear Stud. **12**(3), 659–681 (2012)
39. Lam, N., Lu, G., Tang, H.: Sharp affine and improved Moser-Trudinger-Adams type inequalities on unbounded domains in the spirit of Lions. Preprint (2013)
40. Lam, N., Lu, G., Tang, H.: Sharp subcritical Moser-Trudinger inequalities on Heisenberg groups and subelliptic PDEs. Nonlinear Anal. **95**, 77–92 (2014)
41. Leckband, M.: Moser's inequality on the ball $B^n$ for functions with mean value zero. Commun. Pure Appl. Math. **LVIII**, 0789–0798 (2005)

42. Li, Y.X.: Moser-Trudinger inequality on compact Riemannian manifolds of dimension two. J. Partial Differ. Equ. **14**(2), 163–192 (2001)
43. Li, Y.X.: Extremal functions for the Moser-Trudinger inequalities on compact Riemannian manifolds. Sci. China Ser. A **48**(5), 618–648 (2005)
44. Li, Y.X., Ndiaye, C.: Extremal functions for Moser-Trudinger type inequality on compact closed 4-manifolds. J. Geom. Anal. **17**(4), 669–699 (2007)
45. Li, Y.X., Ruf, B.: A sharp Trudinger-Moser type inequality for unbounded domains in $\mathbb{R}^n$. Indiana Univ. Math. J. **57**(1), 451–480 (2008)
46. Lieb, E.H., Loss, M.: Analysis. Graduate Studies in Mathematics, vol. 14, xxii+346 pp, 2nd edn. American Mathematical Society, Providence (2001)
47. Lin, K.: Extremal functions for Moser's inequality. Trans. Am. Math. Soc. **348**(7), 2663–2671 (1996)
48. Lions, P.L.: The concentration-compactness principle in the calculus of variations. The limit case. II. Rev. Mat. Iberoam. **1**(2), 45–121 (1985)
49. Lu, G., Tang, H.: Best constants for Moser-Trudinger inequalities on high dimensional hyperbolic spaces. Adv. Nonlinear Stud. **13**, 1035–1052 (2013)
50. Lu, G., Tang, H.: Sharp Moser-Trudinger inequalities on hyperbolic spaces with the exact growth condition. J. Geom. Anal. (2015)
51. Lu, G., Yang, Y.: A sharpened Moser-Pohozaev-Trudinger inequality with mean value zero in $R^2$. Nonlinear Anal. **70**(8), 2992–3001 (2009)
52. Lu, G., Yang, Y.: Adams' inequalities for bi-Laplacian and extremal functions in dimension four. Adv. Math. **220**, 1135–1170 (2009)
53. Lu, G., Tang, H., Zhu, M.: Sharp Trudinger-Moser-Adams inequality with the exact growth condition in $R^n$. Adv. Nonlinear Stud. **15**(4) (2015)
54. Masmoudi, N., Sani, F.: Adams' inequality with the exact growth condition in $R^4$. Commun. Pure Appl. Math. **67**(8), 1307–1335 (2014)
55. Moser, J.: A sharp form of an inequality by N. Trudinger. Indiana Univ. Math. J. **20**, 1077–1092 (1970/1971)
56. Ogawa, T.: A proof of Trudinger's inequality and its application to nonlinear Schršdinger equations. Nonlinear Anal. **14**(9), 765–769 (1990)
57. Ozawa, T.: On critical cases of Sobolev's inequalities. J. Funct. Anal. **127**(2), 259–269 (1995)
58. Pohožaev, S.I.: On the eigenfunctions of the equation $\Delta u + \lambda f(u) = 0$. (Russian) Dokl. Akad. Nauk SSSR **165**, 36–39 (1965)
59. Ruf, B.: A sharp Trudinger-Moser type inequality for unbounded domains in $\mathbb{R}^2$. J. Funct. Anal. **219**(2), 340–367 (2005)
60. Ruf, B., Sani, F.: Sharp Adams-type inequalities in $\mathbb{R}^n$. Trans. Am. Math. Soc. **365**(2), 645–670 (2013)
61. Tarsi, C.: Adams' inequality and limiting Sobolev embeddings into Zygmund spaces. Potential Anal. **37**(4), 353–385 (2012)
62. Tian, G., Wang, X.: Moser-Trudinger type inequalities for the Hessian equation. J. Funct. Anal. **259**(8), 1974–2002 (2010)
63. Trudinger, N.S.: On imbeddings into Orlicz spaces and some applications. J. Math. Mech. **17**, 473–483 (1967)
64. Wang, G., Ye, D.: A Hardy-Moser-Trudinger inequality. Adv. Math. **230**(1), 294–320 (2012)
65. Yudovič, V. I.: Some estimates connected with integral operators and with solutions of elliptic equations. (Russian) Dokl. Akad. Nauk SSSR **138**, 805–808 (1961)

# A Quantitative Lusin Theorem for Functions in BV

**András Telcs and Vincenzo Vespri**

*Dedicated to our friend Ermanno Lanconelli on the occasion of his 70th birthday*

**Abstract** We extend to the BV case a measure theoretic lemma previously proved by DiBenedetto et al. (Atti Accad. Naz. Lincei Cl. Sci. Mat. Appl. **9**, 223–225, 2006) in $W^{1,1}_{loc}$. It states that if the set where $u$ is positive occupies a sizable portion of an open set $E$ then the set where $u$ is positive clusters about at least one point of $E$. In this note we follow the proof given in the Appendix of DiBenedetto and Vespri (Arch. Ration. Mech. Anal. **132**, 247–309, 1995) so we are able to use only a 1-dimensional Poincaré inequality.

**Keywords** BV functions • Lusin theorem

**AMS Classification:** Primary: 46E35, Secondary: 28A12

## 1 Introduction

For $\rho > 0$, denote by $K_\rho(y) \subset \mathbb{R}^N$ a cube of edge $\rho$ centered at $y$. If $y$ is the origin on $\mathbb{R}^N$, we write $K_\rho(0) = K_\rho$. For any measurable set $A \subset \mathbb{R}^N$, by $|A|$ we denote its $N$-dimensional Lebesgue measure.

A. Telcs
Department of Quantitative Methods, Faculty of Economics, University of Pannonia, Veszprém, Hungary

Department of Computer Science and Information Theory, Budapest University of Technology and Economic, Budapest, Hungary
e-mail: telcs.szit.bme@gmail.com

V. Vespri (✉)
Dipartimento di Matematica ed Informatica Ulisse Dini, Università degli Studi di Firenze, Firenze, Italy
e-mail: vincenzo.vespri@unifi.it



G. Citti et al. (eds.), *Geometric Methods in PDE's*, Springer INdAM Series 13,
DOI 10.1007/978-3-319-02666-4_4

If $u$ is a continuous function in a domain $E$ and $u(x_0) > 0$ for a point $x_0 \in E$ then there is a $r > 0$ such that $u(x) > 0$ in $K_r(x_0) \cap E$. If $u \in C^1$ then we can quantify the radius $r$ in terms of the $C^1$ norm of $u$.

The Lusin Theorem says that if $u$ is a measurable function in a bounded domain $E$, than for any $\varepsilon > 0$ there is a continuous function $g$ such that $g = u$ in $E$ except in a small set $V \subset E$ such that $|V| \leq \varepsilon$.

In this note we want to generalize the previous property in the case of measurable functions. Very roughly speaking, we prove that if $u \in BV(E)$ and $u(x_0) > 0$ for a point $x_0 \in E$ than for any $\varepsilon > 0$ there is a positive $r$, that can be quantitatively estimated in terms of $\varepsilon$ and the $BV$ norm of $u$, such that $u(x) > 0$ for any $x \in K_r(x_0) \cap E$ except in a small set $V \subset E$ such that $|V| \leq \varepsilon |K_r(x_0)|$. Obviously we will state a more precise result in the sequel.

Such kind of result has natural application in regularity theory for solutions to PDE's (see for instance the monograph [3] for an overview). The case of $W^{1,p}(E)$, with $1 < p < \infty$ , was studied in the Appendix of [1]. It was generalized in the case of $W^{1,1}(E)$ in [2].

Here we combine the proofs of [1, 2] in order to generalize this result in BV spaces. Moreover in this note we use a proof based only on 1-dimensional Poincaré inequality. This approach could be useful in the case anisotropic operators where it is likely that will be necessary to develop a new approach tailored on the structure of the operator (a first step in this direction can be found in [4]). We prove the following Measure Theoretical Lemma.

**Lemma 1.1** *Let $u \in BV(K_\rho)$ satisfy*

$$\|u\|_{BV(K_\rho)} \leq \gamma \rho^{N-1} \quad \text{and} \quad |[u > 1]| \geq \alpha |K_\rho| \tag{1}$$

*for some $\gamma > 0$ and $\alpha \in (0,1)$. Then, for every $\delta \in (0,1)$ and $0 < \lambda < 1$ there exist $x_o \in K_\rho$ and $\eta = \eta(\alpha, \delta, \gamma, \lambda, N) \in (0,1)$, such that*

$$|[u > \lambda] \cap K_{\eta\rho}(x_o)| > (1 - \delta)|K_{\eta\rho}(x_o)|. \tag{2}$$

Roughly speaking the Lemma asserts that if the set where $u$ is bounded away from zero occupies a sizable portion of $K_\rho$, then there exists at least one point $x_o$ and a neighborhood $K_{\eta\rho}(x_o)$ where $u$ remains large in a large portion of $K_{\eta\rho}(x_o)$. Thus the set where $u$ is positive clusters about at least one point of $K_\rho$.

In Sect. 2, we operate a suitable partition of $K_\rho$. In Sect. 3 we prove the result in the case $N = 2$ ( an analogous proof works for $N = 1$. We consider more meaningful to prove the result directly in the less trivial case $N = 2$). In Sect. 4, by an induction argument, we extend the lemma to any dimension.

## 2 Proof: A Partition of the Cube

It suffices to establish the Lemma for $u$ continuous and $\rho = 1$. For $n \in \mathbb{N}$ partition $K_1$ into $n^N$ cubes, with pairwise disjoint interior and each of edge $1/n$. Divide these cubes into two finite subcollections $\mathbf{Q}^+$ and $\mathbf{Q}^-$ by

$$Q_j \in \mathbf{Q}^+ \qquad \Longleftrightarrow \qquad |[u > 1] \cap Q_j| > \frac{\alpha}{2}|Q_j|$$

$$Q_i \in \mathbf{Q}^- \qquad \Longleftrightarrow \qquad |[u > 1] \cap Q_i| \leq \frac{\alpha}{2}|Q_i|$$

and denote by $\#(\mathbf{Q}^+)$ the number of cubes in $\mathbf{Q}^+$. By the assumption

$$\sum_{Q_j \in \mathbf{Q}^+} |[u > 1] \cap Q_j| + \sum_{Q_i \in \mathbf{Q}^-} |[u > 1] \cap Q_i| > \alpha|K_1| = \alpha n^N |Q|$$

where $|Q|$ is the common measure of the $Q_l$. From the definitions of the classes $\mathbf{Q}^\pm$,

$$\alpha n^N < \sum_{Q_j \in \mathbf{Q}^+} \frac{|[u > 1] \cap Q_j|}{|Q_j|} + \sum_{Q_i \in \mathbf{Q}^-} \frac{|[u > 1] \cap Q_i|}{|Q_i|} < \#(\mathbf{Q}^+) + \frac{\alpha}{2}(n^N - \#(\mathbf{Q}^+)).$$

Therefore

$$\#(\mathbf{Q}^+) > \frac{\alpha}{2 - \alpha} n^N.$$

Consider now a subcollection $\bar{\mathbf{Q}}^+$ of $\mathbf{Q}^+$. A cube $Q_j$ belongs to $\bar{\mathbf{Q}}^+$ if $Q_j \in \mathbf{Q}^+$ and $\|u\|_{BV(Q_j)} \leq \dfrac{2\alpha}{(2-\alpha)n^N} \|u\|_{BV(K_1)}$.

Clearly

$$\#(\bar{\mathbf{Q}}^+) > \frac{\alpha}{2(2 - \alpha)} n^N. \tag{3}$$

Fix $\delta, \lambda \in (0, 1)$. The idea of the proof is that an alternative occurs. Either there is a cube $Q_j \in \bar{\mathbf{Q}}^+$ such that there is a subcube $\tilde{Q} \subset Q_j$ where

$$|[u > \lambda] \cap \tilde{Q}| \geq (1 - \delta)|\tilde{Q}| \tag{4}$$

or for any cube $Q_j \in \bar{\mathbf{Q}}^+$ there exists a constant $c = c(\alpha, \delta, \gamma, \eta, N)$ such that

$$\|u\|_{BV(Q_j)} \geq c(\alpha, \delta, \gamma, \lambda, N) \frac{1}{n^{N-1}}. \tag{5}$$

Hence if (4) does not hold for any cube $Q_j \in \bar{\mathbf{Q}}^+$, we can add (5) over all such $Q_j$. Therefore taking into account (3), we have

$$\frac{\alpha}{2-\alpha} c(\alpha, \delta, \gamma, N) n \leq \|u\|_{BV(K_1)} \leq \gamma$$

and for $n$ large enough this fact leads to an evident absurdum.

## 3 Proof of the Lemma 1.1 When $N = 2$

The proof is quite similar to the one of Appendix A.1 of [1] to which we refer the reader for more details. For sake of simplicity we will use the same notation of [1].

Let $K_{\frac{1}{n}}(x_o, y_o) \in \bar{\mathbf{Q}}^+$. Without loss of generality we may assume $(x_o, y_o) = (0, 0)$. Assume that

$$\left|[u > 1] \cap K_{\frac{1}{n}}\right| > \frac{\alpha}{2} |K_{\frac{1}{n}}| \tag{6}$$

$$\|u\|_{BV(K_{\frac{1}{n}})} \leq \frac{2\alpha}{(2-\alpha)n^2} \|u\|_{BV(K_1)}. \tag{7}$$

Denote by $(x, y)$ the coordinates of $\mathbb{R}^2$ and, for $x \in (-\frac{1}{2n}, \frac{1}{2n})$ let $\mathfrak{Y}(x)$ the cross section of the set $[u > 1] \cap K_{\frac{1}{n}}$ with lines parallel to $y$-axis, through the abscissa $x$, i.e.

$$\mathfrak{Y}(x) \equiv \{y \in (-\frac{1}{2n}, \frac{1}{2n}) \text{ such that } u(x, y) > 1\}.$$

Therefore

$$|[u > 1] \cap K_{\frac{1}{n}}| \equiv \int_{-\frac{1}{2n}}^{\frac{1}{2}n} |\mathfrak{Y}(x)| dx.$$

Since, by (6), $|[u > 1] \cap K_{\frac{1}{n}}| > \frac{\alpha}{2} |K_{\frac{1}{n}}|$,

there exists some $\tilde{x} \in (-\frac{1}{2n}, \frac{1}{2n})$ such that

$$|\mathfrak{Y}(\tilde{x})| \geq \frac{\alpha}{4n}. \tag{8}$$

Define

$$A_{\tilde{x}} \equiv \{y \in \mathfrak{Y}(\tilde{x}) \text{ such that } \exists x \in (-\frac{1}{2n}, \frac{1}{2n}) \text{ such that } u(x,y) \leq \frac{(1+\lambda)}{2}\}.$$

Note that for any $y \in A_{\tilde{x}}$ the variation along the $x$ direction is at least $\frac{(1-\lambda)}{2}$. If $|A_{\tilde{x}}| \geq \frac{\alpha}{8n}$, we have that the BV norm of $u$ in $K_{\frac{1}{n}}$ is at least $\frac{\alpha(1-\lambda)}{16n}$ and therefore (5) holds.

If $|A_{\tilde{x}}| \leq \frac{\alpha}{8n}$, we have that there exists at least a $\tilde{y} \in \mathfrak{Y}(\tilde{x})$ such that u(x,$\tilde{y}$) $\geq \frac{(1+\lambda)}{2}$ for any $x \in (-\frac{1}{2n}, \frac{1}{2n})$.

Define

$$A_{\tilde{y}} \equiv \{x \in (-\frac{1}{2n}, \frac{1}{2n}) \text{ such that } \exists y \in (-\frac{1}{2n}, \frac{1}{2n}) \text{ such that } u(x,y) \leq \lambda\}.$$

Note that for any $x \in A_{\tilde{y}}$ the variation along the $y$ direction is at least $\frac{(1-\lambda)}{2}$.

If $|A_{\tilde{y}}| \geq \frac{\delta}{n}$ we have that the BV norm of $u$ in $K_{\frac{1}{n}}$ is at least $\frac{\delta(1-\lambda)}{2n}$ and therefore (5) holds.

If $|A_{\tilde{y}}| \leq \frac{\delta}{n}$ we have that $|[u > \lambda] \cap K_{\frac{1}{n}}| \geq (1-\delta)|K_{\frac{1}{n}}|$ and therefore (4) holds.

Summarizing either (4) or (5) hold. Therefore the alternative occurs and the case $N = 2$ is proved.

## 4 Proof of the Lemma 1.1 When $N > 2$

Assume that Lemma 1.1 is proved in the case $N = m$ and let us prove it when $N = m + 1$.

Let $z$ a point of $\mathbb{R}^{m+1}$. To make to notation easier, write $z = (x, y)$ where $x \in \mathbb{R}$ and $y \in \mathbb{R}^m$.

Let $K_{\frac{1}{n}}(z) \in \bar{\mathbf{Q}}^+$. Without loss of generality we may assume $z = (0, 0)$. Assume that

$$\left|[u > 1] \cap K_{\frac{1}{n}}\right| > \frac{\alpha}{2}|K_{\frac{1}{n}}| \tag{9}$$

$$\|u\|_{BV(K_{\frac{1}{n}})} \leq \frac{2\alpha}{(2-\alpha)n^{m+1}}\|u\|_{BV(K_1)}. \tag{10}$$

For any $x \in (-\frac{1}{2n}, \frac{1}{2n})$ consider the $m$ -dimensional cube centered in $(x, 0)$, orthogonal to the $x$-axis and with edge $\frac{1}{n}$ and denote this cube with $\bar{K}_{\frac{1}{n}}(x)$. Define $\bar{A}$ as the set of the $x \in (-\frac{1}{2n}, \frac{1}{2n})$ such that

$$\left|[u > 1] \cap \bar{K}_{\frac{1}{n}}(x)\right| > \frac{\alpha}{4}|\bar{K}_{\frac{1}{n}}(x)|$$

and

$$\|u\|_{BV(\bar{K}_{\frac{1}{n}}(x))} \leq \frac{16}{(2-\alpha)n^m}\|u\|_{BV(K_1)}.$$

It is possible to prove that

$$|\bar{A}| \geq \frac{\alpha}{8n}.$$

Let $\bar{x} \in \bar{A}$ and apply Lemma 1.1 to $\bar{K}_{\frac{1}{n}}(\bar{x})$ (we can do so because $\bar{K}_{\frac{1}{n}}(\bar{x})$ is a $m$-dimensional set).

So we get the existence of a constant $\eta_0 > 0$ and a point $y_o \in \bar{K}_{\frac{1}{n}}(\bar{x})$ such that if we define the set

$$A \equiv \{(\bar{x}, y) \in \bar{K}_{\frac{\eta_0}{n}}(\bar{x}, y_0) \text{ such that } u(\bar{x}, y) \geq \frac{(1+\lambda)}{2}\}$$

where $\bar{K}_{\frac{\eta_0}{n}}(\bar{x}, y_0)$ denotes the $m$-dimensional cube of edge $\frac{\eta_0}{n}$, centered in $(\bar{x}, y_0)$ and orthogonal to the $x$-axis, we have

$$|A| \geq (1 - \frac{\delta}{2})(\frac{\eta_0}{n})^m. \tag{11}$$

Define

$$B \equiv \{y \in A \text{ such that } \exists x \in (-\frac{1}{2n}, \frac{1}{2n}) \text{ such that } u(x, y) \leq \lambda\}.$$

Note that for any $y \in B$ the variation along the $x$ direction is at least $\frac{(1-\lambda)}{2}$.

If $|B| \geq \frac{\delta}{2}(\frac{\eta_0}{n})^m$, we have that the BV norm of $u$ in $K_{\frac{1}{n}}$ is at least $\frac{\delta(1-\lambda)}{4}(\frac{\eta_0}{n})^m$ and therefore (5) holds.

If $|B| \geq \frac{\delta}{2}(\frac{\eta_0}{n})^m$, taking in account (11) we have that in the cylinder $(-\frac{1}{2n}, \frac{1}{2n}) \times \bar{K}_{\frac{\eta_0}{n}}(0, y_0)$ the measure of the set where $u(x, y) \geq \lambda$ is greater than $(1-\delta)\frac{\eta_o^m}{n^{m+1}}$. Therefore (4) holds in a suitable subcube of $K_{\frac{1}{n}}$.

Summarizing either (4) or (5) hold. Therefore the alternative occurs and the case $N > 2$ is proved.

**Acknowledgements** This research was supported by the Italian-Hungarian executive project HU11MO10 "NonLinear Diffusion Processes and Mathematical Modelling in Finance."

## References

1. DiBenedetto, E., Vespri, V.: On the singular equation $\beta(u)_t = \Delta u$. Arch. Ration. Mech. Anal. **132**, 247–309 (1995)
2. DiBenedetto, E., Gianazza, U., Vespri, V.: Local clustering of the non-zero set of functions in $W^{1,1}(E)$. Atti Accad. Naz. Lincei Cl. Sci. Mat. Appl. **9**, 223–225 (2006)
3. DiBenedetto, E., Gianazza, U., Vespri, V.: Harnack's inequality for degenerate and singular parabolic equations. In: Springer Monographs in Mathematics. Springer, New York (2011)
4. Düzgün, F.G., Marcellini, P., Vespri, V.: An alternative approach to the Hölder continuity of solutions to some elliptic equations. NonLinear Anal. **94**, 133–141 (2014)

# $X$-Elliptic Harmonic Maps

**Sorin Dragomir**

*Dedicated to Ermanno Lanconelli on the occasion of his 70th birthday*

**Abstract** We study $X$-elliptic harmonic maps of an open set $\mathscr{U} \subset \mathbb{R}^N$ endowed with a family of vector fields $X = \{X_1, \cdots, X_m\}$ into a Riemannian manifold $S$ i.e. $C^\infty$ solutions $\phi : \mathscr{U} \to S$ to the nonlinear system $-L\phi^\alpha + a^{ij}(\Gamma^\alpha_{\beta\gamma} \circ \phi)(\partial_{x^i}\phi^\beta)(\partial_{x^j}\phi^\gamma) = 0$ where $L = \sum_{i,j=1}^N \partial_{x_i}(a^{ij}(x)\, \partial_{x_j} u)$ is an uniformly $X$-elliptic operator. We establish a Solomon type (cf. Solomon, J Differ Geom 21:151–162, 1985) result for $X$-elliptic harmonic maps $\phi : \mathscr{U} \to S^M \setminus \Sigma$ with values into a sphere and omitting a codimension two totally geodesic submanifold $\Sigma \subset S^M$. As an application of Harnack inequality (for positive solutions to $Lu = 0$) in Gutiérrez and Lanconelli (Commun Partial Differ Equ 28:1833–1862, 2003) we prove openness of $X$-elliptic harmonic morphisms.

**Keywords** Harmonic maps • X-elliptic vector fields

**AMS Classification:** Primary: 53C43, Secondary: 35H20

## 1 Statement of Main Results

Let $\mathscr{U} \subset \mathbb{R}^N$ be an open set and let $X = \{X_1, \cdots, X_m\} \subset \mathfrak{X}(\mathscr{U})$ be a family of vector fields

$$X_a = \sum_{i=1}^N b^i_a(x)\, \partial_{x^i}\,, \quad 1 \le a \le m,$$

S. Dragomir (✉)
Dipartimento di Matematica, Informatica ed Economia, Università degli Studi della Basilicata, Potenza, Italy
e-mail: sorin.dragomir@unibas.it



G. Citti et al. (eds.), *Geometric Methods in PDE's*, Springer INdAM Series 13,
DOI 10.1007/978-3-319-02666-4_5

with $b^i_a \in C^\infty(\mathscr{U})$. Let $L$ be the second order differential operator

$$Lu = \sum_{i,j=1}^{N} \partial_{x_j}(a^{ij}(x)\, \partial_{x_j} u), \tag{1}$$

where $a^{ij} = a^{ji}$ are measurable functions on $\mathscr{U}$. Let us set $A = [a^{ij}]$ and let $\Omega \subset \mathscr{U}$ be an open subset. $L$ is *$X$-elliptic* in $\Omega$ if there is a constant $\lambda > 0$ such that

$$\lambda \sum_{a=1}^{m} \langle X_a(x), \xi\rangle^2 \leq \langle A(x)\xi, \xi\rangle\,, \quad \xi \in \left(\mathbb{R}^N\right)^*\,, \quad x \in \Omega, \tag{2}$$

where $\langle A(x)\xi, \xi\rangle = \sum_{i,j=1}^{N} a^{ij}(x)\xi_i\xi_j$. Also $L$ is *uniformly $X$-elliptic* in $\Omega$ if $L$ is $X$-elliptic in $\Omega$ and there is a constant $\Lambda > 0$ such that

$$\langle A(x)\xi, \xi\rangle \leq \Lambda \sum_{a=1}^{m} \langle X_a(x), \xi\rangle^2\,, \quad \xi \in \left(\mathbb{R}^N\right)^*\,, \quad x \in \Omega. \tag{3}$$

$X$-Elliptic operators were introduced by E. Lanconelli and A.E. Kogoi, [8] (cf. also [4]) although the notion is implicit in [10] and goes back as far as the work by E. Lanconelli, [7], for particular systems of vector fields $X = \{\lambda^j\, \partial_{x^j} : 1 \leq j \leq N\}$. In this paper we consider nonlinear PDEs systems of variational origin

$$-L\phi^\alpha + a^{ij}(\Gamma^\alpha_{\beta\gamma} \circ \phi)(\partial_{x^i}\phi^\beta)(\partial_{x^j}\phi^\gamma) = 0 \tag{4}$$

whose principal part is an $X$-elliptic operator. $C^\infty$ solutions $\phi : \mathscr{U} \to S$ to (4) are termed *$X$-elliptic harmonic maps* and are the main object of study in this work. Here $S$ is a Riemannian manifold, with the metric tensor $h$, and $\Gamma^\alpha_{\beta\gamma}$ are the Christoffel symbols of $h_{\alpha\beta} = h(\partial_\alpha\,, \partial_\beta)$ with respect to a local coordinate system $(V, y^\alpha)$ on $S$ (also $\partial_\alpha \equiv \partial/\partial y^\alpha$). The vector fields $X = \{X_1, \cdots, X_m\}$ are requested to satisfy the structural assumptions in [4] (i.e. the assumptions **(D)**, **(S)**, **(LT)**, **(P)** and **(I)** in Sect. 2). Exploiting the variational structure of (4) we establish

**Theorem 1.1** *Let $L$ be uniformly $X$-elliptic on $\mathscr{U}$. Let $\phi : \mathscr{U} \to S^M$ be an $X$-elliptic harmonic map. Let $\Sigma \subset S^M$ be a totally geodesic codimension two submanifold. Then either (i) $\phi$ meets $\Sigma$ i.e. $\phi(\mathscr{U}) \cap \Sigma \neq \emptyset$ or (ii) $\phi : \mathscr{U} \to S^M \setminus \Sigma$ is homotopically nontrivial, or (iii) if $\phi : \mathscr{U} \to S^M \setminus \Sigma$ is null-homotopic then for every smoothly bounded domain $\Omega \subset \mathbb{R}^N$ such that $\overline{\Omega} \subset \mathscr{U}$, $X$ is a Hörmander system on $\Omega$, and $\partial\Omega$ is characteristic relative to $X$, the map $\phi : \Omega \to S^M \setminus \Sigma$ is constant.*

If $X = \{\partial_{x^i} : 1 \leq i \leq N\}$ and $a^{ij}$ are the (reciprocal) coefficients of a Riemannian metric $a$ on $\mathscr{U}$ an $X$-elliptic harmonic map $\phi : \mathscr{U} \to S$ is an ordinary harmonic map (in the sense of [9]) among the Riemannian manifolds $(\mathscr{U}, a)$ and $(S, h)$ hence Theorem 1.1 is an analog to a result in [11]. When $S$ is the sphere $S^M \subset \mathbb{R}^{M+1}$ we may (as well as in [11]) exploit the fact that $S^M \setminus \Sigma$ is isometric to a warped product manifold $S_+^{M-1} \times_w S^1$, to consider a special variation $(F, u + t\varphi)$ of $\phi = (F, u)$ with respect to which (4) reduces to a *single* $X$-elliptic equation and the maximum principle (as devised in [4]) applies. However it doesn't directly yield constancy of $u$ and the conclusion in Lemma 4.1 requires (unlike the elliptic case in [11]) an additional argument working only on domains with characteristic boundary and provided $X$ is a Hörmander system.

A map $\phi \in C^\infty(\mathscr{U}, S)$ is an *X-elliptic harmonic morphism* if for every local harmonic function $v : V \to \mathbb{R}$ (i.e. $V \subset S$ is an open subset and $\Delta_h v = 0$ in $V$) one has $L(v \circ \phi) = 0$ in $U = \phi^{-1}(V)$. Here $\Delta_h$ is the Laplace-Beltrami operator of the Riemannian manifold $(S, h)$. We establish the following Fuglede-Ishihara type (cf. [2] and [5]) result

**Theorem 1.2** *Let $L$ be $X$-elliptic on $\mathscr{U}$. Then (i) any $X$-elliptic harmonic morphism $\phi : \mathscr{U} \to S$ is an $X$-elliptic harmonic map and there is a continuous function $\lambda_\phi : \mathscr{U} \to [0, +\infty)$ such that $\lambda_\phi^2$ is $C^\infty$ and*

$$\big\langle A(x) D\phi^\alpha(x) \, , \, D\phi^\beta(x) \big\rangle = \lambda_\phi(x)^2 \, \delta^{\alpha\beta} \, , \quad 1 \leq \alpha, \beta \leq M, \tag{5}$$

*for any $x \in \mathscr{U}$ and any normal local coordinate system $(V, y^\alpha)$ on $S$ centered at $x$ (here $\phi^\alpha = y^\alpha \circ \phi$). Conversely (ii) any $X$-elliptic harmonic map satisfying (5) is an $X$-elliptic harmonic morphism. (iii) If $M > N$ then $\phi_* X_a = 0$ for any $1 \leq a \leq m$. In particular if $X = \{X_a : 1 \leq a \leq m\}$ is a Hörmander system on $\mathscr{U}$ then there are no nonconstant $X$-elliptic harmonic morphisms $\phi : \mathscr{U} \to S$. (iv) If $M \leq N$ then for every nonconstant $X$-elliptic harmonic morphism $\phi : \mathscr{U} \to S$ and for every $x \in \mathscr{U}$ such that $\lambda_\phi(x) \neq 0$ there is an open set $U \subset \mathscr{U}$ such that $x \in U$ and $\phi : U \to S$ is a submersion. (v) For every $X$-elliptic harmonic morphism $\phi : \mathscr{U} \to S$ and any $f \in C^2(S)$*

$$L(f \circ \phi) = \lambda_\phi^2 \, (\Delta_h f) \circ \phi. \tag{6}$$

In Sect. 5 we apply the Harnack inequality (as established in [4]) to prove

**Theorem 1.3** *Let $L$ be uniformly $X$-elliptic on $\mathscr{U}$. Let $\phi : \mathscr{U} \to S$ be an $X$-elliptic harmonic morphism whose dilation $\lambda_\phi$ has at most isolated zeros. Then $\phi : \mathscr{U} \to S$ is an open map i.e. for every open set $U \subset \mathscr{U}$ its image $\phi(U)$ is open in $S$.*

This is the $X$-elliptic analog to a result by B. Fuglede (cf. Theorem 4.3.8 in [1], p. 112).

## 2 $X$-Elliptic Operators: Structural Assumptions

Let $\mathfrak{L}$ be the second order differential operator

$$\mathfrak{L} u = Lu + \sum_{i=1}^{N} a^i(x) \, \frac{\partial u}{\partial x^i} \, , \tag{7}$$

where $L$ is given by (1) and $a^i$ are measurable functions on $\mathscr{U}$. We set $a = (a^1, \cdots, a^N)$. Let $\Omega \subset \mathscr{U}$ be an open set. $\mathfrak{L}$ is *X-elliptic* in $\Omega$ if (i) $L$ is $X$-elliptic in $\Omega$ (i.e. (2) in Sect. 1 holds for some $\lambda > 0$) and (ii) there is a function $\gamma(x) \geq 0$ such that

$$\langle a(x)\,,\,\xi\rangle^2 \leq \gamma(x) \sum_{a=1}^{m} \langle X_a(x)\,,\,\xi\rangle^2\,, \quad \xi \in \mathbb{R}^N\,, \quad x \in \Omega\,. \tag{8}$$

Also $\mathfrak{L}$ is *uniformly X-elliptic* in $\Omega$ if $\mathfrak{L}$ is $X$-elliptic in $\Omega$ and $L$ is uniformly $X$-elliptic in $\Omega$ (i.e. (3) in Sect. 1 holds for some $\Lambda > 0$). A piecewise $C^1$ curve $\gamma : [0, 1] \to \mathscr{U}$ is an *X-path* if

$$\dot{\gamma}(t) = \sum_{a=1}^{m} f_a(t)\, X_a(\gamma(t))$$

for some functions $f_a(t)$ and for a.e. $t \in [0, 1]$. One sets

$$\ell(\gamma) = \sup_{0 \leq t \leq 1} \left( \sum_{a=1}^{m} f_a(t)^2 \right)^{1/2}.$$

Let $\Gamma(x, y)$ be the set of all $X$-paths connecting $x, y \in \mathscr{U}$. The *control distance* $d = d_X$ is given by

$$d(x, y) = \inf\{\ell(\gamma) : \gamma \in \Gamma(x, y)\}\,.$$

As a fundamental assumption on the system of vector fields $X = (X_1, \cdots, X_m)$ that we adopt through this paper, the control distance relative to $X$ is well defined, continuous in the Euclidean topology, and the following *doubling condition* is satisfied:

**(D)** For every compact subset $K \subset \mathscr{U}$ there exist constants $C_d > 1$ and $R_0 > 0$ such that

$$0 < |B_{2r}| \leq C_d\, |B_r| \tag{9}$$

for every $d$-ball $B_r$ centered at a point of $K$ and of radius $r \leq R_0$. Here $|E| = \mu(E)$ is the Lebesgue measure of $E \subset \mathbb{R}^N$.

Let $\Omega \subset \mathscr{U}$ be a bounded open set. As another fundamental assumption on the given vector fields and the set $\Omega$, we set $Xu = (X_1 u, \cdots, X_m u)$ and postulate that the following *Sobolev inequality* holds good.

**(S)** There exist constants $q = q(\Omega) > 2$ and $S = S(\Omega) > 0$ such that

$$\|u\|_{L^q(\Omega)} \leq S\, \|Xu\|_{L^2(\Omega)} \tag{10}$$

for every $u \in C^1_0(\Omega)$.

We adopt the notation $Q = 2q/(q-2)$. The maximum principle for $X$-elliptic operators $\mathfrak{L}$ as established in [4] also relies on the following assumption on the lower order term $a$.

**(LT)** There is $Q/2 < p < \infty$ such that $\gamma \in L^{2p}(\Omega)$ [where $\gamma$ is the function appearing in (8)].

Moreover the given set $\Omega$ is assumed to support the *Poincaré inequality* i.e.

**(P)** For each compact subset $K \subset \mathscr{U}$ there is a constant $C_P > 0$ such that

$$\frac{1}{|B_r|}\int_{B_r} |u - u_r|\, d\mu \leq \frac{C_P r}{|B_{2r}|}\int_{B_{2r}} |Xu|\, d\mu, \quad u \in C^1(\overline{\Omega}), \tag{11}$$

for any $d$-ball $B_r(x)$ with center $x \in K$ and radius $r \leq R_0$. Here $u_r = (1/|B_r|)\int_{B_r} u\, d\mu$.

Finally the following *dilation invariance* property should hold good. Let $\alpha_1, \cdots, \alpha_N \in \mathbb{N}$ be positive integers such that $Q = \alpha_1 + \cdots + \alpha_N$ and let us set

$$\delta_R x = (R^{\alpha_1} x_1, \cdots, R^{\alpha_N} x_N), \quad R > 0, \quad x = (x_1, \cdots, x_N).$$

The system of vector fields $X$ is required to be *dilation invariant* i.e.

**(I)** For every $R > 0$ and $x \in \mathscr{U}$

$$X_a (\delta_R u)_x = R\,(X_j u)_{\delta_R(x)} \tag{12}$$

for any smooth function $u$. Here $\delta_R u$ is given by $(\delta_R u)(x) = u(\delta_R x)$.

## 3 First Variation Formula

We start with the differential operator $L$ given by (7). Through this paper we work under the assumption that $a^{ij} \in C^\infty(\mathscr{U})$. For each $\epsilon > 0$ we set

$$(g_\epsilon)^{ij} = a^{ij} + \epsilon\,\delta^{ij}, \quad \left[(g_\epsilon)_{ij}\right] = \left[(g_\epsilon)^{ij}\right]^{-1},$$

so that $g_\epsilon = (g_\epsilon)_{ij}\, dx^i \otimes dx^j$ is a Riemannian metric on $\mathscr{U}$. Let $(S,h)$ be a Riemannian manifold, with the Riemannian metric $h$ and for each $\phi \in C^\infty(\mathscr{U}, S)$ let $\|d\phi\|_\epsilon : U \to \mathbb{R}$ be the Hilbert-Schmidt norm of $d\phi$ i.e. for every local coordinate system $(V, y^\alpha)$ on $S$ and every $x \in U = \phi^{-1}(V)$

$$\|d\phi\|_\epsilon (x) = \left\{(g_\epsilon)^{ij}(x)\,\frac{\partial \phi^\alpha}{\partial x^i}(x)\,\frac{\partial \phi^\beta}{\partial x^j}(x)\,h_{\alpha\beta}(\phi(x))\right\}^{1/2},$$

$$\phi^\alpha = y^\alpha \circ \phi, \quad h_{\alpha\beta} = h(\partial_\alpha, \partial_\beta), \quad \partial_\alpha = \partial/\partial y^\alpha.$$

Let $\Omega \subset\subset \mathscr{U}$ be a relatively compact domain. The energy of $\phi$ as a map of $(\mathscr{U}, g_\epsilon)$ into $(S, h)$ is

$$E_{\Omega,\epsilon}(\phi) = \frac{1}{2}\int_\Omega \|d\phi\|_\epsilon^2 \, d\mu$$

where $\mu$ is the Lebesgue measure on $\mathbb{R}^N$. Let $x \in \mathscr{U}$ and let $(V, y^\alpha)$ be a local coordinate system on $S$ such that $\phi(x) \in V$. We comply with the philosophy in [6] and separate the metric on $\mathscr{U}$ (here $g_\epsilon$) from the measure of integration (here $d\mu$, rather than the canonical Riemannian measure of $(\mathscr{U}, g_\epsilon)$). Let us set

$$e_A(\phi)(x) = a^{ij}(x)\,\frac{\partial\phi^\alpha}{\partial x^i}(x)\,\frac{\partial\phi^\beta}{\partial x^j}(x)\,h_{\alpha\beta}(\phi(x)).$$

The definition of $e_A(\phi)(x)$ doesn't depend on the choice of local coordinates $(V, y^\alpha)$ about $\phi(x)$. Note that $\|d\phi\|_\epsilon^2 \to e_A(\phi)$ as $\epsilon \to 0$ thus prompting the energy functional

$$E_A(\phi) = \frac{1}{2}\int_\Omega e_A(\phi)\, d\mu. \tag{13}$$

A map $\phi \in C^\infty(\mathscr{U}, S)$ is said to be *X-elliptic harmonic* if

$$\frac{d}{dt}\{E_A(\phi_t)\}_{t=0} = 0$$

for any $\Omega \subset\subset \mathscr{U}$ and any smooth 1-parameter variation $\{\phi_t\}_{|t|<\epsilon} \subset C^\infty(\mathscr{U}, S)$ of $\phi$ (i.e. $\phi_0 = \phi$) supported in $\Omega$ i.e. $\mathrm{Supp}(V) \subset \Omega$ where

$$V \in C^\infty\left(\phi^{-1}T(S)\right), \quad V_x = (d_{(x,0)}\Phi)(\partial/\partial t)_{(x,0)},$$

$$\Phi : \mathscr{U} \times (-\epsilon, \epsilon) \to S, \quad \Phi(x,t) = \phi_t(x), \quad x \in \mathscr{U}, \quad |t| < \epsilon.$$

We proceed by deriving the first variation formula for the functional (13). One has

$$e_A(\phi_t)(x) = a^{ij}(x)\frac{\partial\Phi^\alpha}{\partial x^i}(x,t)\frac{\partial\Phi^\beta}{\partial x^j}(x,t)\,h_{\alpha\beta}(\phi_t(x)) \tag{14}$$

for any $x \in U = \phi^{-1}(V)$ and $|t| < \delta$. Let us consider the function $f : \mathscr{U} \times (-\delta, \delta) \to \mathbb{R}$ given by $f(x,t) = e_A(\phi_t)(x)$ for any $x \in \mathscr{U}$ and $|t| < \delta$. Then [by (14)]

$$\frac{\partial f}{\partial t}(x,t) = 2a^{ij}(x)\frac{\partial^2\Phi^\alpha}{\partial t\,\partial x^i}(x,t)\frac{\partial\Phi^\beta}{\partial x^j}(x,t)\,h_{\alpha\beta}(\phi_t(x)) +$$

$$+a^{ij}(x)\frac{\partial\Phi^\alpha}{\partial x^i}(x,t)\frac{\partial\Phi^\beta}{\partial x^j}(x,t)\,\frac{\partial h_{\alpha\beta}}{\partial y^\gamma}(\phi_t(x))\frac{\partial\Phi^\gamma}{\partial t}(x,t)$$

hence

$$\frac{\partial f}{\partial t}(x,0) = 2a^{ij}(x)\frac{\partial^2 \Phi^\alpha}{\partial t\,\partial x^i}(x,0)\frac{\partial \phi^\beta}{\partial x^j}(x)\,h_{\alpha\beta}(\phi(x)) +$$
$$+a^{ij}(x)\frac{\partial \phi^\alpha}{\partial x^i}(x)\frac{\partial \phi^\beta}{\partial x^j}(x)\,\frac{\partial h_{\alpha\beta}}{\partial y^\gamma}(\phi(x))\frac{\partial \Phi^\gamma}{\partial t}(x,0)$$

or

$$\frac{\partial f}{\partial t}(\cdot\,,\,0) =$$
$$= 2\frac{\partial}{\partial x^i}\left(a^{ij}V^\alpha\frac{\partial \phi^\beta}{\partial x^j}(h_{\alpha\beta}\circ\phi)\right) - 2\,V^\alpha\frac{\partial}{\partial x^i}\left(a^{ij}\frac{\partial \phi^\beta}{\partial x^j}(h_{\alpha\beta}\circ\phi)\right) +$$
$$+a^{ij}\frac{\partial \phi^\alpha}{\partial x^i}\frac{\partial \phi^\beta}{\partial x^j}\left(\frac{\partial h_{\alpha\beta}}{\partial y^\gamma}\circ\phi\right)V^\gamma =$$
$$= 2\,\mathrm{div}_0\left(a^{ij}V^\alpha\frac{\partial \phi^\beta}{\partial x^j}(h_{\alpha\beta}\circ\phi)\,\frac{\partial}{\partial x^i}\right) -$$
$$-V^\gamma\left\{2\,\frac{\partial}{\partial x^i}\left(a^{ij}\frac{\partial \phi^\beta}{\partial x^j}\right)(h_{\beta\gamma}\circ\phi) + 2\,a^{ij}\frac{\partial \phi^\beta}{\partial x^j}\left(\frac{\partial h_{\beta\gamma}}{\partial y^\alpha}\circ\phi\right)\frac{\partial \phi^\alpha}{\partial x^i} -\right.$$
$$\left. -a^{ij}\frac{\partial \phi^\alpha}{\partial x^i}\frac{\partial \phi^\beta}{\partial x^j}\left(\frac{\partial h_{\alpha\beta}}{\partial y^\gamma}\circ\phi\right)\right\}$$

where $\mathrm{div}_0$ is the Euclidean divergence operator and

$$V^\alpha(x) = \frac{\partial \Phi^\alpha}{\partial t}(x,0), \quad x \in U.$$

Moreover if

$$\Gamma_{\alpha\beta\gamma} = \frac{1}{2}\left(\frac{\partial h_{\alpha\gamma}}{\partial y^\beta} - \frac{\partial h_{\beta\gamma}}{\partial y^\alpha} - \frac{\partial h_{\alpha\beta}}{\partial y^\gamma}\right), \quad \Gamma^\nu_{\alpha\beta} = h^{\nu\gamma}\,\Gamma_{\alpha\beta\gamma}\,,$$

then

$$\frac{\partial f}{\partial t}(\cdot\,,\,0) \equiv$$
$$\equiv V^\gamma\left\{-2(L\phi^\beta)(h_{\beta\gamma}\circ\phi) + a^{ij}\frac{\partial \phi^\alpha}{\partial x^i}\frac{\partial \phi^\beta}{\partial x^j}\left(2\frac{\partial h_{\beta\gamma}}{\partial y^\alpha} - \frac{\partial h_{\alpha\beta}}{\partial y^\gamma}\right)\circ\phi\right\} \equiv$$
$$\equiv 2V^\gamma(h_{\gamma\nu}\circ\phi)\left\{-L\phi^\nu + a^{ij}\frac{\partial \phi^\alpha}{\partial x^i}\frac{\partial \phi^\beta}{\partial x^j}\left(\Gamma^\nu_{\alpha\beta}\circ\phi\right)\right\}, \quad \mathrm{mod}\ \mathrm{div}_0$$

where $L$ is given by (7). Summing up

$$\frac{d}{dt}\{E_A(\phi_t)\}_{t=0} = \int_\Omega h^\phi(V, \tau_A(\phi))\, d\mu \tag{15}$$

(the *first variation formula* seek after) and $\tau_A(\phi) \in C^\infty\left(\phi^{-1}T(S)\right)$ is locally given by

$$\tau_A(\phi)^\alpha = -L\phi^\alpha + \left(\Gamma^\alpha_{\beta\gamma} \circ \phi\right) \frac{\partial \phi^\beta}{\partial x^i} \frac{\partial \phi^\gamma}{\partial x^j} a^{ij}.$$

Also $h^\phi = \phi^{-1}h$ is the pullback of $h$ by $\phi$ (a Riemannian bundle metric on $\phi^{-1}T(S) \to \mathbb{R}^N$). Finally $\int_\Omega (\partial f/\partial t)(x, 0)\, d\mu = 0$ yields

$$-L\phi^\alpha + \left(\Gamma^\alpha_{\beta\gamma} \circ \phi\right) \frac{\partial \phi^\beta}{\partial x^i} \frac{\partial \phi^\gamma}{\partial x^j} a^{ij} = 0 \tag{16}$$

which is the *X-elliptic harmonic map system.*

## 4 X-Elliptic Harmonic Maps into Spheres

Let $S^M = \{(x_1, \cdots, x_{M+1}) \in \mathbb{R}^{M+1} : x_1^2 + \cdots + x_{M+1}^2 = 1\}$. If $S = S^M$ then

$$h_{\alpha\beta} = \delta_{\alpha\beta} + \frac{y^\alpha y^\beta}{1 - |y|^2}, \quad \Gamma^\alpha_{\beta\gamma} = y^\alpha h_{\beta\gamma},$$

hence (16) becomes $L\phi^\alpha + e_A(\phi)\phi^\alpha = 0$. Due to the constraint $\sum_{K=1}^{M+1} \Phi_K^2 = 1$ the $X$-elliptic harmonic map system for $S^M$-valued maps $\Phi = (\Phi_1, \cdots, \Phi_{M+1})$ is

$$-L\Phi + \sum_{K=1}^{M+1} a^{ij} \frac{\partial \Phi^K}{\partial x^i} \frac{\partial \Phi^K}{\partial x^j} \Phi = 0. \tag{17}$$

**Lemma 4.1** *Let $L$ be uniformly elliptic in $\mathcal{U}$. Let $\phi : \mathcal{U} \to S^M_+$ be an $X$-elliptic harmonic map where $S^M_+ = \{y \in S^M : y^{M+1} > 0\}$. Then $\phi_* X_a = 0$ in $\Omega$, $1 \le a \le m$, for every smoothly bounded domain $\Omega \subset \mathcal{U}$ such that $\overline{\Omega} \subset \mathcal{U}$ and whose boundary $\partial\Omega$ is characteristic relative to $X$. In particular if $X$ is a Hörmander system then $\phi$ is constant on $\Omega$.*

*Proof* By Green's lemma

$$\int_\Omega L\Phi_{M+1}\, d\mu = \int_{\partial\Omega} a^{ij} \frac{\partial \Phi_{M+1}}{\partial x^j} \nu_i\, d\sigma$$

where $\nu = \nu^i \partial/\partial x^i$ and $d\sigma$ are the outward unit normal and "area" measure on $\partial\Omega$. On the other hand (by uniform $X$-ellipticity)

$$\left| a^{ij}(x) \frac{\partial \Phi_{M+1}}{\partial x^j} \nu_i \right| \leq \Lambda \langle A(x) D\Phi_{M+1}, D\Phi_{M+1} \rangle \sum_{a=1}^{M} \langle X_a(x), \nu \rangle^2 = 0$$

for every $x \in \partial\Omega$, provided that each $X_a$ is tangent to $\partial\Omega$. Finally one may integrate in (17) over $\Omega$ so that

$$0 = \sum_{K=1}^{M+1} \int_\Omega a^{ij} \frac{\partial \Phi^K}{\partial x^i} \frac{\partial \Phi^K}{\partial x^j} \Phi_{M+1} \, d\mu \geq$$

$$\geq \lambda \sum_{K=1}^{M+1} \sum_{a=1}^{m} \int_\Omega \langle X_a, D\Phi^K \rangle^2 \Phi_{M+1} \, d\mu$$

hence (by $\Phi_{M+1} > 0$) one obtains $X_a(\Phi^K) = 0$ in $\Omega$. □

Let $\phi : \mathscr{U} \to S^M$ be a continuous map. Let $\Sigma \subset S^M$ be a codimension two totally geodesic submanifold, so that $\Sigma = \{x \in S^M : x_1 = x_2 = 0\}$ up to a coordinate transformation. We say $\phi$ *meets* $\Sigma$ if $\phi(\mathscr{U}) \cap \Sigma \neq \emptyset$. If $\phi$ doesn't meet $\Sigma$ then $\phi$ *links* $\Sigma$ when $\phi : \mathscr{U} \to S^M \setminus \Sigma$ is not null-homotopic. By a result of B. Solomon, [11], a harmonic map of a compact Riemannian manifold into $S^M$ either meets or links $\Sigma$. To prove Theorem 1.1 we need some preparation. We establish

**Theorem 4.1** *Let $\phi : \mathscr{U} \to S$ be an $X$-elliptic harmonic map. Let $S = P \times_w \mathbb{R}$ be a warped product Riemannian manifold, where $(P, g_P)$ is a $(M-1)$-dimensional Riemannian manifold and $w : S \to (0, +\infty)$ is a $C^\infty$ function, endowed with the Riemannian metric*

$$h = \pi_1^* g_P + w^2 \pi_2^* ds \otimes ds.$$

*Let $F = \pi_1 \circ \phi$ and $u = \pi_2 \circ \phi$. Then*

$$(w \circ \phi) Lu + 2a^{ij} \frac{\partial u}{\partial x^i} \frac{\partial}{\partial x^j} (w \circ \phi) = \left( \frac{\partial w}{\partial s} \circ \phi \right) a^{ij} \frac{\partial u}{\partial x^i} \frac{\partial u}{\partial x^j} . \tag{18}$$

*In particular if $w \in C^\infty(P)$ then*

$$Lu + 2a^{ij} \frac{\partial u}{\partial x^i} \frac{\partial}{\partial x^j} \{\log(w \circ F)\} = 0. \tag{19}$$

*Consequently for every bounded open subset $\Omega \subset \mathscr{U}$ on which $L$ is uniformly $X$-elliptic*

$$\sup_\Omega u^+ \leq \sup_{\partial\Omega} u^+ . \tag{20}$$

*If $\Omega \subset \mathbb{R}^N$ is a smoothly bounded domain such that $\overline{\Omega} \subset \mathcal{U}$, its boundary $\partial\Omega$ is characteristic relative to X, and L is uniformly X-elliptic on $\mathcal{U}$, then $X_a(u) = 0$ for every $1 \leq a \leq m$. If additionally X is a Hörmander system on $\Omega$ then $\phi(\Omega) \subset P \times \{t_\phi\}$ for some $t_\phi \in \mathbb{R}$.*

Here $\pi_1 : S \to P$ and $\pi_2 : S \to \mathbb{R}$ are the natural projections. Also the notations in (20) will be explained shortly. To prove Theorem 4.1 let $\Omega \subset \mathcal{U}$ be a bounded open subset and $\varphi \in C_0^\infty(\Omega)$. Correspondingly we consider the 1-parameter variation

$$\phi_t(x) = (F(x)\,,\, u(x) + t\,\varphi(x)), \quad x \in \mathcal{U}, \quad |t| < \epsilon.$$

Then

$$e_A\,(\phi_t) = e_A(F) + (w \circ \phi_t)^2\, a^{ij} \frac{\partial u}{\partial x^i} \frac{\partial u}{\partial x^j} + \tag{21}$$

$$+2t\,(w \circ \phi_t)^2\, a^{ij} \frac{\partial u}{\partial x^i} \frac{\partial \varphi}{\partial x^j} + O(t^2).$$

Differentiation with respect to $t$ in (21) gives

$$\frac{d}{dt}\,\{e_A\,(\phi_t)\} = 2\,(w \circ \phi_t)^2\, a^{ij}\, \frac{\partial u}{\partial x^i}\, \frac{\partial \varphi}{\partial x^j} +$$

$$+2\varphi\,(w \circ \phi_t)\left(\frac{\partial w}{\partial s} \circ \phi_t\right) a^{ij}\, \frac{\partial u}{\partial x^i}\, \frac{\partial u}{\partial x^j} + O(t)$$

hence

$$\frac{d}{dt}\,\{E_A\,(\phi_t)\}_{t=0} = \int_\Omega (w \circ \phi)^2\, a^{ij}\, \frac{\partial u}{\partial x^i}\, \frac{\partial \varphi}{\partial x^j}\, d\mu + \tag{22}$$

$$+ \int_\Omega \varphi\,(w \circ \phi)\left(\frac{\partial w}{\partial s} \circ \phi\right) a^{ij}\, \frac{\partial u}{\partial x^i}\, \frac{\partial u}{\partial x^j}\, d\mu.$$

Next one observes that

$$(w \circ \phi)^2 a^{ij} \frac{\partial u}{\partial x^i} \frac{\partial \varphi}{\partial x^j} = -\varphi\, \frac{\partial}{\partial x^j}\left((w \circ \phi)^2\, a^{ij}\, \frac{\partial u}{\partial x^i}\right) +$$

$$+\mathrm{div}_0\left((w \circ \phi)^2\, a^{ij}\, \frac{\partial u}{\partial x^i}\, \varphi\, \frac{\partial}{\partial x^j}\right)$$

and integrates by parts [in the right hand side of (22)]. Hence

$$\frac{d}{dt}\{E_A(\phi_t)\}_{t=0} = -\int_\Omega \varphi \left\{\frac{\partial}{\partial x^j}\left((w\circ\phi)^2 a^{ij}\frac{\partial u}{\partial x^i}\right) - \right. \tag{23}$$

$$\left. -(w\circ\phi)\left(\frac{\partial w}{\partial s}\circ\phi\right) a^{ij}\frac{\partial u}{\partial x^i}\frac{\partial u}{\partial x^j}\right\} d\mu.$$

Yet $\phi : \mathscr{U} \to P \times_w \mathbb{R}$ is $X$-elliptic harmonic so that $\{dE_A(\phi_t)/dt\}_{t=0} = 0$ for every $\varphi \in C_0^\infty(\Omega)$. Hence (23) yields (18). Also when $\partial w/\partial s = 0$ (i.e. $w \in C^\infty(P)$ and $S$ is endowed with the Riemannian metric $\pi_1^* g_P + (w\circ\pi_1)^2 \pi_2^* ds^2$) Eq. (18) yields (19). To prove (20) in Theorem 4.1 we need to recall the maximum principle for $X$-elliptic operators (cf. Theorem 3.1 in [4], p. 1840) as it applies to the situation at hand.

Let $\Omega \subset \mathscr{U}$ be a bounded open set and let us assume that the structure condition **(S)** is satisfied. Also let us assume that $\mathfrak{L}$ is uniformly $X$-elliptic in $\Omega$ with $\gamma$ obeying to **(LT)**. Let $Q/2 < p < \infty$ and $f \in L^p(\Omega)$. Then there is a constant

$$C = C\left(\lambda,\ S(\Omega),\ p,\ Q,\ |\Omega|,\ \int_\Omega \gamma(x)^{2p}\, d\mu(x)\right) > 0$$

such that for every weak subsolution $u \in W^1(\Omega, X)$ to $\mathfrak{L}u = f$ one has

$$\sup_\Omega u^+ \le \sup_{\partial\Omega} u^+ + C\,\|f\|_{L^p(\Omega)}. \tag{24}$$

As to the notations in (24) we set $u^+ = \max\{u,\ 0\}$. Also if $u \in W^1(\Omega, X)$ and $\ell \in \mathbb{R}$ then one says that $u \le \ell$ on $\partial\Omega$ if $(u-\ell)^+ \in W_0^1(\Omega, X)$ and sets by definition

$$\sup_{\partial\Omega} u^+ = \inf\{\ell : u^+ \le \ell \ \text{ on } \ \partial\Omega\}.$$

As to the function spaces we use, by the Sobolev inequality (10) the function $u \mapsto \|Xu\|_{L^2(\Omega)}$ is a norm in $C_0^1(\Omega)$ and one takes $W_0^1(\Omega, X)$ to be the closure of $C_0^1(\Omega)$ in this norm. Also $W^1(\Omega, X)$ is the space of all $u \in L^2(\Omega)$ admitting weak derivatives $X_a u \in L^2(\Omega)$ for every $1 \le a \le m$. Finally we need to recall the notion of weak (sub)solution to $\mathfrak{L}u = f$. To this end one considers the bilinear form

$$B_{\mathfrak{L}}(u, v) = \int_\Omega \{\langle A(x)Du\,,\, Dv\rangle - \langle a(x)\,,\, Du\rangle\, v\}\, d\mu(x)$$

with $u \in C^1(\Omega)$ and $v \in C_0^1(\Omega)$. For every $x \in \Omega$ we set

$$\langle \xi, \eta\rangle_{A(x)} = a^{ij}(x)\xi_i\eta_j\,, \quad \xi, \eta \in \left(\mathbb{R}^N\right)^*.$$

By the $X$-ellipticity condition (2)

$$\langle \xi + t\eta, \xi + t\eta \rangle_{A(x)} \geq 0, \quad \xi, \eta \in \left(\mathbb{R}^N\right)^*, \quad t \in \mathbb{R},$$

hence

$$\left| \langle \xi, \eta \rangle_{A(x)} \right| \leq \left[ \langle \xi, \xi \rangle_{A(x)} \right]^{\frac{1}{2}} \left[ \langle \eta, \eta \rangle_{A(x)} \right]^{\frac{1}{2}}. \tag{25}$$

Uniform $X$-ellipticity of $\mathfrak{L}$ in $\Omega$ yields

$$|B_{\mathfrak{L}}(u, v)| \leq \Lambda \int_{\Omega} |Xu|\, |Xv|\, d\mu + \int_{\Omega} (|Xu|\, |v| + |Xv|\, |u|)\, \gamma\, d\mu \tag{26}$$

hence $B_{\mathfrak{L}}$ is well defined. Then (by (26) and the structure assumption (**LT**))

$$|B_{\mathfrak{L}}(u, v)| \leq \Lambda \|Xu\|_{L^2(\Omega)} \|Xv\|_{L^2(\Omega)} +$$
$$+ \left( \|Xv\|_{L^2(\Omega)} \|u\|_{L^r(\Omega)} + \|Xu\|_{L^2(\Omega)} \|v\|_{L^r(\Omega)} \right) \|\gamma\|_{L^{2p}(\Omega)}$$

where $1/r = 1/2 - 1/(2p)$. Moreover (by (10), boundedness of $\Omega$, and $p > Q/2$)

$$|B_{\mathfrak{L}}(u, v)| \leq C \left( \|Xu\|_{L^2(\Omega)} + \|u\|_{L^r(\Omega)} \right) \|Xv\|_{L^2(\Omega)} \tag{27}$$

where $C = C\left(\Omega,\ S(\Omega),\ \Lambda + \|\gamma\|_{L^{2p}(\Omega)}\right) > 0$. As a consequence of (27) the map $(u, v) \mapsto B(u, v)$ may be extended continuously to a bilinear form

$$B_{\mathfrak{L}} : \left( W^1(\Omega, X) \cap L^r(\Omega) \right) \times W_0^1(\Omega, X) \to \mathbb{R}$$

and the following definitions are legitimate. Let $f \in L^1_{\text{loc}}(\Omega)$. A function $u \in W^1(\Omega, X)$ is a *weak solution* to $\mathfrak{L}u = f$ if

$$B_{\mathfrak{L}}(u, v) = -\int_{\Omega} f\, v\, d\mu, \quad v \in C_0^1(\Omega).$$

If in turn

$$B_{\mathfrak{L}}(u, v) \leq -\int_{\Omega} f\, v\, d\mu, \quad v \in C_0^1(\Omega), \quad v \geq 0,$$

then $u$ is a *weak subsolution* to $\mathfrak{L}u = f$.

Let us go back to the proof of Theorem 4.1. Let us consider the differential operator

$$\mathscr{L}u = Lu + \sum_{i=1}^{N} a^i \frac{\partial u}{\partial x^i}, \quad a^i = 2a^{ij} \frac{\partial \rho}{\partial x^i}, \quad \rho = \log(w \circ F).$$

Since $L$ is uniformly $X$-elliptic (i.e. the matrix $A$ satisfies requirements (2) and (3), while (8) is identically satisfied because $L$ has no lower order terms) the operator $\mathfrak{L}$ is uniformly $X$-elliptic in $\Omega$, as well. Indeed for the lower order term of $\mathfrak{L}$ one has the estimate

$$\langle a(x), \xi \rangle^2 = \left[ 2\, a^{ij}(x) \xi_i \frac{\partial \rho}{\partial x^j}(x) \right]^2 \leq$$

$$\leq 4 \left[ a^{ij}(x) \xi^i \xi^j \right] \left[ a^{ij}(x) \frac{\partial \rho}{\partial x^i}(x) \frac{\partial \rho}{\partial x^j}(x) \right] \leq$$

[by (3)]

$$\leq \gamma(x) \sum_{a=1}^{m} \langle X_a(x)\, ,\, \xi \rangle^2$$

for every $\xi \in \mathbb{R}^N$ and $x \in \Omega$, where we have set

$$\gamma(x) = 4\Lambda\, a^{ij}(x) \frac{\partial \rho}{\partial x^i}(x)\, \frac{\partial \rho}{\partial x^j}(x).$$

Since $u = \pi_2 \circ \phi \in C(\overline{\Omega})$ and $\Omega$ is bounded one has $u \in L^2(\Omega) \cap C^\infty(\Omega) \subset W^1(\Omega, X)$. Finally as $u$ is a strong solution to $Lu + \langle a, Du \rangle = 0$ one may take the $L^2$ inner product with $v \in C^1_0(\Omega)$ and integrate by parts in $\int_\Omega (Lu) v \, d\mu$ to show that $B_{\mathscr{L}}(u, v) = 0$ i.e. $u$ is also a weak solution to $\mathscr{L} u = 0$. Finally one may apply (24) with $f = 0$ to conclude that (20) holds. To prove the last statement in Theorem 4.1 one starts from

$$\frac{\partial}{\partial x^i} \left( (w \circ F)^2 a^{ij} \frac{\partial u}{\partial x^j} \right) = 0.$$

Then

$$\frac{\partial}{\partial x^i} \left( (w \circ \phi)^2\, a^{ij}\, u \,\frac{\partial u}{\partial x^j} \right) = (w \circ F)^2\, a^{ij} \frac{\partial u}{\partial x^i} \frac{\partial u}{\partial x^j}$$

so that (by Green's lemma)

$$\int_{\partial \Omega} u (w \circ F)^2 a^{ij} \frac{\partial u}{\partial x^j}\, \nu_i \, d\sigma = \int_\Omega (w \circ F)^2 \langle A(x) Du, Du \rangle \, d\mu(x)$$

and (as $L$ is uniformly $X$-elliptic in $\mathscr{U}$)

$$\left| a^{ij}(x) \frac{\partial u}{\partial x^j} \nu_i \right| \leq \Lambda \langle A(x) Du, Du \rangle \sum_{a=1}^{m} \langle X_a(x), \nu \rangle^2 = 0, \quad x \in \partial \Omega.$$

Consequently $0 = \langle A(x)Du, Du\rangle \geq \lambda \sum_a \langle X_a(x), Du\rangle^2$ for every $x \in \Omega$, thus yielding $X_a(u) = 0$ for any $1 \leq a \leq m$. Q.e.d.

Let $\phi : \mathscr{U} \to S^M$ be an $X$-elliptic harmonic map. Let $\Omega \subset \mathbb{R}^N$ be a smoothly bounded domain such that $\overline{\Omega} \subset \mathscr{U}$ and let us assume that $\phi(\Omega) \cap \Sigma = \emptyset$. We shall show that $\phi : \Omega \to S^M \setminus \Sigma$ is homotopically nontrivial, provided that $\partial\Omega$ is characteristic relative to the system of vector fields $X$. To this end we need to recall that $S^M \setminus \Sigma$ is isometric to a warped product $S_+^{M-1} \times_w S^1$. Let $S_+^{M-1} \subset \mathbb{R}^M$ be the open upper hemisphere

$$S_+^{M-1} = \{(y_1, \cdots, y_M) \in \mathbb{R}^M : \sum_{i=1}^M y_i^2 = 1, \quad y_M > 0\}.$$

The map

$$I : S_+^{M-1} \times S^1 \to S^M \setminus \Sigma, \quad I(y, \zeta) = (y_M\, u,\, y_M\, v,\, y'),$$
$$y \in S_+^{M-1}, \quad \zeta = u + iv \in S^1, \quad y' = (y_1, \cdots, y_{M-1}),$$

is an isometry of $S_+^{M-1} \times_w S^1$ onto $(S^M \setminus \Sigma, h_M)$. Here $h_M$ denotes the standard Riemannian metric on the sphere $S^M$. Also $S_+^{M-1} \times S^1$ is organized as a warped product manifold with the metric

$$\pi_1^* h_{M-1} + (w \circ \pi_1)^2\, \pi_2^* h_1\,,$$
$$w \in C^\infty(S_+^{M-1}), \quad w(y) = y_M\,, \quad y \in S_+^{M-1}.$$

**Lemma 4.2** *Let $\mathscr{U} \subset \mathbb{R}^N$ be a domain. If $\phi : \mathscr{U} \to S^M \setminus \Sigma$ is $X$-elliptic harmonic then $\tilde{\psi} = I^{-1} \circ \phi : \mathscr{U} \to S_+^{M-1} \times_w S^1$ is $X$-elliptic harmonic. Consequently if $\phi : \mathscr{U} \to S^M \setminus \Sigma$ is null-homotopic then $\phi$ lifts to an $X$-elliptic harmonic map $\psi : \mathscr{U} \to S_+^{M-1} \times_w \mathbb{R}$.*

*Proof* The warped product $S_+^{M-1} \times_w \mathbb{R}$ carries the Riemannian metric

$$\Pi_1^* h_{M-1} + (w \circ \Pi_1)^2\, \Pi_2^*\, dt \otimes dt.$$

Let $f : \mathbb{R} \to S^1$ be the natural covering map i.e. $f(t) = \exp(2\pi i t)$. The map

$$\left(1_{S_+^{M-1}}, f\right) : S_+^{M-1} \times_w \mathbb{R} \to S_+^{M-1} \times_w S^1$$

is a local isometry. The property that a map is $X$-elliptic harmonic is invariant with respect to (local) isometries of the target manifold. Let us set

$$F = \pi_1 \circ \tilde{\psi}\,, \quad \tilde{u} = \pi_2 \circ \tilde{\psi}\,.$$

Let $x_0 \in \mathscr{U}$ and $\zeta_0 = \tilde{u}(x_0)$. Let also $t_0 \in \mathbb{R}$ such that $f(t_0) = \zeta_0$. Since $\phi$ is null-homotopic, so does $\tilde{\psi}$. Hence $\tilde{u}_* \pi_1(\mathscr{U}, x_0) = 0$. Since $\mathscr{U}$ is connected and locally connected by arcs one may apply standard homotopy theory to conclude that there is a unique continuous map $u : \mathscr{U} \to \mathbb{R}$ such that $u(x_0) = t_0$ and $f \circ u = \tilde{u}$. Then $\psi = (F, u) : \mathscr{U} \to S_+^{M-1} \times_w \mathbb{R}$ is the $X$-elliptic harmonic lift of $\tilde{\psi}$ claimed in Lemma 4.2. □

To prove Theorem 1.1 let $\phi : \mathscr{U} \to S^M \setminus \Sigma$ be a null-homotopic $X$-elliptic harmonic map. By applying first Lemma 4.2 and then Lemma 4.1 for any domain $\Omega \subset \mathbb{R}^N$ such that $\overline{\Omega} \subset \mathscr{U}$ and $\partial\Omega$ is characteristic, it follows that $\psi(\Omega) \subset S_+^{M-1} \times \{t_\psi\}$ for some $t_\psi \in \mathbb{R}$. Let $\rho$ be a defining function for $\Omega$ i.e. $\Omega = \{x \in \mathscr{U} : \rho(x) < 0\}$ and $D\rho(x) \neq 0$ for every $x \in \partial\Omega$. There is $\epsilon_0 > 0$ such that $M_\epsilon = \{x \in \Omega : \rho(x) = -\epsilon\}$ is a smooth hypersurface for every $0 < \epsilon \leq \epsilon_0$. Also $X$ is a Hörmander system on $\Omega_\epsilon = \{x \in \Omega : \rho(x) + \epsilon < 0\}$ and $M_\epsilon = \partial\Omega_\epsilon$ is characteristic. Then Lemma 4.1 applies (with $\Omega$ replaced by $\Omega_\epsilon$) to the $X$-elliptic harmonic map $p \circ \psi : \Omega \to S_+^{M-1}$ (where $p(x,t) = x$) i.e. $p \circ \psi$ is constant on $\Omega_\epsilon$ and then, by passing to the limit with $\epsilon \to 0$, constant on the whole of $\Omega$. Q.e.d.

## 5 $X$-Elliptic Harmonic Morphisms

For every $C^\infty$ function $v : V \to \mathbb{R}$

$$L(v \circ \phi) = \left( \frac{\partial v}{\partial y^\alpha} \circ \phi \right) \tau_A(\phi)^\alpha + a^{ij} \frac{\partial \phi^\alpha}{\partial x^i} \frac{\partial \phi^\beta}{\partial x^j} \left\{ \left( \nabla_\alpha v_\beta \right) \circ \phi \right\} \tag{28}$$

where

$$\nabla_\alpha v_\beta = \frac{\partial^2 v}{\partial y^\alpha \, \partial y^\beta} - \Gamma_{\alpha\beta}^\gamma \frac{\partial v}{\partial y^\gamma} .$$

We need the following result (referred hereafter as *Ishihara's lemma*)

**Lemma 5.1** *Let $C_\alpha$, $C_{\alpha\beta} \in \mathbb{R}$, $1 \leq \alpha, \beta \leq M$, such that $C_{\alpha\beta} = C_{\beta\alpha}$ and $\sum_{\alpha=1}^M C_{\alpha\alpha} = 0$. Let $y_0 \in S$ and let $(V, y^\alpha)$ a local system of normal coordinates on $S$ centered at $y_0$ such that $y^\alpha(y_0) = 0$. There is a harmonic function $v : V \to \mathbb{R}$ such that*

$$\frac{\partial v}{\partial y^\alpha}(y_0) = C_\alpha , \quad \left( \nabla_\alpha v_\beta \right)(y_0) = C_{\alpha\beta} , \quad 1 \leq \alpha, \beta \leq M.$$

Cf. [5]. Let us prove Theorem 1.2. Let $\phi : \mathbb{R}^N \to S$ be an $X$-elliptic harmonic morphism. Let $\alpha_0 \in \{1, \cdots, M\}$ be a fixed index. Let $x_0 \in \mathbb{R}^N$ and let us consider a normal coordinate system $(V, y^\alpha)$ on $S$ at $y_0 = \phi(x_0)$. By Ishihara's lemma applied for the constants $C_\alpha = \delta_{\alpha\alpha_0}$ and $C_{\alpha\beta} = 0$ there is a harmonic function $v : V \to \mathbb{R}$

such that $(\partial v/\partial y^\alpha)(y_0) = \delta_{\alpha\alpha_0}$ and $(\nabla_\alpha v_\beta)(y_0) = 0$ hence (by (28) and $L(v \circ \phi)(x_0) = 0$) one has $\tau_A(\phi)^{\alpha_0}(x_0) = 0$.

Let $C_{\alpha\beta} \in \mathbb{R}$ such that $C_{\alpha\beta} = C_{\beta\alpha}$ and $\sum_{\alpha=1}^M C_{\alpha\alpha} = 0$. By Ishihara's lemma there is a harmonic function $v : V \to \mathbb{R}$ such that $(\partial v/\partial y^\alpha)(y_0) = 0$ and $(\nabla_\alpha v_\beta)(y_0) = C_{\alpha\beta}$. Then [by (28)]

$$a^{ij}(x_0)\frac{\partial\phi^\alpha}{\partial x^i}(x_0)\frac{\partial\phi^\beta}{\partial x^j}(x_0)C_{\alpha\beta} = 0. \tag{29}$$

Let us set $X^{\alpha\beta} = a^{ij}(\partial\phi^\alpha/\partial x^i)(\partial\phi^\beta/\partial x^j)$. Then (29) may be written

$$\sum_{\alpha\neq\beta} C_{\alpha\beta}X^{\alpha\beta}(x_0) + \sum_\alpha C_{\alpha\alpha}\left[X^{\alpha\alpha}(x_0) - X^{11}(x_0)\right] = 0. \tag{30}$$

If $\alpha_0 \in \{2, \cdots, M\}$ is a fixed index then let $C_{\alpha\beta} \in \mathbb{R}$ be given by

$$\alpha \neq \beta \Longrightarrow C_{\alpha\beta} = 0, \quad C_{\alpha\alpha} = \begin{cases} 1, & \alpha = \alpha_0\,, \\ -1 & \alpha = 1, \\ 0, & \text{otherwise.} \end{cases}$$

Identity (30) then yields $X^{\alpha_0\alpha_0}(x_0) - X^{11}(x_0) = 0$ hence $X^{11}(x_0) = \cdots = X^{MM}(x_0)$ and (30) becomes

$$\sum_{\alpha\neq\beta} C_{\alpha\beta}X^{\alpha\beta}(x_0) = 0.$$

Finally let $\alpha_0, \beta_0 \in \{1, \cdots, M\}$ be arbitrary fixed indices such that $\alpha_0 \neq \beta_0$ and let $C_{\alpha\beta} \in \mathbb{R}$ be given by

$$C_{\alpha\beta} = \begin{cases} 1, & \alpha = \alpha_0 \text{ and } \beta = \beta_0\,, \\ 0, & \text{otherwise,} \end{cases}$$

so that to obtain $X^{\alpha_0\beta_0}(x_0) = 0$. Gathering the information got so far

$$X^{\alpha\beta}(x_0) = X^{11}(x_0)\,\delta^{\alpha\beta}\,. \tag{31}$$

Let us set $X^{11} = \lambda_U^2$ so that $\lambda_U \in C(U)$ and $\lambda_U^2 \in C^\infty(U)$ where $U = \phi^{-1}(V)$. Contraction of $\alpha$ and $\beta$ in (31) furnishes

$$M\,\lambda_U(x_0)^2 = \sum_{\alpha=1}^M a^{ij}(x_0)\frac{\partial\phi^\alpha}{\partial x^i}(x_0)\frac{\partial\phi^\alpha}{\partial x^j}(x_0).$$

On the other hand if $(V', y'^{\alpha})$ is another normal coordinate system centered at $y_0 = \phi(x_0)$ then $y'^{\alpha} = a^{\alpha}_{\beta} y^{\beta}$ for some orthogonal matrix $\left[a^{\alpha}_{\beta}\right] \in \mathrm{O}(M)$ hence

$$\sum_{\alpha} a^{ij}(x_0) \frac{\partial \phi'^{\alpha}}{\partial x^i}(x_0) \frac{\partial \phi'^{\alpha}}{\partial x^j}(x_0) = \sum_{\alpha} a^{ij}(x_0) \frac{\partial \phi^{\alpha}}{\partial x^i}(x_0) \frac{\partial \phi^{\alpha}}{\partial x^j}(x_0)$$

and the functions $\lambda_U$ glue up to a globally defined function $\lambda_{\phi} : \mathbb{R}^N \to [0, +\infty)$ such that $\lambda_{\phi}\big|_U = \lambda_U$. Clearly (31) may be written as (5) in Theorem 1.2.

To prove (ii) in Theorem 1.2 let $\phi : \mathbb{R}^N \to S$ be an $X$-elliptic harmonic map satisfying (5) and let $v : V \to \mathbb{R}$ with $V \subset S$ open and $\Delta_h v = 0$ in $V$. Then (by (28) and $\tau_A(\phi) = 0$)

$$L(v \circ \phi)(x_0) = a^{ij}(x_0) \frac{\partial \phi^{\alpha}}{\partial x^i}(x_0) \frac{\partial \phi^{\beta}}{\partial x^j}(x_0) \frac{\partial^2 v}{\partial y^{\alpha} \partial y^{\beta}}(y_0) =$$

[by (5)]

$$= \lambda_{\phi}(x_0)^2 \delta^{\alpha\beta} \frac{\partial^2 v}{\partial y^{\alpha} \partial y^{\beta}}(y_0) = -\lambda_{\phi}(x_0)^2 (\Delta_h v)(y_0) = 0$$

because of

$$\Delta_h v = -g^{\alpha\beta} \left( \frac{\partial^2 v}{\partial y^{\alpha} \partial y^{\beta}} - \Gamma^{\gamma}_{\alpha\beta} \frac{\partial v}{\partial y^{\gamma}} \right),$$

$$g^{\alpha\beta}(x_0) = \delta^{\alpha\beta}, \quad \Gamma^{\gamma}_{\alpha\beta}(y_0) = 0.$$

Therefore $\phi$ is an $X$-elliptic harmonic morphism.

To prove (iii)–(iv) in Theorem 1.2 let $\phi : \mathbb{R}^N \to S$ be an $X$-elliptic harmonic morphism. Let $x_0 \in \mathbb{R}^N$ be an arbitrary point and let $(V, y^{\alpha})$ be a normal coordinate system on $S$ centered at $y_0 = \phi(x_0)$. If $\phi^{\alpha} = y^{\alpha} \circ \phi$ and $\xi^{\alpha} = D\phi^{\alpha}(x_0) \in \mathbb{R}^N$, $1 \leq \alpha \leq M$, then [by (5)]

$$\left\langle A(x_0)\xi^{\alpha}, \xi^{\beta} \right\rangle = \lambda_{\phi}(x_0)^2 \delta^{\alpha\beta}. \tag{32}$$

**Lemma 5.2** *If the vectors $\{\xi^{\alpha} : 1 \leq \alpha \leq M\}$ are linearly dependent then* $\lambda_{\phi}(x_0)=0$.

*Proof* Assume there is $\alpha_0 \in \{1, \cdots, M\}$ such that $\xi^{\alpha_0} = \sum_{\alpha \neq \alpha_0} a_{\alpha} \xi^{\alpha}$ for some $a_{\alpha} \in \mathbb{R}$. Then formula (32) for $\alpha = \beta = \alpha_0$ gives

$$\lambda_{\phi}(x_0)^2 = \langle A(x_0)\xi^{\alpha_0}, \xi^{\alpha_0} \rangle = \sum_{\alpha \neq \alpha_0} \langle A(x_0)\xi^{\alpha_0}, \xi^{\alpha} \rangle = 0$$

by applying (32) once more. □

Let us assume that $M > N$. Then $\{\xi^\alpha : 1 \le \alpha \le M\}$ are linearly dependent so that (by Lemma 5.2) $\lambda_\phi(x_0) = 0$. Consequently [by (5)]

$$\big\langle A(x_0)D\phi^\alpha(x_0)\,,\, D\phi^\beta(x_0)\big\rangle = 0.$$

In particular for $\beta = \alpha$ [by the $X$-ellipticity condition (2)]

$$\langle X_a(x_0)\,,\, D\phi^\alpha(x_0)\rangle = 0$$

or $X_a(\phi^\alpha)_{x_0} = 0$ for every $1 \le a \le m$ and $1 \le \alpha \le M$. Let $(W, w^\alpha)$ be an arbitrary local coordinate system on $S$ such that $y_0 \in W$. Let $f^\alpha(y^1, \cdots, y^M)$ be the transition functions relative to the pair of local coordinate systems $(V, y^\alpha)$ and $(W, w^\alpha)$. Then $w^\alpha \circ \phi = f^\alpha(\phi^1, \cdots, \phi^M)$ on $\phi^{-1}(V \cap W)$ and

$$X_a\,(w^\alpha \circ \phi)_{x_0} = \frac{\partial f^\alpha}{\partial \xi^\beta}(\phi^1(x_0), \cdots, \phi^M(x_0))\, X_a(\phi^\beta)_{x_0} = 0$$

or $\phi_* X_a = 0$ for every $1 \le a \le m$. Let us define $X^{(k)} \subset \mathfrak{X}(\mathbb{R}^N)$ by recurrence by setting

$$X^{(1)} = \{X_a : 1 \le a \le m\},$$

$$X^{(k+1)} = \big\{[X_a\,,\, Y] : 1 \le a \le m,\ \ Y \in X^{(k)}\big\}, \quad k \ge 1.$$

Then $\phi_* X^{(k)} = 0$ for every $k \ge 1$. In particular if $X = X^{(1)}$ is a Hörmander system on $\mathbb{R}^N$ (i.e. there is $k \ge 1$ such that $\{Y_x : Y \in X^{(k)}\}$ spans the tangent space $T_x(\mathbb{R}^N)$ at any $x \in \mathbb{R}^N$) then $\phi$ is constant.

Let now $M \le N$ and let $x_0 \in \mathbb{R}^N$ such that $\lambda_\phi(x_0) \neq 0$. Then (again by Lemma 5.2) the vectors $\{\xi^\alpha : 1 \le \alpha \le M\}$ are linearly independent. Yet the $N \times M$ matrix $\left[(\xi^1)^T\,,\ \cdots\,,\ (\xi^M)^T\right]$ is the Jacobian of $\phi$ at $x_0$ hence rank $(d_{x_0}\phi) = M$ i.e. $\phi$ is a submersion on some neighborhood of $x_0$. Formula (6) follows from (28) and (5). Q.e.d.

Let us prove Theorem 1.3. Given an open connected subset $U \subset \mathbb{R}^N$ we shall show that $V = \phi(U) \subset S$ is an open set as well. Proof is by contradiction. Let us assume that $V \setminus \mathring{V} \neq \emptyset$ and consider $y_0 \in V \setminus \mathring{V}$. Then $B(y_0\,,\,1/j) \setminus V \neq \emptyset$ for any $j \ge 1$, because otherwise $y_0$ would be an interior point of $V$. Here $B(x, r) \subset S$ is the ball of radius $r > 0$ and center $x \in S$, relative to the distance function associated to the Riemannian metric $h$. Let $y_j \in B(y_0\,,\,1/j) \setminus V$ for any $j \ge 1$. This amounts to having chosen a sequence of points $y_j \in S \setminus V$ such that $y_j \to y_0$ as $j \to \infty$. As $S$ is Riemannian, there is an open set $W \subset S$ such that $y_0 \in W$ and there is a fundamental solution $G$ to $\Delta_h$ on $W \times W$, which is $C^\infty$ off the diagonal, and has the property that for every $y \in W$ the function $x \longmapsto G(x, y)$ is strictly positive and $\Delta_h G(\cdot\,,\,y) = 0$ in $W \setminus \{y\}$. Also if $D = \{(x, y) \in W \times W : x = y\}$ is the diagonal then $G(x, y) \to +\infty$

as $(x, y) \to D$. We may assume without loss of generality that $U \subset \phi^{-1}(W)$. Next let us consider the sequence of functions

$$v_j : W \setminus \{y_j\} \to (0, +\infty),$$
$$v_j(y) = G(y, y_j), \quad y \in W, \quad y \neq y_j .$$

Then $\Delta_h\left(v_j\right) = 0$ in $W \setminus \{y_j\}$ for every $j \geq 1$. The set $W \setminus \{y_j\}$ is open in $S$. Hence, as $\phi$ is an $X$-elliptic harmonic morphism, the function

$$u_j = v_j \circ \phi : \phi^{-1}\left(W \setminus \{y_j\}\right) \to (0, +\infty)$$

must satisfy $L(u_j) = 0$ in $\phi^{-1}\left(W \setminus \{y_j\}\right)$ and in particular in $U$. Let $x_0 \in U$ such that $\phi(x_0) = y_0$. Then

$$u_j(x_0) = v_j(y_0) = G(y_0\,,\, y_j) \to +\infty, \quad j \to \infty.$$

To end the proof of Theorem 1.3 we need to recall the Harnack inequality for uniformly $X$-elliptic operators $L$ (cf. Theorem 4.1 in [4], p. 1848).

Let $\Omega \subset \mathbb{R}^N$ be a bounded open subset. By a result in [4] nonnegative solutions to $Lu = 0$ in $\Omega$ satisfy an invariant Harnack inequality provided that the structure assumptions **(D)** and **(P)** are satisfied. Let $K \subset \mathbb{R}^N$ be a fixed compact subset containing the closure of $\Omega$ i.e. $\overline{\Omega} \subset K$. By Theorem 1.15 in [3] there is a constant $r_0(K) > 0$ such that the following Sobolev inequality

$$\|u\|_{L^q_*(B_r)} \leq C\, r\, \|Xu\|_{L^2_*(B_r)}\,, \quad u \in C^1_0(B_r), \tag{33}$$

$$q = \frac{2Q}{Q-2}\,, \quad Q = \log_2 C_d\,,$$

holds for every control $d$-ball $B_r$ whose center lies in $K$ and having radius $0 < r \leq r_0(K)$. Here one adopted the notation

$$\|u\|_{L^s_*(B_r)} = \left(\frac{1}{|B_r|}\int_{B_r} |u|^s\, dx\right)^{1/s}.$$

It is worth mentioning that (33) follows from the doubling condition **(D)** and Poincaré inequality **(P)**. The result from [4] that we need is that given a nonnegative solution $u \in W^1_{\rm loc}(\Omega, X)$ to $Lu = 0$ in $\Omega$ and $0 < r \leq r_0(K)/4$ then for any control $d$-ball $B_{4r} \subset \Omega$

$$\sup_{B_r} u \leq C \inf_{B_r} u \tag{34}$$

for some constant $C = C(\lambda, \Lambda, C_d, C_P, , \alpha) > 0$ where $\alpha$ is a Lipschitz constant for the vector fields $\{X_a : 1 \leq a \leq m\}$ on $\Omega$. Harnack inequality (34) for nonnegative solutions to $Lu = 0$ was proved in [8] and successively generalized to a larger class of uniformly $X$-elliptic operators (including equations with lower order terms $\mathfrak{L}u = 0$) in [4]. To keep notation unitary we however make explicit references to [4] alone. Let us go back to the proof of Theorem 1.3. Since $U$ is open there exist a bounded open set $\Omega$ and a compact set $K$ such that $x_0 \in \Omega$ and $\overline{\Omega} \subset K \subset U$. Let $0 < r < r_0(K)/4$ be sufficiently small such that $x_0 \in B_r$ and $B_{4r} \subset \Omega$. As $Lu_j = 0$ and $u_j > 0$ in $\Omega$ one may apply Harnack inequality to get

$$u_j(x_0) \leq \sup_{B_r} u_j \leq C \inf_{B_r} u_j$$

hence on one hand $\lim_{j\to\infty} \inf_{B_r} u_j = \infty$. On the other hand

$$B_r \setminus \phi^{-1}(y_0) \neq \emptyset. \tag{35}$$

Indeed if (35) is not true then $\phi$ is constant on $B_r$ hence $\lambda_\phi(x) = 0$ for every $x \in B_r$ [as a consequence of (5)], a situation excluded by the hypothesis of Theorem 1.3. Let then $x \in B_r \setminus \phi^{-1}(y_0)$ and let $y = \phi(x)$. It follows that

$$\inf_{B_r} u_j \leq u_j(x) = v_j(y) = G(y, y_j) \to G(y, y_0) < \infty, \quad j \to \infty,$$

a contradiction.

## References


1. Baird, P., Wood, J.C.: Harmonic Morphisms Between Riemannian Manifolds. Oxford Science Publications, London Mathematical Society Monographs, New Series, vol. 29. Clarendon Press, Oxford (2003)
2. Fuglede, B.: Harmonic morphisms between semi-Riemannian manifolds. Ann. Acad. Sci. Fenn. **21**, 31–50 (1996)
3. Garofalo, N., Nhieu, D.-M.: Isoperimetric and Sobolev inequalities for Carnot-Carathéodory spaces and the existence of minimal surfaces. Commun. Pure Appl. Math. **49**, 1081–1144 (1996)
4. Gutiérrez, C.E., Lanconelli, E.: Maximum principle, nonhomogeneous Harnack inequality, and Liouville theorems for $X$-elliptic operators. Commun. Partial Differ. Equ. **28**, 1833–1862 (2003)
5. Ishihara, T.: A mapping of Riemannian manifolds which preserves harmonic functions. J. Math. Kyoto Univ. **19**, 215–29 (1979)
6. Jost, J., Xu, C.-J.: Subelliptic harmonic maps. Trans. Am. Math. Soc. **350**, 4633–4649 (1988)
7. Lanconelli, E.: Regolarità Hölderiana delle soluzioni deboli di certe equazioni ellittiche fortemente degeneri. Seminario di Analisi Matematica, Istituto Matematico dell'Università di Bologna, 22–29 April 1982
8. Lanconelli, E., Kogoi, A.E.: $X$-elliptic operators and $X$-elliptic distances. Ricerche Mat. **49**, 222–243 (2000)

9. Moser, R.: Partial Regularity for Harmonic Maps and Related Problems. World Scientific, New Jersey, London, Singapore, Beijing, Shanghai, Hong Kong, Taipei, Chennai (2005)
10. Parmeggiani, A., Xu, C.-J.: The Dirichlet problem for sub-elliptic second order equations. Ann Mat. Pura Appl. **173**, 233–243 (1997)
11. Solomon, B.: Harmonic maps to spheres. J. Differ. Geom. **21**, 151–162 (1985)

# Sum Operators and Fefferman–Phong Inequalities

**Giuseppe Di Fazio, Maria Stella Fanciullo, and Pietro Zamboni**

*Dedicated to Ermanno Lanconelli on the occasion of his 70th birthday*

**Abstract** We define Stummel-Kato type classes in a quasimetric homogeneous setting using sum operators introduced in (J. Fourier Anal. Appl. **9**(5), 511–540, 2003) by Franchi, Perez and Wheeden. Then we prove an embedding inequality of Fefferman–Phong type. As an application we give a unique continuation result for non negative solutions of some subelliptic equations.

**Keywords** Stummel spaces • Unique continuation

**Mathematics Subject Classification 2000:** 46E35, 35B05

## 1 Introduction

In his celebrated paper [22] C. Fefferman proved the following inequality

$$\int_B |u(x) - u_B|^p |V(x)|\, dx \le c \int_B |\nabla u(x)|^p\, dx \qquad \forall u \in C^\infty$$

in the case $p = 2$ assuming the potential $V$ to belong in the Morrey class $L^{r,n-2r}$, $1 < r \le n/2$. Later, Chiarenza and Frasca [4] extended Fefferman result - with different proof - assuming $V$ in $L^{r,n-pr}$, $1 < r \le n/p$, $1 < p < n$ (see also [35, 37]).

Danielli, Garofalo and Nhieu in [9] and Danielli [8] introduced a suitable version of Morrey spaces adapted to the Carnot-Carathéodory metric and proved

G. Di Fazio (✉) • M.S. Fanciullo • P. Zamboni
Dipartimento di Matematica e Informatica, Università di Catania, Catania, Italy
e-mail: difazio@dmi.unict.it; fanciullo@dmi.unict.it; zamboni@dmi.unict.it



G. Citti et al. (eds.), *Geometric Methods in PDE's*, Springer INdAM Series 13,
DOI 10.1007/978-3-319-02666-4_6

the following inequality assuming $V$ in a Morrey space defined by using a system of locally Lipschitz vector fields $X = (X_1, X_2, \ldots, X_q)$

$$\int_B |u(x) - u_B|^p |V(x)|\, dx \leq c \int_B |Xu(x)|^p\, dx \qquad \forall u \in C^\infty. \tag{1}$$

In [10] inequality (1) was proved assuming $V$ in a Stummel-Kato class defined by using the Carnot-Carathéodory metric $d$ induced by the system $X$ of locally Lipschitz vector fields.

In the Euclidean case the Stummel-Kato class was introduced by Aizenman and Simon in [1] (see also [6, 7]).

The fundamental tool of the proof of (1) is the following integral representation formula (see [33])

$$|u(x) - u_B| \leq c \int_B |Xu(y)| \frac{d(x,y)}{|B(x, d(x,y))|} dy\,, \quad x \in B. \tag{2}$$

The Fefferman-Phong inequality (1) has been used by many Authors to obtain regularity results for solutions of quasilinear equations (see [8, 9, 11–13, 15, 36, 37]). Moreover a Fefferman-Phong inequality type inequality has been used also for regularity of solutions of linear and quasilinear degenerate elliptic equations with a degeneracy of $A_2$ type or strong $A_\infty$ type (see [14, 16–20, 32, 34, 38]).

This paper is the first step in a project whose aim is to obtain regularity properties of generalized solutions of subelliptic equations in a particular setting (see [21]).

Let $X = (X_1, \ldots, X_q)$ be a system of locally Lipschitz vector fields in $\mathbf{R}^n$ that turns $\mathbf{R}^n$ into a space of homogeneous type.

We assume that $X$ satisfies the Poincaré inequality

$$\frac{1}{|B|} \int_B |f(x) - f_B|\, dx \leq c r(B) \left( \frac{1}{|B|} \int_B |Xf(x)|^p\, dx \right)^{1/p} \quad p > 1 \tag{3}$$

for any smooth $f$. Unfortunately, in our setting inequality (3) does not imply $(1,1)$ Poincaré-Sobolev inequality (see examples in [30, 31]).

It is known (see [24]) that the integral representation formula (2) is equivalent to the following $(1,1)$ Poincaré inequality

$$\int_{B_r} |u(x) - u_{B_r}|\, dx \leq c r \int_{B_r} |Xu(x)|\, dx\,.$$

This implies that we cannot use the representation formula (2).

In [27] Franchi, Perez and Wheeden showed that the Poincaré inequality (3) implies the following representation formula

$$|u(x) - u_{B_0}| \leq c \sum_{j=0}^{\infty} r(B_j(x)) \left( \frac{1}{|B_j(x)|} \int_{B_j(x)} |Xu|^p dy \right)^{1/p}$$

where $\{B_j\}$ is a suitable sequence of Carnot-Carathéodory metric balls (see Sect. 2).

This representation formula allows us to introduce a Stummel-Kato type class modeled to our setting and prove an embedding result of Fefferman-Phong type.

As application, we prove a result of unique continuation for non negative solutions of linear degenerate subelliptic equations in divergence form with a potential belonging to the Stummel-Kato class. In other settings similar results can be found in [5, 11, 16, 28, 35, 37].

## 2 Preliminaries

Let $(\mathscr{S}, \rho)$ be a quasimetric space, in the sense that $\rho : \mathscr{S} \times \mathscr{S} \to \mathbf{R}$ satisfies

(1) $\rho(x, y) \geq 0$ for all $x, y \in \mathscr{S}$ and $\rho(x, y) = 0$ iff $x = y$.
(2) $\rho(x, y) = \rho(y, x)$ for all $x, y \in \mathscr{S}$.
(3) $\rho(x, y) \leq K[\rho(x, z) + \rho(z, y)]$ for all $x, y, z \in \mathscr{S}$.

We denote by $B(x, r)$ the ball centered at $x \in \mathscr{S}$ of radius $r$. Let $(\mathscr{S}, \rho)$ be endowed with a Borel measure $\nu$ satisfying the following doubling property

(4) There exists $A > 0$ such that for all $x \in \mathscr{S}$ and $r > 0$

$$\nu(B(x, 2r)) \leq A\nu(B(x, r)) .$$

Such a quasimetric space $(\mathscr{S}, \rho, \nu)$ is called space of homogeneous type.

Let $B_0$ be a fixed ball in $\mathscr{S}$. We assume the following geometric hypotheses: for all $x \in B_0$ there exists a chain of balls $\{B_i\} = \{B_i(x)\}_{i=1}^{\infty}$, of radius $r(B_i)$, such that

(H1) $B_i \subset B_0$ for all $i \geq 0$.
(H2) $r(B_i) \sim 2^{-i} r(B_0)$ for all $i \geq 0$.
(H3) $\rho(B_i, x) \leq cr(B_i)$ for all $i \geq 0$.
(H4) For all $i \geq 0$, $B_i \cap B_{i+1}$ contains a ball $S_i$ with $r(S_i) \sim r(B_i)$.

The balls $B_j(x)$ may or may not contain $x$, but the sequence $\{B_i(x)\}_{i=1}^{\infty}$ depends on $x$. Any positive constant that depends only on $K, A$ and the constants in (H2) and (H3) will be called a geometric constant.

Now we give the definition of sum operator in $(\mathscr{S}, \rho, \nu)$.

**Definition 2.1** Let $a : B \to a(B)$ be a nonnegative functional defined on balls $B \subset B_0$. If $x \in B_0$, let

$$T(x) = \sum_{i=0}^{\infty} a(B_i(x)),$$

where $\{B_i(x)\}_{i=1}^{\infty}$ is a sequence of balls satisfying (H1), (H2), (H3), and $B_0(x) = B_0$ for all $x \in B_0$. The operator $T(x)$ is called a sum operator associated with the functional $a(B)$.

The meaning of $T(x)$ lies on the following pointwise representation formula (see [27]).

**Theorem 2.1** *Let $B_0$ be a ball and (H1), (H2), (H3), (H4) hold true. Let $u \in L^1(B_0, \nu)$ be such that for any ball $B \subset B_0$*

$$\frac{1}{\nu(B)} \int_B |u - u_B| d\nu \le c a(B) \tag{4}$$

*where $u_B = \frac{1}{\nu(B)} \int_B u d\nu$. Then for $\nu - a.e. x \in B_0$*

$$|u(x) - u_{B_0}| \le c\, T(x),$$

*where c is a geometric constant which also depends on the constant in (4).*

We define Stummel-Kato type spaces in the space of homogeneous type $(\mathscr{S}, \rho, \nu)$.

**Definition 2.2** Let $1 \le p < +\infty$. Let $B(x_0, r)$ be a ball of $\mathscr{S}$ and let us $\{B_j(x)\}_{j=1}^{\infty}$ be a chain of balls related to $x \in B(x_0, r)$ satisfying (H1), (H2), (H3) and (H4) and $B_0(x) = B(x_0, r)$. We say that $V \in L^1_{loc}(\mathscr{S}, \nu)$ belongs to the space $\tilde{S}_p(\mathscr{S})$ if

$$\eta_V(r) \equiv \sup_{x_0 \in \mathscr{S}} \sup_{y \in B(x_0, r)} \int_{B(x_0, r)} \sum_{j=0}^{\infty} \frac{r^p(B_j(x))|V(x)|}{\nu(B_j(x))} \chi_{B_j(x)}(y) d\nu(x)$$

is finite for all $r > 0$.

We say that $V \in \tilde{S}_p(\mathscr{S})$ belongs to $S_p(\mathscr{S})$ if $\lim_{r \to 0} \eta_V(r) = 0$.

## 3 Embedding Inequality of Fefferman-Phong Type

Let $X = (X_1, X_2, \ldots, X_q)$ be a system of locally Lipschitz vector fields in $\mathbf{R}^n$. Denoted by $d$ the associated Carnot-Carathéodory distance defined by means of subunit curves, we assume that $d$ is finite for each pair of points $x, y \in \mathbf{R}^n$ (more details can be found in [2, 23]). From now on we denote by $B(x, r)$ the Carnot-Carathéodory ball centered at $x \in \mathbf{R}^n$ of radius $r$. Throughout the paper, we shall assume the following:

$(A1)$ $d$ is continuous with respect to the Euclidean distance in $\mathbf{R}^n$.

$(A2)$ The Lebesgue measure is globally doubling with doubling constant $A$, that is for all $x \in \mathbf{R}^n$ and $r > 0$

$$|B(x, 2r)| \leq A|B(x, r)| .$$

$(A3)$ The $(1, p)$ Poincaré inequality holds true: let $p > 1$ and $B_0$ be a ball in $\mathbf{R}^n$, then there exists a positive constant $c$ such that $\forall\ B = B(x, r) \subset B_0$, and $\forall u \in C^\infty(\overline{B}_0)$

$$\frac{1}{|B|}\int_B |u - u_B| dy \leq cr\left(\frac{1}{|B|}\int_B |Xu|^p dy\right)^{1/p},$$

where $u_B = \frac{1}{|B|}\int_B u dy$.

We call $Q = \log_2 A$ the homogeneous dimension.

*Remark 3.1* We recall that the Carnot-Carathéodory metric satisfies the segment property, i.e. for every pair of points $x, y \in \mathbf{R}^n$ there is a continuous curve $\gamma : [0, T] \to \mathbf{R}^n$ connecting $x$ and $y$ such that $d(\gamma(t), \gamma(s)) = |t - s|$ for all $s, t \in [0, T]$ (see [26], Remark 2.6). Then in Carnot-Carathéodory space $\forall x \in B_0 = B(x_0, r)$ there exists a chain of balls satisfying (H1), (H2), (H3) and (H4). An example is $\{B(x, 2^{-i} d(x, \partial B_0))\}$. In this model case

$$\eta_V(r) = \sup_{x_0 \in \mathbf{R}^n} \sup_{y \in B(x_0, r)} \int_{B(x_0, r)} |V(x)| \left[\frac{1}{(d(x, y))^{Q-p}} - \frac{1}{(d(x, \partial B_0))^{Q-p}}\right] dx$$

(see e.g. [29]).

In the sequel we will use the following Sobolev space.

**Definition 3.1** Let $\Omega$ be a bounded domain in $\mathbf{R}^n$. We say that $u$ belongs to $W^{1,2}(\Omega)$ if $u, X_i u \in L^2(\Omega)$, for $i = 1, \ldots q,$. We denote by $W_0^{1,2}(\Omega)$ the closure of the smooth and compactly supported functions in $W^{1,2}(\Omega)$ with respect to the norm

$$\|u\|_{W^{1,2}(\Omega)} = \|u\|_{L^2(\Omega)} + \sum_{i=1}^{q} \|X_i u\|_{L^2(\Omega)} .$$

We say that $u$ belongs to $W^{1,2}_{loc}(\Omega)$ if $u \in W^{1,2}(\Omega')$ for any $\Omega' \subset\subset \Omega$.

Let $u \in W^{1,2}(\Omega)$, we denote by $Xu$ the vector $(X_1 u, X_2 u, \ldots, X_q u)$. We prove the following embedding theorem and some useful corollaries (see also [4, 10, 14, 35–37]).

**Theorem 3.1** *Let $B_0 = B(x_0, r)$ be a ball of $\mathbf{R}^n$ and $1 < p < Q$. Let $V$ be a function in $\tilde{S}_p(\mathbf{R}^n)$. Then there exists a positive constant $c$ such that*

$$\int_{B_0} |V(x)|\, |u(x) - u_{B_0}|^p \, dx \leq \, c\, \eta_V(r) \int_{B_0} |Xu(x)|^p \, dx$$

*for any smooth function $u$ in $B_0$.*

*Proof* Let $u$ be a smooth function in $B_0$. Choosing the functional

$$a(B(z,s)) = s \left( \frac{1}{|B(z,s)|} \int_{B(z,s)} |Xu|^p dy \right)^{1/p},$$

from $(A3)$, the Theorem 2.1 yields the following representation formula for $u$

$$|u(x) - u_{B_0}| \leq c \sum_{j=0}^{\infty} r(B_j(x)) \left( \frac{1}{|B_j(x)|} \int_{B_j(x)} |Xu|^p dy \right)^{1/p}, \tag{5}$$

for a.e. $x \in B_0$.

Now from (5) and Hölder inequality

$$\int_{B_0} |V(x)||u(x) - u_{B_0}|^p dx$$

$$\leq \int_{B_0} |V(x)||u(x) - u_{B_0}|^{p-1} \sum_{j=0}^{\infty} r(B_j(x)) \left[ \frac{1}{|B_j(x)|} \int_{B_j(x)} |Xu(y)|^p dy \right]^{1/p} dx$$

$$\leq \left[ \int_{B_0} |V(x)||u(x) - u_{B_0}|^p dx \right]^{1/p'} \left[ \int_{B_0} \sum_{j=0}^{\infty} |V(x)| \frac{r^p(B_j(x))}{|B_j(x)|} \int_{B_j(x)} |Xu(y)|^p dy dx \right]^{1/p}$$

$$\leq \left[ \int_{B_0} |V(x)||u(x) - u_{B_0}|^p dx \right]^{1/p'} \left[ \int_{B_0} \sum_{j=0}^{\infty} |V(x)| \frac{r^p(B_j(x))}{|B_j(x)|} \int_{B_0} |Xu(y)|^p \chi_{B_j(x)}(y) dy dx \right]^{1/p}$$

$$\leq \left[ \int_{B_0} |V(x)||u(x) - u_{B_0}|^p dx \right]^{1/p'} \left[ \int_{B_0} |Xu(y)|^p \int_{B_0} \sum_{j=0}^{\infty} |V(x)| \frac{r^p(B_j(x))}{|B_j(x)|} \chi_{B_j(x)}(y) dx dy \right]^{1/p}$$

$$\leq \left[ \int_{B_0} |V(x)||u(x) - u_{B_0}|^p dx \right]^{1/p'} \eta_V^{1/p}(r) \left[ \int_{B_0} |Xu(y)|^p dy \right]^{1/p},$$

from which

$$\int_{B_0} |V(x)||u(x) - u_{B_0}|^p dx \le c\eta_V(r) \int_{B_0} |Xu(y)|^p dy .$$

**Corollary 3.1** *Let* $1 < p < Q$ *and* $V$ *be a function in* $\tilde{S}_p(\mathbf{R}^n)$*. Then there exists a positive constant* $c$ *such that*

$$\int_{\mathbf{R}^n} |V(x)|\, |u(x)|^p\, dx \le c\, \eta_V(r) \int_{\mathbf{R}^n} |Xu(x)|^p\, dx$$

*for any compactly supported smooth function* $u$ *in* $\mathbf{R}^n$*.*

**Corollary 3.2** *Let* $\Omega$ *be a bounded open set* $\Omega \subset \mathbf{R}^n$ *and* $1 < p < Q$*. Let* $V$ *be a function in* $S_p(\Omega)$*. For any* $\epsilon > 0$ *there exists a positive function* $K(\epsilon) \sim \frac{\epsilon}{[\eta_V^{-1}(\epsilon)]^{Q+p}}$ *(where* $\eta_V^{-1}$ *is the inverse function of* $\eta_V$*), such that*

$$\int_{\Omega} |V(x)|\, |u(x)|^p\, dx \le \epsilon \int_{\Omega} |Xu(x)|^p\, dx + K(\epsilon) \int_{\Omega} |u(x)|^p\, dx \tag{6}$$

*for any compactly supported smooth function* $u$ *in* $\Omega$*.*

*Proof* Let $\epsilon > 0$. Let $r$ be a positive number that we will choose later. Let $\{\alpha_i^p\}$, $i = 1, 2, \ldots N(r)$, be a finite partition of unity of $\overline{\Omega}$, such that $\text{spt}\alpha_i \subset B(x_i, r)$ with $x_i \in \overline{\Omega}$ (for the construction of the cut off functions $\alpha_i$ see [25]).

From Corollary 3.1 we obtain

$$\begin{aligned}
\int_{\Omega} |V(x)|\, |u(x)|^p\, dx &\le \int_{\Omega} |V(x)|\, |u(x)|^p \sum_{i=1}^{N(r)} \alpha_i^p(x)\, dx \\
&= \sum_{i=1}^{N(r)} \int_{\Omega} |V(x)|\, |u(x)|^p \alpha_i^p(x) dx \\
&\le c \sum_{i=1}^{N(r)} \eta_V(r) \left( \int_{\Omega} |Xu(x)|^p \alpha_i^p(x)\, dx + \int_{\Omega} |X\alpha_i(x)|^p |u(x)|^p\, dx \right) \\
&\le c\eta_V(r) \left( \int_{\Omega} |Xu(x)|^p\, dx + \frac{N(r)}{r^p} |u(x)|^p\, dx \right) .
\end{aligned}$$

Now we choose $r$ such that $c\eta_V(r) = \epsilon$ and since $N(r) \sim r^{-Q}$ we get (6).

*Remark 3.2* Our setting is not comparable to Euclidean one. Indeed the $(1-1)$ Poincaré inequality is not known to be true.

# 4 Unique Continuation

Let $\Omega$ be a bounded domain in $\mathbf{R}^n$. In $\Omega$ we consider the following linear elliptic equation

$$- X_i^*(a_{ij}(x)X_ju) + Vu = 0 \tag{7}$$

where the coefficients satisfy the following condition

$$\exists \lambda > 0: \ \lambda|\xi|^2 \leq a_{ij}(x)\xi_i\xi_j \leq \lambda^{-1}|\xi|^2 \, , \forall \xi \in \mathbf{R}^q \, , \text{a.e.}\, x \in \Omega \, ,$$

and $V \in \tilde{S}_2(\Omega)$.

**Definition 4.1** A function $u \in W^{1,2}_{loc}(\Omega)$ is a local weak solution of (7) if $\forall \phi \in W^{1,2}_0(\Omega)$

$$\int_\Omega [a_{ij}X_iuX_j\phi + Vu\phi]dx = 0\,.$$

Now we prove the unique continuation property for solutions of (7) (for similar results see also [5, 11, 16, 35, 37]). First we recall the definition of zero of infinite order.

**Definition 4.2** A function $u \in L^1_{loc}(\Omega)$ such that $u(x) \geq 0$ a.e. $x \in \Omega$, is said to vanish of infinite order at $x_0 \in \Omega$ if $\forall k > 0$

$$\lim_{\sigma \to 0} \frac{\int_{B(x_0,\sigma)} u(x)dx}{|B(x_0,\sigma)|^k} = 0\,.$$

**Theorem 4.1** *Let $u \in W^{1,2}_{loc}(\Omega)$, $u \geq 0$, $u \not\equiv 0$, be a solution of (7). Then u has no zero of infinite order in $\Omega$.*

*Proof* Let $x_0 \in \Omega$, $B(x_0, r)$ a ball such that $B(x_0, 2r) \subset \Omega$. Consider a ball $B = B(y, h)$ contained in $B(x_0, r)$. Let $\eta$ be a non negative smooth function with support in $B(y, 2h)$. We take $\phi = \eta^2 u^{-1}$[1] as test function and we obtain

$$\int_\Omega |X \log u(x)|^2 \eta^2(x)dx \leq c\left\{\int_\Omega |X\eta(x)|^2dx + \int_\Omega |V(x)|\eta^2(x)dx\right\}\,.$$

[1] We should take $u + \epsilon$ ($\epsilon > 0$) which is positive in $\Omega$ and, after obtaining estimates independing of $\epsilon$, go to the limit for $\epsilon \to 0$.

From Corollary 3.1 we have

$$\int_{\Omega} |X \log u(x)|^2 \eta^2(x) dx \leq c \int_{\Omega} |X\eta(x)|^2 dx .$$

Choosing $\eta$ such that $\eta = 1$ in $B$ and $X\eta \leq \frac{c}{h}$ we obtain

$$\int_{B} |X \log u(x)|^2 dx \leq c \frac{|B|}{h^2} .$$

Then from John-Nirenberg Lemma (see [3])

$$\int_{B} u^{\delta}(x) dx \int_{B} u^{-\delta}(x) dx \leq c|B|^2 ,$$

that is $u^{\delta}$ is a $A_2$ weight for some $\delta > 0$. Then $u^{\delta}$ satisfies a doubling property from which $u^{\delta}$ (and then also $u$) has no zero of infinite order in $\Omega$.

## References

1. Aizenman, M., Simon, B.: Brownian motion and Harnack inequality for Schrödinger operators. Commun. Pure Appl. Math. **35**(2), 209–273 (1982)
2. Bonfiglioli, A., Lanconelli, E., Uguzzoni, F.: Stratified Lie Groups and Potential Theory for Their Sub-Laplacians. Springer Monographs in Mathematics. Springer, Berlin (2007)
3. Buckley, S.M.: Inequalities of John-Nirenberg type in doubling spaces. J. Anal. Math. **79**, 215–240 (1999)
4. Chiarenza, F., Frasca, M.: A remark on a paper by C. Fefferman. Proc. Am. Math. Soc. **108**, 407–409 (1990)
5. Chiarenza, F., Garofalo, N.: Unique continuation for non-negative solutions of Schrödinger operators. Institute for Mathematics and Its Applications, University of Minnesota. Preprint Series n. 122 (1984)
6. Chiarenza, F., Fabes, E., Garofalo, N.: Harnack's inequality for Schrödinger operators and the continuity of solutions. Proc. Am. Math. Soc. **98**(3), 415–425 (1986)
7. Citti, G., Garofalo, N., Lanconelli, E.: Harnack's inequality for sum of squares of vector fields plus a potential. Am. J. Math. **115**(3), 699–734 (1993)
8. Danielli, D.: A Fefferman–Phong inequality and applications to quasilinear subelliptic equations. Potential Anal. **11**, 387–413 (1999)
9. Danielli, D., Garofalo, N., Nhieu, D.M.: Trace inequalities for Carnot-Caratheodory spaces and applications. Ann. Scuola Norm. Sup. Pisa **XXVII**, 195–252 (1998)
10. Di Fazio, G., Zamboni, P.: A Fefferman-Poincaré type inequality for Carnot-Carathéodory vector fields. Proc. Am. Math. Soc. **130**(9), 2655–2660 (2002)
11. Di Fazio, G., Zamboni, P.: Unique continuation of non negative solutions to quasilinear subelliptic equations in Carnot Caratheodory spaces. Nonlinear elliptic and parabolic equations and systems. Commun. Appl. Nonlinear Anal. **10**(2), 97–105 (2003)
12. Di Fazio, G., Zamboni, P.: Hölder continuity for quasilinear subelliptic equations in Carnot Carathodory spaces. Math. Nachr. **272**, 310 (2004)
13. Di Fazio, G., Zamboni, P.: Fefferman Poincaré inequality and regularity for quasilinear subelliptic equations. Lect. Notes Sem. Int. Matematica **3**, 103–122 (2004)

14. Di Fazio, G., Zamboni, P.: Regularity for quasilinear degenerate elliptic equations. Math. Z. **253**(4), 787–803 (2006)
15. Di Fazio, G., Zamboni, P.: Local regularity of solutions to quasilinear subelliptic equations in Carnot Caratheodory spaces. Boll. Unione Mat. Ital. Sez. B Artic. Ric. Mat. (8) **9**(2), 485–504 (2006)
16. Di Fazio, G., Zamboni, P.: Unique continuation for positive solutions of degenerate elliptic equations. Math. Nachr. **283**(7), 994–999 (2010)
17. Di Fazio, G., Fanciullo, M.S., Zamboni, P.: Harnack inequality and smoothness for quasilinear degenerate elliptic equations. J. Differ. Equ. **245**(10), 2939–2957 (2008)
18. Di Fazio, G., Fanciullo, M.S., Zamboni, P.: Harnack inequality and regularity for degenerate quasilinear elliptic equations. Math. Z. **264**(3), 679–695 (2010)
19. Di Fazio, G., Fanciullo, M.S., Zamboni, P.: Harnack inequality for a class of strongly degenerate elliptic operators formed by Hörmander vector fields. Manuscripta Math. **135**(3–4), 361–380 (2011)
20. Di Fazio, G., Fanciullo, M.S., Zamboni, P.: Regularity for a class of strongly degenerate quasilinear operators. J. Differ. Equ. **255**(11), 3920–3939 (2013)
21. Di Fazio, G., Fanciullo, M.S., Zamboni, P.: Harnack inequality for degenerate elliptic equations and sum operators. Commun. Pure Appl. Anal. **14**(6) (2015)
22. Fefferman, C.: The uncertainty principle. Bull. Am. Math. Soc. **9**, 129–206 (1983)
23. Franchi, B., Lanconelli, E.: Une métrique associée à une classe d'opérateurs elliptiques égénérés. Rend. Sem. Mat. Univ. Politec. Torino (Special Issue), 105–114 (1983)
24. Franchi, B., Lu, G., Wheeden, R.L.: A relationship between Poincaré-type inequalities and representation formulas in spaces of homogeneous type. Int. Math. Res. Not. **1996**(1), 1–14
25. Franchi, B., Serapioni, R., Cassano, F.S.: Approximation and imbedding theorems for weighted Sobolev spaces associated with Lipschitz continuous vector fields. Boll. Un. Mat. Ital. B (7) **11**(1), 83–117 (1997)
26. Franchi, B., Perez, C., Wheeden, R.L.: Self-improving properties of John-Nirenberg and Poincaré inequalities on spaces of homogeneous type. J. Funct. Anal. **153**, 108–146 (1998)
27. Franchi, B., Perez, C., Wheeden, R.L.: A sum operator with applications to self-improving properties of Poincaré inequalities in metric spaces. J. Fourier Anal. Appl. **9**(5), 511–540 (2003)
28. Garofalo, N., Lanconelli, E.: Zero-order perturbations of the subelliptic Laplacian on the Heisenberg group and their uniqueness properties. Bull. Am. Math. Soc. **23**(2), 501–512 (1990)
29. Gutierrez, C.E.: Harnack's inequality for degenerate Schrödinger operators. Trans. AMS **312**, 403–419 (1989)
30. Hajlasz, P., Koskela, P.: Sobolev met Poincaré. Mem. Am. Math. Soc. **145**(688), x+101 (2000)
31. Heinonen, J., Koskela, P.: Quasiconformal maps on metric spaces with controlled geometry. Acta Math. **181**, 1–61 (1998)
32. Lu, G.: On Harnack's inequality for a class of strongly degenerate Schrödinger operators formed by vector fields. Differ. Integr. Equ. **7**(1), 73–100 (1994)
33. Lu, G., Wheeden, R.L.: An optimal representation formula for Carnot- Carathéodory vector fields. **30**(6), 578–584 (1998)
34. Vitanza, C., Zamboni, P.: Necessary and sufficient conditions for Hölder continuity of solutions of degenerate Schrödinger operators. Le Matematiche **52**, 393–409 (1997)
35. Zamboni, P.: Some function spaces and elliptic partial differential equations. Le Matematiche (Catania) **42**(1–2), 171–178 (1987)
36. Zamboni, P.: The Harnack inequality for quasilinear elliptic equations under minimal assumptions. Manuscripta Math. **102**(30), 311–323 (2000)
37. Zamboni, P.: Unique continuation for non-negative solutions of quasilinear elliptic equations. Bull. Aust. Math. Soc. **64**(1), 149–156 (2001)
38. Zamboni, P.: Hölder continuity for solutions of linear degenerate elliptic equations under minimal assumptions. J. Differ. Equ. **182**(1), 121–140 (2002)

# $L^p$-Parabolic Regularity and Non-degenerate Ornstein-Uhlenbeck Type Operators

**Enrico Priola**

*Dedicated to Ermanno Lanconelli on the occasion of his 70th birthday*

**Abstract** We prove $L^p$-parabolic a-priori estimates for $\partial_t u + \sum_{i,j=1}^d c_{ij}(t)\partial^2_{x_i x_j} u = f$ on $\mathbb{R}^{d+1}$ when the coefficients $c_{ij}$ are locally bounded functions on $\mathbb{R}$. We slightly generalize the usual parabolicity assumption and show that still $L^p$-estimates hold for the second spatial derivatives of $u$. We also investigate the dependence of the constant appearing in such estimates from the parabolicity constant. The proof requires the use of the stochastic integral when $p$ is different from 2. Finally we extend our estimates to parabolic equations involving non-degenerate Ornstein-Uhlenbeck type operators.

**Keywords** A-priori $L^p$-estimates • Ornstein-Uhlenbeck operators • Parabolic equations

**Mathematics Subject Classification:** 35K10, 35B65, 35R05

## 1 Introduction and Basic Notations

In this paper we deal with global a-priori $L^p$-estimates for solutions $u$ to second order parabolic equations like

$$u_t(t,x) + \sum_{i,j=1}^{d} c_{ij}(t) u_{x_i x_j}(t,x) = f(t,x), \quad (t,x) \in \mathbb{R}^{d+1}, \tag{1}$$

$d \geq 1$, with locally bounded coefficients $c_{ij}(t)$. Here $u_t$ and $u_{x_i x_j}$ denote respectively the first partial derivative with respect to $t$ and the second partial derivative with

E. Priola (✉)
Dipartimento di Matematica “Giuseppe Peano”, Università di Torino, Torino, Italy
e-mail: enrico.priola@unito.it



G. Citti et al. (eds.), *Geometric Methods in PDE's*, Springer INdAM Series 13,
DOI 10.1007/978-3-319-02666-4_7

respect to $x_i$ and $x_j$. We slightly generalize the usual parabolicity assumption and show that still $L^p$-estimates hold for the second spatial derivatives of $u$. We also investigate the dependence of the constant appearing in such estimates from the symmetric $d \times d$-matrix $c(t) = \big(c_{ij}(t)\big)_{i,j=1,\dots,d}$. In the final section we treat more general equations involving Ornstein-Uhlenbeck type operators and show that the previous a-priori estimates are still true.

The $L^p$-estimates we are interested in are the following: for any $p \in (1,\infty)$, there exists $\tilde{M} > 0$ such that, for any $u \in C_0^\infty(\mathbb{R}^{d+1})$ which solves (1), we have

$$\|u_{x_i x_j}\|_{L^p(\mathbb{R}^{d+1})} \le \tilde{M}\, \|f\|_{L^p(\mathbb{R}^{d+1})}, \quad i,j = 1,\dots,d, \tag{2}$$

where the $L^p$-spaces are considered with respect to the $d+1$-dimensional Lebesgue measure. Usually, in the literature such a-priori estimates are stated requiring that there exists $\lambda$ and $\Lambda > 0$ such that

$$\lambda|\xi|^2 \le \sum_{i,j=1}^{d} c_{ij}(t)\xi_i\xi_j \le \Lambda|\xi|^2, \quad t \in \mathbb{R}, \xi \in \mathbb{R}^d, \tag{3}$$

where $|\xi|^2 = \sum_{i=1}^d \xi_i^2$. We refer to Chap. 4 in [16], Appendix in [24], Sect. VII.3 in [19], which also assumes that $c_{ij}$ are uniformly continuous, and Chap. 4 in [15]. The proofs are based on parabolic extensions of the Calderon-Zygmund theory for singular integrals (cf. [8, 11]). This theory was originally used to prove a-priori Sobolev estimates for the Laplace equation (see [5]). In the above mentioned references, it is stated that $\tilde{M}$ depends not only on $d, p, \lambda$ (the parabolicity constant) but also on $\Lambda$. An attempt to determine the explicit dependence of $\tilde{M}$ from $\lambda$ and $\Lambda$ has been done in Theorem A.2.4 of [24] finding a quite complicate constant.

The fact that $\tilde{M}$ is actually *independent* of $\Lambda$ is mentioned in Remark 2.5 of [14]. This property follows from a general result given in Theorem 2.2 of [13]. Once this independence from $\Lambda$ is proved one can use a rescaling argument (cf. Corollary 2.1) to show that we have

$$\tilde{M} = \frac{M_0}{\lambda}, \tag{4}$$

for a suitable positive constant $M_0$ depending only on $d$ and $p$.

In Theorem 2.1 and Corollary 2.1 we generalize the parabolicity condition by requiring that the symmetric $d \times d$ matrix $c(t) = \big(c_{ij}(t)\big)$ is non-negative definite, for any $t \in \mathbb{R}$, and, moreover, that there exists and integer $p_0$, $1 \le p_0 \le d$, and $\lambda \in (0,\infty)$ such that

$$\lambda \sum_{j=1}^{p_0} \xi_j^2 \le \sum_{i,j=1}^{d} c_{ij}(t)\xi_i\xi_j, \quad t \in \mathbb{R}, \ \xi \in \mathbb{R}^d \tag{5}$$

(cf. Hypothesis 2.1 in Sect. 2). We show that (5) is enough to get estimates like (2) for $i, j = 1, \dots, p_0$, with a constant $\tilde{M}$ as in (4) (now $M_0$ depends on $p, d$ and $p_0$). An example in which (5) holds is

$$u_t(t,x,y) + u_{xx}(t,x,y) + t u_{xy}(t,x,y) + t^2 u_{yy}(t,x,y) = f(t,x,y), \tag{6}$$

$(t,x,y) \in \mathbb{R}^3$ (see Example 2.1). In this case we have an a-priori estimates for $\|u_{xx}\|_{L^p}$.

We will first provide a purely analytic proof of Theorem 2.1 in the case of $L^2$-estimates. This is based on Fourier transform techniques. Then we provide the proof for the general case $1 < p < \infty$ in Sect. 2.2. This proof is inspired by the one of Theorem 2.2 in [13] and requires the concept of stochastic integral with respect to the Wiener process. In Sect. 2.2.1 we recall basic properties of the stochastic integral. It is not clear how to prove Theorem 2.1 for $p \neq 2$ in a purely analytic way. One possibility could be to follow step by step the proof given in Appendix of [24] trying to improve the constants appearing in the various estimates.

In Sect. 3 we will extend our estimates to more general equations like

$$u_t(t,x) + \sum_{i,j=1}^{d} c_{ij}(t) u_{x_i x_j}(t,x) + \sum_{i,j=1}^{d} a_{ij} x_j \, u_{x_i}(t,x) = f(t,x), \tag{7}$$

where $A = (a_{ij})$ is a given real $d \times d$-matrix. If (5) holds with $p_0 = d$ then we show that estimate (2) is still true with $M_0 = M_0(d, p, T, A) > 0$ for any solution $u \in C_0^\infty((-T,T) \times \mathbb{R}^d)$ of (7) (see Theorem 3.1 for a more general statement).

An interesting case of (7) is when $c(t)$ is constant, i.e., $c(t) = Q$, $t \in \mathbb{R}$. Then Eq. (7) becomes

$$u_t + \mathscr{A} u = f,$$

where $\mathscr{A}$ is the Ornstein-Uhlenbeck operator, i.e.,

$$\mathscr{A} v(x) = \mathrm{Tr}(Q D^2 v(x)) + \langle Ax, Dv(x) \rangle, \quad x \in \mathbb{R}^d, \quad v \in C_0^\infty(\mathbb{R}^d). \tag{8}$$

The operator $\mathscr{A}$ and its parabolic counterpart $\mathscr{L} = \mathscr{A} - \partial_t$, which is also called Kolmogorov-Fokker-Planck operator, have recently received much attention (see, for instance, [3, 4, 6, 7, 9, 17, 20, 23] and the references therein). The operator $\mathscr{A}$ is the generator of the Ornstein-Uhlenbeck process which solves a linear stochastic differential equation (SDE) describing the random motion of a particle in a fluid (see [21]). Several interpretations in physics and finance for $\mathscr{A}$ and $\mathscr{L}$ are explained in the survey papers [18, 22]. From the a-priori estimates for the parabolic equation (7) one can deduce elliptic estimates like

$$\|v_{x_i x_j}\|_{L^p(\mathbb{R}^d)} \le C_1 \left( \|\mathscr{A} v\|_{L^p(\mathbb{R}^d)} + \|v\|_{L^p(\mathbb{R}^d)} \right), \tag{9}$$

with $C_1 = \frac{M_2(d,p,A)}{\lambda}$, assuming that $\mathscr{A}$ is non-degenerate (i.e., $Q$ is positive definite; see Corollary 3.1). Similar estimates have been already obtained in [20]. Here we can show in addition the precise dependence of the constant $C_1$ from the matrix $Q$.

More generally, estimates like (9) hold for possibly degenerate hypoelliptic Ornstein-Uhlenbeck operators $\mathscr{A}$ (see [3]); a typical example in $\mathbb{R}^2$ is $\mathscr{A}v = qv_{xx} + xv_y$ with $q > 0$ (cf. Example 43). In this case we have

$$\|v_{xx}\|_{L^p(\mathbb{R}^2)} \leq C_1\big(\|qv_{xx} + xv_y\|_{L^p(\mathbb{R}^2)} + \|v\|_{L^p(\mathbb{R}^2)}\big). \tag{10}$$

Estimates as (10) have been deduced in [3] by corresponding parabolic estimates for $\mathscr{A} - \partial_t$, using that such operator is left invariant with respect to a suitable Lie group structure on $\mathbb{R}^{d+1}$ (see [17]). We also mention [4] which contains a generalization of [3] when $Q$ may also depend on $x$ and [23] where the results in [3] are used to study well-posedness of related SDEs. Finally, we point out that in the degenerate hypoelliptic case considered in [3] it is not clear how to prove the precise dependence of the constant appearing in the a-priori $L^p$-estimates from the matrix $Q$.

We denote by $|\cdot|$ the usual Euclidean norm in any $\mathbb{R}^k$, $k \geq 1$. Moreover, $\langle \cdot, \cdot \rangle$ indicates the usual inner product in $\mathbb{R}^k$.

We denote by $L^p(\mathbb{R}^k)$, $k \geq 1$, $1 < p < \infty$ the usual Banach spaces of measurable real functions $f$ such that $|f|^p$ is integrable on $\mathbb{R}^k$ with respect to the Lebesgue measure. The space of all $L^p$-functions $f : \mathbb{R}^k \to \mathbb{R}^j$ with $j > 1$ is indicated with $L^p(\mathbb{R}^k; \mathbb{R}^j)$. Let $H$ be an open set in $\mathbb{R}^k$; $C_0^\infty(H)$ stands for the vector space of all real $C^\infty$-functions $f : H \to \mathbb{R}$ which have compact support.

Let $d \geq 1$. Given a regular function $u : \mathbb{R}^{d+1} \to \mathbb{R}$, we denote by $D_x^2 u(t,x)$ the $d \times d$ Hessian matrix of $u$ with respect to the spatial variables at $(t,x) \in \mathbb{R}^{d+1}$, i.e., $D_x^2 u(t,x) = (u_{x_i x_j}(t,x))_{i,j=1,\dots,d}$. Similarly we define the gradient $D_x u(t,x) \in \mathbb{R}^d$ with respect to the spatial variables.

Given a real $k \times k$ matrix $A$, $\|A\|$ denotes its operator norm and $Tr(A)$ its trace.

Let us recall the notion of *Gaussian measure* (see, for instance, Sect. 1.2 in [2] or Chap. 1 in [7] for more details). Let $d \geq 1$. Given a symmetric non-negative definite $d \times d$ matrix $Q$, the symmetric Gaussian measure $N(0,Q)$ is the unique Borel probability measure on $\mathbb{R}^d$ such that its Fourier transform is

$$\int_{\mathbb{R}^d} e^{i\langle x,\xi\rangle}\, N(0,Q)(dx) = e^{-\langle \xi, Q\xi\rangle}, \quad \xi \in \mathbb{R}^d; \tag{11}$$

$N(0,Q)$ is the Gaussian measure with mean 0 and covariance matrix $2Q$. If in addition $Q$ is positive definite than $N(0,Q)$ has the following density $f$ with respect to the $d$-dimensional Lebesgue measure

$$f(x) = \frac{1}{\sqrt{(4\pi)^d \det(Q)}} e^{-\frac{1}{4}\langle Q^{-1}x, x\rangle}, \quad x \in \mathbb{R}^d. \tag{12}$$

Given two Borel probability measures $\mu_1$ and $\mu_2$ on $\mathbb{R}^d$ the convolution $\mu_1 * \mu_2$ is the Borel probability measure defined as

$$\mu_1 * \mu_2(B) = \int_{\mathbb{R}^d}\int_{\mathbb{R}^d} 1_B(x+y)\mu_1(dx)\mu_2(dy) = \int_{\mathbb{R}^d}\mu_1(dx)\int_{\mathbb{R}^d} 1_B(x+y)\mu_2(dy),$$

for any Borel set $B \subset \mathbb{R}^d$. Here $1_B$ is the indicator function of $B$ (i.e., $1_B(x) = 1$ if $x \in B$ and $1_B(x) = 0$ if $x \notin B$). It can be easily verified that

$$N(0,Q) * N(0,R) = N(0,Q+R), \tag{13}$$

where $Q + R$ is the sum of the two symmetric non-negative definite matrices $Q$ and $R$.

## 2 A-priori $L^p$-Estimates

In this section we consider parabolic equations like (1).

We always assume that the coefficients $c_{ij}(t)$ *of the symmetric* $d \times d$ *matrix* $c(t)$ *appearing in* (1) *are (Borel) measurable and locally bounded on* $\mathbb{R}$ *and, moreover, that* $\langle c(t)\xi, \xi\rangle \geq 0$, $t \in \mathbb{R}$, $\xi \in \mathbb{R}^d$. Moreover, we will consider the symmetric non-negative $d \times d$ matrix

$$C_{sr} = \int_s^r c(t)dt, \quad s \leq r, \quad s,r \in \mathbb{R}. \tag{14}$$

We start with a simple representation formula for solutions to Eq. (1). This formula is usually obtained assuming that $c(t)$ is uniformly positive. However there are no difficulties to prove it even in the present case when $c(t)$ is only non-negative definite.

**Proposition 2.1** *Let* $u \in C_0^\infty(\mathbb{R}^{d+1})$ *be a solution to* (1). *Then we have, for* $(s,x) \in \mathbb{R}^{d+1}$,

$$u(s,x) = -\int_s^\infty dr \int_{\mathbb{R}^d} f(r, x+y)N(0,C_{sr})(dy). \tag{15}$$

*Proof* Let us denote by $\hat{u}(t,\cdot)$ the Fourier transform of $u(t,\cdot)$ in the space variable $x$. Applying such partial Fourier transform to both sides of (1) we obtain

$$\hat{u}_t(s,\xi) - \sum_{i,j=1}^d c_{ij}(s)\xi_i\xi_j\hat{u}(s,\xi) = \hat{f}(s,\xi),$$

i.e., we have

$$\hat{u}(s,\xi) = -\int_s^\infty e^{-\langle C_{sr}\xi,\xi\rangle}\hat{f}(r,\xi)dr, \quad (s,\xi) \in \mathbb{R}^{d+1}. \tag{16}$$

It follows that

$$\hat{u}(s,\xi) = -\int_s^\infty \Big(\int_{\mathbb{R}^d} e^{i\langle x,\xi\rangle} N(0, C_{sr})(dx)\Big)\hat{f}(r,\xi)dr.$$

By some straightforward computations, using also the uniqueness property of the Fourier transform, we get (15).

Alternatively, starting from (16) one can directly follow the computations of pages 48 in [15] and obtain (15). These computations use that there exists $\epsilon > 0$ such that $\langle c(t)\xi,\xi\rangle \ge \epsilon|\xi|^2$, $\xi \in \mathbb{R}^d$. We write, for $\epsilon > 0$, using the Laplace operator,

$$u_t(t,x) + \sum_{i,j=1}^d c_{ij}(t) u_{x_i x_j}(t,x) + \epsilon \Delta u(t,x) = f(t,x) + \epsilon \Delta u(t,x),$$

$(t,x) \in \mathbb{R}^{d+1}$; since $c(t) + \epsilon I$ is uniformly positive, following [15] we find

$$u(s,x) = -\int_s^\infty dr \int_{\mathbb{R}^d} f(r, y+x) N(0, C_{sr} + \epsilon(r-s)I)(dy)$$
$$-\epsilon \int_s^\infty dr \int_{\mathbb{R}^d} \Delta u(r, y+x) N(0, C_{sr} + \epsilon(r-s)I)(dy).$$

Using also (13) we get

$$u(s,x) = -\int_s^\infty dr \int_{\mathbb{R}^d} N(0,(r-s)I)(dz) \int_{\mathbb{R}^d} f(r, x+y+\sqrt{\epsilon}\, z) N(0, C_{sr})(dy)$$
$$-\epsilon \int_s^\infty dr \int_{\mathbb{R}^d} N(0,(r-s)I)(dz) \int_{\mathbb{R}^d} \Delta u(r, x+y+\sqrt{\epsilon}\, z) N(0, C_{sr})(dy).$$

Now we can pass to the limit as $\epsilon \to 0^+$ by the Lebesgue theorem and get (15). □

The next assumption is a slight generalization of the usual parabolicity condition which corresponds to the case $p_0 = d$.

**Hypothesis 2.1** *The coefficients $c_{ij}$ are locally bounded on $\mathbb{R}$ and the matrix $c(t) = \big(c_{ij}(t)\big)$ is symmetric non-negative definite, $t \in \mathbb{R}$. In addition, there exists an integer $p_0$, $1 \le p_0 \le d$, and $\lambda \in (0,\infty)$ such that*

$$\langle c(t)\xi,\xi\rangle = \sum_{i,j=1}^d c_{ij}(t)\xi_i\xi_j \ge \lambda \sum_{j=1}^{p_0} \xi_j^2, \quad t \in \mathbb{R},\ \xi \in \mathbb{R}^d. \tag{17}$$

A possible generalization of this hypothesis is given in Remark 2.1. Note that if we introduce the orthogonal projection

$$I_0 : \mathbb{R}^d \to F_{p_0}, \tag{18}$$

where $F_{p_0}$ is the subspace generated by $\{e_1, \dots, e_{p_0}\}$ (here $\{e_i\}_{i=1,\dots,d}$ denotes the canonical basis in $\mathbb{R}^d$) then (19) can be rewritten as

$$\langle c(t)\xi, \xi\rangle \geq \lambda |I_0 \xi|^2, \quad t \in \mathbb{R},\ \xi \in \mathbb{R}^d. \tag{19}$$

**Lemma 2.1** *Let $g : \mathbb{R}^{d+1} \to \mathbb{R}$ be Borel, bounded, with compact support and such that $g(t, \cdot) \in C_0^\infty(\mathbb{R}^d)$, $t \in \mathbb{R}$. Fix $i, j \in \{1, \dots, p_0\}$ and consider*

$$w_{ij}(s,x) = -\int_s^\infty dr \int_{\mathbb{R}^d} g_{x_i x_j}(r, x+y) N(0, I_0(r-s))(dy),\ \ (s,x) \in \mathbb{R}^{d+1},$$

*where $I_0$ is defined in* (18). *For any $p \in (1, \infty)$, there exists $M_0 = M_0(d, p, p_0) > 0$, such that*

$$\|w_{ij}\|_{L^p(\mathbb{R}^{d+1})} \leq M_0 \|g\|_{L^p(\mathbb{R}^{d+1})}. \tag{20}$$

*Proof* If $p_0 = d$ the estimate is classical. In such case we are dealing with the heat equation

$$\partial_t u + \Delta u = g$$

on $\mathbb{R}^{d+1}$ and $w_{ij}$ coincides with the second partial derivative with respect to $x_i$ and $x_j$ of the heat potential applied to $g$ (see, for instance, page 288 in [16] or Appendix in [24]). If $p_0 < d$ we write $x = (x', x'')$ for $x \in \mathbb{R}^d$, where $x' \in \mathbb{R}^{p_0}$ and $x'' \in \mathbb{R}^{d-p_0}$. We get

$$w_{ij}(s, x', x'') = -\int_s^\infty dr \int_{\mathbb{R}^{p_0}} g_{x_i x_j}(r, x' + y', x'') N(0, I_{p_0}(r-s))(dy'),$$

where $I_{p_0}$ is the identity matrix in $\mathbb{R}^{p_0}$. Let us fix $x'' \in \mathbb{R}^{d-p_0}$ and consider the function $l(t, x') = g(t, x', x'')$ defined on $\mathbb{R} \times \mathbb{R}^{p_0}$. By classical estimates for the heat equation $\partial_t u + \Delta u = l$ on $\mathbb{R}^{p_0+1}$ we obtain

$$\int_{\mathbb{R}^{p_0+1}} |w_{ij}(s, x', x'')|^p ds dx' \leq M_0^p \int_{\mathbb{R}^{p_0+1}} |g(s, x', x'')|^p ds dx'.$$

Integrating with respect to $x''$ we get the assertion. □

In the sequel we also consider the differential operator $L$

$$Lu(t,x) = \sum_{i,j=1}^{d} c_{ij}(t) u_{x_i x_j}(t,x), \quad (t,x) \in \mathbb{R}^{d+1}, \quad u \in C_0^\infty(\mathbb{R}^{d+1}). \tag{21}$$

The next regularity result when $p_0 = d$ follows by a general result given in Theorem 2.2 of [13] (cf. Remark 2.5 in [14]).

In the next two sections we provide the proof. First we give a direct and self-contained proof in the case $p = 2$ by Fourier transform techniques (see Sect. 2.1). Then in Sect. 2.2 we consider the general case. The proof for $1 < p < \infty$ is inspired by the one of Theorem 2.2 in [13] and uses also a probabilistic argument. This argument is used to "decompose" a suitable Gaussian measure in order to apply successfully the Fubini theorem (cf. (30) and (31)).

We stress again that in the case of $d = p_0$, usually, the next result is stated under the stronger assumption that (17) holds with $\lambda = 1$ and also that $c_{ij}$ are *bounded*, i.e., assuming (3) with $\lambda = 1$ and $\Lambda \geq 1$ (see, for instance, Appendix in [16, 24]).

**Theorem 2.1** *Assume Hypothesis 2.1 with* $\lambda = 1$ *in* (17). *Then, for* $p \in (1,\infty)$, *there exists a constant* $M_0 = M_0(d,p,p_0)$ *such that, for any* $u \in C_0^\infty(\mathbb{R}^{d+1})$, $i,j = 1,\ldots,p_0$, *we have*

$$\|u_{x_i x_j}\|_{L^p(\mathbb{R}^{d+1})} \leq M_0 \|u_t + Lu\|_{L^p(\mathbb{R}^{d+1})}. \tag{22}$$

As a consequence of the previous result we obtain

**Corollary 2.1** *Assume Hypothesis 2.1. Then, for any* $u \in C_0^\infty(\mathbb{R}^{d+1})$, $p \in (1,\infty)$, $i,j = 1,\ldots,p_0$, *we have (see* (21)*)*

$$\|u_{x_i x_j}\|_{L^p(\mathbb{R}^{d+1})} \leq \frac{M_0}{\lambda} \|u_t + Lu\|_{L^p(\mathbb{R}^{d+1})}, \tag{23}$$

*where* $M_0 = M_0(d,p,p_0)$ *is the same constant appearing in* (22).

*Proof* Let us define $w(t,y) = u(t,\sqrt{\lambda}\, y)$. Set $f = u_t + Lu$; since $u(t,x) = w(t,\frac{x}{\sqrt{\lambda}})$, we find

$$f(t,\sqrt{\lambda}\, y) = w_t(t,y) + \frac{1}{\lambda} Lw(t,y).$$

Now the matrix $(\frac{1}{\lambda} c_{ij})$ satisfies $\frac{1}{\lambda}\sum_{i,j=1}^{d} c_{ij}(t)\xi_i\xi_j \geq \sum_{j=1}^{p_0} \xi_j^2$, $t \in \mathbb{R}$, $\xi \in \mathbb{R}^d$. Applying Theorem 2.1 to $w$ we find

$$\|w_{x_i,x_j}\|_{L^p} \leq M_0 \lambda^{-\frac{d}{2p}} \|f\|_{L^p}$$

and so

$$\lambda^{1-\frac{d}{2p}} \|u_{x_i x_j}\|_{L^p} \le M_0 \lambda^{-\frac{d}{2p}} \|f\|_{L^p}$$

which is the assertion. □

*Example 2.1* Equation (6) verifies the assumptions of Corollary 2.1 with $p_0 = 1$ and $\lambda = 3/4$ since

$$\sum_{i,j=1}^{2} c_{ij}(t)\xi_i\xi_j = \xi_1^2 + t\xi_1\xi_2 + t^2\xi_2^2 \ge \frac{3}{4}\xi_1^2, \quad (t, \xi_1, \xi_2) \in \mathbb{R}^3.$$

Hence there exists $M_0 > 0$ such that if $u \in C_0^\infty(\mathbb{R}^3)$ solves (6) then

$$\|u_{xx}\|_{L^p(\mathbb{R}^3)} \le \frac{M_0}{\lambda} \|f\|_{L^p(\mathbb{R}^3)}.$$

*Remark 2.1* One can easily generalize Hypothesis 2.1 as follows:

the coefficients $c_{ij}$ are locally bounded on $\mathbb{R}$ and, moreover, there exists an orthogonal projection $I_0 : \mathbb{R}^d \to \mathbb{R}^d$ and $\lambda > 0$ such that, for any $t \in \mathbb{R}$,

$$\langle c(t)\xi, \xi\rangle \ge \lambda |I_0\xi|^2, \quad \xi \in \mathbb{R}^d. \tag{24}$$

Theorem 2.1 and Corollary 2.1 continue to hold under this assumption.

Indeed if (24) holds then by a suitable linear change of variables in Eq. (1) we may assume that $I_0(\mathbb{R}^d)$ is the linear subspace generated by $\{e_1, \ldots, e_{p_0}\}$ for some $p_0$ with $1 \le p \le d$ and so apply Theorem 2.1.

Under hypothesis (24) assertion (22) in Theorem 2.1 becomes

$$\|\langle D_x^2 u(\cdot)h, k\rangle\|_{L^p(\mathbb{R}^{d+1})} \le M_0 \|u_t + Lu\|_{L^p(\mathbb{R}^{d+1})},$$

where $h, k \in I_0(\mathbb{R}^d)$.

## 2.1 *Proof of Theorem 2.1 When p = 2*

This proof is inspired by the one of Lemma A.2.2 in [24]. Note that such lemma has $p_0 = d$ and, moreover, it assumes the stronger condition (3). In Lemma A.2.2 the constant $M_0$ appearing in (22) is $2\sqrt{\Lambda}$.

We start from (16) with

$$f = u_t + Lu.$$

Recall that for $g : \mathbb{R}^{d+1} \to \mathbb{R}$, $\hat{g}(s,\xi)$ denotes the Fourier transform of $g(s,\cdot)$ with respect to the $x$-variable ($s \in \mathbb{R}$, $\xi \in \mathbb{R}^d$) assuming that $g(s,\cdot) \in L^1(\mathbb{R}^d)$.

Let us fix $s \in \mathbb{R}$. Let $i,j = 1,\dots,p_0$. We easily compute the Fourier transform of $u_{x_i x_j}(s,\cdot)$ (the matrix $C_{sr}$ is defined in (14)):

$$\hat{u}_{x_i x_j}(s,\xi) = -\xi_i \xi_j \, \hat{u}(s,\xi) = \xi_i \xi_j \int_s^\infty e^{-\langle C_{sr}\xi,\xi\rangle} \hat{f}(r,\xi)dr, \quad \xi \in \mathbb{R}^d.$$

Since $|I_0\xi|^2 = \sum_{i=1}^{p_0} |\xi_i|^2$, we get

$$2|\hat{u}_{x_i x_j}(s,\xi)| \le |I_0\xi|^2 \int_s^\infty e^{-\left(\langle C_{0r}\xi,\xi\rangle - \langle C_{0s}\xi,\xi\rangle\right)} |\hat{f}(r,\xi)|dr = G_\xi(s).$$

Now we fix $\xi \in \mathbb{R}^d$, such that $|I_0\xi| \neq 0$, and define

$$g_\xi(r) = \langle C_{0r}\xi,\xi\rangle = \int_0^r \langle c(p)\xi,\xi\rangle dp, \quad r \in \mathbb{R}.$$

Changing variable $t = g_\xi(r)$, we get

$$G_\xi(s) = |I_0\xi|^2 \int_{g_\xi(s)}^\infty e^{(g_\xi(s)-t)} \, |\hat{f}(g_\xi^{-1}(t),\xi)| \frac{1}{\langle c(g_\xi^{-1}(t))\xi,\xi\rangle} dt.$$

Let us introduce $\varphi(t) = e^t \cdot 1_{(-\infty,0)}(t)$, $t \in \mathbb{R}$, and

$$F_\xi(t) = |I_0\xi|^2 \, |\hat{f}(g_\xi^{-1}(t),\xi)| \frac{1}{\langle c(g_\xi^{-1}(t))\xi,\xi\rangle}.$$

Using the standard convolution for real functions defined on $\mathbb{R}$ we find

$$G_\xi(s) = (\varphi * F_\xi)\,(g_\xi(s)).$$

Therefore (recall (19) with $\lambda = 1$)

$$\|G_\xi\|^2_{L^2(\mathbb{R})} = \int_{\mathbb{R}} |(\varphi * F_\xi)(t)|^2 \frac{1}{\langle c(g_\xi^{-1}(t))\xi,\xi\rangle} dt \le \frac{1}{|I_0\xi|^2} \|\varphi * F_\xi\|^2_{L^2(\mathbb{R})} \tag{25}$$

which implies $\|G_\xi\|_{L^2(\mathbb{R})} \le \frac{1}{|I_0\xi|} \|\varphi * F_\xi\|_{L^2(\mathbb{R})}$. On the other hand, using the Young inequality, we find, for any $\xi \in \mathbb{R}^d$ with $|I_0\xi| \neq 0$,

$$\begin{aligned}\|\varphi * F_\xi\|_{L^2(\mathbb{R})} &\le \|\varphi\|_{L^1(\mathbb{R})} \, \|F_\xi\|_{L^2(\mathbb{R})} = \|F_\xi\|_{L^2(\mathbb{R})} \\ &= |I_0\xi|^2 \Big( \int_{\mathbb{R}} |\hat{f}(g_\xi^{-1}(t),\xi)|^2 \frac{1}{(\langle c(g_\xi^{-1}(t))\xi,\xi\rangle)^2} dt \Big)^{1/2}\end{aligned}$$

$$= |I_0\xi|^2\Big(\int_{\mathbb{R}} |\hat{f}(r,\xi)|^2 \frac{1}{(\langle c(r)\xi,\xi\rangle)^2}\langle c(r)\xi,\xi\rangle dr\Big)^{1/2}$$
$$\leq \frac{|I_0\xi|^2}{|I_0\xi|}\Big(\int_{\mathbb{R}} |\hat{f}(r,\xi)|^2\Big)^{1/2} = |I_0\xi|\cdot \|\hat{f}(\cdot,\xi)\|_{L^2(\mathbb{R})}.$$

Using also (25) we obtain, for any $\xi \in \mathbb{R}^d$, $|I_0\xi| \neq 0$,

$$2\|\hat{u}_{x_ix_j}(\cdot,\xi)\|_{L^2(\mathbb{R})} \leq \|G_\xi\|_{L^2(\mathbb{R})} \leq \|\hat{f}(\cdot,\xi)\|_{L^2(\mathbb{R})}.$$

From the previous inequality, integrating with respect to $\xi$ over $\mathbb{R}^d$ we find

$$4\int_{\mathbb{R}} ds \int_{\mathbb{R}^d} |\hat{u}_{x_ix_j}(s,\xi)|^2 d\xi \leq \int_{\mathbb{R}} ds \int_{\mathbb{R}^d} |\hat{f}(s,\xi)|^2 d\xi.$$

By using the Plancherel theorem in $L^2(\mathbb{R}^d)$ we easily obtain (22) with $M_0 = 1/2$. The proof is complete. □

## 2.2 *Proof of Theorem 2.1 When* $1 < p < \infty$

The proof uses the concept of stochastic integral in a crucial point (see (30) and (31)). Before starting the proof we collect some basic properties of the stochastic integral with respect to the Wiener process which are needed (see, for instance, Chap. 4 in [1] or Sect. 4.3 in [24] for more details).

### 2.2.1 The Stochastic Integral

Let $W = (W_t)_{t\geq 0}$ be a standard $d$-dimensional Wiener process defined on a probability space $(\Omega, \mathscr{F}, \mathbb{P})$. Denote by $\mathbb{E}$ the expectation with respect to $\mathbb{P}$.

Consider a function $F \in L^2([a,b];\mathbb{R}^d \otimes \mathbb{R}^d)$ (here $0 \leq a \leq b$ and $\mathbb{R}^d \otimes \mathbb{R}^d$ denotes the space of all real $d \times d$-matrices).

Let $(\pi_n)$ be any sequence of partitions of $[a,b]$ such that $|\pi_n| \to 0$ as $n \to \infty$ (given a partition $\pi = \{t_0 = a, \dots, t_N = b\}$ we set $|\pi| = \sup_{t_k,t_{k+1}\in\pi} |t_{k+1} - t_k|$). One defines the stochastic integral $\int_a^b F(s)dW_s$ as the limit in $L^2(\Omega,\mathbb{P};\mathbb{R}^d)$ of

$$J_n = \sum_{t_k^n,t_{k+1}^n\in\pi_n} F(t_k^n)\,(W_{t_{k+1}^n} - W_{t_k^n}),$$

as $n \to \infty$ (recall that the previous formula means $J_n(\omega) = \sum_{t_k^n,t_{k+1}^n\in\pi_n} F(t_k^n)$ $(W_{t_{k+1}^n}(\omega) - W_{t_k^n}(\omega))$, for any $\omega \in \Omega$). One can prove that the previous limit is independent of the choice of $(\pi_n)$. Moreover, we have, $\mathbb{P}$-a.s.,

$$\int_a^b F(s)dW_s = \int_0^b F(s)dW_s - \int_0^a F(s)dW_s. \qquad (26)$$

Set $\Gamma_{ab} = \int_a^b F(s)F^*(s)ds$ where $F^*(s)$ denotes the adjoint matrix of $F(s)$. Clearly, $\Gamma_{ab}$ is a $d \times d$ symmetric non-negative definite matrix. Moreover, we have (see, for instance, page 77 in [1])

$$\mathbb{E}\big[e^{i\sqrt{2}\langle \int_a^b F(s)dW_s, \xi\rangle}\big] = \int_\Omega e^{i\sqrt{2}\langle \left(\int_a^b F(s)dW_s\right)(\omega)\,,\,\xi\rangle}\,\mathbb{P}(d\omega) \qquad (27)$$

$$= \int_{\mathbb{R}^d} e^{i\langle x,\xi\rangle}\, N(0, \Gamma_{ab})(dx) = e^{-\langle \xi, \Gamma_{ab}\xi\rangle}, \quad \xi \in \mathbb{R}^d.$$

Formula (27) is equivalent to require that for any Borel and bounded $f : \mathbb{R}^d \to \mathbb{R}$,

$$\mathbb{E}\Big[f\Big(\sqrt{2}\int_a^b F(s)dW_s\Big)\Big] = \int_{\mathbb{R}^d} f(y)N(0, \Gamma_{ab})(dy). \qquad (28)$$

Equivalently, one can say that the distribution (or image measure) of $\sqrt{2}\int_a^b F(s)dW_s$ is $N(0, \Gamma_{ab})$.

### 2.2.2 Proof of the Theorem

It is convenient to suppose that $u(t, \cdot) = 0$ if $t \le 0$ so that $u \in C_0^\infty([0, \infty) \times \mathbb{R}^d)$.

Indeed if $u(t, \cdot) = 0$, $t \le T$, for some $T \in \mathbb{R}$, then we can introduce $v(t, x) = u(t + T, x)$ which belongs to $u \in C_0^\infty([0, \infty) \times \mathbb{R}^d)$; from the a-priori estimate for $v_{x_ix_j}$ it follows (22) since $\|v_{x_ix_j}\|_{L^p(\mathbb{R}^{d+1})} = \|u_{x_ix_j}\|_{L^p(\mathbb{R}^{d+1})}$.

We know that, for $s \ge 0$, $x \in \mathbb{R}^d$,

$$u(s, x) = -\int_s^\infty dr \int_{\mathbb{R}^d} f(r, x + y)N(0, C_{sr})(dy),$$

where $f = u_t + Lu$ is bounded, with compact support on $\mathbb{R}^{d+1}$ and such that $f(t, \cdot) \in C_0^\infty(\mathbb{R}^d)$, $t \ge 0$. Let us fix $i, j \in \{1, \dots, p_0\}$.

Differentiating under the integral sign it is not difficult to prove that

$$u_{x_ix_j}(s, x) = -\int_s^\infty dr \int_{\mathbb{R}^d} f_{x_ix_j}(r, x + y)N(0, C_{sr})(dy).$$

Let us fix $s$ and $r$, $0 \le s \le r$, and consider

$$C_{sr} = A_{sr} + (r-s)I_0, \quad \text{where } A_{sr} = \int_s^r (c(t) - I_0)dt.$$

By (13) we know that $N(0, C_{sr}) = N(0, A_{sr}) * N(0, (r-s)I_0)$ and so

$$\int_{\mathbb{R}^d} f_{x_i x_j}(r, x+y) N(0, C_{sr})(dy) \tag{29}$$

$$= \int_{\mathbb{R}^d} N(0, A_{sr})(dz) \int_{\mathbb{R}^d} f_{x_i x_j}(r, x+y+z) N(0, (r-s)I_0)(dy).$$

Now we introduce a standard $d$-dimensional Wiener process $W = (W_t)_{t \ge 0}$ on a probability space $(\Omega, \mathscr{F}.\mathbb{P})$ (see Sect. 2.2.1). Consider the symmetric $d \times d$ square root $\sqrt{c(t) - I_0}$ of $c(t) - I_0$ and define the stochastic integral

$$\Lambda_{sr} = \sqrt{2} \int_s^r \sqrt{c(t) - I_0}\, dW_t.$$

By (26) we know that

$$\Lambda_{sr} = b_r - b_s, \quad \text{where } b_t = \sqrt{2} \int_0^t \sqrt{c(p) - I_0}\, dW_p,$$

$t \ge 0$, and $b_t = 0$ if $t \le 0$. Moreover (cf. (28)) for any Borel and bounded $g : \mathbb{R}^d \to \mathbb{R}$, we have

$$\mathbb{E}[g(b_r - b_s)] = \int_\Omega g\big(b_r(\omega) - b_s(\omega)\big)\mathbb{P}(d\omega) = \int_{\mathbb{R}^d} g(y) N(0, A_{sr})(dy). \tag{30}$$

Using this fact and the Fubini theorem we get from (29)

$$\int_{\mathbb{R}^d} f_{x_i x_j}(r, x+y) N(0, C_{sr})(dy)$$

$$= \mathbb{E}\Big[\int_{\mathbb{R}^d} f_{x_i x_j}(r, x+y+\Lambda_{rs})\, N(0, (r-s)I_0)(dy)\Big]$$

$$= \mathbb{E}\Big[\int_{\mathbb{R}^d} f_{x_i x_j}(r, x+y+b_r - b_s)\, N(0, (r-s)I_0)(dy)\Big]. \tag{31}$$

Therefore we find

$$u_{x_i x_j}(s, x) = -\mathbb{E}\Big[\int_s^\infty dr \int_{\mathbb{R}^d} f_{x_i x_j}(r, x+y+b_r - b_s) N(0, (r-s)I_0)(dy)\Big]. \tag{32}$$

Now we estimate the $L^p$-norm of $u_{x_ix_j}$. To simplify the notation in the sequel we set $N\big(0,(r-s)I_0\big) = \mu_{sr}$. Using the Jensen inequality and the Fubini theorem we get

$$\int_{\mathbb{R}_+} ds \int_{\mathbb{R}^d} |u_{x_ix_j}(s,x)|^p dx$$
$$= \int_{\mathbb{R}_+} ds \int_{\mathbb{R}^d} \Big| \mathbb{E}\Big[ \int_s^\infty dr \int_{\mathbb{R}^d} f_{x_ix_j}(r, x+y+b_r-b_s)\mu_{sr}(dy) \Big] \Big|^p dx$$
$$\le \mathbb{E}\Big[ \int_{\mathbb{R}_+} ds \int_{\mathbb{R}^d} \Big| \int_s^\infty dr \int_{\mathbb{R}^d} f_{x_ix_j}(r, x+y+b_r-b_s)\mu_{sr}(dy) \Big|^p dx \Big].$$

Now in the last line of the previous formula we change variable in the integral over $\mathbb{R}^d$ with respect to the $x$-variable; we obtain

$$\int_{\mathbb{R}_+} ds \int_{\mathbb{R}^d} |u_{x_ix_j}(s,x)|^p dx \tag{33}$$
$$\le \mathbb{E}\Big[ \int_{\mathbb{R}_+} ds \int_{\mathbb{R}^d} \Big| \int_s^\infty dr \int_{\mathbb{R}^d} f_{x_ix_j}(r, z+y+b_r)\mu_{sr}(dy) \Big|^p dz \Big].$$

To estimate the last term we fix $\omega \in \Omega$ and consider the function

$$g_\omega(t,x) = f(t, x+b_t(\omega)), \quad (t,x) \in \mathbb{R}^{d+1}.$$

The function $g_\omega$ is bounded, with compact support on $\mathbb{R}^{d+1}$ and such that $g_\omega(t,\cdot) \in C_0^\infty(\mathbb{R}^d)$, $t \in \mathbb{R}$.

By Lemma 2.1 we know that there exists $M_0 = M_0(d,p,p_0) > 0$ such that, for any $\omega \in \Omega$,

$$\int_{\mathbb{R}_+} ds \int_{\mathbb{R}^d} \Big| \int_s^\infty dr \int_{\mathbb{R}^d} f_{x_ix_j}\big(r, z+y+b_r(\omega)\big)\mu_{sr}(dy) \Big|^p dz$$
$$= \int_{\mathbb{R}_+} ds \int_{\mathbb{R}^d} \Big| \int_s^\infty dr \int_{\mathbb{R}^d} \partial^2_{x_ix_j} g_\omega\big(r, z+y\big)\mu_{sr}(dy) \Big|^p dz \le M_0^p \, \|g_\omega\|^p_{L^p}.$$

Using also (33) we find

$$\int_{\mathbb{R}_+} ds \int_{\mathbb{R}^d} |u_{x_ix_j}(s,x)|^p dx \le M_0^p \, \mathbb{E}\Big[ \int_{\mathbb{R}} ds \int_{\mathbb{R}^d} |g_\omega(s,x)|^p dx \Big]$$
$$= M_0^p \, \mathbb{E}\Big[ \int_{\mathbb{R}} ds \int_{\mathbb{R}^d} |f(s, x+b_s)|^p dx \Big]$$
$$= M_0^p \int_{\mathbb{R}} ds \int_{\mathbb{R}^d} |f(s,z)|^p dz.$$

The proof is complete. □

## 3 $L^p$-Estimates Involving Ornstein-Uhlenbeck Operators

Let $A = (a_{ij})$ be a given real $d \times d$-matrix. We consider the following Ornstein-Uhlenbeck type operator

$$L_0 u(t,x) = \sum_{i,j=1}^{d} c_{ij}(t) u_{x_i x_j}(t,x) + \sum_{i,j=1}^{d} a_{ij} x_j \, u_{x_i}(t,x)$$

$$= \mathrm{Tr}(c(t) D_x^2 u(t,x)) + \langle Ax, D_x u(t,x) \rangle,$$

$(t,x) \in \mathbb{R}^{d+1}$, $u \in C_0^\infty(\mathbb{R}^{d+1})$. This is a kind of perturbation of $L$ given in (21) by the first order term $\langle Ax, D_x u(t,x) \rangle$ which has linear coefficients.

We will extend Corollary 2.1 to cover the parabolic equation

$$u_t(t,x) + L_0 u(t,x) = f(t,x) \tag{34}$$

on $\mathbb{R}^{d+1}$. We will assume Hypothesis 2.1 and also

**Hypothesis 3.1** *Let $p_0$ as in Hypothesis 2.1. Define $F_{p_0} \simeq \mathbb{R}^{p_0}$ as the linear subspace generated by $\{e_1, \ldots, e_{p_0}\}$. Let $F^{p_0}$ be the linear subspace generated by $\{e_{p_0+1}, \ldots, e_d\}$ if $p_0 < d$ (when $p_0 = d$, $F^{p_0} = \{0\}$). We suppose that*

$$A(F_{p_0}) \subset F_{p_0}, \quad A(F^{p_0}) \subset F^{p_0}. \tag{35}$$

Recall that given a $d \times d$-matrix $B$, $\|B\|$ and $Tr(B)$ denote, respectively, the operator norm and the trace of $B$. In the next result we will use that there exists $\omega > 0$ and $\eta > 0$ such that

$$\|e^{tA}\| \le \eta e^{\omega|t|}, \quad t \in \mathbb{R}, \tag{36}$$

where $e^{tA}$ is the exponential matrix of $tA$. Note that the constant $M_0$ below is the same given in (22).

**Theorem 3.1** *Assume Hypotheses 2.1 and 3.1. Let $T > 0$ and set $S_T = (-T,T) \times \mathbb{R}^d$. Suppose that $u \in C_0^\infty(S_T)$. For any $p \in (1,\infty)$, $i,j = 1, \ldots, p_0$,*

$$\|u_{x_i x_j}\|_{L^p(\mathbb{R}^{d+1})} \le \frac{M_1(T)}{\lambda} \|u_t + L_0 u\|_{L^p(\mathbb{R}^{d+1})}; \quad M_1(T) = c(d) M_0 \eta^4 e^{4T\omega + \frac{2T}{p}|Tr(A)|}. \tag{37}$$

*Proof* We fix $T > 0$ and use a change of variable similar to that used in page 100 of [6]. Define $v(t,y) = u(t, e^{tA}y)$, $(t,y) \in \mathbb{R}^{d+1}$. We have $v \in C_0^\infty(\mathbb{R}^{d+1})$,

$u(t,x) = v(t, e^{-tA}x)$ and

$$\begin{aligned} &u_t(t,x) + L_0 u(t,x) \\ &= v_t(t, e^{-tA}x) - \langle D_y v(t, e^{-tA}x), Ae^{-tA}x\rangle + \mathrm{Tr}\big(e^{-tA}c(t)e^{-tA^*} D_y^2 v(t, e^{-tA}x)\big) \\ &\quad + \langle D_y v(t, e^{-tA}x), Ae^{-tA}x\rangle \\ &= v_t(t, e^{-tA}x) + \mathrm{Tr}\big(e^{-tA}c(t)e^{-tA^*} D_y^2 v(t, e^{-tA}x)\big). \end{aligned}$$

It follows that

$$u_t(t, e^{tA}y) + L_0 u(t, e^{tA}y) = v_t(t,y) + \mathrm{Tr}\big(e^{-tA}c(t)e^{-tA^*} D_y^2 v(t,y)\big). \tag{38}$$

Now we have to check Hypothesis 2.1. We first define $c_0(t)$, $t \in \mathbb{R}$,

$$c_0(t) = e^{-tA}c(t)e^{-tA^*}, \quad t \in [-T, T], \tag{39}$$

$$c_0(t) = e^{-TA}c(T)e^{-TA^*}, \ \ t \geq T, \ \ c_0(t) = e^{TA}c(-T)e^{TA^*}, \ \ t \leq -T.$$

Since $v \in C_0^\infty(S_T)$ we have on $\mathbb{R}^{d+1}$

$$v_t(t,y) + \mathrm{Tr}\big(e^{-tA}c(t)e^{-tA^*} D_y^2 v(t,y)\big) = v_t(t,y) + \mathrm{Tr}\big(c_0(t) D_y^2 v(t,y)\big)$$

and so it is enough to check that $c_0(t)$ verifies (19). Moreover, by (39) it is enough to verify (19) for $t \in [-T, T]$. We have

$$\langle c_0(t)\xi, \xi\rangle = \langle c(t)e^{-tA^*}\xi, e^{-tA^*}\xi\rangle \geq \lambda\, |I_0 e^{-tA^*}\xi|^2.$$

By (35) we deduce that $F_{p_0}$ and $F^{p_0}$ are both invariant for $A^*$. It follows easily that

$$I_0 e^{sA^*}\xi = e^{sA^*} I_0 \xi, \quad \xi \in \mathbb{R}^d, \ \ s \in \mathbb{R}. \tag{40}$$

Using this fact we find for $t \in [-T, T]$, $\xi \in \mathbb{R}^d$,

$$|I_0\xi|^2 = |I_0 e^{tA^*} e^{-tA^*}\xi|^2 = |e^{tA^*} I_0 e^{-tA^*}\xi|^2 \leq \eta^2 e^{2T\omega}\, |I_0 e^{-tA^*}\xi|^2$$

and so

$$\lambda |I_0\xi|^2 \leq \lambda \eta^2 e^{2T\omega}\, |I_0 e^{-tA^*}\xi|^2 \leq \eta^2 e^{2T\omega}\, \langle c_0(t)\xi, \xi\rangle,$$

which implies $\lambda\eta^{-2}e^{-2T\omega}\,|I_0\xi|^2 \le \langle c_0(t)\xi,\xi\rangle$. By Corollary 2.1 and (38) we get, for any $i,j = 1,\dots p_0$,

$$\|v_{y_iy_j}\|_{L^p} = \|\langle D_y^2 v(\cdot)e_i, e_j\rangle\|_{L^p} \le \frac{M_0\,\eta^2 e^{2T\omega}}{\lambda}\|v_t + \mathrm{Tr}(c_0(t)D_y^2 v)\|_{L^p} \tag{41}$$

$$= \frac{M_0\,\eta^2 e^{2T\omega}}{\lambda}\|u_t(\cdot, e^{\cdot A}\cdot) + L_0 u(\cdot, e^{\cdot A}\cdot)\|_{L^p} \le \frac{M_0\eta^2 e^{2T\omega}}{\lambda} e^{\frac{T}{p}|\mathrm{Tr}(A)|}\,\|u_t + L_0 u\|_{L^p}.$$

Note that

$$\langle D_y^2 v(t,y) I_0 e_i, I_0 e_j\rangle = \langle D_y^2 v(t,y)e_i, e_j\rangle = \langle e^{tA^*} D_x^2 u(t, e^{tA}y)e^{tA}e_i, e_j\rangle$$

and so $I_0 D_y^2 v(t,y) I_0 = e^{tA^*} I_0 D_x^2 u(t, e^{tA}y) I_0 e^{tA}$, $t \in \mathbb{R}$, $y \in \mathbb{R}^d$. Indicating by $\mathbb{R}^{p_0} \otimes \mathbb{R}^{p_0}$ the space of all real $p_0 \times p_0$-matrices, we find

$$\|I_0 D_y^2 v\, I_0\|_{L^p(\mathbb{R}^{d+1};\mathbb{R}^{p_0}\otimes\mathbb{R}^{p_0})} \ge e^{-\frac{T}{p}|\mathrm{Tr}(A)|}\,\|e^{\cdot A^*} I_0 D_x^2 u\, I_0 e^{\cdot A}\|_{L^p(\mathbb{R}^{d+1};\mathbb{R}^{p_0}\otimes\mathbb{R}^{p_0})}.$$

Since, for $(t,x) \in \mathbb{R}^{d+1}$,

$$\|I_0 D_x^2 u(t,x)\, I_0\| \le \eta^2 e^{2T\omega}\,\|e^{tA^*} I_0 D_x^2 u(t,x)\, I_0 e^{tA}\|$$

by (41) we deduce

$$\|I_0 D_x^2 u\, I_0\|_{L^p(\mathbb{R}^{d+1};\mathbb{R}^{p_0}\otimes\mathbb{R}^{p_0})} \le c(d)\,\frac{M_0}{\lambda}\,\eta^4 e^{4T\omega}\, e^{\frac{2T}{p}|\mathrm{Tr}(A)|}\,\|u_t + L_0 u\|_{L^p}$$

which gives (37). The proof is complete. □

*Example 3.1* The equation

$$u_t(t,x,y) + (1+e^t)u_{xx}(t,x,y) + t u_{xy}(t,x,y) + t^2 u_{yy}(t,x,y) + y u_y(t,x,y) = f(t,x,y), \tag{42}$$

$(t,x,y) \in \mathbb{R}^3$, verifies the assumptions of Theorem 3.1 with $p_0 = 1$ and so estimate (37) holds for $u_{xx}$.

*Remark 3.1* Assumption (35) does not hold for the degenerate hypoelliptic operators considered in [3]. To see this let us consider the following classical example of hypoelliptic operator (cf. [10, 12])

$$u_t(t,x,y) + u_{xx}(t,x,y) + x u_y(t,x,y) = f(t,x,y), \tag{43}$$

$(t,x,y) \in \mathbb{R}^3$. In this case $p_0 = 1$ and $A = \begin{pmatrix} 0 & 0 \\ 1 & 0 \end{pmatrix}$. It is clear that (35) does not hold in this case. Indeed we can not recover the $L^p$-estimates in [3].

As an application of the previous theorem we obtain elliptic estimates for non-degenerate Ornstein-Uhlenbeck operators $\mathscr{A}$. These estimates have been first proved in [20]. Differently with respect to [20] in the next result we can show the explicit dependence of the constant $C_1$ in (46) from the ellipticity constant $\lambda$.

Let

$$\mathscr{A} u(x) = \mathrm{Tr}(Q D^2 u(x)) + \langle Ax, Du(x) \rangle, \tag{44}$$

$x \in \mathbb{R}^d$, $u \in C_0^\infty(\mathbb{R}^d)$, where $A$ is a $d \times d$ matrix and $Q$ is a symmetric positive define $d \times d$-matrix such that

$$\langle Q\xi, \xi \rangle \geq \lambda |\xi|^2, \quad \xi \in \mathbb{R}^d, \tag{45}$$

for some $\lambda > 0$.

**Corollary 3.1** *Let us consider* (44) *under assumption* (45). *For any* $w \in C_0^\infty(\mathbb{R}^d)$, $p \in (1, \infty)$, $i, j = 1, \ldots, d$, *we have (the constant* $M_1(1)$ *is given in* (37)*)*

$$\|w_{x_i x_j}\|_{L^p(\mathbb{R}^d)} \leq \frac{c(p)\, M_1(1)}{\lambda} \big( \|\mathscr{A} w\|_{L^p(\mathbb{R}^d)} + \|w\|_{L^p(\mathbb{R}^d)} \big). \tag{46}$$

*Proof* We will deduce (46) from (37) in $S_1 = (-1, 1) \times \mathbb{R}^d$ with $p_0 = d$.

Let $\psi \in C_0^\infty(-1, 1)$ with $\int_{-1}^1 \psi(t) dt > 0$. We define, similarly to Sect. 1.3 of [3],

$$u(t, x) = \psi(t) w(x).$$

Since $u_t + L_0 u = \psi'(t) w(x) + \psi(t) \mathscr{A} w(x)$, applying (37) to $u$ we easily get (46). □

## References


1. Arnold, L.: Stochastic Differential Equations, Theory and Applications. Wiley, New York (1974)
2. Bogachev V. I.: Gaussian Measures. Mathematical Surveys and Monographs, vol. 62. American Mathematical Society, Providence (1998)
3. Bramanti, M., Cupini, G., Lanconelli, E., Priola, E.: Global $L^p$ estimates for degenerate Ornstein-Uhlenbeck operators. Math. Z. **266**, 789–816 (2010)
4. Bramanti, M., Cupini, G., Lanconelli, E., Priola, E.: Global $L^p$ estimates for degenerate Ornstein-Uhlenbeck operators with variable coefficients. Math. Nachr. **286**, 1087–1101 (2013)
5. Calderon, A.P., Zygmund, A.: On the existence of certain singular integrals. Acta Math. **88**, 85–139 (1952)
6. Da Prato, G., Lunardi, A.: On the Ornstein-Uhlenbeck operator in spaces of continuous functions. J. Funct. Anal. **131**, 94–114 (1995)
7. Da Prato, G., Zabczyk, J.: Second Order Partial Differential Equations in Hilbert Spaces. London Mathematical Society Lecture Note Series, vol. 293. Cambridge University Press, Cambridge (2002)

8. Fabes, E.B., Rivière, N.M.: Singular integrals with mixed homogeneity. Stud. Math. **27**, 19–38 (1966)
9. Farkas, A. Lunardi: Maximal regularity for Kolmogorov operators in $L^2$ spaces with respect to invariant measures. J. Math. Pures Appl. (9) **86**(4), 310–321 (2006)
10. Hörmander, L.: Hypoelliptic second order differential equations. Acta Math. **119**, 147–171 (1967)
11. Frank Jones, B., Jr.: A class of singular integrals. Am. J. Math. **86**, 441–462 (1964)
12. Kolmogoroff, A.: Zufällige Bewegungen (zur Theorie der Brownschen Bewegung). Ann. Math. **35**, 116–117 (1934)
13. Krylov, N.V.: A parabolic Littlewood-Paley inequality with applications to parabolic equations. Topol. Methods Nonlinear Anal. **4**, 355–364 (1994)
14. Krylov, N.V.: Parabolic equations in $L_p$-spaces with mixed norms. Algebra i Analiz. **14** 91–106 (2002, in Russian). English translation in St. Petersburg Math. J. **14**, 603–614 (2003)
15. Krylov, N.V.: Lectures on Elliptic and Parabolic Equations in Sobolev Spaces. American Mathematical Society, Providence (2008)
16. Ladyzhenskaya, O.A., Solonnikov, V.A., Ural'tseva, N.N.: Linear and Quasi-Linear Parabolic Equations. Nauka, Moscow (1967, in Russian). English translation: American Mathematical Society, Providence (1968)
17. Lanconelli, E., Polidoro, S.: On a class of hypoelliptic evolution operators. Partial differential equations, II (Turin, 1993). Rend. Sem. Mat. Univ. Politec. Torino **52**, 29–63 (1994)
18. Lanconelli, E., Pascucci, A., Polidoro, S.: Linear and nonlinear ultraparabolic equations of kolmogorov type arising in diffusion theory and in finance. In: Nonlinear Problems in Mathematical Physics and Related Topics. In honor of Professor O.A. Ladyzhenskaya, International Mathematical Series, vol. II, pp. 243–265. Kluwer, New York (2002)
19. Lieberman, G.M.: Second Order Parabolic Differential Equations. World Scientific, River Edge (1996)
20. Metafune, G., Prüss, J., Rhandi, A., Schnaubelt, R.: The domain of the Ornstein-Uhlenbeck operator on an $L^p$-space with invariant measure. Ann. Sc. Norm. Super. Pisa Cl. Sci. **1**, 471–485 (2002)
21. Ornstein, L.S., Uhlenbeck, G.E.: On the theory of the Brownian motion. Phys. Rev. **36**(3), 823–841 (1930). This paper is also contained In: Wax, N. (ed.) Selected Papers on Noise and Stochastic Processes. Dover, New York (2003)
22. Priola, E.: On weak uniqueness for some degenerate SDEs by global $L^p$ estimates. Potential Anal. **42**, 247–281 (2015). doi:10.1007/s11118-014-9432-7
23. Priola, E.: On weak uniqueness for some degenerate SDEs by global $L^p$ estimates. Potential Anal. (to appear). doi:10.1007/s11118-014-9432-7 [Preprint arXiv.org http://arxiv.org/abs/1305.7174]
24. Stroock, D.W., Varadhan, S.R.S.: Multidimensional Diffusion Processes. Grundlehren der Mathematischen Wissenschaften, vol. 233. Springer, Berlin (1979)

# Local Solvability of Nonsmooth Hörmander's Operators

**Marco Bramanti**

*Dedicated to Ermanno Lanconelli on the occasion of his 70th birthday*

**Abstract** This note describes the results of a joint research with L. Brandolini, M. Manfredini and M. Pedroni, contained in Bramanti et al. [Fundamental solutions and local solvability of nonsmooth Hörmander's operators. Mem. Am. Math. Soc., in press. http://arxiv.org/abs/1305.3398], with some background. We consider operators of the form $L = \sum_{i=1}^{n} X_i^2 + X_0$ in a bounded domain of $\mathbb{R}^p$ ($p \geq n+1$) where $X_0, X_1, \ldots, X_n$ are *nonsmooth* Hörmander's vector fields of step $r$, such that the highest order commutators are only $C^{1,\alpha}$. Applying Levi's parametrix method we construct a local fundamental solution $\gamma$ for $L$, provide growth estimates for $\gamma$ and its first and second order derivatives with respect to the vector fields and deduce the local solvability of $L$ in $C_X^{2,\beta}$ spaces (for any $\beta < \alpha$).

**Keywords** Fundamental solution • Hölder estimates • Nonsmooth Hörmander's vector fields • Solvability

**Mathematics Subject Classification:** 35A08, 35A17, 35H20

## 1 Introduction and Main Results

In the study of elliptic-parabolic degenerate partial differential operators, an important class is represented by *Hörmander's operators*

$$L = \sum_{i=1}^{n} X_i^2 + X_0$$

M. Bramanti (✉)
Dipartimento di Matematica, Politecnico di Milano, Milano, Italy
e-mail: marco.bramanti@polimi.it



G. Citti et al. (eds.), *Geometric Methods in PDE's*, Springer INdAM Series 13,
DOI 10.1007/978-3-319-02666-4_8

built on real smooth vector fields

$$X_i = \sum_{j=1}^{p} b_{ij}(x)\, \partial_{x_j}$$

$(i = 0, 1, 2, \ldots, n;\, n < p)$ defined in a domain $\Omega \subset \mathbb{R}^p$.

Let us assume for the moment that $b_{ij} \in C^\infty(\Omega)$ and the vector fields $X_i$ satisfy *Hörmander's condition*, i.e: if we define the commutator of two vector fields

$$[X, Y] = XY - YX,$$

then the system consisting in the $X_i$'s, their commutators, the commutators of the $X_i$'s with their commutators, and so on up to some step $r$, generates $\mathbb{R}^p$ at any point of $\Omega$.

A famous theorem by Hörmander [10] states that under these assumptions the operator $L$ is *hypoelliptic* in $\Omega$:

$$Lu = f \text{ in } \Omega \text{ (in distributional sense)}, A \subset \Omega, f \in C^\infty(A) \Longrightarrow u \in C^\infty(A).$$

After this result, several fundamental properties have been proved, both about *systems of Hörmander's vector fields and their metric*, and about second order *Hörmander's operators.* In the first group of results, let us quote the doubling property of the Lebesgue measure with respect to metric balls (Nagel et al. [17]); Poincaré's inequality with respect to the vector fields (Jerison [9]). In the second group of results, we recall the a priori estimates (in $L^p$ or $C^\alpha$) on $X_iX_ju$, in terms of $Lu$ and $u$ proved by Folland [7], Rothschild and Stein [19]; estimates about the fundamental solution of $L$ or $\partial_t - L$ ([17], Sanchez-Calle [20]).

It is natural to ask whether part of this theory still holds for a family of vector fields possessing just the right number of derivatives required to check Hörmander's condition. Actually, a quite extensive literature exists, by now, regarding the geometry of nonsmooth Hörmander's vector fields. However, if we restrict our attention to the research about systems of nonsmooth Hörmander's vector fields of general structure, only supposed to satisfy Hörmander's condition at some step $r$, the literature becomes much narrower. In this note I am going to describe some results of this type obtained jointly with Brandolini et al. in [6]. This paper is the third step in a larger project started by three of us in [4, 5], and the first one devoted to the study of second order Hörmander's operators built with nonsmooth vector fields of general form, and also allowing the presence of a drift term of weight two. Other results in this direction of research have been obtained in some papers by Montanari and Morbidelli [14–16] and by Karmanova-Vodopyanov (see [11, 21] and the references therein).

Our framework is the following. Let $X_0, X_1, \ldots, X_n$ be a system of real vector fields, defined in a bounded domain $\Omega \subset \mathbb{R}^p$ such that for some integer $r \geq 2$ and $\alpha \in (0, 1]$ the coefficients of $X_1, X_2, \ldots, X_n$ belong to $C^{r-1,\alpha}(\Omega)$ while the coefficients of $X_0$ belong to $C^{r-2,\alpha}(\Omega)$. Here $C^{k,\alpha}(\Omega)$ stands for the classical space of functions with Hölder continuous Cartesian derivatives up to the order $k$. If $r = 2$, we take $\alpha = 1$. We assume that $X_0, X_1, \ldots, X_n$ satisfy Hörmander's condition of *weighted step* $r$ in $\Omega$, where the weight of a commutator

$$\left[X_{i_1}, \left[X_{i_2}, \left[\ldots \left[X_{i_{j-1}}, X_{i_j}\right]\right]\right]\right]$$

is defined as the sum of the weights of the vector fields $X_{i_k}$ ($k = 1, 2, \ldots, j$), with $X_1, X_2, \ldots, X_n$ having weight one while $X_0$ has weight two.

In a few words, our main results are the following. We show how to build a local fundamental solution $\gamma$ for the operator $L$, prove natural bounds on the growth of $\gamma$, $X_k\gamma$, $X_iX_j\gamma$ ($i, j = 1, 2, \ldots, n$, $k = 0, 1, \ldots, n$) and the local Hölder continuity of $X_iX_j\gamma$, and we use these facts to prove a local solvability result: every point $\overline{x} \in \Omega$ possesses a neighborhood $U(\overline{x})$ such that for every $\beta \in (0, \alpha)$ and $f \in C_X^{\beta}(U)$ there exists a solution $u \in C_{X,loc}^{2,\beta}(U)$ to $Lu = f$ in $U$. For some results of ours we require the stronger assumption $X_i$ ($i = 1, 2, \ldots, n$) in $C^{r,\alpha}(\Omega)$ and $X_0$ in $C^{r-1,\alpha}(\Omega)$. (See Remark 4.1).

## 2 The Classical Framework

To describe the general line and strategy of this paper it is necessary to recall first the strategy which has been followed in the classical case, as well as some of the results proved in the nonsmooth case in the previous two papers [4, 5], which the paper [6] under discussion is built on.

### 2.1 Homogeneous Groups

The simplest situation in which a priori estimates on $X_iX_ju$ have been proved for classical Hörmander's operators is that of homogeneous groups.

A *homogeneous group in* $\mathbb{R}^N$ is a Lie group $\mathbb{G} = \left(\mathbb{R}^N, \circ\right)$ (which we think as "translations") for which 0 is the identity and the inverse of $u$ is $-u$, endowed with a family of automorphisms ("dilations")

$$D(\lambda)(u_1, u_2, \ldots, u_N) = \left(\lambda^{\alpha_1} u_1, \lambda^{\alpha_2} u_2, \ldots, \lambda^{\alpha_N} u_N\right).$$

We can define in $\mathbb{G}$ a *homogeneous norm* $\|\cdot\|$ letting

$$\|u\| = r \quad \Leftrightarrow \quad \left|D\left(\frac{1}{r}\right)u\right| = 1,$$

where $|\cdot|$ is the Euclidean norm and $\|0\| = 0$. The Lebesgue measure in $\mathbb{R}^N$ is biinvariant with respect to $\circ$ and

$$|B(u,r)| = |B(u,1)|\, r^Q \quad \forall u \in \mathbb{G}, r > 0,$$

where $Q = \sum_{i=1}^N \alpha_i$ is called *homogeneous dimension of the group.*

We assume there exists a basis $Y_0, Y_1, \ldots, Y_q$ of the Lie algebra of left invariant vector fields, $Y_0$ homogeneous of degree 2, the $Y_i$'s of degree 1. Under these assumptions, the operator

$$L = \sum_{i=1}^n Y_i^2 + Y_0$$

is hypoelliptic by Hörmander's theorem, left invariant and homogeneous of degree two. Its transpose is simply

$$L^* = \sum_{i=1}^n Y_i^2 - Y_0,$$

and shares the same properties of $L$. Thanks to these facts, Folland proved in [7] the existence of a left invariant fundamental solution $\Gamma$, homogeneous of degree $2 - Q$, such that

$$f(v) = \int_{\mathbb{R}^N} \Gamma\left(u^{-1} \circ v\right) Lf(u)\, du = L \int_{\mathbb{R}^N} \Gamma\left(u^{-1} \circ v\right) f(u)\, du$$

for any $f \in C_0^\infty\left(\mathbb{R}^N\right), v \in \mathbb{R}^N$.

The good properties of the kernel $\Gamma$ allow to apply a suitable general theory of singular integrals in a way which is strictly analogous to the one used for classical elliptic operators, proving local a priori $L^p$ estimates on $X_iX_ju$ and $X_0u$ in terms of $u, Lu, X_iu$ $(i,j = 1,2,\ldots,n)$.

## 2.2 *Lifting and Approximation*

Let us now come to the case of general Hörmander's vector fields, that is without an underlying structure of homogeneous groups. This case has been dealt by Rothschild and Stein in [19]. The goal is to approximate locally, in a suitable sense, a general family of Hörmander's vector fields by another family which is homogeneous and left invariant with respect to a structure of homogeneous group, in order to exploit the previous theory developed by Folland. However, it turns out that this approximation is not possible for a completely general system of Hörmander's vector fields, but it is possible for a family satisfying an extra property. To bypass this problem without losing generality, Rothschild and Stein then implement a two-step procedure. First the vector fields $X_i$ are "lifted" (in a neighborhood of any fixed point) to a higher dimensional space $\mathbb{R}^{p+m}$ getting new vector fields $\tilde{X}_i$, which project on the original $X_i$, that is

$$\tilde{X}_i = X_i + \sum_{j=1}^{m} u_{ij}(x, h) \frac{\partial}{\partial h_j};$$

the $\tilde{X}_i$'s still fulfil Hörmander's condition, and also satisfy the required extra property: they are *free up to step* $r$; which means that all their commutators up to step $r$ do not satisfy linear relations others than those which follow from anticommutativity and Jacobi identity.

As a second step, the $\tilde{X}_i$'s are locally approximated by homogeneous left invariant vector fields $Y_i$ on a suitable homogeneous group $\mathbb{G}$ of the kind we have described above. The description of this approximation is quite technical, but we cannot avoid it completely:

**Theorem 2.1 (Approximation)** *There exist a homogeneous group $\mathbb{G}$ on $\mathbb{R}^N$, $N = p + m$, a family of homogeneous left invariant Hörmander's vector fields $Y_0, Y_1, Y_2, \ldots, Y_n$ on $\mathbb{G}$ and a neighborhood $V$ of $(x_0, 0)$ in $\mathbb{R}^N$ such that for any $\eta \in V$ there exists a smooth diffeomorphism $\Theta_\eta$ from a neighborhood of $\eta$ onto a neighborhood of the origin in $\mathbb{G}$, smoothly depending on $\eta$, and for any smooth function $f : \mathbb{G} \to \mathbb{R}, i = 0, 1, \ldots, n$*

$$\tilde{X}_i (f \circ \Theta_\eta)(\xi) = (Y_i f + R_i^\eta f)(\Theta_\eta(\xi)) \ \forall \xi, \eta \in V$$

*where $R_i^\eta$ are smooth vector fields, smoothly depending on $\eta$, of weight $\geq 0$ for $i = 1, 2, \ldots, n$, $\geq -1$ for $i = 0$.*

The assertion about the *weight* means that, for small $u$,

$$\left|R_i^\eta \Gamma(u)\right| \leq \frac{c}{\|u\|^{Q-2}} \text{ while } |Y_i \Gamma(u)| \leq \frac{c}{\|u\|^{Q-1}} \text{ for } i = 1, 2, \ldots, n$$

$$\left|R_0^\eta \Gamma(u)\right| \leq \frac{c}{\|u\|^{Q-1}} \text{ while } |Y_0 \Gamma(u)| \leq \frac{c}{\|u\|^{Q}}.$$

Moreover, the map $\Theta_\eta(\xi)$ enjoys suitable properties, which however we shall not recall here. The line followed by Rothschild-Stein is then the following: they use the function $\Gamma(\Theta_\eta(\xi))$ as a *parametrix* of the lifted operator: exploiting this kernel, which is a good local approximation of the fundamental solution of the lifted operator

$$\tilde{L} = \sum_{i=1}^{n} \tilde{X}_i^2 + \tilde{X}_0,$$

they write representation formulas of $u$ and $\tilde{X}_i\tilde{X}_j u$ in terms of $\tilde{L}u$, $u$ and $\tilde{X}_i u$. From these representation formulas and a suitable abstract theory of singular integrals, $L^p$ a priori estimates are derived for $\tilde{X}_i\tilde{X}_j u$ and then, by projection on the original space of variables, for $X_iX_ju$.

## 3 Attacking the Problem: The Nonsmooth Background

In [4] we have built in the nonsmooth context an analogous kit of tools consisting in "lifting, approximation and basic properties of the map $\Theta$". Namely: the lifting theorem is perfectly analogous to the smooth one, with the new set of vector fields

$$\tilde{X}_i = X_i + \sum_{j=1}^{m} u_{ij}(x,h)\frac{\partial}{\partial h_j}$$

still satisfying Hörmander's condition at step $r$; the $\tilde{X}_i$'s have the same regularity of the $X_i$'s; the approximation formula takes the same form

$$\tilde{X}_i\left(f\circ\Theta_\eta\right)(\xi) = \left(Y_i f + R_i^\eta f\right)\left(\Theta_\eta(\xi)\right)\ \forall\xi,\eta\in V$$

where the $Y_i$ are homogeneous left invariant (smooth) vector fields on $\mathbb{G}$; the $R_i^\eta$ are $C^{r-p_i,\alpha}$ vector fields of weight $\geq \alpha - p_i$ (where $p_0 = 2, p_i = 1$ for $i = 1, 2, \ldots, n$), that is

$$\left|R_i^\eta\Gamma(u)\right| \leq \frac{c}{\|u\|^{Q-1-\alpha}} \text{ while } \left|Y_i\Gamma(u)\right| \leq \frac{c}{\|u\|^{Q-1}}$$
$$\left|R_0^\eta\Gamma(u)\right| \leq \frac{c}{\|u\|^{Q-\alpha}} \text{ while } \left|Y_0\Gamma(u)\right| \leq \frac{c}{\|u\|^{Q}}.$$

Moreover, the coefficients of the $R_i^\eta$'s and their first order derivatives depend on $\eta$ in a $C^\alpha$ way; the map $\Theta_\eta(\xi)$ is smooth in $\xi$ and $C^\alpha$ in $\eta$, and satisfies suitable other properties which we do not recall here. The important difference with respect to the classical situation is therefore the lack of smoothness in the dependence on $\eta$ of both the map $\Theta_\eta(\xi)$ and the "remainder vector fields" $R_i^\eta$. This fact will have an important drawback, as we will see.

Let us consider the function $\Gamma\left(\Theta_\eta(\xi)\right)$, where $\Gamma$ is Folland's homogeneous fundamental solution of the left invariant operator $\sum_{i=1}^n Y_i^2 + Y_0$. This kernel should be a parametrix for the operator $\tilde{L}$. Note however that this function is now smooth in $\xi$ but just $C^\alpha$ in $\eta$: a strong asymmetry in the roles of $\xi, \eta$. On the other hand, Rothschild-Stein's procedure to write by this parametrix good representation formulas for $\tilde{X}_i\tilde{X}_j u$ heavily relies on the possibility of differentiating the parametrix with respect to *both* variables, what we *cannot* do, due to the lack of regularity of $\Theta_\eta(\xi)$ with respect to the "bad" variable $\eta$.

So we apparently are stuck: after rebuilding Rothschild-Stein's set of tools in the nonsmooth case, we cannot use it as they do. The idea is then to exploit a different technique, compatible with the different degree of regularity of $\Gamma\left(\Theta_\eta(\xi)\right)$ with respect to $\xi$ and $\eta$. This technique is the *parametrix method* devised by Levi [13] to build fundamental solutions for elliptic equations or systems, and later extended to uniformly parabolic operators (see e.g. [8]). We recall that this method was first adapted to hypoelliptic ultraparabolic operators of Kolmogorov-Fokker-Planck type in [18], exploiting the knowledge of an explicit expression for the fundamental solution of the "frozen" operator, which had been constructed in [12]. It was later adapted in [1] to a general class of operators structured on homogeneous left invariant (smooth) vector fields on Carnot groups, for which no explicit fundamental solution is known in general, and in [3] to the more general context of arbitrary (smooth) Hörmander's vector fields.

Our strategy is the following: instead of using the parametrix for writing a representation formula for second order derivatives (what we cannot do in the nonsmooth case) and then using the representation formula for proving a priori estimates, we use the parametrix to build an exact (local) fundamental solution, and then use this to produce a local solution to the equation having a natural degree of smoothness.

However, if we applied the method starting with $\Gamma\left(\Theta_\eta(\xi)\right)$, we would build a local fundamental solution $\tilde{\gamma}$ for $\tilde{L}$, *in the space of lifted variables*; but starting with $\tilde{\gamma}$ there is no easy way to build a local fundamental solution for $L$: integrating $\tilde{\gamma}$ with respect to the lifted variables just produces a *parametrix* for $L$.

So, we have to reverse the order of the procedures: first, starting with $\Gamma\left(\Theta_\eta(\xi)\right)$ we define the kernel

$$P(x,y) = \int_{\mathbb{R}^m}\left(\int_{\mathbb{R}^m}\Gamma\left(\Theta_{(y,k)}(x,h)\right)\varphi(h)\,dh\right)\varphi(k)\,dk,$$

where $\varphi \in C_0^\infty(\mathbb{R}^m)$ is a fixed cutoff function, $= 1$ near 0. Then, starting with this $P$, "candidate parametrix" for $L$, we want to implement the Levi method *in the space of the original variables*, to produce a local fundamental solution for $L$.

This however is not an easy task, because in this space we do not have a simple geometry like in the space of lifted variables: the volume of metric balls does not behave like a fixed power of the radius, there is not a number having the meaning of "homogeneous dimension". In order to measure the growth of the kernels involved in the procedure, we have to use the control distance induced by the vector fields, we need to know an estimate for the volume of metric balls, and a comparison between volumes of balls in the lifted and original space. Briefly: we need a nonsmooth analog of the theory developed by Nagel et al. in [17] for general Hörmander's vector fields with drift. This theory is contained in the paper [5]. Let us recall a few facts from this paper.

The subelliptic metric, analogous to that introduced by Nagel-Stein-Wainger, is defined as follows:

**Definition 3.1** For any $\delta > 0$, let $C(\delta)$ be the class of absolutely continuous mappings $\varphi : [0,1] \longrightarrow \Omega$ such that

$$\varphi'(t) = \sum_{|I| \le r} a_I(t) \left(X_{[I]}\right)_{\varphi(t)} \text{ a.e.}$$

with $a_I : [0,1] \to \mathbb{R}$ measurable functions, $|a_I(t)| \le \delta^{|I|}$.Then:

$$d(x,y) = \inf\{\delta > 0 : \exists \varphi \in C(\delta) \text{ with } \varphi(0) = x, \varphi(1) = y\}.$$

The following basic result can be proved also in this case:

**Theorem 3.1 (Doubling Condition)** *Under the previous assumptions, for any domain $\Omega' \Subset \Omega$, there exist positive constants $c, \rho_0$, depending on $\Omega, \Omega'$ and the $X_i$'s, such that*

$$|B(x, 2\rho)| \le c\,|B(x,\rho)| \quad \forall x \in \Omega', \rho < \rho_0.$$

We also prove a sharp estimate on the volume of metric balls. A consequence of this estimate is the couple of bounds:

$$c_1 \left(\frac{R}{r}\right)^p \le \frac{|B(x,R)|}{|B(x,r)|} \le c_2 \left(\frac{R}{r}\right)^Q \quad \forall r, R \text{ with } \rho_0 > R > r > 0.$$

The *dimensional gap* between $p$ (Euclidean dimension) and $Q$ (homogeneous dimension of the locally approximating group) causes a lot of problems in implementing the Levi method. In contrast with this, note that

$$c_1 r^Q \le \left|\tilde{B}(\xi, r)\right| \le c_2 r^Q.$$

We can now also give the definitions of the function spaces defined by the vector fields, which will be used in the following.

**Definition 3.2** For any $\beta \in (0,1)$ we will denote by $C_X^\beta(U)$ the space of Hölder continuous functions of order $\beta$ with respect to the distance $d$.

By known results the following relations between Euclidean and subelliptic Hölder spaces holds:

$$f \in C^\alpha(U) \Rightarrow f \in C_X^\alpha(U)$$

$$f \in C_X^\alpha(U) \Rightarrow f \in C^{\alpha/r}(U)$$

where $r$ is the step of the Lie algebra generated by the $X_i$'s (i.e., $r$ is the integer appearing in our assumptions).

**Definition 3.3** Let $C_X^2(U)$ be the space of continuous functions on $U$ possessing in $U$ continuous derivatives with respect to $X_i$ $(i = 0, 1, 2, \dots, n)$ and second derivatives $X_iX_j$ $(i, j = 1, 2, \dots, n)$. For any $\beta \in (0, 1)$, let $C_X^{2,\beta}(U)$ be the space of $C_X^2(U)$ functions $u$ such that $X_k u$, $X_iX_j u$ $(k = 0, 1, \dots, n,\ i, j = 1, 2, \dots, n)$ belong to $C_X^\beta(U)$. We will also use the symbols $C_{X,0}^\beta(U)$, $C_{X,0}^{2,\beta}(U)$ for the analogous spaces of compactly supported functions.

## 4 The Parametrix Method

Let us recall what the parametrix method is. We look for a fundamental solution $\gamma$ for $L$ (that is, such that $L(\gamma(\cdot, y)) = -\delta_y$) of the form

$$\gamma(x, y) = P(x, y) + J(x, y), \text{ with}$$
$$J(x, y) = \int_U P(x, z)\, \Phi(z, y)\, dz$$

and $\Phi(z, y)$ to be determined. In other words, the fundamental solution is sought as a small integral perturbation of the parametrix.

Computing, for $x \neq y$,

$$0 = L(\gamma(\cdot, y)) = L(P(\cdot, y)) + L(J(\cdot, y))$$

and letting $Z_1(x, y) = L(P(\cdot, y))(x)$, we find

$$-Z_1(x, y) = L(J(\cdot, y))(x) = L\int_U P(x, z)\, \Phi(z, y)\, dz$$
$$= \int_U Z_1(x, z)\, \Phi(z, y)\, dz - \Phi(x, y),$$

whence $\Phi(z, y)$ satisfies the integral equation

$$\Phi(x, y) = Z_1(x, y) + \int_U Z_1(x, z)\, \Phi(z, y)\, dz$$

and can be therefore computed by Neumann series,

$$\Phi(x, y) = \sum_{j=1}^{\infty} Z_j(x, y), \text{ with}$$
$$Z_{j+1}(x, y) = \int_U Z_1(x, z)\, Z_j(z, y)\, dz.$$

So far, all this stuff is purely formal. To make this procedure rigorous we need, in sequence:

(a) To compute $Z_1(x, y) = L(P(\cdot, y))(x)$ and find an upper bound.
(b) To bound iteratively $Z_{j+1}(x, y)$.
(c) To bound $\Phi(z, y)$ (getting the series converge for $U$ small enough).
(d) To bound $J(x, y)$.
(e) To bound $\gamma(x, y)$.

Let me give just an idea of the proof of step (a) in this procedure, to show how several tools previously developed in the nonsmooth context are needed.

Let us start noting the following bounds:

$$\left|\left(Y_iY_j\Gamma\right)\left(\Theta_\eta(\xi)\right)\right| \leq \frac{c}{\left\|\Theta_\eta(\xi)\right\|^Q}, \text{ while}$$

$$\left|\left(R_i^\eta R_j^\eta\Gamma\right)\left(\Theta_\eta(\xi)\right)\right| \leq \frac{c}{\left\|\Theta_\eta(\xi)\right\|^{Q-\alpha}}.$$

Then, for $x \neq y$, the approximation theorem allows to write

$$Z_1(x, y) = LP(x, y) = \int_{\mathbb{R}^m}\int_{\mathbb{R}^m} \tilde{L}\left[\Gamma\left(\Theta_{(y,k)}(x, h)\right)\varphi(h)\right]dh\,\varphi(k)\,dk$$

$$= \int_{\mathbb{R}^m}\int_{\mathbb{R}^m}\left[-\delta_{(y,k)} + (\text{remainders})\right]\left(\Theta_{(y,k)}(x, h)\right)\varphi(h)\,dh\,\varphi(k)\,dk$$

since, for $x \neq y$, $\Theta_{(y,k)}(x, h) \neq 0$, hence $\delta_{(y,k)}\left(\Theta_{(y,k)}(x, h)\right) = 0$,

$$= \int_{\mathbb{R}^m}\int_{\mathbb{R}^m}\left[(\text{remainders})\left(\Theta_{(y,k)}(x, h)\right)\right]\varphi(h)\,dh\,\varphi(k)\,dk, \text{ where:}$$

$$\left|(\text{remainders})\left(\Theta_{(y,k)}(x, h)\right)\right| \leq \frac{c}{\left\|\Theta_\eta(\xi)\right\|^{Q-\alpha}}.$$

Therefore

$$|Z_1(x, y)| = |LP(x, y)| \leq \int_{\mathbb{R}^m}\int_{\mathbb{R}^m}\frac{\psi(h)}{\left\|\Theta_\eta(\xi)\right\|^{Q-\alpha}}dh\,\varphi(k)\,dk.$$

So far, we have used the (nonsmooth) Rothschild-Stein-like machinery. To bound the last integral we have to use the (nonsmooth) Nagel-Stein-Weinger-like machinery. The local doubling property and the comparison between the volumes of lifted and unlifted balls give the following:

**Lemma 4.1** *For every $\beta \in \mathbb{R}$ there exists $c > 0$ such that for any $x, y \in U(x_0)$,*

$$\int_{\mathbb{R}^m}\int_{\mathbb{R}^m} \frac{\psi(h)}{\left\|\Theta_{(y,k)}(x,h)\right\|^{Q-\beta}} dh\, \varphi(k)\, dk \le c \int_{d(x,y)}^{R} \frac{r^{\beta-1}}{|B(x,r)|} dr.$$

The function at the right hand side of the last inequality plays a central role in our estimates:

$$\phi_\beta(x,y) = \int_{d(x,y)}^{R} \frac{r^{\beta-1}}{|B(x,r)|} dr.$$

To visualize its size, note that we have the following bounds (which however are not equivalences!):

$$\phi_\beta(x,y) \le \begin{cases} c\dfrac{d(x,y)^\beta}{|B(x,d(x,y))|} & \text{for } \beta < p \\ c\dfrac{d(x,y)^p}{|B(x,d(x,y))|} R^{\beta-p} & \text{for } \beta > p. \end{cases}$$

For instance,

$$|Z_1(x,y)| = |LP(x,y)| \le \int_{\mathbb{R}^m}\int_{\mathbb{R}^m} \frac{\psi(h)}{\left\|\Theta_\eta(\xi)\right\|^{Q-\alpha}} dh\, \varphi(k)\, dk$$

$$\le c\phi_\alpha(x,y) \le c\frac{d(x,y)^\alpha}{|B(x,d(x,y))|} \text{ which is locally integrable,}$$

while, by comparison, we have

$$|P(x,y)| \le c\phi_2(x,y);$$
$$|X_iP(x,y)| \le c\phi_1(x,y) \text{ for } i = 1,2,\ldots,n;$$
$$\left|X_jX_iP(x,y)\right|, |X_0P(x,y)| \le c\phi_0(x,y) \text{ for } i,j = 1,2,\ldots,n.$$

Having bounded $|Z_1(x,y)|$ with a locally integrable kernel satisfying a quantitative size condition is the starting point for iterative computations which allow to carry out the parametrix method. Skipping many other facts, let us now jump to the statement of the first main result of ours:

**Theorem 4.1 (Existence of Fundamental Solution)** *The functions $\gamma(x,y) = P(x,y) + J(x,y)$ and $X_i\gamma(x,y)$ $(i = 1,2,\ldots,n)$ are well defined and continuous in the joint variables $x, y \in U, x \neq y$, and satisfy the following bounds:*

$$|\gamma(x,y)| \le c\phi_2(x,y);$$
$$|X_i\gamma(x,y)| \le c\phi_1(x,y).$$

*Moreover,* $\gamma(\cdot, y)$ *is a weak solution to* $L\gamma(\cdot, y) = -\delta_y$, *that is:*

$$\int_U \gamma(x, y) L^* \psi(x)\, dx = -\psi(y) \quad \forall \psi \in C_0^\infty(U), y \in U.$$

*Finally, if* $X_0 \equiv 0$, *then* $\exists \varepsilon > 0$ *such that*

$$\gamma(x, y) > 0 \text{ for } d(x, y) < \varepsilon.$$

*Remark 4.1* All the results proved so far hold under the following assumptions $X_i \in C^{r-1,\alpha}(\Omega), X_0 \in C^{r-2,\alpha}(\Omega)$ for some $\alpha \in (0, 1]$.

Henceforth we will make instead the stronger assumptions $X_i \in C^{r,\alpha}(\Omega), X_0 \in C^{r-1,\alpha}(\Omega)$ (if $r = 2$ then $\alpha = 1$).

The results stated in the previous theorem are not yet satisfying: we want to know further regularity results for the fundamental solution. Our next step is the following:

**Theorem 4.2 (Second Derivatives of the Fundamental Solution)** *For* $i, j = 1, 2, \ldots, n$ *and* $x, y \in U, x \neq y$, *the following assertions hold true:*

*(a)* $\exists\, X_jX_iJ(x, y)$, $X_0J(x, y)$, $X_iX_j\gamma(x, y)$, $X_0\gamma(x, y)$ *continuous in the joint variables for* $x \neq y$; *in particular,*

$$\gamma(\cdot, y) \in C_X^2(U \setminus \{y\}) \text{ for any } y \in U.$$

*(b)* $\forall \varepsilon \in (0, \alpha)$, $U' \Subset U\ \exists c > 0$ *such that* $\forall x \in U'$ *and* $y \in U$,

$$\left|X_jX_iJ(x, y)\right|, \left|X_0J(x, y)\right| \leq c \frac{d(x, y)^{\alpha-\varepsilon}}{|B(x, d(x, y))|}$$

$$\left|X_jX_i\gamma(x, y)\right|, \left|X_0\gamma(x, y)\right| \leq c \frac{1}{|B(x, d(x, y))|}.$$

Let us give just an idea of the problems involved in the proof. Recall that

$$\begin{aligned}\gamma(x, y) &= P(x, y) + J(x, y) \\ &= P(x, y) + \int_U P(x, z)\, \Phi(z, y)\, dz.\end{aligned}$$

Hence, in order to compute and bound $X_iX_j\gamma$, the problem is the computation of the singular integral

$$X_i \int_U X_jP(x, z)\, \Phi(z, y)\, dz.$$

This computation is made possible by a Hölder type estimate on the function $\Phi(\cdot, y)$:

**Theorem 4.3** *For every $\varepsilon \in (0, \alpha)$ there exists $c > 0$ such that*

$$|\Phi(x_1, y) - \Phi(x_2, y)| \leq c d(x_1, x_2)^{\alpha-\varepsilon} \phi_\varepsilon(x_1, y)$$

*for any $x_1, x_2, y \in U$ with $x_1 \neq x_2, d(x_1, y) \geq 3d(x_1, x_2)$.*

Exploiting this Hölder bound we are able to prove the following:

$$X_j X_i J(x, y) = \int_U \int_{\mathbb{R}^m} \int_{\mathbb{R}^m} (Y_j D_1 \Gamma)(\Theta_{(z,k)}(x, h)) a_0(h) b_0(k) dh dk \times$$
$$\times [\Phi(z, y) - \Phi(x, y)] dz + c(x) \Phi(x, y) + \int_U R_2(x, z) \Phi(z, y) dz$$

with $R_2$ locally integrable kernel. This representation formula is the starting point for several computations.

In turn, proving the Hölder bound on $\Phi(\cdot, y)$ requires a more regular dependence on the parameter for the "remainder vector fields" appearing in Rothschild-Stein-type approximation formula, and better properties of the map $\Theta_\eta(\xi)$ with respect to the "bad variable" $\eta$. This is made possible by a careful analysis of the properties of this map $\Theta_\eta(\xi)$.

We can now refine the previous analysis of the second derivatives of our local fundamental solution and prove a sharp bound of Hölder type on $X_i X_j \gamma$:

**Theorem 4.4** *For every $\varepsilon \in (0, \alpha)$ and $U' \Subset U$ there exists $c > 0$ such that for any $x_1, x_2 \in U'$, $y \in U$ with $d(x_1, y) \geq 2d(x_1, x_2)$, $i, j = 1, 2, \ldots, n$,*

$$\left|X_i X_j P(x_1, y) - X_i X_j P(x_2, y)\right| \leq c \frac{d(x_1, x_2)}{d(x_1, y)} \frac{1}{|B(x_1, d(x_1, y))|}$$

$$\left|X_i X_j J(x_1, y) - X_i X_j J(x_2, y)\right| \leq c \left(\frac{d(x_1, x_2)}{d(x_1, y)}\right)^{\alpha-\varepsilon} \frac{d(x_1, y)^{\alpha-\varepsilon}}{|B(x_1, d(x_1, y))|}$$

$$\left|X_i X_j \gamma(x_1, y) - X_i X_j \gamma(x_2, y)\right| \leq c \left(\frac{d(x_1, x_2)}{d(x_1, y)}\right)^{\alpha-\varepsilon} \frac{1}{|B(x_1, d(x_1, y))|}$$

$$|X_0 \gamma(x_1, y) - X_0 \gamma(x_2, y)| \leq c \left(\frac{d(x_1, x_2)}{d(x_1, y)}\right)^{\alpha-\varepsilon} \frac{1}{|B(x_1, d(x_1, y))|}.$$

*In particular, for every $\varepsilon \in (0, \alpha)$ and $y \in U$,*

$$\gamma(\cdot, y) \in C^{2,\alpha-\varepsilon}_{X,loc}(U \setminus \{y\}).$$

## 5 Local Solvability

We can now come to another of the main results in the paper:

**Theorem 5.1 (Local Solvability in $C_X^2$)** *The function $\gamma$ is a solution to the equation*

$$L\gamma(\cdot, y) = 0 \text{ in } U \setminus \{y\}, \forall y \in U.$$

*Moreover, $\forall \beta > 0, f \in C_X^{\beta}(U)$, the function*

$$w(x) = -\int_U \gamma(x, y) f(y)\, dy \tag{1}$$

*is a $C_X^2(U)$ solution to the equation $Lw = f$ in $U$.*

*Also, if $X_0 \equiv 0$, choosing $U$ small enough, we have the following positivity property: if $f \in C_X^{\beta}(U), f \leq 0$ in $U$, then the equation $Lw = f$ has at least a $C_X^2(U)$ solution $w \geq 0$ in $U$.*

The proof amounts to showing that the function $w$ assigned by (1) is actually $C_X^2(U)$, since this easily implies it solves the equation.

The by-product of the proof of this theorem is a convenient representation formula for the derivatives $X_iX_jw$, which is the starting point for the proof of Hölder continuity of $X_iX_jw$. In order to simplify this presentation, I write this formula in a simplified, schematic, way:

**Corollary 5.1** *$\forall x \in B\left(\overline{x}, \frac{R}{2}\right)$ and $i,j = 1, 2, \dots, n$, we have:*

$$X_jX_iw(x) = c(x)f(x) + \int_U k_0(x, y) f(y)\, dy +$$
$$+ \int_{B(\overline{x},R)} k_1(x, z)\, b(z) f(z)\, dz + \int_{B(\overline{x},R)} k_2(x, z)\, [f(z) - f(x)]\, b(z)\, dz$$

*where $c \in C_X^{\alpha}\left(B\left(\overline{x}, \frac{R}{2}\right)\right)$, $k_0$ is a bounded kernel, $k_1$ is a fractional integral kernel, $k_2$ is a singular integral kernel, $b$ a cutoff function.*

Then, a careful application of suitable abstract theories of singular and fractional integrals on locally doubling spaces allows to prove Hölder continuity of $X_iX_jw$. We will give some more details about this in the next section. Let us first conclude the description of the main line of the paper. The final goal is to prove the following:

**Theorem 5.2** *For every $\beta \in (0, \alpha)$ and $f \in C_X^{\beta}(U)$, let $w \in C_X^2(U)$ be the solution to $Lw = f$ in $U$ assigned by (1). Then $w \in C_{X,loc}^{2,\beta}(U)$. More precisely, for every $U' \Subset U$ there exists $c > 0$ such that*

$$\|w\|_{C_X^{2,\beta}(U')} \leq c\, \|f\|_{C_X^{\beta}(U)}.$$

**Corollary 5.2** (**$C_X^{2,\beta}$ Local Solvability**) *For every $\beta \in (0,\alpha)$ the operator L is locally $C_X^{2,\beta}$ solvable in $\Omega$ in the following senses:*

$\forall \overline{x} \in \Omega\ \exists U(\overline{x})$ such that $\forall f \in C_X^{\beta}(U)$ there exists a solution $u \in C_{X,loc}^{2,\beta}(U)$ to $Lu = f$ in $U$.

$\forall \overline{x} \in \Omega\ \exists U(\overline{x})$ such that $\forall f \in C_{X,0}^{\beta}(U)$ there exists a solution $u \in C_X^{2,\beta}(U)$ to $Lu = f$ in $U$.

# 6 Real Analysis Estimates

Let us now spend a few words about the real analysis machinery we need to use to get the $C^{2,\beta}$ estimates described above.

An important starting remark is the following. Our construction lives in a fixed neighborhood $U$, where the distance $d$ induced by the vector fields $X_i$ is defined; then the Lebesgue measure is *locally* doubling, while we cannot assure the validity of a *global* doubling condition in $U$, which should mean:

$$|B(x,2r) \cap U| \leqslant c\,|B(x,r) \cap U| \text{ for any } x \in U, r > 0. \tag{2}$$

Actually, even for the Carnot-Carathéodory distance induced by smooth Hörmander's vector fields, condition (2) is known when $U$ is for instance a metric ball and the drift term $X_0$ is lacking; in presence of a drift, however, the validity of a condition (2) on some reasonable $U$ seems to be an open problem. This means that in our situation $(U, d, dx)$ is not a space of homogeneous type in the sense of Coifman-Weiss. However, $(U, d, dx)$ fits the assumptions of *locally homogeneous spaces* as defined in [2]. We apply some results proved in [2] which assure the local $C^\alpha$ continuity of singular and fractional integrals defined by a kernel of the kind

$$a(x)\,k(x,y)\,b(y)$$

(with $a, b$ smooth cutoff functions) provided that the kernel $k$ satisfies natural assumptions which never involve integration over domains of the kind $B(x,r) \cap U$, but only over balls $B(x,r) \Subset U$, which makes our local doubling condition usable.

**Definition 6.1** We say that a kernel $k(x,y)$ satisfies the *standard estimates of fractional integrals* with (positive) exponents $\nu, \beta$ in $B(\overline{x}, R)$ if

$$|k(x,y)| \le c\frac{d(x,y)^{\nu}}{|B(x,d(x,y))|} \quad \forall x,y \in B(\overline{x},R), \text{ and}$$

$$|k(x,y) - k(x_0,y)| \le c\frac{d(x_0,y)^{\nu}}{|B(x_0,d(x_0,y))|}\left(\frac{d(x_0,x)}{d(x_0,y)}\right)^{\beta}$$

$\forall x_0, x, y \in B(\overline{x}, R)$ such that $d(x_0,y) \ge M d(x_0,x)$ for suitable $M > 1$.

We say that $k(x,y)$ satisfies the *standard estimates of singular integrals* if the previous estimates hold with $\nu = 0$ and some $\beta > 0$.

As already stated, some of the terms in the representation formula of $X_iX_jw$ are fractional integrals while another is a multiplicative term by a Hölder function; the singular integral part is the following term:

$$Tf(x) = \int_{B(\overline{x},R)} k_2(x,z)\,[f(z) - f(x)]\,b(z)\,dz, \quad \text{where}$$

$$k_2(x,z) = \int_{\mathbb{R}^m}\int_{\mathbb{R}^m} Y_iY_j\Gamma\left(\Theta_{(z,k)}(x,h)\right)a(h)\,b(k)\,dhdk,$$

$k_2$ satisfying the standard estimates of singular integrals.

In order to deduce an Hölder estimate for $Tf$ we need in particular to establish a suitable *cancellation property* for $k_2$:

**Theorem 6.1** *There exists $C > 0$ such that for a.e. $x \in B(\overline{x},R)$ and $0 < \varepsilon_1 < \varepsilon_2 < \infty$*

$$\left|\int_{\varepsilon_1 < d(x,y) < \varepsilon_2} a(x)k_2(x,y)\,b(y)\,dy\right| \leq C.$$

As far as I know this is the first case of a priori estimate for PDEs when a singular integral operator is directly handled in a context where there is no kind of exact or approximate homogeneity, and the measure of a ball does not behave like a fixed power of the radius.

## References

1. Bonfiglioli, A., Lanconelli, E., Uguzzoni, F.: Fundamental solutions for non-divergence form operators on stratified groups. Trans. Am. Math. Soc. **356**(7), 2709–2737 (2004)
2. Bramanti, M., Zhu, M.: Local real analysis in locally doubling spaces. Manuscripta Math. **138**(3–4), 477–528 (2012)
3. Bramanti, M., Brandolini, L., Lanconelli, E., Uguzzoni, F.: Non-divergence equations structured on Hörmander vector fields: heat kernels and Harnack inequalities. Mem. Am. Math. Soc. **204**(961), 1–136 (2010)
4. Bramanti, M., Brandolini, L., Pedroni, M.: On the lifting and approximation theorem for nonsmooth vector fields. Indiana Univ. Math. J. **59**(6), 1889–1934 (2010)
5. Bramanti, M., Brandolini, L., Pedroni, M.: Basic properties of nonsmooth Hörmander's vector fields and Poincaré's inequality. Forum Mathematicum **25**(4), 703–769 (2013)
6. Bramanti, M., Brandolini, L., Manfredini, M., Pedroni, M.: Fundamental solutions and local solvability of nonsmooth Hörmander's operators. Mem. Am. Math. Soc., in press. http://arxiv.org/abs/1305.3398
7. Folland, G.B.: Subelliptic estimates and function spaces on nilpotent Lie groups. Arkiv Mat. **13**, 161–207 (1975)

8. Friedman, A.: Partial Differential Equations of Parabolic Type. Prentice-Hall, Englewood Cliffs (1964)
9. Jerison, D.: The Poincaré inequality for vector fields satisfying Hörmander's condition. Duke Math. J. **53**(2), 503–523 (1986)
10. Hörmander, L.: Hypoelliptic second order differential equations. Acta Math. **119**, 147–171 (1967)
11. Karmanova, M., Vodopyanov, S.: Geometry of Carnot-Carathéodory spaces, differentiability, coarea and area formulas. Anal. Math. Phys. Trends Math. 233–335 (2009)
12. Lanconelli, E., Polidoro, S.: On a class of hypoelliptic evolution operators. Partial differential equations, II (Turin, 1993). Rend. Sem. Mat. Univ. Politec. Torino **52**(1), 29–63 (1994)
13. Levi, E.E.: Sulle equazioni lineari totalmente ellittiche alle derivate parziali. Rend. Circ. Mat. Palermo **24**, 312–313 (1907)
14. Montanari, A., Morbidelli, D.: Nonsmooth Hörmander vector fields and their control balls. Trans. Am. Math. Soc. **364**, 2339–2375 (2012)
15. Montanari, A., Morbidelli, D.: Generalized Jacobi identities and ball-box theorem for horizontally regular vector fields. J. Geom. Anal **24**(2), 687–720 (2014)
16. Montanari, A., Morbidelli, D.: Steps involutive families of vector fields, their orbits and the Poincaré inequality. J. Math. Pures Appl. (9) **99**(4), 375–394 (2013)
17. Nagel, A., Stein, E.M., Wainger, S.: Balls and metrics defined by vector fields I: Basic properties. Acta Math. **155**, 130–147 (1985)
18. Polidoro, S.: On a class of ultraparabolic operators of Kolmogorov-Fokker-Planck type. Le Matematiche **49**, 53–105 (1994)
19. Rothschild, L.P., Stein, E.M.: Hypoelliptic differential operators and nilpotent groups. Acta Math. **137**, 247–320 (1976)
20. Sánchez-Calle, A.: Fundamental solutions and geometry of sum of squares of vector fields. Inv. Math. **78**, 143–160 (1984)
21. Vodopyanov, S.K., Karmanova, M.B.: Sub-Riemannian geometry for vector fields of minimal smoothness. Dokl. Akad. Nauk **422**(5), 583–588 (2008, Russian). Translation in Dokl. Math. **78**(2), 737–742 (2008)

# Multiple Solutions for an Eigenvalue Problem Involving Non-local Elliptic $p$-Laplacian Operators

**Patrizia Pucci and Sara Saldi**

*Dedicated to Ermanno Lanconelli on the occasion of his 70th birthday, with great feelings of esteem and affection*

**Abstract** In this paper we establish the existence of two nontrivial weak solutions of a one parameter non-local eigenvalue problem under homogeneous Dirichlet boundary conditions in bounded domains, involving a general non-local elliptic $p$-Laplacian operator.

**Keywords** Existence and multiplicity of solutions • Fractional elliptic Dirichlet problems • Fractional Sobolev spaces • Variational methods

**Mathematics Subject Classification:** 35R11, 35J60, 35S15, 47G20

## 1 Introduction

In the paper [1] *Arcoya* and *Carmona* extend to a wide class of functionals the three critical point theorem of *Pucci* and *Serrin* in [12] (see also [11]) and applied it to a one-parameter family of functionals $J_\lambda$, $\lambda \in I \subset \mathbb{R}$. Under suitable assumptions, they locate an open subinterval of values $\lambda$ in $I$ for which $J_\lambda$ possesses at least three

The manuscript was sent on November 7, 2013.

P. Pucci (✉)
Dipartimento di Matematica e Informatica, Università degli Studi di Perugia, Perugia, Italy
e-mail: patrizia.pucci@unipg.it

S. Saldi
Dipartimento di Matematica e Informatica "U. Dini", Università degli Studi di Firenze, Firenze, Italy

Dipartimento di Matematica e Informatica, Università degli Studi di Perugia, Perugia, Italy
e-mail: saldi@math.unifi.it; sara.saldi@dmi.unipg.it



G. Citti et al. (eds.), *Geometric Methods in PDE's*, Springer INdAM Series 13,
DOI 10.1007/978-3-319-02666-4_9

critical points. Recently, a slight variant of the main abstract theorem of [1] has been proposed in [4]. In both papers [1, 4] several interesting applications to quasilinear boundary value problems are given.

In this paper, taking inspiration of [4], we establish the existence of two nontrivial weak solutions of a one parameter eigenvalue problem under homogeneous Dirichlet boundary conditions in an open bounded subset $\Omega$ of $\mathbb{R}^N$, with Lipschitz boundary. More precisely, we consider the problem

$$\begin{cases} \mathscr{L}_K u = \lambda[a(x)|u|^{p-2}u + f(x,u)], & \text{in } \Omega, \\ u = 0, & \text{in } \mathbb{R}^N \setminus \Omega, \end{cases} \tag{1}$$

where $\mathscr{L}_K$ is an integro-differential non-local operator, defined pointwise by

$$\mathscr{L}_K\varphi(x) = -\int_{\mathbb{R}^N} |\varphi(x) - \varphi(y)|^{p-2}[\varphi(x) - \varphi(y)] \cdot K(x-y)dy,$$

along any function $\varphi \in C_0^\infty(\Omega)$, and the *weight* $K : \mathbb{R}^N \setminus \{0\} \to \mathbb{R}^+$ satisfies the natural restriction. *There exists* $s \in (0,1)$, *with* $N > ps$, *such that*

($\mathscr{K}$) $\epsilon \le K(x)|x|^{N+ps} \le \delta$ *for all* $x \in \mathbb{R}^N \setminus \{0\}$ *and some suitable numbers* $\epsilon$, $\delta$, *with* $0 < \epsilon \le \delta$;

*holds.*

Clearly, when $K(x) = |x|^{-(N+ps)}$, the operator $\mathscr{L}_K$ reduces to the more familiar fractional $p$-Laplacian operator $(-\Delta)_p^s$, which up to a multiplicative constant depending only on $N$, $s$ and $p$ is defined by

$$(-\Delta)_p^s\varphi(x) = -\int_{\mathbb{R}^N} \frac{|\varphi(x) - \varphi(y)|^{p-2}[\varphi(x) - \varphi(y)]}{|x-y|^{N+ps}}dy,$$

along any function $\varphi \in C_0^\infty(\Omega)$.

In (1) we assume that the coefficient *a is a positive weight of class* $L^\alpha(\Omega)$, with $\alpha > N/ps$, and that the perturbation $f : \overline{\Omega} \times \mathbb{R} \to \mathbb{R}$ *is a Carathéodory function, with* $f \not\equiv 0$, *satisfying the main assumption* ($\mathscr{F}$) *of Sect. 2.*

In Sect. 3 we determine precisely the intervals of $\lambda$'s for which problem (1) admits only the trivial solution and for which (1) has at least two nontrivial solutions. More precisely, we study problem (1) by a slight variant of the *Arcoya* and *Carmona* result in [1], as proved in Theorem 2.1 of [4].

## 2 Preliminaries and Auxiliary Results

Throughout the paper $\Omega$ is an open bounded subset of $\mathbb{R}^N$, with Lipschitz boundary, $a$ is a positive weight of class $L^\alpha(\Omega)$, with $\alpha > N/ps$, and $K$ satisfies ($\mathscr{K}$) of the Introduction in $\mathbb{R}^N \setminus \{0\}$. Since $s \in (0,1)$ and $N > ps$ we denote by $p^*$ the critical Sobolev exponent for $W_0^{s,p}(\Omega)$, that is $p^* = pN/(N-ps)$. We recall that $D_0^{s,p}(\Omega) = \overline{C_0^\infty(\Omega)}^{\|\cdot\|_\Omega}$, where $\|\cdot\|_\Omega$ is the standard fractional Gagliardo norm, given by

$$\|u\|_\Omega = \left(\iint_{\Omega\times\Omega} |u(x)-u(y)|^p|x-y|^{-(N+ps)}dxdy\right)^{1/p}$$

for all $u \in W_0^{s,p}(\Omega)$. Furthermore, $D^{s,p}(\mathbb{R}^N)$ denotes the fractional Beppo–Levi space, that is the completion of $C_0^\infty(\mathbb{R}^N)$, with respect to the norm

$$\|u\|_s = \left(\iint_{\mathbb{R}^{2N}} |u(x)-u(y)|^p|x-y|^{-(N+ps)}dxdy\right)^{1/p}.$$

Moreover, by Theorems 1 and 2 of [10]

$$\begin{aligned} \|u\|^p_{L^{p^*}(\mathbb{R}^N)} &\le c_{N,p}\frac{s(1-s)}{(N-ps)^{p-1}}\|u\|_s^p, \\ \int_{\mathbb{R}^N}|u(x)|^p\frac{dx}{|x|^{ps}} &\le c_{N,p}\frac{s(1-s)}{(N-ps)^p}\|u\|_s^p \end{aligned} \tag{2}$$

for all $u \in D^{s,p}(\mathbb{R}^N)$, where $c_{N,p}$ is a positive constant depending only on $N$ and $p$. Hence

$$D^{s,p}(\mathbb{R}^N) = \{u \in L^{p^*}(\mathbb{R}^N) \,:\, |u(x)-u(y)|\cdot|x-y|^{-(s+N/p)} \in L^p(\mathbb{R}^{2N})\}.$$

Following [8], we put

$$\tilde{D}^{s,p}(\Omega) = \{u \in L^{p^*}(\Omega) \,:\, \tilde{u} \in D^{s,p}(\mathbb{R}^N)\},$$

with the norm $\|u\|_{\tilde{s}} = \|\tilde{u}\|_s$, where $\tilde{u}$ is the natural extension of $u$ in the entire $\mathbb{R}^N$, with value 0 in $\mathbb{R}^N \setminus \Omega$. Clearly,

$$\|u\|_{\tilde{s}} = \left(\|u\|^p_\Omega + 2\int_\Omega |u(x)|^p dx \int_{\mathbb{R}^N\setminus\Omega}|x-y|^{-(N+ps)}dy\right)^{1/p} \ge \|u\|_\Omega.$$

Since here $\Omega$ is regular, an application of Theorem 1.4.2.2 of [8] shows that $\tilde{D}^{s,p}(\Omega) = \overline{C_0^\infty(\Omega)}^{\|\cdot\|_{\tilde{s}}}$. Finally, since $\Omega$ is bounded and regular, by (2) there exists a constant $c_\Omega > 0$ such that

$$c_\Omega\|\tilde{u}\|_{W^{s,p}(\mathbb{R}^N)} \le \|\tilde{u}\|_s = \|u\|_{\tilde{s}} \le \|\tilde{u}\|_{W^{s,p}(\mathbb{R}^N)}$$

for all $u \in \tilde{D}^{s,p}(\Omega)$, and so, using also Corollary 1.4.4.10 of [8], we have the main property

$$\begin{aligned}\tilde{D}^{s,p}(\Omega) &= \{u \in W_0^{s,p}(\Omega) \, : \, u\, d(\cdot, \partial\Omega)^{-s} \in L^p(\Omega)\} \\ &= \{u \in D^{s,p}(\mathbb{R}^N) \, : \, u = 0 \text{ a.e. in } \mathbb{R}^N \setminus \Omega\} \\ &= \{u \in W^{s,p}(\Omega) \, : \, \tilde{u} \in W^{s,p}(\mathbb{R}^N)\},\end{aligned}$$

where $d(x, \partial\Omega)$ is the distance from $x$ to the boundary $\partial\Omega$ of $\Omega$.

It is not hard to see that $\tilde{D}^{s,p}(\Omega)$ is a closed subspace of $D^{s,p}(\mathbb{R}^N)$. Hence also $\tilde{D}^{s,p}(\Omega) = \left(\tilde{D}^{s,p}(\Omega), \|\cdot\|_s\right)$ is a reflexive Banach space. For simplicity and abuse of notation, in the following we still denote by $u$ the extension of every function $u \in \tilde{D}^{s,p}(\Omega)$, by setting $u = 0$ in $\mathbb{R}^N \setminus \Omega$.

From now on we endow $\tilde{D}^{s,p}(\Omega)$ with the weighted *Gagliardo* norm

$$\|u\| = \left(\iint_{\mathbb{R}^{2N}} |u(x) - u(y)|^p K(x-y)dxdy\right)^{1/p},$$

equivalent to the norm $\|\cdot\|_s$ by virtue of $(\mathscr{K})$. Indeed, $(\mathscr{K})$ implies at once that $mK \in L^1(\mathbb{R}^N)$, where $m(x) = \min\{1, |x|^p\}$, so that in particular $\|\varphi\| < \infty$ for all $\varphi \in C_0^2(\Omega)$.

In conclusion, also *the natural solution space* $\tilde{D}^{s,p}(\Omega) = \left(\tilde{D}^{s,p}(\Omega), \|\cdot\|\right)$ *of* (1) *is a reflexive Banach space.*

Note that by Corollary 7.2 of [5] the embedding $\tilde{D}^{s,p}(\Omega) \hookrightarrow L^{\alpha' p}(\Omega)$ is compact, being $\alpha' p < p^*$ by the assumption that $\alpha > N/ps$. Moreover, the embedding $L^{\alpha' p}(\Omega) \hookrightarrow L^p(\Omega, a)$ is continuous, since $\|u\|_{p,a}^p \leq \|a\|_\alpha \|u\|_{\alpha' p}^p$ for all $u \in L^{\alpha' p}(\Omega)$ by Hölder's inequality. Hence,

$$\textit{the embedding } \tilde{D}^{s,p}(\Omega) \hookrightarrow L^p(\Omega, a) \textit{ is compact.} \tag{3}$$

Let $\lambda_1$ be *the first eigenvalue* of the problem

$$\begin{cases} \mathscr{L}_K u = \lambda\, a(x)|u|^{p-2}u, & \text{in } \Omega, \\ u = 0, & \text{in } \mathbb{R}^N \setminus \Omega, \end{cases} \tag{4}$$

in $\tilde{D}^{s,p}(\Omega)$, that is $\lambda_1$ is defined by the Rayleigh quotient

$$\lambda_1 = \inf_{u \in \tilde{D}^{s,p}(\Omega),\ u \neq 0} \frac{\iint_{\mathbb{R}^{2N}} |u(x) - u(y)|^p K(x-y)dxdy}{\int_\Omega a(x)|u|^p dx}. \tag{5}$$

By Lemma 2.1 of [7] (see also Theorem 5 of [9] for the fractional $p$-Laplacian first eigenvalue) *the infimum in* (5) *is achieved and* $\lambda_1 > 0$, *when* $a \equiv 1$. We refer also to [6] for the special linear case of the fractional Laplacian and $a \in Lip(\overline{\Omega})$. For sake of completeness we prove the result for the general weight $a$, using a completely different argument.

**Proposition 2.1** *The infimum $\lambda_1$ in* (5) *is positive and attained at a certain function $u_1 \in \tilde{D}^{s,p}(\Omega)$, with $\|u_1\|_{p,a} = 1$ and $\|u_1\|^p = \lambda_1 > 0$. Moreover, $u_1$ is a solution of* (4) *when $\lambda = \lambda_1$.*

*Proof* For any $u \in \tilde{D}^{s,p}(\Omega)$ define the functionals $\mathscr{I}(u) = \|u\|^p$ and $\mathscr{J}(u) = \|u\|^p_{p,a}$. Let $\lambda_0 = \inf\{\mathscr{I}(u)/\mathscr{J}(u) : u \in \tilde{D}^{s,p}(\Omega) \setminus \{0\}, \ \|u\|_{p,a} \le 1\}$. Observe that $\mathscr{I}$ and $\mathscr{J}$ are continuously Fréchet differentiable and convex in $\tilde{D}^{s,p}(\Omega)$. Clearly $\mathscr{I}'(0) = \mathscr{J}'(0) = 0$. Moreover, $\mathscr{J}'(u) = 0$ implies $u = 0$. In particular, $\mathscr{I}$ and $\mathscr{J}$ are weakly lower semi-continuous on $\tilde{D}^{s,p}(\Omega)$. Actually, $\mathscr{J}$ is weakly sequentially continuous on $\tilde{D}^{s,p}(\Omega)$. Indeed, if $(u_n)_n$ and $u$ are in $\tilde{D}^{s,p}(\Omega)$ and $u_n \rightharpoonup u$ in $\tilde{D}^{s,p}(\Omega)$, then $u_n \to u$ in $L^p(\Omega, a)$ by (3). This implies at once that $\mathscr{J}(u_n) = \|u_n\|^p_{p,a} \to \|u\|^p_{p,a} = \mathscr{J}(u)$, as claimed.

Now, either $W = \{u \in \tilde{D}^{s,p}(\Omega) : \mathscr{J}(u) \le 1\}$ is bounded in $\tilde{D}^{s,p}(\Omega)$, or not. In the first case we are done, while in the latter $\mathscr{I}$ is coercive in $W$, being coercive in $\tilde{D}^{s,p}(\Omega)$. Therefore, all the assumptions of Theorem 6.3.2 of [2] are fulfilled, being $\tilde{D}^{s,p}(\Omega)$ a reflexive Banach space, so that $\lambda_0$ is attained at a point $u_1 \in \tilde{D}^{s,p}(\Omega)$, with $\|u_1\|_{p,a} = 1$. We claim now that $\lambda_0 = \lambda_1$. Indeed,

$$\lambda_1 = \inf_{u \in \tilde{D}^{s,p}(\Omega)\setminus\{0\}} \left\| \frac{u}{\|u\|_{p,a}} \right\|^p = \inf_{\substack{u \in \tilde{D}^{s,p}(\Omega) \\ \|u\|_{p,a}=1}} \|u\|^p \ge \inf_{\substack{u \in \tilde{D}^{s,p}(\Omega) \\ 0<\|u\|_{p,a}\le 1}} \frac{\|u\|^p}{\|u\|^p_{p,a}} = \lambda_0 \ge \lambda_1.$$

In particular, $\lambda_1 = \|u_1\|^p > 0$ and $\mathscr{I}'(u_1) = \lambda_1 \mathscr{J}'(u_1)$ again by Theorem 6.3.2 of [2]. Hence $u_1$ is a solution of (4) when $\lambda = \lambda_1$. □

From the proof of Proposition 2.1 it is also evident that

$$\lambda_1 = \inf_{\substack{u \in \tilde{D}^{s,p}(\Omega) \\ \|u\|_{p,a}=1}} \|u\|^p.$$

Moreover Proposition 2.1 gives at once that

$$\lambda_1 \|u\|^p_{p,a} \le \|u\|^p \quad \text{for every } u \in \tilde{D}^{s,p}(\Omega). \tag{6}$$

In the following we put $c^p_{p,a} = 1/\lambda_1$.

On the perturbation $f$ we assume condition

($\mathscr{F}$) *Let $f : \overline{\Omega}\times\mathbb{R} \to \mathbb{R}$ be a Carathéodory function, $f \not\equiv 0$, satisfying the following properties.*

(a) *There exist two measurable functions $f_0$, $f_1$ on $\Omega$ and an exponent $q \in (1, p)$, such that $0 \le f_0(x) \le C_f a(x)$, $0 \le f_1(x) \le C_f a(x)$ a.e. in $\Omega$ and some appropriate constant $C_f > 0$, and*

$$|f(x,u)| \le f_0(x) + f_1(x)|u|^{q-1} \text{ for a.a. } x \in \Omega \text{ and all } u \in \mathbb{R}.$$

(b) *There exists* $\gamma \in (p, p^*/\alpha')$ *such that* $\limsup_{u\to 0} \dfrac{|f(x,u)|}{a(x)|u|^{\gamma-1}} < \infty$, *uniformly a.e. in* $\Omega$.

(c) $\int_\Omega F(x, u_1(x))dx > 0$, *where* $F(x,u) = \int_0^u f(x,v)dv$ *and* $u_1$ *is the first normalized eigenfunction given in Proposition* 2.1.

Note that, in the more familiar and standard setting in the literature, as e.g. in [6, 7, 9], in which $a \in L^\infty(\Omega)$, the exponent $\gamma$ in $(\mathscr{F})$–$(b)$ belongs to the open interval $(p, p^*)$. In any case $p < p^*/\alpha'$, since $\alpha > N/ps$.

As shown in [4], it is clear from $(\mathscr{F})$–$(a)$ and $(b)$ that problem (1) admits always the trivial solution since $f(x,0) = 0$ a.e. in $\Omega$, and that *the quantity*

$$S_f = \operatorname*{ess\,sup}_{u\neq 0,\, x\in\Omega} \frac{|f(x,u)|}{a(x)|u|^{p-1}} \tag{7}$$

*is a finite positive number*. In particular,

$$\operatorname*{ess\,sup}_{u\neq 0,\, x\in\Omega} \frac{|F(x,u)|}{a(x)|u|^{p}} \le \frac{S_f}{p} \tag{8}$$

and *the positive number*

$$\lambda_\star = \frac{\lambda_1}{1 + S_f} \tag{9}$$

is well defined and positive.

The main result of the section is proved by using the energy functional $J_\lambda$ associated to (1), which is given by $J_\lambda(u) = \Phi(u) + \lambda\Psi(u)$, where

$$\begin{gathered}\Phi(u) = \frac{1}{p}\|u\|^p, \quad \Psi(u) = -\mathscr{H}(u), \quad \mathscr{H}(u) = \mathscr{H}_1(u) + \mathscr{H}_2(u),\\ \mathscr{H}_1(u) = \frac{1}{p}\|u\|^p_{p,a}, \quad \mathscr{H}_2(u) = \int_\Omega F(x,u(x))dx.\end{gathered} \tag{10}$$

It is easy to see that the functional $J_\lambda$ is well defined in $\tilde{D}^{s,p}(\Omega)$ and of class $C^1$ in $\tilde{D}^{s,p}(\Omega)$. Furthermore, for all $u, \varphi \in \tilde{D}^{s,p}(\Omega)$,

$$\begin{aligned}\langle J_\lambda'(u), \varphi\rangle = &\iint_{\mathbb{R}^{2N}} |u(x)-u(y)|^{p-2}[u(x)-u(y)]\cdot[\varphi(x)-\varphi(y)]\cdot K(x-y)dxdy\\ &-\lambda\int_\Omega \left\{a(x)|u(x)|^{p-2}u(x) + f(x,u(x))\right\}\varphi(x)dx,\end{aligned}$$

where $\langle\cdot,\cdot\rangle$ denotes the duality pairing between $\tilde{D}^{s,p}(\Omega)$ and its dual space $\tilde{D}^{-s,p'}(\Omega)$. Therefore, the critical points $u \in \tilde{D}^{s,p}(\Omega)$ of the functional $J_\lambda$ are exactly the weak solutions of problem (1).

**Lemma 2.1** *The functional $\Phi : \tilde{D}^{s,p}(\Omega) \to \mathbb{R}$ is convex, weakly lower semicontinuous and of class $C^1$ in $\tilde{D}^{s,p}(\Omega)$.*
*Moreover, $\Phi' : \tilde{D}^{s,p}(\Omega) \to \tilde{D}^{-s,p'}(\Omega)$ verifies the $(\mathscr{S}_+)$ condition, i.e., for every sequence $(u_n)_n \subset \tilde{D}^{s,p}(\Omega)$ such that $u_n \rightharpoonup u$ weakly in $\tilde{D}^{s,p}(\Omega)$ and*

$$\limsup_{n\to\infty}\,\langle\Phi'(u_n), u_n - u\rangle \le 0,$$

$$\begin{aligned}\langle\Phi'(u_n), u_n - u\rangle = \iint_{\mathbb{R}^{2N}} &|u_n(x) - u_n(y)|^{p-2}[u_n(x) - u_n(y)]\cdot \\ &\times [u_n(x) - u(x) - u_n(y) + u(y)]\cdot K(x-y)dxdy,\end{aligned} \tag{11}$$

*then $u_n \to u$ strongly in $\tilde{D}^{s,p}(\Omega)$.*

*Proof* A simple calculation shows that the functional $\Phi$ is convex and of class $C^1$ in $\tilde{D}^{s,p}(\Omega)$. Hence, in particular $\Phi$ is weakly lower semicontinuous in $\tilde{D}^{s,p}(\Omega)$, see Corollary 3.9 of [3].

Let $(u_n)_n$ be a sequence in $\tilde{D}^{s,p}(\Omega)$ as in the statement. Then $\Phi(u) \le \liminf_{n\to\infty}\Phi(u_n)$, being $\Phi$ weakly lower semicontinuous in $\tilde{D}^{s,p}(\Omega)$. Furthermore, the linear functional $\langle\Phi'(u),\cdot\rangle : \tilde{D}^{s,p}(\Omega) \to \mathbb{R}$ is in $\tilde{D}^{-s,p'}(\Omega)$, since $(x,y) \mapsto |u(x) - u(y)|^{p-1}|x-y|^{-(N+ps)/p'} \in L^{p'}(\mathbb{R}^{2N})$, so that also $(x,y) \mapsto |u(x) - u(y)|^{p-1}K(x-y)^{1/p'} \in L^{p'}(\mathbb{R}^{2N})$ by $(\mathscr{K})$. Hence, since $u_n \rightharpoonup u$ in $\tilde{D}^{s,p}(\Omega)$ as $n \to \infty$,

$$\langle\Phi'(u), u_n - u\rangle = o(1) \quad \text{as } n \to \infty. \tag{12}$$

Therefore, $0 \le \limsup_{n\to\infty}\langle\Phi'(u_n) - \Phi'(u), u_n - u\rangle \le 0$ by convexity and (11). In other words,

$$\lim_{n\to\infty}\langle\Phi'(u_n) - \Phi'(u), u_n - u\rangle = 0. \tag{13}$$

Combining (12) with (13), we get

$$\lim_{n\to\infty}\langle\Phi'(u_n), u_n - u\rangle = 0. \tag{14}$$

By the convexity of $\Phi$ we have $\Phi(u) + \langle\Phi'(u_n), u_n - u\rangle \ge \Phi(u_n)$ for all $n$, so that $\Phi(u) \ge \limsup_{n\to\infty}\Phi(u_n)$ by (14). In conclusion,

$$\Phi(u) = \lim_{n\to\infty}\Phi(u_n). \tag{15}$$

Furthermore (13) implies that the sequence

$$n \mapsto \mathscr{U}_n(x,y) = \{|u_n(x)-u_n(y)|^{p-2}[u_n(x)-u_n(y)] - |u(x)-u(y)|^{p-2}[u(x)-u(y)]\}\cdot \\ \times [u_n(x)-u(x)-u_n(y)+u(y)]\cdot K(x-y) \geq 0$$

converges to 0 in $L^1(\mathbb{R}^{2N})$.

Fix now a subsequence $(u_{n_k})_k$ of $(u_n)_n$. Hence, up to a further subsequence if necessary, $\mathscr{U}_{n_k}(x,y) \to 0$ a.e. in $\mathbb{R}^{2N}$, and so $u_{n_k}(x) - u_{n_k}(y) \to u(x) - u(y)$ for a.a. $(x,y) \in \mathbb{R}^{2N}$. Indeed, fixing $x, y \in \mathbb{R}^N$, with $x \neq y$ and $\mathscr{U}_{n_k}(x,y) \to 0$, and putting $u_{n_k}(x) - u_{n_k}(y) = \xi_k$ and $u(x) - u(y) = \xi$, we get

$$\left(|\xi_k|^{p-2}\xi_k - |\xi|^{p-2}\xi\right)\cdot(\xi_k - \xi) \to 0, \tag{16}$$

since $K > 0$ by $(\mathscr{K})$. Hence $(\xi_k)_k$ is bounded in $\mathbb{R}$. Otherwise, up to a subsequence,

$$\left(|\xi_k|^{p-2}\xi_k - |\xi|^{p-2}\xi\right)\cdot(\xi_k - \xi) \sim |\xi_k|^p \to \infty,$$

which is obviously impossible. Therefore, $(\xi_k)_k$ is bounded and possesses a subsequence $(\xi_{k_j})_j$, which converges to some $\eta \in \mathbb{R}$. Thus (16) implies at once that $\left(|\eta|^{p-2}\eta_k - |\xi|^{p-2}\xi\right)\cdot(\eta - \xi) = 0$ and the strict convexity of $t \mapsto |t|^p$ yields $\eta = \xi$. This also shows that actually the entire sequence $(\xi_k)_k$ converges to $\xi$.

Consider the sequence $(g_{n_k})_k$ in $L^1(\mathbb{R}^{2N})$ defined pointwise by

$$g_{n_k}(x,y) = \left\{\frac{1}{2}\left(|u_{n_k}(x) - u_{n_k}(y)|^p + |u(x)-u(y)|^p\right) \right.\\ \left. - \left|\frac{u_{n_k}(x) - u_{n_k}(y) - u(x) + u(y)}{2}\right|^p\right\} K(x-y).$$

By convexity $g_{n_k} \geq 0$ and $g_{n_k}(x,y) \to |u(x)-u(y)|^p K(x-y)$ for a.a. $(x,y) \in \mathbb{R}^{2N}$ as $k \to \infty$. Therefore, by the Fatou lemma and (15) we have

$$p\,\Phi(u) \leq \liminf_{k\to\infty} \iint_{\mathbb{R}^{2N}} g_{n_k}(x,y)dxdy \\ = p\,\Phi(u) - \frac{1}{2^p}\limsup_{k\to\infty} \iint_{\mathbb{R}^{2N}} |u_{n_k}(x) - u_{n_k}(y) - u(x) + u(y)|^p K(x-y)dxdy.$$

Hence, $\limsup_{k\to\infty} \|u_{n_k} - u\| \leq 0$, that is $\lim_{k\to\infty} \|u_{n_k} - u\| = 0$. Since $(u_{n_k})_k$ is an arbitrary subsequence of $(u_n)_n$, this shows that actually the entire sequence $(u_n)_n$ converges strongly to $u$ in $\tilde{D}^{s,p}(\Omega)$, as required. □

If $\Psi(v) < 0$ at some $v \in \tilde{D}^{s,p}(\Omega)$, that is $\Psi^{-1}(I_0)$ is non-empty, where $I_0 = (-\infty, 0) = \mathbb{R}^-$, then *the crucial positive number*

$$\lambda^\star = \inf_{u\in\Psi^{-1}(I_0)} -\frac{\Phi(u)}{\Psi(u)} \tag{17}$$

is well defined.

**Lemma 2.2** *If* $(\mathscr{F})$*–(a), (b) and (c) hold, then* $\Psi^{-1}(I_0)$ *is non-empty and moreover* $\lambda_\star \le \lambda^\star < \lambda_1$.

*Proof* By $(\mathscr{F})$–$(c)$ it follows that

$$\mathscr{H}(u_1) > \frac{1}{p}, \quad \text{i.e. } u_1 \in \Psi^{-1}(I_0).$$

Hence, $\lambda^\star$ is well defined. Again by $(\mathscr{F})$–$(c)$ and Proposition 2.1

$$\lambda^\star = \inf_{u\in\Psi^{-1}(I_0)} -\frac{\Phi(u)}{\Psi(u)} \le \frac{\Phi(u_1)}{\mathscr{H}(u_1)} = \frac{\|u_1\|^p}{p\mathscr{H}(u_1)} < \|u_1\|^p = \lambda_1,$$

as required. Finally, by $(\mathscr{F})$–$(a)$, $(b)$, (6), (8) and (10), for all $u \in \tilde{D}^{s,p}(\Omega)$, with $u \neq 0$, we have

$$\frac{\Phi(u)}{|\Psi(u)|} \ge \frac{\|u\|^p}{(1+S_f)\|u\|^p_{p,a}} \ge \frac{\lambda_1}{1+S_f} = \lambda_\star.$$

Hence, in particular $\lambda^\star \ge \lambda_\star$. □

**Lemma 2.3** *If* $(\mathscr{F})$*–(a) holds, then* $\mathscr{H}_1'$, $\mathscr{H}_2'$, $\Psi' : \tilde{D}^{s,p}(\Omega) \to \tilde{D}^{-s,p'}(\Omega)$ *are compact and* $\mathscr{H}_1$, $\mathscr{H}_2$, $\Psi$ *are sequentially weakly continuous in* $\tilde{D}^{s,p}(\Omega)$.

*Proof* Since $\Psi = -\mathscr{H}$, it is enough to prove the lemma for $\mathscr{H}$. Of course, $\mathscr{H}' = \mathscr{H}_1' + \mathscr{H}_2'$, where

$$\langle \mathscr{H}_1'(u), v\rangle = \int_\Omega a(x)|u|^{p-2}uv\,dx \quad \text{and} \quad \langle \mathscr{H}_2'(u), v\rangle = \int_\Omega f(x,u)v\,dx,$$

for all $u, v \in \tilde{D}^{s,p}(\Omega)$. Since $\mathscr{H}_1'$ and $\mathscr{H}_2'$ are continuous, thanks to the reflexivity of $\tilde{D}^{s,p}(\Omega)$ it is sufficient to show that $\mathscr{H}_1'$ and $\mathscr{H}_2'$ are weak-to-strong sequentially continuous, i.e. if $(u_n)_n$, $u$ are in $\tilde{D}^{s,p}(\Omega)$ and $u_n \rightharpoonup u$ in $\tilde{D}^{s,p}(\Omega)$ as $n \to \infty$, then $\|\mathscr{H}_1'(u_n) - \mathscr{H}_1'(u)\|_{\tilde{D}^{-s,p'}(\Omega)} \to 0$ and $\|\mathscr{H}_2'(u_n) - \mathscr{H}_2'(u)\|_{\tilde{D}^{-s,p'}(\Omega)} \to 0$ as $n \to \infty$. To this aim, fix $(u_n)_n \subset \tilde{D}^{s,p}(\Omega)$, with $u_n \rightharpoonup u$ in $\tilde{D}^{s,p}(\Omega)$.

From the fact that $u_n \to u$ in $L^p(\Omega, a)$ by (3), then $\|u_n\|_{p,a} \to \|u\|_{p,a}$, or equivalently, $\|v_n\|_{p',a} \to \|v\|_{p',a}$, where $v_n = |u_n|^{p-2}u_n$ and similarly $v = |u|^{p-2}u$. We claim that $v_n \to v$ in $L^{p'}(\Omega, a)$. Indeed, fix any subsequence $(v_{n_k})_k$ of $(v_n)_n$. The related subsequence $(u_{n_k})_k$ of $(u_n)_n$ converges in $L^p(\Omega, a)$ and admits a subsequence, say $(u_{n_{k_j}})_j$, converging to $u$ a.e. in $\Omega$. Hence, the corresponding subsequence $(v_{n_{k_j}})_j$ of $(v_{n_k})_k$ converges to $v$ a.e. in $\Omega$. Therefore, being $1 < p' < \infty$, by the Clarkson and Mil'man theorems it follows that $v_{n_{k_j}} \rightharpoonup v$ in $L^{p'}(\Omega, a)$, since the sequence $(\|v_n\|_{p',a})_n$ is bounded, and so by Radon's theorem we get that $v_{n_{k_j}} \to v$ in $L^{p'}(\Omega, a)$, since $\|v_n\|_{p',a} \to \|v\|_{p',a}$. This shows the claim, since the subsequence $(v_{n_k})_k$ of $(v_n)_n$ is arbitrary.

Now, for all $\varphi \in \tilde{D}^{s,p}(\Omega)$, with $\|\varphi\| = 1$, by Hölder's inequality,

$$\begin{aligned}|\langle \mathscr{H}_1'(u_n) - \mathscr{H}_1'(u), \varphi\rangle| &\le \int_\Omega a^{1/p'}|v_n - v| \cdot a^{1/p}|\varphi| dx \le \|v_n - v\|_{p',a}\|\varphi\|_{p,a} \\ &\le c_{p,a}\|v_n - v\|_{p',a},\end{aligned}$$

where $c_{p,a}^p = 1/\lambda_1$ is the Sobolev constant for the embedding $\tilde{D}^{s,p}(\Omega) \hookrightarrow L^p(\Omega, a)$ by (5) and (6). Hence, $\|\mathscr{H}_1'(u_n) - \mathscr{H}_1'(u)\|_{\tilde{D}^{-s,p'}(\Omega)} \to 0$ as $n \to \infty$ and $\mathscr{H}_1'$ is compact.

Similarly, $u_n \to u$ in $L^q(\Omega, a)$, since the embedding $\tilde{D}^{s,p}(\Omega) \hookrightarrow L^q(\Omega, a)$ is compact, being $L^p(\Omega, a) \hookrightarrow L^q(\Omega, a)$ continuous, since $1 < q < p$ by $(\mathscr{F})$–$(a)$. Indeed $\|v\|_{q,a} \le \|a\|_1^{1/q-1/p}\|v\|_{p,a}$ for all $v \in L^p(\Omega, a)$ by Hölder's inequality and the fact that $a \in L^\alpha(\Omega) \subset L^1(\Omega)$, $\alpha > N/ps > 1$ and $\Omega$ is bounded. Clearly, the Nemitskii operator $N_f : L^q(\Omega, a) \to L^{q'}(\Omega, a^{1/(1-q)})$ given by $N_f(u) = f(\cdot, u(\cdot))$ for all $u \in L^q(\Omega, a)$ is well defined thanks to $(\mathscr{F})$–$(a)$. We assert that $N_f(u_n) \to N_f(u)$ in $L^{q'}(\Omega, a^{1/(1-q)})$ as $n \to \infty$. Indeed, fix a subsequence $(u_{n_k})_k$ of $(u_n)_n$. Hence, there exists a subsequence, still denoted by $(u_{n_k})_k$, such that $u_{n_k} \to u$ a.e. in $\Omega$ and $|u_{n_k}| \le h$ a.e. in $\Omega$ for all $k \in \mathbb{N}$ and some $h \in L^q(\Omega, a)$. In particular, $|N_f(u_{n_k}) - N_f(u)|^{q'} a^{1/(1-q)} \to 0$ a.e. in $\Omega$, being $f(x, \cdot)$ continuous for a.a. $x \in \Omega$. Furthermore, $|N_f(u_{n_k}) - N_f(u)|^{q'} a^{1/(1-q)} \le \kappa a(1 + h^q) \in L^1(\Omega)$, $\kappa = (2C_f)^{q'} 2^{q'-1}$, by $(\mathscr{F})$–$(a)$, being $a \in L^\alpha(\Omega) \subset L^1(\Omega)$, since $\alpha > N/ps > 1$ and $\Omega$ is bounded. This shows the assertion, since $1 < q < p$ by $(\mathscr{F})$–$(a)$. Hence, by the dominated convergence theorem, we have $N_f(u_{n_k}) \to N_f(u)$ in $L^{q'}(\Omega, a^{1/(1-q)})$. Therefore the entire sequence $N_f(u_n) \to N_f(u)$ in $L^{q'}(\Omega, a^{1/(1-q)})$ as $n \to \infty$.

Finally, for all $\varphi \in \tilde{D}^{s,p}(\Omega)$, with $\|\varphi\| = 1$, we have by Hölder's inequality,

$$\begin{aligned}|\langle \mathscr{H}_2'(u_n) - \mathscr{H}_2'(u), \varphi\rangle| &\le \int_\Omega a^{-1/q}|N_f(u_n) - N_f(u)| \cdot a^{1/q}|\varphi| dx \\ &\le \|N_f(u_n) - N_f(u)\|_{q',a^{1/(1-q)}}\|\varphi\|_{q,a} \\ &\le c_{p,a}\|a\|_1^{1/q-1/p}\|N_f(u_n) - N_f(u)\|_{q',a^{1/(1-q)}},\end{aligned}$$

where $c_{p,a}$ is given in (6). Thus, $\|\mathscr{H}_2'(u_n) - \mathscr{H}_2'(u)\|_{\tilde{D}^{-s,p'}(\Omega)} \to 0$ as $n \to \infty$, that is $\mathscr{H}_2'$ is compact.

Since by the above steps $\mathscr{H}' = \mathscr{H}_1' + \mathscr{H}_2'$ is compact, then $\mathscr{H}$ is sequentially weakly continuous by Zeidler [13, Corollary 41.9], being $\tilde{D}^{s,p}(\Omega)$ reflexive. □

**Lemma 2.4** *Under the assumption* $(\mathscr{F})$–$(a)$ *the functional* $J_\lambda(u) = \Phi(u) + \lambda\Psi(u)$ *is coercive for every* $\lambda \in (-\infty, \lambda_1)$.

*Proof* Fix $\lambda \in (-\infty, \lambda_1)$. Then by (6) and $(\mathscr{F})$–$(a)$

$$\begin{aligned} J_\lambda(u) &\geq \frac{1}{p}\left(1 - \frac{\lambda}{\lambda_1}\right)\|u\|^p - |\lambda| \int_\Omega |F(x,u)|dx \\ &\geq \frac{1}{p}\left(1 - \frac{\lambda}{\lambda_1}\right)\|u\|^p - |\lambda| \int_\Omega f_0(x)dx - |\lambda| \int_\Omega \left[f_0(x) + \frac{f_1(x)}{q}\right]|u|^q dx \\ &\geq \frac{1}{p}\left(1 - \frac{\lambda}{\lambda_1}\right)\|u\|^p - |\lambda| C_1 - |\lambda|\, C_2\|u\|^q, \end{aligned} \tag{18}$$

where $C_1 = \|f_0\|_1$, $C_2 = c_{\alpha' q}^q \|f_0 + f_1/q\|_\alpha$ and $c_{\alpha' q}$ denotes the Sobolev constant of the compact embedding $\tilde{D}^{s,p}(\Omega) \hookrightarrow L^{\alpha' q}(\Omega)$, being $\alpha' q < p^*$. Note that $C_1 < \infty$, since $f_0 \in L^\alpha(\Omega) \subset L^1(\Omega)$, by $(\mathscr{F})$–$(a)$, being $\alpha > N/ps > 1$ and $\Omega$ bounded. This shows the assertion, since $1 < q < p$ by $(\mathscr{F})$–$(a)$.

## 3 The Main Result

In this section we prove an existence theorem for (1) as an application of the principle abstract Theorem 2.1-(*ii*), Part (*a*) in [4], which represents the differential version of the *Arcoya* and *Carmona* Theorem 3.4 in [1]. In order to simplify the notation let us introduce the main auxiliary functions

$$\begin{aligned} \varphi_1(r) &= \inf_{u \in \Psi^{-1}(I_r)} \frac{\inf_{v \in \Psi^{-1}(r)} \Phi(v) - \Phi(u)}{\Psi(u) - r}, \quad I_r = (-\infty, r), \\ \varphi_2(r) &= \sup_{u \in \Psi^{-1}(I^r)} \frac{\inf_{v \in \Psi^{-1}(r)} \Phi(v) - \Phi(u)}{\Psi(u) - r}, \quad I^r = (r, \infty), \end{aligned} \tag{19}$$

which are well-defined for all $r \in \left(\inf_{u \in \tilde{D}^{s,p}(\Omega)} \Psi(u), \sup_{u \in \tilde{D}^{s,p}(\Omega)} \Psi(u)\right)$, see [1, 4].

**Theorem 3.1** *Assume* $(\mathscr{F})$–$(a)$ *and* $(b)$.

(*i*) *If* $\lambda \in [0, \lambda_\star)$, *where* $\lambda_\star$ *is defined in* (9), *then* (1) *has only the trivial solution.*
(*ii*) *If* $f$ *satisfies also* $(\mathscr{F})$–$(c)$, *then problem* (1) *admits at least two nontrivial solutions for every* $\lambda \in (\lambda^\star, \lambda_1)$, *where* $\lambda^\star < \lambda_1$ *is given in* (17).

*Proof* Let $u \in \tilde{D}^{s,p}(\Omega)$ be a nontrivial weak solution of the problem (1), then

$$\iint_{\mathbb{R}^{2N}} |u(x)-u(y)|^{p-2}[u(x)-u(y)]\cdot[\varphi(x)-\varphi(y)]\cdot K(x-y)dxdy$$
$$= \lambda \int_{\Omega} \{a(x)|u|^{p-2}u + f(x,u)\}\varphi\, dx$$

for all $\varphi \in \tilde{D}^{s,p}(\Omega)$. Take $\varphi = u$ and put $\Omega_0 = \{x \in \Omega\, :\, u(x) \neq 0\}$, so that

$$\lambda_1\|u\|^p = \lambda_1\lambda\left(\|u\|_{p,a}^p + \int_{\Omega_0}\frac{f(x,u)}{a(x)|u|^{p-1}}a(x)|u|^p dx\right) \leq \lambda_1\lambda\left(1+S_f\right)\|u\|_{p,a}^p$$
$$\leq \lambda\left(1+S_f\right)\|u\|^p$$

by (6) and (7). Therefore $\lambda \geq \lambda_\star$ by (9), as required.

(*ii*) The functional $\Phi$ is clearly convex, $\Phi$ is also weakly lower semicontinuous in $\tilde{D}^{s,p}(\Omega)$ and $\Phi'$ verifies condition $(\mathscr{S}_+)$, as already proved in Lemma 2.1. Furthermore, $\Psi' : \tilde{D}^{s,p}(\Omega) \to \tilde{D}^{-s,p'}(\Omega)$ is compact and $\Psi$ is sequentially weakly continuous in $\tilde{D}^{s,p}(\Omega)$ by Lemma 2.3. The functional $J_\lambda$ is coercive for every $\lambda \in I$, where $I = (-\infty, \lambda_1)$, thanks to Lemma 2.4.

We claim that $\Psi(\tilde{D}^{s,p}(\Omega)) \supset \mathbb{R}_0^-$. Indeed, $\Psi(0) = 0$ and by $(\mathscr{F})$–$(a)$

$$\Psi(u) \leq -\frac{1}{p}\|u\|_{p,a}^p + \int_\Omega |F(x,u)|dx \leq -\frac{1}{p}\|u\|_{p,a}^p + \|f_0\|_1 + 2C_f\int_\Omega a(x)|u|^q dx$$
$$\leq -\frac{1}{p}\|u\|_{p,a}^p + \|f_0\|_1 + 2C_f\|a\|_1^{(p-q)/p}\|u\|_{p,a}^q,$$

since $a \in L^\alpha(\Omega) \subset L^1(\Omega)$, being $\alpha > N/ps > 1$ and $\Omega$ bounded. Therefore,

$$\lim_{u\in\tilde{D}^{s,p}(\Omega),\ \|u\|_{p,a}\to\infty} \Psi(u) = -\infty,$$

being $q < p$. Hence, the claim follows by the continuity of $\Psi$.

Thus, $(\inf\Psi, \sup\Psi) \supset \mathbb{R}_0^-$. For every $u \in \Psi^{-1}(I_0)$ we have

$$\varphi_1(r) \leq \frac{\Phi(u)}{r-\Psi(u)} \quad \text{for all } r \in (\Psi(u), 0),$$

so that

$$\limsup_{r\to 0^-}\varphi_1(r) \leq -\frac{\Phi(u)}{\Psi(u)} \quad \text{for all } u \in \Psi^{-1}(I_0).$$

In other words, by (17) and (19)

$$\limsup_{r\to 0^-} \varphi_1(r) \leq \varphi_1(0) = \lambda^\star. \tag{20}$$

From $(\mathscr{F})$–$(b)$ and L'Hôpital's rule

$$\limsup_{u\to 0} \frac{|F(x,u)|}{a(x)|u|^\gamma} < \infty \quad \text{uniformly a.e. in } \Omega,$$

so that using also $(\mathscr{F})$–$(a)$ and again $(b)$, that is (7), it follows the existence of a positive real number $L > 0$ such that

$$|F(x,u)| \leq La(x)|u|^\gamma \quad \text{for a.a. } x \in \Omega \text{ and all } u \in \mathbb{R}. \tag{21}$$

The embedding $\tilde{D}^{s,p}(\Omega) \hookrightarrow L^\gamma(\Omega, a)$ is continuous, since $\gamma \in (p, p^*/\alpha')$ by $(\mathscr{F})$–$(b)$. Indeed, by Hölder's inequality $\|u\|_{\gamma,a}^\gamma \leq |\Omega|^{1/\wp}\|a\|_\alpha \|u\|_{p^*}^\gamma \leq c\|u\|^\gamma$ for all $u \in \tilde{D}^{s,p}(\Omega)$, where $c = c_{p^*}^\gamma |\Omega|^{1/\wp}\|a\|_\alpha$ and $c_{p^*}$ is the Sobolev constant of the continuous embedding $\tilde{D}^{s,p}(\Omega) \hookrightarrow L^{p^*}(\Omega)$ and $\wp$ is the crucial exponent

$$\wp = \frac{\alpha' p^*}{p^* - \gamma\alpha'} > 1,$$

being $\gamma \in (p, p^*/\alpha')$ by $(\mathscr{F})$–$(b)$. Hence, by (21)

$$|\Psi(u)| \leq \frac{1}{p\lambda_1}\|u\|^p + C\|u\|^\gamma, \tag{22}$$

for every $u \in \tilde{D}^{s,p}(\Omega)$, where $C = cL$. Therefore, given $r < 0$ and $v \in \Psi^{-1}(r)$, we get

$$r = \Psi(v) \geq -\frac{1}{p\lambda_1}\|v\|^p - C\|v\|^\gamma = -\frac{1}{\lambda_1}\Phi(v) - \kappa\Phi(v)^{\gamma/p}, \tag{23}$$

where $\kappa = Cp^{\gamma/p}$. Since the functional $\Phi$ is bounded below, coercive and lower semicontinuous on the reflexive Banach space $\tilde{D}^{s,p}(\Omega)$, it is easy to see that $\Phi$ is also coercive on the sequentially weakly closed non-empty set $\Psi^{-1}(r)$, see Lemma 2.3. Therefore, by Theorem 6.1.1 of [2], there exists an element $u_r \in \Psi^{-1}(r)$ such that $\Phi(u_r) = \inf_{v\in\Psi^{-1}(r)} \Phi(v)$. Taking $u \equiv 0 \in \Psi^{-1}(I^r)$ in (19), we have

$$\varphi_2(r) \geq -\frac{1}{r}\inf_{v\in\Psi^{-1}(r)} \Phi(v) = \frac{\Phi(u_r)}{|r|}.$$

Hence (23), evaluated at $v = u_r$ and divided by $r < 0$, gives

$$1 \le \frac{1}{\lambda_1} \cdot \frac{\Phi(u_r)}{|r|} + \kappa |r|^{\gamma/p-1} \left( \frac{\Phi(u_r)}{|r|} \right)^{\gamma/p} \le \frac{\varphi_2(r)}{\lambda_1} + \kappa |r|^{\gamma/p-1} \varphi_2(r)^{\gamma/p}.$$

There are now two possibilities to be considered: either $\varphi_2$ is locally bounded at $0^-$, so that the above inequality shows at once that

$$\liminf_{r \to 0^-} \varphi_2(r) \ge \lambda_1,$$

being $\gamma > p$ by $(\mathscr{F})$–$(b)$, or $\limsup_{r \to 0^-} \varphi_2(r) = \infty$. In both cases (20) and Lemma 2.2 yield

$$\limsup_{r \to 0^-} \varphi_1(r) \le \lambda^\star < \lambda_1 \le \limsup_{r \to 0^-} \varphi_2(r).$$

Hence for all integers $n \ge n^\star = 1 + [2/(\lambda_1 - \lambda^\star)]$ there exists a number $r_n < 0$ so close to zero that $\varphi_1(r_n) < \lambda^\star + 1/n < \lambda_1 - 1/n < \varphi_2(r_n)$. In particular,

$$[\lambda^\star + 1/n, \lambda_1 - 1/n] \subset (\varphi_1(r_n), \varphi_2(r_n)) \cap I = (\varphi_1(r_n), \varphi_2(r_n)) \tag{24}$$

for all $n \ge n^\star$. Therefore, since all the assumptions of Theorem 2.1-(*ii*), Part (*a*), in [4] are satisfied and $u \equiv 0$ is a critical point of $J_\lambda$, problem (1) admits at least two nontrivial solutions for all $\lambda \in (\varphi_1(r_n), \varphi_2(r_n))$ and for all $n \ge n^\star$. In conclusion, problem (1) admits at least two nontrivial solutions for all $\lambda \in (\lambda^\star, \lambda_1)$ as required, being

$$(\lambda^\star, \lambda_1) = \bigcup_{n=n^\star}^{\infty} [\lambda^\star + 1/n, \lambda_1 - 1/n] \subset \bigcup_{n=n^\star}^{\infty} (\varphi_1(r_n), \varphi_2(r_n))$$

by (24). □

Taking inspiration from [4], also in this new setting we can derive an interesting consequence from the main Theorem 3.1 for a simpler problem. Let us therefore replace $(\mathscr{F})$–$(c)$ by the next condition much easier to verify.

$(\mathscr{F})$–$(c')$ *Assume there exist* $x_0 \in \Omega$, $t_0 \in \mathbb{R}$ *and* $r_0 > 0$ *so small that the closed ball* $B_0 = \{x \in \mathbb{R}^N : |x - x_0| \le r_0\}$ *is contained in* $\Omega$ *and*

$$\operatorname*{ess\,inf}_{B_0} F(x, |t_0|) = \mu_0 > 0, \qquad \operatorname*{ess\,sup}_{B_0} \max_{|t| \le |t_0|} |F(x,t)| = M_0 < \infty.$$

Clearly, when $f$ does not depend on $x$, condition $(\mathscr{F})$–$(c')$ simply reduces to the request that $F(t_0) > 0$ at a point $t_0 \in \mathbb{R}$.

**Corollary 3.1** *Assume that* $f : \overline{\Omega} \times \mathbb{R} \to \mathbb{R}$ *satisfies* $(\mathscr{F})$–$(a)$, $(b)$. *Consider the problem*

$$\begin{cases} \mathscr{L}_K u = \lambda f(x,u), & \text{in } \Omega, \\ u = 0, & \text{in } \mathbb{R}^N \setminus \Omega. \end{cases} \tag{25}$$

(*i*) *If* $\lambda \in [0, \ell_\star)$, *where* $\ell_\star = \lambda_1/S_f$, *then* (25) *has only the trivial solution.*
(*ii*) *If furthermore* $f$ *satisfies* $(\mathscr{F})$–$(c')$, *then there exists* $\ell^\star \geq \ell_\star$ *such that* (25) *admits at least two nontrivial solutions for all* $\lambda \in (\ell^\star, \infty)$.

*Proof* The energy functional $J_\lambda$ associated to problem (25) is simply given by $J_\lambda(u) = \Phi(u) + \lambda \Psi_2(u)$, where as before $\Psi_2(u) = -\displaystyle\int_\Omega F(x, u(x))dx$, see (10). First, note that $J_\lambda$ is coercive for every $\lambda \in \mathbb{R}$. Indeed, by (18)

$$J_\lambda(u) \geq \frac{1}{p}\|u\|^p - |\lambda| \int_\Omega |F(x,u)|dx \geq \frac{1}{p}\|u\|^p - |\lambda|\, C_1 - |\lambda|\, C_2\|u\|^q,$$

where $C_1$ and $C_2$ are as in (18). Hence $J_\lambda(u) \to \infty$ as $\|u\| \to \infty$, since $1 < q < p$ by $(\mathscr{F})$–$(a)$. In conclusion, here $I = \mathbb{R}$, as claimed.

The part (*i*) of the statement is proved using the same argument produced for the proof of Theorem 3.1-(*i*), being

$$\lambda_1\|u\|^p = \lambda_1\lambda \int_\Omega f(x,u)udx \leq \lambda_1\lambda S_f\|u\|^p_{p,a} \leq \lambda S_f\|u\|^p$$

by (6) and (7). Thus, if $u$ is a nontrivial weak solution of (25), then necessarily $\lambda \geq \ell_\star = \lambda_1/S_f$, as required.

In order to prove (*ii*), we first show that there exists $u_0 \in \tilde{D}^{s,p}(\Omega)$ such that $\Psi_2(u_0) < 0$, so that the crucial number

$$\ell^\star = \varphi_1(0) = \inf_{u \in \Psi_2^{-1}(I_0)} -\frac{\Phi(u)}{\Psi_2(u)}, \quad I_0 = (-\infty, 0) = \mathbb{R}^-,$$

is well defined. Indeed, in this special subcase (19) simply reduces to

$$\begin{aligned} \varphi_1(r) &= \inf_{u \in \Psi_2^{-1}(I_r)} \frac{\displaystyle\inf_{v \in \Psi_2^{-1}(r)} \Phi(v) - \Phi(u)}{\Psi_2(u) - r}, \quad I_r = (-\infty, r), \\ \varphi_2(r) &= \sup_{u \in \Psi_2^{-1}(I^r)} \frac{\displaystyle\inf_{v \in \Psi_2^{-1}(r)} \Phi(v) - \Phi(u)}{\Psi_2(u) - r}, \quad I^r = (r, \infty). \end{aligned} \tag{26}$$

Clearly $t_0 \neq 0$ in $(\mathscr{F})$–$(c')$. Now take $\sigma \in (0,1)$ and put

$$B = \{x \in \mathbb{R}^N : |x - x_0| \leq \sigma r_0\}, \quad B_1 = \left\{x \in \mathbb{R}^N : |x - x_0| \leq r_1\right\},$$

where $r_1 = (1+\sigma)r_0/2$. Hence $B \subset B_1 \subset B_0$. Set $v_0(x) = |t_0|\chi_{B_1}(x)$ and denote by $\rho_\varepsilon$ the convolution kernel of fixed radius $\varepsilon$, with $0 < \varepsilon < (1-\sigma)r_0/2$. Define

$$u_0(x) = \rho_\varepsilon * v_0(x),$$

so that $u_0(x) = |t_0|$ for all $x \in B$, $0 \leq u_0(x) \leq |t_0|$ for all $x \in \Omega$, $u_0 \in C_0^\infty(\Omega)$ and $\operatorname{supp} u_0 \subset B_0$. Therefore, $u_0 \in \tilde{D}^{s,p}(\Omega)$ by $(\mathscr{K})$. From $(\mathscr{F})$–$(c')$ we also have

$$\begin{aligned}\Psi_2(u_0) &= -\int_B F(x,|t_0|)dx - \int_{B_0\setminus B} F\left(x, u_0(x)\right)dx \leq M_0 \int_{B_0\setminus B} dx - \mu_0 \int_B dx \\ &\leq \omega_N r_0^N \left[M_0(1-\sigma^N) - \mu_0\sigma^N\right].\end{aligned}$$

Hence, taking $\sigma \in (0,1)$ so large that $\sigma^N > M_0/(\mu_0 + M_0)$, we get that $\Psi_2(u_0) < 0$, as claimed.

Furthermore, by (6), (8) and (10), for all $u \in \tilde{D}^{s,p}(\Omega)$, with $u \not\equiv 0$, we have

$$\frac{\Phi(u)}{|\Psi_2(u)|} \geq \frac{\|u\|^p}{S_f\|u\|_{p,a}^p} \geq \frac{\lambda_1}{S_f} = \ell_\star.$$

Thus, $\ell^\star \geq \ell_\star$.

In particular, for $\varphi_1$ given now by (26) and for all $u \in \Psi_2^{-1}(I_0)$, we get

$$\varphi_1(r) \leq \frac{\Phi(u)}{r - \Psi_2(u)} \quad \text{for all } r \in (\Psi_2(u), 0).$$

Therefore,

$$\limsup_{r\to 0^-} \varphi_1(r) \leq \varphi_1(0) = \ell^\star,$$

which is the analog of (20).

Also in this setting (21) holds and (22) simply reduces to

$$|\Psi_2(u)| \leq C\|u\|^\gamma.$$

Taken $r < 0$ and $v \in \Psi_2^{-1}(r)$, we obtain

$$r = \Psi_2(v) \geq -C\|v\|^\gamma \geq -C\left[p\Phi(v)\right]^{\gamma/p}.$$

Therefore, by (26), since $u \equiv 0 \in \Psi_2^{-1}(I^r)$,

$$\varphi_2(r) \geq \frac{1}{|r|} \inf_{v \in \Psi_2^{-1}(r)} \Phi(v) \geq \kappa |r|^{p/\gamma - 1},$$

where $\kappa = C^{-p/\gamma}/p$. This implies that $\lim_{r \to 0^-} \varphi_2(r) = \infty$, being $\gamma > p$ by $(\mathscr{F})$–$(b)$.

In conclusion, we have proved that

$$\limsup_{r \to 0^-} \varphi_1(r) \leq \varphi_1(0) = \ell^\star < \lim_{r \to 0^-} \varphi_2(r) = \infty.$$

This shows that for all integers $n \geq n^\star = 2 + [\ell^\star]$ there exists $r_n < 0$ so close to zero that $\varphi_1(r_n) < \ell^\star + 1/n < n < \varphi_2(r_n)$. Hence, all the assumptions of Theorem 2.1-($ii$), Part ($a$) are satisfied and, being $u \equiv 0$ a critical point of $J_\lambda$ and $I = \mathbb{R}$, problem (25) admits at least two nontrivial solutions for all

$$\lambda \in \bigcup_{n=n^\star}^{\infty} (\varphi_1(r_n), \varphi_2(r_n)) \supset \bigcup_{n=n^\star}^{\infty} [\ell^\star + 1/n, n] = (\ell^\star, \infty),$$

as stated. □

**Acknowledgements** The first author was partially supported by the Italian MIUR Project *Aspetti variazionali e perturbativi nei problemi differenziali nonlineari* (201274FYK7) and is a member of the *Gruppo Nazionale per l'Analisi Matematica, la Probabilità e le loro Applicazioni* (GNAMPA) of the *Istituto Nazionale di Alta Matematica* (INdAM). The manuscript was realized within the auspices of the INdAM–GNAMPA Project 2015 titled *Modelli ed equazioni nonlocali di tipo frazionario* (Prot_2015_000368).

We thank the anonymous referee for the careful reading of our manuscript and the useful comments.

## References

1. Arcoya, D., Carmona, J.: A nondifferentiable extension of a theorem of Pucci and Serrin and applications. J. Differ. Equ. **235**, 683–700 (2007)
2. Berger, M.S.: Nonlinearity and Functional Analysis. Lectures on Nonlinear Problems in Mathematical Analysis. Pure and Applied Mathematics. Academic, New York/London (1977)
3. Brezis, H.: Functional Analysis, Sobolev Spaces and Partial Differential Equations. Universitext. Springer, New York (2011)
4. Colasuonno, F., Pucci P., Varga, Cs.: Multiple solutions for an eigenvalue problem involving $p$-Laplacian type operators. Nonlinear Anal. **75**, 4496–4512 (2012)
5. Di Nezza, E., Palatucci, G., Valdinoci, E.: Hitchhiker's guide to the fractional Sobolev spaces. Bull. Sci. Math. **136**, 521–573 (2012)
6. Fiscella, A., Servadei, R., Valdinoci, E., A resonance problem for non-local elliptic operators. Z. Anal. Anwend. **32**, 411–431 (2013)
7. Franzina, G., Palatucci, G.: Fractional $p$-eigenvalues. Riv. Math. Univ. Parma (N.S.) **5**, 373–386 (2014)

8. Grisvard, P.: Elliptic Problems in Nonsmooth Domains. Monographs and Studies in Mathematics, vol. 24. Pitman, Boston (1985)
9. Lindgren, E., Lindqvist, P.: Fractional eigenvalues. Calc. Var. Partial Differ. Equ. **49**, 795–826 (2013)
10. Maz'ya, V., Shaposhnikova, T.: On the Bourgain, Brezis, and Mironescu theorem concerning limiting embeddings of fractional Sobolev spaces. J. Funct. Anal. **195**, 230–238 (2002)
11. Pucci, P., Serrin, J.: Extensions of the mountain pass theorem. J. Funct. Anal. **59**, 185–210 (1984)
12. Pucci, P., Serrin, J.: A mountain pass theorem. J. Differ. Equ. **60**, 142–149 (1985)
13. Zeidler, E.: Nonlinear Functional Analysis and Its Applications. III: Variational Methods and Optimization. Springer, New York (1985)

# Uniqueness of Solutions of a Class of Quasilinear Subelliptic Equations

**Lorenzo D'Ambrosio and Enzo Mitidieri**

*Dedicated to Ermanno Lanconelli on the occasion of his 70th birthday*

**Abstract** We study the uniqueness problem of the equation,

$$-\Delta_{L,p}u + |u|^{q-1}u = h \quad on \quad \mathbb{R}^N,$$

where $q > p - 1 > 0$. and $N > p$. Uniqueness results proved in this paper hold for equations associated to the mean curvature type operators as well as for more general quasilinear coercive subelliptic problems.

**Keywords** A priori estimates • Carnot groups • Comparison and uniqueness • Quasilinear elliptic inequalities

**AMS Subject Classifications:** 35B45, 35B51, 35B53, 35J62, 35J70, 35R03

## 1 Introduction

Nonlinear elliptic problems of coercive type is still an interesting subject for scholars of nonlinear partial differential equations.

---

L. D'Ambrosio
Dipartimento di Matematica, Università degli Studi di Bari, Bari, Italy
e-mail: dambros@dm.uniba.it

E. Mitidieri (✉)
Dipartimento di Matematica e Geoscienze, Università degli Studi di Trieste, Trieste, Italy
e-mail: mitidier@units.it

 
G. Citti et al. (eds.), *Geometric Methods in PDE's*, Springer INdAM Series 13,
DOI 10.1007/978-3-319-02666-4_10

In [3] the authors studied, among other things, one of the simplest canonical quasilinear coercive problem with non regular data, namely,

$$-\Delta_p u + |u|^{q-1}u = h \quad on \quad \mathbb{R}^N, \tag{1}$$

where $q > p - 1 > 0$ and $h \in L^1_{loc}(\mathbb{R}^N)$.

An earlier contribution to this problem in the case $p = 2$ was obtained in [5]. Among other things in [5] it was proved that for the semilinear equation (1), for any $h \in L^1_{loc}(\mathbb{R}^N)$ there exists a unique distributional solution $u \in L^q_{loc}(\mathbb{R}^N)$.

Later on in [3] the Authors studied the general case $p > 1$.

By using an approximation procedure they proved that if $q > p - 1$ and $p > 2 - \frac{1}{N}$, then for any $h \in L^1_{loc}(\mathbb{R}^N)$ the Eq. (1) possesses a solution belonging to the space

$$X = W^{1,1}_{loc}(\mathbb{R}^N) \cap W^{1,p-1}_{loc}(\mathbb{R}^N) \cap L^q_{loc}(\mathbb{R}^N).$$

No general results about uniqueness of solutions were claimed in that paper.

In this work, we shall study the uniqueness problem of solutions of general quasilinear equations of the type

$$-\operatorname{div}_L(\mathscr{A}(x, u(x), \nabla_L u(x))) + \psi^\ell |u|^{q-1}u = h \quad on \quad \mathbb{R}^N, \tag{2}$$

and related qualitative properties in the subelliptic setting (see Sect. 2 for details). The main goal of this paper is to show that the ideas introduced in [10] and developed [11] apply to this more general setting as well.

In this regards we observe that the Eq. (2) contains a weight function $\psi$ which is related to subellipticity of the operator appearing in (2) and may vanish on some unbounded negligible set. Problems containing this kind of degeneracy were not studied in [11].

By using the notations introduced in Sect. 2, we shall prove the uniqueness of solutions of (2) in the space

$$W^{1,p}_{L,loc}(\mathbb{R}^N) \cap L^q_{loc}(\mathbb{R}^N) = \{u \in L^p_{loc}(\mathbb{R}^N) \cap L^q_{loc}(\mathbb{R}^N) : |\nabla_L u| \in L^p_{loc}(\mathbb{R}^N)\}.$$

To this end, first we set up two essential tools which are of independent interest. Namely, the regularity of weak solutions of (2) in the space $W^{1,p}_{L,loc}(\mathbb{R}^N) \cap L^q_{loc}(\mathbb{R}^N)$ and comparison principles on $\mathbb{R}^N$. Further we shall derive some properties of the solutions of the problems under consideration.

Our efforts here is to apply an approach that can be useful when dealing with more general operators and related equations or inequalities.

Canonical cases of the main results proved in this paper are the following.

**Theorem 1.1** *Let* $1 < p < 2$, $0 \le \ell \le p$, $q \ge 1$, $h \in L^1_{loc}(\mathbb{R}^N)$, *then the problem*

$$-\operatorname{div}_L\left(|\nabla_L u|^{p-2}\nabla_L u\right) + \psi^\ell |u|^{q-1}u = h \quad on \quad \mathbb{R}^N,$$

*has at most one weak solution* $v \in W^{1,p}_{L,loc}(\mathbb{R}^N) \cap L^q_{loc}(\mathbb{R}^N)$. *Moreover,*

$$\inf_{\mathbb{R}^N} \frac{h}{\psi^\ell} \le |v|^{q-1}v \le \sup_{\mathbb{R}^N} \frac{h}{\psi^\ell}.$$

In the semilinear case we have,

**Theorem 1.2** *Let* $0 \le \ell \le 2$, $q > 1$, $h \in L^1_{loc}(\mathbb{R}^N)$, *then the problem*

$$-\operatorname{div}_L(\nabla_L u) + \psi^\ell |u|^{q-1}u = h \quad on \quad \mathbb{R}^N,$$

*has at most one weak solution* $v \in W^{1,2}_{L,loc}(\mathbb{R}^N) \cap L^q_{loc}(\mathbb{R}^N)$. *Moreover,*

$$\inf_{\mathbb{R}^N} \frac{h}{\psi^\ell} \le |v|^{q-1}v \le \sup_{\mathbb{R}^N} \frac{h}{\psi^\ell}.$$

**Theorem 1.3** *Let* $q \ge 1$, $0 \le \ell \le 1$, $h \in L^1_{loc}(\mathbb{R}^N)$ *then the problem,*

$$-\operatorname{div}_L\left(\frac{\nabla_L u}{\sqrt{1+|\nabla_L u|^2}}\right) + \psi^\ell |u|^{q-1}u = h \quad on \quad \mathbb{R}^N,$$

*has at most one weak solution* $v \in W^{1,1}_{L,loc}(\mathbb{R}^N) \cap L^q_{loc}(\mathbb{R}^N)$. *Moreover,*

$$\inf_{\mathbb{R}^N} \frac{h}{\psi^\ell} \le |v|^{q-1}v \le \sup_{\mathbb{R}^N} \frac{h}{\psi^\ell}.$$

When considering the case $\ell > 1$, we need to look at solutions that belong to a functional space which is smaller than $W^{1,1}_{L,loc}(\mathbb{R}^N) \cap L^q_{loc}(\mathbb{R}^N)$.

We have the following.

**Theorem 1.4** *Let* $1 < \ell \le 2$, $q \ge 1$, $q > \ell - 1$, $h \in L^1_{loc}(\mathbb{R}^N)$ *then the problem,*

$$-\operatorname{div}_L\left(\frac{\nabla_L u}{\sqrt{1+|\nabla_L u|^2}}\right) + \psi^\ell |u|^{q-1}u = h \quad on \quad \mathbb{R}^N,$$

*has at most one weak solution* $v \in W^{1,\ell}_{L,loc}(\mathbb{R}^N) \cap L^q_{loc}(\mathbb{R}^N)$. *Moreover,*

$$\inf_{\mathbb{R}^N} \frac{h}{\psi^\ell} \le |v|^{q-1}v \le \sup_{\mathbb{R}^N} \frac{h}{\psi^\ell}.$$

Our uniqueness results concern solutions that belong to the class $W^{1,p}_{L,loc}(\mathbb{R}^N) \cap L^q_{loc}(\mathbb{R}^N)$. Of course, this set in the canonical Euclidean case is contained in the space $X$ considered in [3]. We point out that when dealing with uniqueness results additional regularity is required by several Authors. See for instance [1]. Indeed, in that work the Authors obtain the existence of solutions of problem (1) belonging to a certain space $T^{1,p}_0$. Uniqueness of solutions proved in [1] concerns entropy solutions.

The paper is organized as follow. In the next section we describe the setting and the notations. In Sect. 3 we prove some general a priori estimates on the solutions of the problems under consideration.

In Sect. 4 we prove some comparison results and derive some consequences.

Finally in Sect. 5 we discuss an open question and we point out its solution in a special case.

In this paper an important role is played by the **M**-$p$-**C** operators (see below for the definition). For easy reference, in Sect. 6 we recall some inequalities proved in [11]. These inequalities are of independent interest and will be used throughout the paper when checking that an operator satisfies the **M**-$p$-**C** property.

## 2 Notations and Definitions

In this paper $\nabla$ and $|\cdot|$ stand respectively for the usual gradient in $\mathbb{R}^N$ and the Euclidean norm.

Let $\mu \in \mathscr{C}(\mathbb{R}^N;\mathbb{R}^l)$ be a matrix $\mu := (\mu_{ij})$, $i = 1,\ldots,l$, $j = 1,\ldots,N$ and assume that for any $i = 1,\ldots,l$, $j = 1,\ldots,N$ the derivative $\frac{\partial}{\partial x_j}\mu_{ij} \in \mathscr{C}(\mathbb{R}^N)$. For $i = 1,\ldots,l$, let $X_i$ and its formal adjoint $X_i^*$ be defined as

$$X_i := \sum_{j=1}^{N} \mu_{ij}(\xi)\frac{\partial}{\partial \xi_j}, \qquad X_i^* := -\sum_{j=1}^{N} \frac{\partial}{\partial \xi_j}\left(\mu_{ij}(\xi)\cdot\right), \tag{3}$$

and let $\nabla_L$ be the vector field defined by

$$\nabla_L := (X_1,\ldots,X_l)^T = \mu\nabla,$$

and

$$\nabla_L^* := (X_1^*,\ldots,X_l^*)^T.$$

For any vector field $h = (h_1,\ldots,h_l)^T \in \mathscr{C}^1(\mathbb{R}^N,\mathbb{R}^l)$, we shall use the following notation $\mathrm{div}_L(h) := \mathrm{div}(\mu^{T}h)$, that is

$$\mathrm{div}_L(h) = -\sum_{i=1}^{l} X_i^* h_i = -\nabla_L^* \cdot h.$$

We suppose that the vector fields satisfy the following assumption. Let $\delta := (\delta_1, \ldots, \delta_N)$ be an $N$-uple of positive real number. We shall denote by $\delta_R$ the function $\delta_R : \mathbb{R}^N \to \mathbb{R}^N$ defined by

$$\delta_R(x) = \delta_R(x_1, \ldots, x_N) := (R^{\delta_1} x_1, \ldots, R^{\delta_N} x_N). \tag{4}$$

We require that $\nabla_L$ *is* $\delta_R$*-homogeneous*, that is, there exists $\delta = (\delta_1, \ldots, \delta_N)$ such that $\nabla_L$ is pseudo homogeneous of degree 1 with respect to dilation $\delta_R$, namely

$$\nabla_L(\phi(\delta_R(\cdot))) = R(\nabla_L\phi)(\delta_R(\cdot)) \textit{ for } R > 0 \textit{ and } \phi \in \mathscr{C}^1(\mathbb{R}^N).$$

Notice that in the Euclidean framework we have $\mu = I_N$, the identity matrix on $\mathbb{R}^N$. Examples of vector fields satisfying our assumptions are the usual gradient acting on $l(\leq N)$ variables, vector fields related to Bouendi–Grushin operator, Heisenberg–Kohn sub-Laplacian, Heisenberg–Greiner operator, sub-Laplacian on Carnot Groups.

A nonnegative continuous function $S : \mathbb{R}^N \to \mathbb{R}_+$ is called a $\delta_R$*-homogeneous norm* on $\mathbb{R}^N$, if $S(\xi^{-1}) = S(\xi)$, $S(\xi) = 0$ if and only if $\xi = 0$, and it is homogeneous of degree 1 with respect to $\delta_R$ (i.e. $S(\delta_R(\xi)) = RS(\xi)$).

An example of smooth homogeneous norm is

$$S(\xi) := \left( \sum_{i=1}^{N} (\xi_i^r)^{\frac{d}{\delta_i}} \right)^{\frac{1}{rd}}, \tag{5}$$

where $d := \delta_1 \delta_2 \cdots \delta_N$ and $r$ is the lowest even integer such that $r \geq \max\{\delta_1/d, \ldots, \delta_N/d\}$.

Notice that if $S$ is a homogeneous norm differentiable a.e., then $|\nabla_L S|$ is homogeneous of degree 0 with respect to $\delta_R$; hence $|\nabla_L S|$ is bounded.

Throughout this paper we assume that $|\cdot|_L \in \mathscr{C}^1(\mathbb{R}^N \setminus \{0\})$ is a general, however fixed, homogeneous norm.

We denote by $B_R$ the open ball generated by $|\cdot|_L$, that is $B_R := \{\xi \in \mathbb{R}^N : |\xi|_L < R\}$. Since the Jacobian of the map $\delta_R$ is $J(\delta_R) = R^Q$ with $Q := \delta_1 + \delta_2 + \ldots \delta_N$, we have $|B_R| = R^Q |B_1|$,

We define $\psi := |\nabla_L |\xi|_L|$ and assume that the set where $\psi$ vanishes is negligible.

The function $\psi$ is bounded and may vanish at some point. For instance in the Euclidean setting, if $|\cdot|_L$ is the Euclidean norm, then $\psi \equiv 1$. If we endow $\mathbb{R}^N$ with the Heisenberg group structure with $\mathbb{R}^N \approx \mathbb{H}^n = \mathbb{R}^n_x \times \mathbb{R}^n_y \times \mathbb{R}_t$, $\nabla_L$ is the Heisenberg gradient and $|\cdot|_L$ is the gauge of the canonical sublaplacian, then $\psi^2(\xi) = (|x|^2 + |y|^2)/|\xi|_L^2$ with $\xi = (x, y, t)$.

In what follows we shall assume that $\mathscr{A} : \mathbb{R}^N \times \mathbb{R} \times \mathbb{R}^l \to \mathbb{R}^l$ is a Carathéodory function, that is for each $t \in \mathbb{R}$ and $\xi \in \mathbb{R}^l$ the function $\mathscr{A}(\cdot, t, \xi)$ is measurable; and for a.e. $x \in \mathbb{R}^N$, $\mathscr{A}(x, \cdot, \cdot)$ is continuous.

We consider operators $L$ "generated" by $\mathscr{A}$, that is

$$L(u)(x) = \operatorname{div}_L\left(\mathscr{A}(x, u(x), \nabla_L u(x))\right). \tag{6}$$

Our canonical model cases are the $p$-Laplacian operator, the mean curvature operator and some related generalizations. See Examples 2.1 below.

**Definition 2.1** Let $\mathscr{A} : \mathbb{R}^N \times \mathbb{R} \times \mathbb{R}^l \to \mathbb{R}^l$ be a Carathéodory function. The function $\mathscr{A}$ is called *weakly elliptic* if it generates a weakly elliptic operator $L$ i.e.

$$\begin{gathered}\mathscr{A}(x, t, \xi) \cdot \xi \geq 0 \ \textit{for each}\ x \in \mathbb{R}^N,\ t \in \mathbb{R},\ \xi \in \mathbb{R}^l, \\ \mathscr{A}(x, 0, \xi) = 0 \ \ \textit{or} \ \ \mathscr{A}(x, t, 0) = 0.\end{gathered} \tag{WE}$$

Let $p \geq 1$, the function $\mathscr{A}$ is called **W**-$p$-**C** (weakly-$p$-coercive) (see [2]), if $\mathscr{A}$ is (WE) and it generates a weakly-p-coercive operator $L$, i.e. if there exists a constant $k_2 > 0$ such that

$$(\mathscr{A}(x, t, \xi) \cdot \xi)^{p-1} \geq k_2^p |\mathscr{A}(x, t, \xi)|^p \text{ for each } x \in \mathbb{R}^N,\ t \in \mathbb{R},\ \xi \in \mathbb{R}^l. \tag{\textbf{W}-$p$-\textbf{C}}$$

Let $p > 1$, the function $\mathscr{A}$ is called **S**-$p$-**C** (strongly-$p$-coercive) (see [2, 13, 14]), if there exist $k_1, k_2 > 0$ constants such that

$$(\mathscr{A}(x, t, \xi) \cdot \xi) \geq k_1 |\xi|^p \geq k_2^{p'} |\mathscr{A}(x, t, \xi)|^{p'} \text{ for each } x \in \mathbb{R}^N,\ t \in \mathbb{R},\ \xi \in \mathbb{R}^l. \tag{\textbf{S}-$p$-\textbf{C}}$$

We look for solution in the space $W^{1,p}_{L,loc}(\Omega)$ defined as

$$W^{1,p}_{L,loc}(\Omega) := \{u \in L^p_{loc}(\Omega) : |\nabla_L u| \in L^p_{loc}(\Omega)\}.$$

**Definition 2.2** Let $\Omega \subset \mathbb{R}^N$ be an open set and let $f : \Omega \times \mathbb{R} \times \mathbb{R}^l \to \mathbb{R}$ be a Carathéodory function. Let $p \geq 1$. We say that $u \in W^{1,p}_{L,loc}(\Omega)$ is a *weak solution* of

$$\operatorname{div}_L\left(\mathscr{A}(x, u, \nabla_L u)\right) \geq f(x, u, \nabla_L u) \qquad on\ \ \Omega,$$

if $\mathscr{A}(\cdot, u, \nabla_L u) \in L^{p'}_{loc}(\Omega)$, $f(\cdot, u, \nabla_L u) \in L^1_{loc}(\Omega)$, and for any nonnegative $\phi \in \mathscr{C}^1_0(\Omega)$ we have

$$-\int_\Omega \mathscr{A}(x, u, \nabla_L u) \cdot \nabla_L \phi \geq \int_\Omega f(x, u, \nabla_L u)\phi.$$

*Example 2.1*

1. Let $p > 1$. The $p$-Laplacian operator defined on suitable functions $u$ by,

$$\Delta_p u = \operatorname{div}_L\left(|\nabla_L u|^{p-2} \nabla_L u\right)$$

is an operator generated by $\mathscr{A}(x, t, \xi) := |\xi|^{p-2}\xi$ which is **S**-$p$-**C**.

2. If $\mathscr{A}$ is of mean curvature type, that is $\mathscr{A}$ can be written as $\mathscr{A}(x,t,\xi) := A(|\xi|)\xi$ with $A : \mathbb{R} \to \mathbb{R}$ a positive bounded continuous function (see [2, 12]), then $\mathscr{A}$ is **W-2-C**.
3. The mean curvature operator in non parametric form

$$Tu := \operatorname{div}_L\left(\frac{\nabla_L u}{\sqrt{1+|\nabla_L u|^2}}\right),$$

   is generated by $\mathscr{A}(x,t,\xi) := \frac{\xi}{\sqrt{1+|\xi|^2}}$. In this case $\mathscr{A}$ is **W**-$p$-**C** with $1 \le p \le 2$ and of mean curvature type but it is not **S-2-C**.
4. Let $m > 1$. The operator

$$T_m u := \operatorname{div}_L\left(\frac{|\nabla_L u|^{m-2}\nabla_L u}{\sqrt{1+|\nabla_L u|^m}}\right)$$

   is **W**-$p$-**C** for $m \ge p \ge m/2$.

**Definition 2.3** Let $\mathscr{A} : \mathbb{R}^N \times \mathbb{R}^l \to \mathbb{R}^l$ be a Charateodory function. We say that $\mathscr{A}$ is monotone if

$$(\mathscr{A}(x,\xi) - \mathscr{A}(x,\eta)) \cdot (\xi - \eta) \ge 0 \qquad \textit{for } \xi, \eta \in \mathbb{R}^l. \tag{7}$$

Let $p \ge 1$. We say that $\mathscr{A}$ is **M**-$p$-**C** (monotone $p$-coercive) if $\mathscr{A}$ is monotone and if there exists $k_2 > 0$ such that

$$((\mathscr{A}(x,\xi) - \mathscr{A}(x,\eta)) \cdot (\xi - \eta))^{p-1} \ge k_2^p |\mathscr{A}(x,\xi) - \mathscr{A}(x,\eta)|^p. \tag{8}$$

*Example 2.2*

1. Let $1 < p \le 2$ the function $\mathscr{A}(\xi) := |\xi|^{p-2}\xi$ is **M**-$p$-**C** (see Sect. 6 for details).
2. The mean curvature operator is **M**-$p$-**C** with $1 \le p \le 2$ (see Sect. 6).

In what follows we shall use a special family of test functions that we call *cut-off functions*. More precisely, let $\varphi_1 \in \mathscr{C}_0^1(\mathbb{R})$ be such that $0 \le \varphi_1 \le 1$, $\varphi_1(t) = 0$ if $|t| \ge 2$ and $\varphi_1(t) = 1$ if $|t| \le 1$. Next, for $R > 0$ by *cut-off function* we mean the function $\varphi_R$ defined as $\varphi_R(x) = \varphi_1(|x|_L/R)$.

Finally, if not otherwise stated, the integrals are considered on the whole space $\mathbb{R}^N$.

## 3 A Priori Estimates

The following is a slight variation of a result proved in [10]. For easy reference we shall include its detailed proof.

Consider the following inequality,

$$\operatorname{div}_L(\mathscr{A}(x,v,\nabla_L v)) - f \ge \operatorname{div}_L(\mathscr{A}(x,u,\nabla_L u)) - g \qquad \textit{on } \mathbb{R}^N. \tag{9}$$

We have,

**Theorem 3.1** *Let $p \geq 1$ and let $\mathscr{A} : \mathbb{R}^N \times \mathbb{R}^l \to \mathbb{R}^l$ be **M**-p-**C**. Let $f, g \in L^1_{loc}(\mathbb{R}^N)$ and let $(u, v)$ be weak solution of (9). Set $w := (v-u)^+$ and let $s > 0$ and $p \geq \ell \geq 0$. If $(f-g)w \geq 0$ and*

$$w^{s+p-1}\psi^\ell \in L^1(B_{2R} \setminus B_R) \qquad \textit{for } R \textit{ large}, \tag{10}$$

*then*

$$(f-g)w^s, \ (\mathscr{A}(x, \nabla_L v) - \mathscr{A}(x, \nabla_L u)) \cdot \nabla_L w \, w^{s-1} \chi_{\{w>0\}} \in L^1_{loc}(\mathbb{R}^N). \tag{11}$$

*Moreover, for any nonnegative $\phi \in \mathscr{C}^1_0(\mathbb{R}^N)$ we have,*

$$\int (f-g)w^s\phi + c_1 s \int (\mathscr{A}(x, \nabla_L v) - \mathscr{A}(x, \nabla_L u)) \cdot \nabla_L w \, w^{s-1}\phi \leq c_2 s^{1-p} \int w^{s+p-1} \frac{|\nabla_L \phi|^p}{\phi^{p-1}}, \tag{12}$$

*where $c_1 = 1 - \frac{p-1}{p}\left(\frac{\epsilon}{k_2}\right)^{\frac{p}{p-1}} > 0$, $c_2 = \frac{p^p}{p\epsilon^p}$ and $\epsilon > 0$ is sufficiently small for $p > 1$ and $c_1 = 1$ and $c_2 = 1/k_2$ for $p = 1$.*

*Remark 3.1*

i) Notice that from the above result it follows that if $u, v \in W^{1,p}_{L,loc}(\mathbb{R}^N)$ is a weak solution of (9), then $(f-g)\, w \in L^1_{loc}(\mathbb{R}^N)$.
ii) The above lemma still holds if we replace the function $f - g \in L^1_{loc}(\mathbb{R}^N)$ with a regular Borel measure on $\mathbb{R}^N$.
iii) The right hand side in (12) could be divergent since we know only that $w^{s+p-1}\psi^\ell \in L^1_{loc}(\mathbb{R}^N)$.
iv) If in Theorem 3.1 we consider the case $\ell = 0$, then Theorem 3.1 can be restated for inequalities (9) on a open set $\Omega$ by replacing $\mathbb{R}^N$ with $\Omega$ and requiring that $w^{s+p-1} \in L^1_{loc}(\Omega)$.
v) If $(u, v)$ is a weak solution of (9) and $u$ is a constant i.e. $u \equiv const$, then Theorem 3.1 still holds even for **W**-$p$-**C** operators. See the following Lemma 3.1.

**Lemma 3.1** *Let $p \geq 1$ and let $\mathscr{A}$ be **W**-p-**C**. Let $f, g \in L^1_{loc}(\mathbb{R}^N)$ and let $v \in W^{1,p}_{L,loc}(\mathbb{R}^N)$ be a weak solution of*

$$\mathrm{div}_L(\mathscr{A}(x, u, \nabla_L u)) \geq f - g, \qquad \textit{on } \mathbb{R}^N. \tag{13}$$

*Let $k > 0$ and set $w := (v-k)^+$ and let $s > 0$, $p \geq \ell \geq 0$. If $(f-g)w \geq 0$ and (10) holds, then*

$$(f-g)w^s, \ \mathscr{A}(x, v, \nabla_L v) \cdot \nabla_L w \, w^{s-1} \chi_{\{w>0\}} \in L^1_{loc}(\mathbb{R}^N) \tag{14}$$

*and for any nonnegative* $\phi \in \mathscr{C}_0^1(\mathbb{R}^N)$ *we have,*

$$\int (f-g)w^s\phi + c_1 s \int \mathscr{A}(x,v,\nabla_L v)\cdot\nabla_L w\, w^{s-1}\phi \le c_2 s^{1-p}\int w^{s+p-1}\frac{|\nabla_L\phi|^p}{\phi^{p-1}}, \tag{15}$$

*where* $c_1$ *and* $c_2$ *are as in Theorem 3.1.*

The above lemma lies on the following result proved in [10, Theorem 2.7].

**Theorem 3.2 ([10])** *Let* $\mathscr{A} : \Omega \times \mathbb{R} \times \mathbb{R}^N \to \mathbb{R}^N$ *be a monotone Carathéodory function. Let* $f, g \in L^1_{loc}(\Omega)$ *and let* $u, v$ *be weak solution of*

$$\operatorname{div}_L(\mathscr{A}(x,v,\nabla_L v)) - f \ge \operatorname{div}_L(\mathscr{A}(x,u,\nabla_L u)) - g \qquad \text{on } \Omega. \tag{16}$$

*Let* $\gamma \in \mathscr{C}^1(\mathbb{R})$ *be such that* $0 \le \gamma(t), \gamma'(t) \le M$*, then*

$$-\int_\Omega (\mathscr{A}(x,v,\nabla_L v) - \mathscr{A}(x,u,\nabla_L u))\cdot\nabla_L\phi\,\gamma(v-u) \ge \tag{17}$$

$$\ge \int_\Omega \gamma'(v-u)\,(\nabla_L v - \nabla_L u)\cdot(\mathscr{A}(x,v,\nabla_L v) - \mathscr{A}(x,u,\nabla_L u)\,\phi \tag{18}$$

$$+\int_\Omega \phi\gamma(v-u)(f-g) \quad \text{on } \Omega. \tag{19}$$

*Hence*

$$\operatorname{div}_L(\gamma(v-u)(\mathscr{A}(x,v,\nabla_L v) - \mathscr{A}(x,u,\nabla_L u))) \ge \gamma(v-u)(f-g) \quad \text{on } \Omega.$$

*Moreover*[1]

$$\operatorname{div}_L\big(\operatorname{sign}^+(v-u)(\mathscr{A}(x,v,\nabla_L v) - \mathscr{A}(x,u,\nabla_L u))\big) \ge \operatorname{sign}^+(v-u)(f-g) \quad \text{on } \Omega. \tag{20}$$

*Proof (of Theorem 3.1)* Let $\gamma \in \mathscr{C}^1(\mathbb{R})$ be a bounded nonnegative function with bounded nonnegative first derivative and let $\phi \in \mathscr{C}_0^1(\Omega)$ be a nonnegative test function.

For simplicity we shall omit the arguments of $\mathscr{A}$. So we shall write $\mathscr{A}_u$ and $\mathscr{A}_v$ instead of $\mathscr{A}(x,\nabla_L u)$ and $\mathscr{A}(x,\nabla_L v)$ respectively.

[1] We recall that the function $\operatorname{sign}^+$ is defined as $\operatorname{sign}^+(t) := 0$ if $t \le 0$ and $\operatorname{sign}^+(t) := 1$ otherwise.

Applying Lemma 3.2, we obtain

$$\int (f-g)\gamma(w)\phi + \int (\mathscr{A}_v - \mathscr{A}_u)\cdot \nabla_L w\, \gamma'(w)\phi \le -\int (\mathscr{A}_v - \mathscr{A}_u)\cdot \nabla_L \phi\, \gamma(w)$$
$$\le \int |\mathscr{A}_v - \mathscr{A}_u|\, |\nabla_L \phi| \gamma(w). \tag{21}$$

Let $p > 1$. From (21) we have

$$\int (f-g)\gamma(w)\phi + \int (\mathscr{A}_v - \mathscr{A}_u)\cdot \nabla_L w\, \gamma'(w)\phi \le$$
$$\le \left(\int |\mathscr{A}_v - \mathscr{A}_u|^{p'} \gamma'(w)\phi\right)^{1/p'} \left(\int \frac{\gamma(w)^p}{\gamma'(w)^{p-1}} \frac{|\nabla_L\phi|^p}{\phi^{p-1}}\right)^{1/p}$$
$$\le \frac{\epsilon^{p'}}{p' k_2^{p'}} \int (\mathscr{A}_v - \mathscr{A}_u)\cdot \nabla_L w\, \gamma'(w)\phi + \frac{1}{p\epsilon^p}\int \frac{\gamma(w)^p}{\gamma'(w)^{p-1}} \frac{|\nabla_L\phi|^p}{\phi^{p-1}},$$

where $\epsilon > 0$ and all integrals are well defined provided $\frac{\gamma(w)^p}{\gamma'(w)^{p-1}} \in L^1_{loc}(\Omega)$. With a suitable choice of $\epsilon > 0$, for any nonnegative $\phi \in \mathscr{C}^1_0(\Omega)$ and $\gamma \in \mathscr{C}^1(\mathbb{R})$ as above such that $\frac{\gamma(w)^p}{\gamma'(w)^{p-1}} \in L^1_{loc}(\Omega)$, it follows that,

$$\int (f-g)\gamma(w)\phi + c_1 \int (\mathscr{A}_v - \mathscr{A}_u)\cdot \nabla_L w\, \gamma'(w)\phi \le \frac{1}{p\epsilon^p}\int \frac{\gamma(w)^p}{\gamma'(w)^{p-1}} \frac{|\nabla_L\phi|^p}{\phi^{p-1}}. \tag{22}$$

Now for $s > 0$, $1 > \delta > 0$ and $n \ge 1$, define

$$\gamma_n(t) := \begin{cases} (t+\delta)^s & \mathit{if}\ 0 \le t < n-\delta, \\ cn^s - \dfrac{s}{\beta-1} n^{\beta+s-1}(t+\delta)^{1-\beta} & \mathit{if}\ t \ge n-\delta, \end{cases} \tag{23}$$

where $c := \frac{\beta-1+s}{\beta-1}$ and $\beta > 1$ will be chosen later. Clearly $\gamma_n \in \mathscr{C}^1$,

$$\gamma_n'(t) = \begin{cases} s(t+\delta)^{s-1} & \mathit{if}\ 0 \le t < n-\delta, \\ sn^{\beta+s-1}(t+\delta)^{-\beta} & \mathit{if}\ t \ge n-\delta, \end{cases}$$

and $\gamma_n$, $\gamma_n'$ are nonnegative and bounded with $||\gamma_n||_\infty = cn^s$ and $||\gamma_n'||_\infty = sn^{s-1}$. Moreover

$$\frac{\gamma_n(t)^p}{\gamma_n'(t)^{p-1}} = \begin{cases} s^{1-p}(t+\delta)^{s+p-1} & \mathit{for}\ t < n-\delta, \\ \theta(t,n) & \mathit{for}\ t \ge n-\delta, \end{cases}$$

where

$$\theta(t,n) := \frac{(cn^s - \frac{s}{\beta-1} n^{\beta+s-1}(t+\delta)^{1-\beta})^p}{(sn^{\beta+s-1}(t+\delta)^{-\beta})^{p-1}} \leq (cn^s)^p s^{1-p} n^{-(\beta+s-1)(p-1)} \, (t+\delta)^{\beta(p-1)}.$$

Choosing $\beta := \frac{s+p-1}{p-1}$ we have $c = p$, and

$$\theta(t,n) \leq p^p s^{1-p} n^{sp-(\beta+s-1)(p-1)} (t+\delta)^{s+p-1} = p^p s^{1-p} (t+\delta)^{s+p-1}.$$

Therefore, for $t \geq 0$ we have,

$$\frac{\gamma_n(t)^p}{\gamma_n'(t)^{p-1}} \leq p^p s^{1-p} (t+\delta)^{s+p-1}.$$

Since by assumption $w^{s+p-1} \in L^1_{loc}(\Omega)$, from (22) with $\gamma = \gamma_n$, it follows that

$$\int (f-g)\gamma_n(w)\phi + c_1 \int (\mathscr{A}_v - \mathscr{A}_u) \cdot \nabla_L w \, \gamma_n'(w)\phi \leq \frac{p^p s^{1-p}}{p\epsilon^p} \int (w+\delta)^{s+p-1} \frac{|\nabla_L \phi|^p}{\phi^{p-1}}.$$

Now, noticing that $\gamma_n(t) \to (t+\delta)^s$ and $\gamma_n'(t) \to s(t+\delta)^{s-1}$ as $n \to +\infty$, $(f-g)(\gamma_n(w) - \gamma_n(0) \geq 0$ and $\mathscr{A}$ is monotone (that is $(\mathscr{A}_v - \mathscr{A}_u) \cdot \nabla_L w \geq 0$), by Fatou's Lemma theorem we obtain

$$\int (f-g)\,(w+\delta)^s \phi + c_1 s \int (\mathscr{A}_v - \mathscr{A}_u) \cdot \nabla_L w \,(w+\delta)^{s-1} \phi \leq c_2 s^{1-p} \int (w+\delta)^{s+p-1} \frac{|\nabla_L \phi|^p}{\phi^{p-1}}.$$

By letting $\delta \to 0$ in the above inequality, we have the inequality (12).

Next, we choose $R > 0$ large enough and $\phi := \varphi_R^p$ with $\varphi_R$ a cut off function, that is

$$\phi(x) := (\varphi_R(x))^p := (\varphi_1(|x|_L/R))^p.$$

With these choice we have

$$\frac{|\nabla_L \phi|^p}{\phi^{p-1}} = p^p \psi^p R^{-p} |\varphi_1'|^p \left(\frac{|x|_L}{R}\right) \leq p^p ||\psi||^{p-\ell} ||\varphi_1'||_\infty^p R^{-p} \psi^\ell =: c_3 \psi^\ell,$$

and from (12) we deduce

$$\int_{B_R} (f-g)\, w^s + c_1 s \int_{B_R} (\mathscr{A}_v - \mathscr{A}_u) \cdot \nabla_L w \, w^{s-1} \leq c_2 s^{1-p} c_3 \int_{B_{2r} \setminus B_R} w^{s+p-1} \psi^\ell,$$

which completes the proof of the claim in the case $p > 1$.

Let $p = 1$. From (21) and the fact that $\mathscr{A}_v - \mathscr{A}_u$ is bounded, the estimate (22) holds provided we replace $p$ with 1 and $\epsilon$ with $k_2$. The remaining argument is similar to the case $p > 1$, hence we shall omit it.

**Lemma 3.2** *Let $p \geq 1$ and let $\mathscr{A} : \mathbb{R}^N \times \mathbb{R}^l \to \mathbb{R}^l$ be **M**-p-**C**. Let $f, g \in L^1_{loc}(\mathbb{R}^N)$ and let $(u, v)$ be weak solution of (9). Set $w := (v - u)^+$. If $(f - g)w \geq 0$ and $w^q \psi^\ell \in L^1(B_{2R} \setminus B_R)$ for $q > p - 1$, $p \geq \ell \geq 0$ and $R > 0$ large, then*

$$(f - g)w^{q-p+1}, \quad ((\mathscr{A}(x, \nabla_L v) - \mathscr{A}(x, \nabla_L u)) \cdot \nabla_L w \, w^{q-p} \chi_{\{w>0\}} \in L^1_{loc}(\mathbb{R}^N), \tag{24}$$

*and for any $\varphi_R \in \mathscr{C}^1_0(\mathbb{R}^N)$ cut-off function, for R large enough, we have,*

$$\int (f - g) \operatorname{sign}^+(w) \, \varphi_R^\sigma \leq c \left( \int_{B_{2R} \setminus B_R} w^q \psi^\ell \varphi_R^\sigma \right)^{\frac{p-1}{q}} R^{Q(\frac{q-p+1}{q})-p}, \tag{25}$$

*where $c = c(\sigma, k_2, p, q, ||\psi||_\infty, \ell)$ and $\sigma \geq \frac{pq}{q-p+1-s}$, $0 < s < \min\{1, q - p + 1\}$.*

*Proof* The claim (24) follows from Theorem 3.1.

Let $s > 0$ be such that $q \geq s + p - 1$. From Lemma 3.1, for any nonnegative $\phi \in \mathscr{C}^1_0(\mathbb{R}^N)$, we have

$$\int (f - g)w^s \phi + c_1 s \int (\mathscr{A}_v - \mathscr{A}_u) \cdot \nabla_L w \, w^{s-1} \phi \leq c_2 s^{1-p} \int_S w^{s+p-1} \frac{|\nabla_L \phi|^p}{\phi^{p-1}}, \tag{26}$$

where, as in the proof of Theorem 3.1, we write $\mathscr{A}_v$ and $\mathscr{A}_v$ for $\mathscr{A}(x, \nabla_L v)$ and $\mathscr{A}(x, \nabla_L u)$ respectively and $S$ is the support of $|\nabla_L \phi|$.

Next, an application of Theorem 3.2 gives (20). That is

$$\operatorname{div}_L \left( \operatorname{sign}^+(v - u)(\mathscr{A}(x, v, \nabla_L v) - \mathscr{A}(x, u, \nabla_L u)) \right) \geq \operatorname{sign}^+(v - u)(f - g) \quad on\ \mathbb{R}^N. \tag{27}$$

Now we consider the case $p > 1$. Let $0 < s < \min\{1, q - p + 1\}$. By definition of weak solution and Hölder's inequality with exponent $p'$, taking into account that $\mathscr{A}$ is **M**-$p$-**C** and from (26) we get,

$$\int \operatorname{sign}^+ w(f - g) \, \phi \leq \int_S |\mathscr{A}_v - \mathscr{A}_u| |\nabla_L \phi| \operatorname{sign}^+ w \tag{28}$$

$$= \int_S |\mathscr{A}_v - \mathscr{A}_u| w^{\frac{s-1}{p'}} \phi^{\frac{1}{p'}} \, |\nabla_L \phi| w^{\frac{1-s}{p'}} \phi^{-\frac{1}{p'}} \tag{29}$$

$$\leq \frac{1}{k_2} \left( \int_S (\mathscr{A}_v - \mathscr{A}_u) \cdot \nabla_L w \, w^{s-1} \phi \right)^{1/p'} \left( \int_S w^{(1-s)(p-1)} \frac{|\nabla_L \phi|^p}{\phi^{p-1}} \right)^{1/p} \tag{30}$$

$$\leq \frac{1}{k_2} \left( \frac{c_2}{c_1 s^p} \right)^{1/p'} \left( \int_S w^{s+p-1} \frac{|\nabla_L \phi|^p}{\phi^{p-1}} \right)^{1/p'} \left( \int_S w^{(1-s)(p-1)} \frac{|\nabla_L \phi|^p}{\phi^{p-1}} \right)^{1/p}. \tag{31}$$

Since $q > s+p-1$ and $q > p-1$, applying Hölder inequality to (31) with exponents $\chi := \frac{q}{s+p-1}$ and $y := \frac{q}{(1-s)(p-1)}$, we obtain

$$\int \operatorname{sign}^+ w\,(f-g)\phi \le c_3' \left(\int_S w^q \psi^\ell \phi\right)^{\delta} \left(\int_S \frac{|\nabla_L \phi|^{p\chi'}}{\psi^{\chi'-1}\phi^{p\chi'-1}}\right)^{\frac{1}{p'\chi'}} \left(\int_S \frac{|\nabla_L \phi|^{py'}}{\chi^{y'-1}\phi^{py'-1}}\right)^{\frac{1}{py'}}, \tag{32}$$

where

$$\delta := \frac{1}{\chi p'} + \frac{1}{yp} = \frac{p-1}{q}, \qquad c_3' := \left(\frac{c_2}{c_1 s^p}\right)^{1/p'} \frac{1}{k_2}.$$

Next for $\sigma \ge p\chi'$ (notice that $p\chi' > py'$ implies $\sigma > py'$), we choose $\phi := \varphi_R^\sigma$. From (32) it follows that $S = B_{2R} \setminus B_R$ and

$$\begin{aligned}
\int \operatorname{sign}^+ w\,(f-g)\varphi_R^\sigma &\le c_3'\sigma^p \left(\int_S w^q \psi^\ell \varphi_R^\sigma\right)^{\delta} \times \\
&\quad \times \left(\int_S \psi^p R^{-p\chi'} |\varphi_1'|^{p\chi'} (\frac{|x|_L}{R})\right)^{\frac{1}{p'\chi'}} \left(\int_S \psi^p R^{-py'} |\varphi_1'|^{py'} (\frac{|x|_L}{R})\right)^{\frac{1}{py'}} \\
&\le c_3'\sigma^p \left(\int_S w^q \psi^\ell \varphi_R^\sigma\right)^{\delta} ||\psi||_\infty^{\frac{p}{p'\chi'}+\frac{p}{py'}} R^{-\frac{p\chi'}{p'\chi'}-\frac{py'}{py'}} ||\varphi_1'||_\infty^{\frac{p\chi'}{p'\chi'}+\frac{py'}{py'}} |B_{2R} \setminus B_R|^{\frac{1}{p'\chi'}+\frac{1}{py'}} \\
&\le c \left(\int_S w^q \psi^\ell \varphi_R^\sigma\right)^{\delta} R^{Q(1-\delta)-p},
\end{aligned}$$

completing the proof of (25).

Now, we assume that $p = 1$. From (28), with the choice $\phi := \varphi_R^\sigma$, we have

$$\int \operatorname{sign}^+ w\,(f-g)\varphi_R^\sigma \le \frac{\sigma}{k_2} \int_S |\nabla_L \varphi_R| \le cR^{Q-1},$$

which completes the proof.

## 4 Comparison and Uniqueness

In this section we prove a comparison principle and its implication on the uniqueness property.

Consider the following inequality,

$$\operatorname{div}_L(\mathscr{A}(x, \nabla_L v)) - \psi^\ell |v|^{q-1} v \ge \operatorname{div}_L(\mathscr{A}(x, \nabla_L u)) - \psi^\ell |u|^{q-1} u \qquad on\ \mathbb{R}^N. \tag{33}$$

As preliminary result we have the following.

**Lemma 4.1** *Let $p \geq 1$, let $\mathscr{A}$ be **M**-p-**C**, $q \geq 1$ and $q > p-1$.*
*Let $(u, v)$ be weak solution of (33) with $p \geq \ell \geq 0$. Then $((v-u)^+)^r \psi^\ell \in L^1_{loc}(\mathbb{R}^N)$ for any $r < +\infty$.*

*Proof* Let $(u, v)$ be a solution of (33) and set $w := (v-u)^+$. By using the well known inequality

$$|t|^{q-1}t - |s|^{q-1}s \geq c_q(t-s)^q, \quad for\ t \geq s \qquad (q \geq 1), \tag{34}$$

we deduce that $w^q \psi^\ell \in L^1_{loc}(\mathbb{R}^N)$. From this it follows that we are in the position to apply Theorem 3.1, with $s = q - p + 1$ obtaining $w^{q_1}\psi^\ell \in L^1_{loc}(\mathbb{R}^N)$ with $q_1 := 2q - p + 1$. Applying again Theorem 3.1, with $s = q_1 - p + 1$, we get $w^{q_2}\psi^\ell \in L^1_{loc}(\mathbb{R}^N)$ with $q_2 := q_1 + q - p + 1 = q + 2(q - p + 1)$. Iterating $j$ times we have that $w^{q_j}\psi^\ell \in L^1_{loc}(\mathbb{R}^N)$ with $q_j := q + j(q - p + 1)$. By choosing $j$ sufficiently large we get the claim.

**Theorem 4.1** *Let $p \geq 1$, let $\mathscr{A}$ be **M**-p-**C**, $q \geq 1$, $q > p-1$ and $p \geq \ell \geq 0$. Let $(u, v)$ be a weak solution of*

$$\mathrm{div}_L(\mathscr{A}(x, \nabla_L v)) - \psi^\ell |v|^{q-1}v \geq \mathrm{div}_L(\mathscr{A}(x, \nabla_L u)) - \psi^\ell |u|^{q-1}u \qquad on\ \mathbb{R}^N. \tag{35}$$

*Then $v \leq u$ a.e. on $\mathbb{R}^N$.*
*In particular if $(u, v)$ be a weak solution of*

$$\mathrm{div}_L(\mathscr{A}(x, \nabla_L v)) - \psi^\ell |v|^{q-1}v = \mathrm{div}_L(\mathscr{A}(x, \nabla_L u)) - \psi^\ell |u|^{q-1}u \qquad on\ \mathbb{R}^N, \tag{36}$$

*then $u \equiv v$ a.e. on $\mathbb{R}^N$.*

*Proof* Let $(u, v)$ be a solution of (35) and set $w := (v-u)^+$. From Lemma 4.1 we know that $w^r\psi^\ell \in L^1_{loc}(\mathbb{R}^N)$ for any $r$, and hence we are in the position to apply Theorem 3.1 with $s$ large enough. Thus, from (34) and (12) we get $w^{q+s}\psi^\ell \in L^1_{loc}(\mathbb{R}^N)$ and

$$\int w^{q+s}\psi^\ell \phi \leq c(s, q, p) \int w^{s+p-1} \frac{|\nabla_L \phi|^p}{\phi^{p-1}}.$$

Applying the Hölder inequality with exponent $x := \frac{q+s}{s+p-1} > 1$ we have

$$\int w^{q+s}\psi^\ell \phi \leq c(s, q, p) \int \psi^{p(1-x')} \frac{|\nabla_L \phi|^{px'}}{\phi^{px'-1}}.$$

By the same choice of $\phi$ we made in the proof of Theorem 3.1, that is $\phi = \varphi_R$ a cut off functions, it follows that

$$\int_{B_R} w^{q+s}\psi^\ell \le cR^{Q-px'} = cR^{Q-p(q+s)/(q-p+1)}.$$

Choosing $s$ large enough and letting $R \to +\infty$, we have that $w \equiv 0$ a.e. on $\mathbb{R}^N$. This completes the proof.

**Corollary 4.1** *Let $p \ge 1$, let $\mathscr{A}$ be **W**-p-**C** such that $\mathscr{A}(x,0) = 0$. Let $q$ and $\ell$ be as in Theorem 4.1. Let $h \in L^1_{loc}(\mathbb{R}^N)$. Let $v$ be a weak solution of the problem*

$$-\operatorname{div}_L(\mathscr{A}(x, \nabla_L v)) + \psi^\ell |v|^{q-1}v = h. \tag{37}$$

*Then,*

$$\inf_{\mathbb{R}^N} \frac{h}{\psi^\ell} \le |v|^{q-1}v \le \sup_{\mathbb{R}^N} \frac{h}{\psi^\ell}.$$

*In particular, if $h \ge 0$ [resp. $\le 0$], then $v \ge 0$ [resp. $\le 0$] and if $\frac{h}{\psi^\ell} \in L^\infty(\mathbb{R}^N)$, then $v \in L^\infty(\mathbb{R}^N)$.*

*Proof* We shall prove only the estimate

$$|v|^{q-1}v \le \sup_{\mathbb{R}^N} \frac{h}{\psi^\ell},$$

the proof of the other inequality being similar. If $\sup_{\mathbb{R}^N} \frac{h}{\psi^\ell} = +\infty$ there is nothing to prove. Let $M := \sup_{\mathbb{R}^N} \frac{h}{\psi^\ell} < +\infty$. We define $u := \operatorname{sign}(M)|M|^{1/q}$. Then

$$\operatorname{div}_L(\mathscr{A}(x, \nabla_L v)) - \psi^\ell |v|^{q-1}v + h = 0 \ge h - \psi^\ell M = \operatorname{div}_L(\mathscr{A}(x, \nabla_L u)) - \psi^\ell |u|^{q-1}u + h,$$

that is $(u, v)$ satisfy (35) with $u$ constant. In this case all the previous estimates still hold since in this case the operator can be seen as it were **M**-$p$-**C**. See also Remark 3.1 and Lemma 3.1.

Thus the claim follows from Theorem 4.1.

**Corollary 4.2** *Let $p \ge 1$ and let $\mathscr{A}$ be **M**-p-**C**. Let $q$ and $\ell$ be as in Theorem 4.1. Let $h \in L^1_{loc}(\mathbb{R}^N)$. Then the possible weak solution of the problem (37) is unique.*

*Moreover if $\mathscr{A}(x,0) = 0$ and $v$ is a solution of (37), then*

$$\inf_{\mathbb{R}^N} \frac{h}{\psi^\ell} \le |v|^{q-1}v \le \sup_{\mathbb{R}^N} \frac{h}{\psi^\ell}.$$

*Proof* Uniqueness. Let $u$ and $v$ two solutions of (37). Then $(u, v)$ solves

$$\mathrm{div}_L\,(\mathscr{A}(x, \nabla_L v)) - \psi^\ell |v|^{q-1} v = \mathrm{div}_L\,(\mathscr{A}(x, \nabla_L u)) - \psi^\ell |u|^{q-1} u \qquad on\ \mathbb{R}^N,$$

and applying Theorem 4.1 we conclude that $u \equiv v$.

The remaining claim follows from Corollary 4.1.

# 5 Further Applications

## 5.1 Symmetry Results

An application of Theorem 4.1 to the symmetry of solutions is the following.

**Proposition 5.1** *Let $p \geq 1$. Let $\mathscr{A}$ be **M**-p-**C** and Let L be the operator generated by $\mathscr{A}$, see (6). Let q be as in Theorem 4.1.*

*Let $\Phi : \mathbb{R}^N \to \mathbb{R}^N$ be a map which leaves L invariant, that is*

$$\mathrm{div}_L(\mathscr{A}(x, \nabla_L(\phi(\Phi(x))))) = \mathrm{div}_L(\mathscr{A}(\cdot, \nabla_L(\phi(\cdot)))(\Phi(x)) \quad for\ \ any\ \ \phi \in \mathscr{C}^2(\mathbb{R}^N).$$

*i.e.*

$$L(\phi(\Phi(x))) = L(\phi)(\Phi(x)) \quad for\ \ any\ \ \phi \in \mathscr{C}^2(\mathbb{R}^N).$$

*Let $h \in L^1_{loc}(\mathbb{R}^N)$ be a $\Phi$-invariant function, that is $h(\Phi(x)) = h(x)$ for a.e. $x \in \mathbb{R}^N$.*

*If $v$ is a solution of*

$$-\,\mathrm{div}_L\,(\mathscr{A}(x, \nabla_L v)) + |v|^{q-1} v = h, \tag{38}$$

*then $v$ is $\Phi$-invariant.*

*If $\psi$ is $\Phi$-invariant, $p \geq \ell \geq 0$ and $v$ is a solution of*

$$-\,\mathrm{div}_L\,(\mathscr{A}(x, \nabla_L v)) + \psi^\ell |v|^{q-1} v = h, \tag{39}$$

*then $v$ is $\Phi$-invariant.*

*Proof* Set $v_\Phi(x) := v(\Phi(x))$. We have that

$$\begin{aligned} -L(v)(x) + \psi^\ell(x)|v|^{q-1}v(x) &= h(x) = h(\Phi(x)) \\ &= -L(v)(\Phi(x)) + \psi^\ell((\Phi(x))|v|^{q-1}v((\Phi(x)) \\ &= -L(v_\Phi)(x) + \psi^\ell(x)|v_\Phi|^{q-1}v_\Phi(x) \end{aligned}$$

and by the uniqueness of the solution we have the claim.

In the Heisenberg group examples of map which leaves the $p$-laplacian invariant are the following, $\Phi(\xi) = -\xi$, $\Phi(x, y, t) = (-x, y, t)$ and $\Phi(x, y, t) = (2\mu - x, y, -t - 4\mu y)$ for any $\mu \in \mathbb{R}$.

**Proposition 5.2** *Let $q > 1$, $2 \geq \ell \geq 0$ and $h \in L^1_{loc}(\mathbb{R}^N)$. Let $\Delta_H$ be the Heisenberg Laplacian on the Heisenberg group $\mathbb{H}^n$ and let $|\cdot|_L$ the gauge related to $\Delta_H$. Then the problem*

$$-\Delta_H v + \psi^\ell |v|^{q-1} v = h \tag{40}$$

*has at most one solution.*

*Moreover, let $v$ be a solution of (40) we have*

i) *If $h$ is cylindrical, then $v$ is cylindrical.*
ii) *Let $\ell = 0$. If $h$ does not depend on $t$, then $v$ is independent on $t$ and it solves the problem*

$$-\Delta v + |v|^{q-1} v = h \quad on\ \mathbb{R}^{2n}. \tag{41}$$

## 5.2 Some Applications to Systems

Another consequence of Theorem 4.1 is the following.

**Theorem 5.1** *Let $p \geq 1$, let $\mathscr{A}$ be **M**-p-**C** and odd, that is $\mathscr{A}(x, -\xi) = -\mathscr{A}(x, \xi)$ for any $x \in \mathbb{R}^N$ and $\xi \in \mathbb{R}^l$. Let $q \geq 1$, $q > p - 1$ and $p \geq \ell \geq 0$. Let $h_1, h_2 \in L^1_{loc}(\mathbb{R}^N)$. Let $(u, v)$ be a weak solution of*

$$\begin{cases} \operatorname{div}_L(\mathscr{A}(x, \nabla_L u)) \geq \psi^\ell |v|^{q-1} v + h_1 & \text{on } \mathbb{R}^N, \\ \operatorname{div}_L(\mathscr{A}(x, \nabla_L v)) \geq \psi^\ell |u|^{q-1} u + h_2 & \text{on } \mathbb{R}^N. \end{cases} \tag{42}$$

*If $h_1 + h_2 \geq 0$, then $u + v \leq 0$ a.e. on $\mathbb{R}^N$.*

*Moreover, if $(u, v)$ solves also the equation in (42) and $h_1 = -h_2$, then $u = -v$ and $u$ solves*

$$-\operatorname{div}_L(\mathscr{A}(x, \nabla_L u)) = |u|^{q-1} u.$$

*Proof* Let $w := -u$. Summing up the inequalities, we have that $(w, v)$ is a solution of (35). Hence by Theorem 4.1 it follows that $v \leq w$. This completes the first part of the proof.

Now, if $(u, v)$ is a solution of (42) with equality sign, then $(-u, -v)$ solves the same equations. By the first part of this claim we deduce that $-u - v \leq 0$, thereby concluding the proof.

**Corollary 5.1** *Let $p \geq 1$, let $\mathscr{A}$ be **M**-p-**C** and odd. Let $q \geq 1$, $q > p-1$ and $p \geq \ell \geq 0$ and let $(u, v)$ be a weak solution of*

$$\begin{cases} -\operatorname{div}_L(\mathscr{A}(x, \nabla_L u)) = \psi^\ell |v|^{q-1} v & \text{on } \mathbb{R}^N, \\ -\operatorname{div}_L(\mathscr{A}(x, \nabla_L v)) = \psi^\ell |u|^{q-1} u & \text{on } \mathbb{R}^N. \end{cases} \tag{43}$$

*Then $u = v$ a.e. on $\mathbb{R}^N$.*

*Proof* The claim follows by observing that $(-u, v)$ solves the system (42) with equality signs and $h_1 = h_1 \equiv 0$. Hence the claim follows from Theorem 5.1.

The above Theorem 5.1 and Corollary 5.1 were proved in a weaker form by the first author in [7].

## 5.3 An Interesting Question

We the point out the following challenging question.

If $p = 1$ and $q \geq 1$, from the results proved in the preceding sections it follows that uniqueness and comparison principles for problem (37) hold.

A natural open question is whether in the case $0 < q < 1$ the same results hold. In these respects, the following partial results may give some indication that this problem has an affirmative answer.

**Theorem 5.2** *Let $p = 1$, let $\mathscr{A}$ be **M**-p-**C**, $q > 0$ and $p \geq \ell \geq 0$. If $(u, v)$ is a bounded weak solution of*

$$\operatorname{div}_L(\mathscr{A}(x, \nabla_L v)) - \psi^\ell |v|^{q-1} v \geq \operatorname{div}_L(\mathscr{A}(x, \nabla_L u)) - \psi^\ell |u|^{q-1} u \quad \text{on } \mathbb{R}^N, \tag{44}$$

*then $v \leq u$ a.e. on $\mathbb{R}^N$.*

*Proof* It is easy to see that

$$|t|^{q-1}t - |s|^{q-1}s \geq c_q(t-s), \quad \text{for } M \geq t \geq s \geq -M. \tag{45}$$

Therefore by the argument used in the proof of Theorem 4.1, the claim follows.

**Corollary 5.2** *Let $p = 1$, let $\mathscr{A}$ be **M**-p-**C**, $q > 0$, $p \geq \ell \geq 0$ and let $h \in L^1_{loc}(\mathbb{R}^N)$. Then the possible bounded solution of (37) is unique.*

Looking at one of the model case, the $p$-Laplacian, one can easily realize that, for $p > 2$ the $p$-Laplacian operator in not **M**-$p$-**C**. In this direction some efforts have been made in [11]. However, even if the technique developed in the present paper shows that it is possible to study equations associated to general operators satisfying appropriated structural assumptions, the uniqueness problem for the equation

$$-\Delta_p u + |u|^{q-1}u = h \qquad on\ \mathbb{R}^N,$$

for $h \in L^1_{loc}(\mathbb{R}^N)$ and $u \in W^{1,p}_{loc}(\mathbb{R}^N)$, with $q > p-1$ and $p > 2$ remains still open.

Clearly, looking for nonnegative solution with $h \leq 0$ several results are known see [13] for the Euclidean setting and [6] for the degenerate and anisotropic case. The interested reader may refer also to [8–10] and [11].

## 6 Inequalities and M-*p*-C Operators

Here, we recall some fundamental elementary inequalities proved in [11] that we use throughout the paper.

In what follows we shall assume that $\mathscr{A}$ has the form

$$\mathscr{A}(x,\xi) = A(|\xi|)\xi,$$

where $\mathscr{A} : \mathbb{R}_+ \to \mathbb{R}$. We set $\phi(t) := A(t)t$.

**Theorem 6.1** *Let A be nonincreasing and bounded function such that*

$$\phi(0) = 0, \ \ \phi(t) > 0 \text{ for } t > 0, \phi \text{ is nondecreasing.} \tag{46}$$

*Then $\mathscr{A}$ is **M**-p-**C** with $p = 2$.*

**Theorem 6.2** *Let $1 < p \leq 2$. Let $\phi$ be increasing, concave function satisfying (46) and such that there exist positive constants $c_p, c_\phi > 0$ such that*

$$\phi(t) \leq c_p t^{p-1} \tag{47}$$

*and*

$$\phi'(s)s \leq c_\phi \phi(s). \tag{48}$$

*Then $\mathscr{A}$ is **M**-p-**C**.*

*Remark 6.1* We notice that (47) is a necessary condition on $\mathscr{A}$ to be an **M**-*p*-**C** operator. Indeed, if $\mathscr{A}$ is **M**-*p*-**C**, by taking $\eta = 0$, then it follows that $\mathscr{A}$ is **W**-*p*-**C**, and (47) holds by Hölder inequality.

# 7 Examples

*Example 7.1* Let $l \leq N$ be a positive natural number and let $\mu^l \in \mathscr{C}^1(\mathbb{R}^N; \mathbb{R}^l)$ be the matrix defined as

$$\mu^l := \left(I_l\ 0\right).$$

The corresponding vector field $\nabla^l$ is the usual gradient acting only on the first $l$ variables

$$\nabla^l = (\partial_{x_1}, \partial_{x_2}, \ldots, \partial_{x_l}).$$

Clearly $\nabla^N = \nabla$ and $\nabla^l$ is homogeneous with respect to dilation

$$\delta_R(x) = (Rx_1, Rx_2, \ldots, Rx_l, R^{\delta_{l+1}}x_{l+1}, \ldots, R^{\delta_N}x_N)$$

with $\delta_{l+1}, \ldots, \delta_N$ are arbitrary real positive numbers.

*Example 7.2 (Baouendi-Grushin Type Operator)* Let $\xi = (x, y) \in \mathbb{R}^n \times \mathbb{R}^k (= \mathbb{R}^N)$. Let $\gamma \geq 0$ and let $\mu$ be the following matrix

$$\begin{pmatrix} I_n & 0 \\ 0 & |x|^\gamma I_k \end{pmatrix}. \tag{49}$$

The corresponding vector field is given by $\nabla_\gamma = (\nabla_x, |x|^\gamma \nabla_y)^T$ and the linear operator $L = \operatorname{div}_L(\nabla_L \cdot) = \Delta_x + |x|^{2\gamma}\Delta_y$ is the so-called Baouendi-Grushin operator. Notice that if $k = 0$ or $\gamma = 0$, then $L$ coincides with the usual Laplacian operator. The vector field $\nabla_\gamma$ is homogeneous with respect to dilation $\delta_R(x) = (Rx_1, \ldots, Rx_n, R^{1+\gamma}y_1, \ldots, R^{1+\gamma}y_k)$.

*Example 7.3 (Heisenberg-Kohn Operator)* Let $\xi = (x, y, t) \in \mathbb{R}^n \times \mathbb{R}^n \times \mathbb{R} = \mathbb{H}^n$ and let $\mu$ be defined as

$$\begin{pmatrix} I_n & 0 & 2y \\ 0 & I_n & -2x \end{pmatrix}.$$

The corresponding vector field $\nabla_H$ is the Heisenberg gradient on the Heisenberg group $\mathbb{H}^n$. The vector field $\nabla_H$ is homogeneous with respect to $\delta_R(\xi) = (Rx, Ry, R^2t)$ and $Q = 2n + 2$.

In $\mathbb{H}^1$ the corresponding vector fields are $X = \partial_x + 2y\partial_t$, $Y = \partial_y - 2x\partial_t$. In this case $Q = 4$.

In $\mathbb{H}^n$ a canonical homogeneous norm, called *gauge*, is defined as

$$|\xi|_H := \left(\left(\sum_{i=1}^n x_i^2 + y_i^2\right)^2 + t^2\right)^{1/4}.$$

*Example 7.4 (Heisenberg-Greiner Operator)* Let $\xi = (x, y, t) \in \mathbb{R}^n \times \mathbb{R}^n \times \mathbb{R}$, $r := |(x, y)|$, $\gamma \geq 1$ and let $\mu$ be defined as

$$\begin{pmatrix} I_n & 0 & 2\gamma y r^{2\gamma-2} \\ 0 & I_n & -2\gamma x r^{2\gamma-2} \end{pmatrix}. \tag{50}$$

The corresponding vector fields are $X_i = \partial_{x_i} + 2\gamma y_i r^{2\gamma-2}\partial_t$, $Y_i = \partial_{y_i} - 2\gamma x_i r^{2\gamma-2}\partial_t$ for $i = 1, \ldots, n$.

For $\gamma = 1$ $L = \text{div}_L(\nabla_L \cdot)$ is the sub-Laplacian $\Delta_H$ on the Heisenberg group $\mathbb{H}^n$. If $\gamma = 2, 3, \ldots, L$ is a Greiner operator. The vector field associated to $\mu$ is homogeneous with respect to $\delta_R(\xi) = (Rx, Ry, R^{2\gamma}t)$ and $Q = 2n + 2\gamma$.

*Example 7.5* Let $\mathbb{R}^N$ be splitted as

$$\mathbb{R}^N = \mathbb{R}^{n_1} \times \mathbb{R}^{n_2} \times \cdots \times \mathbb{R}^{n_r} \ni (x^1, x^2, \ldots, x^r),$$

and let $\alpha_2, \alpha_3, \ldots, \alpha_r > 0$ be fixed.

Let $g_2 : \mathbb{R}^{n_1} \to \mathbb{R}$ be an homogeneous function of degree $\alpha_2$.

Let $g_3 : \mathbb{R}^{n_1} \times \mathbb{R}^{n_2} \to \mathbb{R}$ be an homogeneous function of degree $\alpha_3$ with respect to dilation $\delta_R(x^1, x^2) = (Rx^1, R^{\alpha_2+1}x^2)$, that is $g_3(Rx^1, R^{\alpha_2+1}x^2) = R^{\alpha_3} g_3(x^1, x^2)$.

Let $g_4 : \mathbb{R}^{n_1} \times \mathbb{R}^{n_2} \times \mathbb{R}^{n_3} \to \mathbb{R}$ be an homogeneous function of degree $\alpha_4$ with respect to dilation $\delta_R(x^1, x^2, x^3) = (Rx^1, R^{\alpha_2+1}x^2, R^{\alpha_3+1}x^3)$.

We iterate the procedure by choosing analogously other homogeneous functions $g_j$ up to $g_r : \mathbb{R}^{n_1} \times \mathbb{R}^{n_2} \times \cdots \times \mathbb{R}^{n_{r-1}} \to \mathbb{R}$ a homogeneous function of degree $\alpha_r$ with respect to dilation $\delta_R(x^1, x^2, \ldots x^{r-1}) = (Rx^1, R^{\alpha_2+1}x^2, \ldots, R^{\alpha_{r-1}+1}x^{r-1})$.

Next we define the matrix $\mu$ as

$$\begin{pmatrix} I_{n_1} & 0 & & & \\ 0 & g_2(x_1)I_{n_2} & 0 & \cdots & \\ \cdots & 0 & g_3(x^1, x^2)I_{n_3} & & \\ & & & & \\ & & \cdots & \cdots & \\ & & 0 & g_r(x^1, x^2, \ldots, x^{r-1})I_{n_r} & \end{pmatrix}. \tag{51}$$

We have that the vector field $\mu\nabla$ satisfies the assumption of Sect. 2. Indeed it is homogeneous with respect to $\delta_R(x) = (Rx^1, R^{\alpha_2+1}x^2, \ldots, R^{\alpha_r+1}x^r)$. This example generalizes the Example 7.2.

*Example 7.6 (Carnot Groups)* On a Carnot group the horizontal gradient can be written in the form $\mu\nabla$ as in Sect. 2 and it satisfies our assumptions. We refer the reader to [4] for more detailed information on this subject. Special examples of Carnot groups are the Euclidean spaces $\mathbb{R}^N$. The simplest nontrivial example of a Carnot group is the Heisenberg group $\mathbb{H}^1 = \mathbb{R}^3$. See Example 7.3. Several other examples can be found in the book [4].

**Acknowledgements** This work is supported by the Italian MIUR National Research Project: Variational and perturbative aspects of nonlinear differential problems.

## References

1. Benilan, Ph., Boccardo, L., Galluet, T., Gariepy, R., Pierre, M., Vazquez, J.L.: An $L^1$ - theory of existence and uniqueness of solutions of nonlinear elliptic equations. Annali della scuola Normale di Pisa **22**, 241–273 (1995)
2. Bidaut-Véron, M.F., Pohozaev, S.I.: Nonexistence results and estimates for some nonlinear elliptic problems. J. Anal. Math. **84**, 1–49 (2001)
3. Boccardo, L., Galluet, T., Vazquez, J.L.: Nonlinear Elliptic Equations in $R^n$ without restriction on the data. J. Differ. Equ. **105**, 334–363 (1993)
4. Bonfiglioli, A., Lanconelli, E., Uguzzoni, F.: Stratified Lie Groups and Potential Theory for Their Sub-Laplacians. Springer Monographs in Mathematics. Springer, Berlin (2007)
5. Brezis, H.: Semilinear equations in $\mathbb{R}^n$ without condition at infinity. Appl. Math. Optim. **12**, 271–282 (1984)
6. D'Ambrosio, L.: Liouville theorems for anisotropic quasilinear inequalities. Nonlinear Anal. **70**, 2855–2860 (2009)
7. D'Ambrosio, L.: A new critical curve for a class of quasilinear elliptic systems. Nonlinear Anal. **78**, 62–78 (2013)
8. D'Ambrosio, L., Mitidieri, E.: A priori estimates, positivity results, and nonexistence theorems for quasilinear degenerate elliptic inequalities. Adv. Math. **224**, 967–1020 (2010)
9. D'Ambrosio, L., Mitidieri, E.: Nonnegative solutions of some quasilinear elliptic inequalities and applications. Math. Sb. **201**, 885–861 (2010)
10. D'Ambrosio, L., Mitidieri, E.: A priori estimates and reduction principles for quasilinear elliptic problems and applications. Adv. Differ. Equ. **17**, 935–1000 (2012)
11. D'Ambrosio, L., Farina, A., Mitidieri, E., Serrin, J.: Comparison principles, uniqueness and symmetry results of solutions of quasilinear elliptic equations and inequalities. Nonlinear Anal. **90**,135–158 (2013)
12. Mitidieri, E., Pohozaev, S.I.: Non existence of positive solutions for quasilinear elliptic problems on $\mathbb{R}^N$. Tr. Mat. Inst. Steklova **227**, 192–222 (1999)
13. Mitidieri, E., Pohozaev, S.I.: A priori estimates and the absence of solutions of nonlinear partial differential equations and inequalities. Tr. Mat. Inst. Steklova **234**, 1–384 (2001)
14. Serrin, J.: Local behavior of solutions of quasi–linear equations. Acta Math. **111**, 247–302 (1964)

# Liouville Type Theorems for Non-linear Differential Inequalities on Carnot Groups

**Luca Brandolini and Marco Magliaro**

*Dedicated to Ermanno Lanconelli on the occasion of his 70th birthday*

**Abstract** We overview some recent results on the existence and non-existence of positive solutions for differential inequalities of the kind

$$\operatorname{div}_0\left(\frac{\varphi(|\nabla_0 u|)}{|\nabla_0 u|}\nabla_0 u\right) \geqslant f(u)\ell(|\nabla_0 u|)$$

in the setting of Carnot groups under the Keller-Osserman condition.

**Keywords** Carnot groups • Keller-Osserman condition

**Mathematical Subject Classification:** Primary: 35R03, Secondary: 35R45, 35B53

## 1 Introduction

A Carnot group $\mathbb{G}$ is a Lie group with underlying manifold $\mathbb{R}^N$ endowed with a one parameter family $\{\delta_\lambda\}_{\lambda>0}$ of group automorphisms of the form

$$\delta_\lambda\left(x^{(1)}, x^{(2)}, \ldots, x^{(r)}\right) = \left(\lambda x^{(1)}, \lambda^2 x^{(2)}, \ldots, \lambda^r x^{(r)}\right)$$

L. Brandolini (✉)
Dipartimento di Ingegneria, Università degli studi di Bergamo, Dalmine (BG), Italy
e-mail: luca.brandolini@unibg.it

M. Magliaro
Dipartimento di Matematica Informatica ed Economia, Università degli Studi della Basilicata, Potenza, Italy
e-mail: marco.magliaro@gmail.com



G. Citti et al. (eds.), *Geometric Methods in PDE's*, Springer INdAM Series 13,
DOI 10.1007/978-3-319-02666-4_11

where $x^{(i)} \in \mathbb{R}^{N_i}$, $N_1 + N_2 + \ldots + N_r = N$ and such that the $N_1$ left-invariant vector fields $X_1, X_2, \ldots, X_{N_1}$ that agree with $\partial/\partial x_i^{(1)}$ at the origin generate the whole Lie algebra of left-invariant vector fields on $\mathbb{G}$.

The vector fields $X_1, \ldots, X_{N_1}$ are homogeneous of degree 1 with respect to the dilations $\delta_\lambda$, that is

$$X_j \left[ f \left( \delta_\lambda \left( \cdot \right) \right) \right] (x) = \lambda X_j f \left( \delta_\lambda x \right).$$

The linear span of the vector fields $X_1, X_2, \ldots, X_{N_1}$ is called the *horizontal layer* of the Lie algebra of $\mathbb{G}$. The canonical sub-Laplacian on $\mathbb{G}$ is the differential operator

$$\Delta_{\mathbb{G}} = \sum_{i=1}^{N_1} X_i^2$$

which is hypoelliptic by Hörmander's theorem (see [15]). We refer the interested readers to [3] for a detailed introduction to Carnot groups.

By a general theorem of Folland (see [13]) the sub-Laplacian $\Delta_{\mathbb{G}}$ admits a smooth fundamental solution $\Gamma$ which is homogeneous of degree $2-Q$ with respect to the dilation $\delta_\lambda$

$$\Gamma \left( \delta_\lambda \left( x \right) \right) = \lambda^{2-Q} \Gamma \left( x \right)$$

where $Q$ is the homogeneous dimension of $\mathbb{G}$ defined by $Q = N_1 + 2N_2 + \cdots + rN_r$. The fundamental solution $\Gamma$ can be used to define on $\mathbb{G}$ a symmetric homogeneous norm by

$$d\left(x\right) = \begin{cases} \Gamma\left(x\right)^{1/(2-Q)} & x \neq 0, \\ 0 & x = 0. \end{cases}$$

More precisely we have $d\left(\delta_\lambda(x)\right) = \lambda d\left(x\right)$ and $d\left(x\right) > 0$ if and only if $x \neq 0$. Setting

$$d\left(x, y\right) = d\left(y^{-1} \circ x\right)$$

one can check that $d\left(x,y\right) = d\left(y,x\right)$, $d\left(x,y\right) = 0$ if and only if $x = y$ and that the pseudo triangle inequality

$$d\left(x,y\right) \leqslant c\left[d\left(x,z\right) + d\left(z,y\right)\right]$$

holds for a suitable $c > 0$ and for every $x, y, z \in \mathbb{G}$.

From now on we will denote by $C_H^1\left(\mathbb{G}\right)$ the space of continuous functions on $\mathbb{G}$ having continuous derivative with respect to the vector fields of the first layer. For $k \in \mathbb{N}$ we define likewise $C_H^k\left(\mathbb{G}\right)$.

Given a function $u \in C^1_H(\mathbb{G})$ we define the *horizontal gradient* $\nabla_0 u$ as the vector field

$$\nabla_0 u = \sum_{i=1}^{N_1} (X_i u) X_i.$$

For a horizontal vector field $W = \sum_{i=1}^{N_1} w_i X_i$ we define the *horizontal divergence*

$$\mathrm{div}_0 W = \sum_{i=1}^{N_1} X_i w_i.$$

If $W = \sum_{i=1}^{N_1} w_i X_i$ and $Z = \sum_{i=1}^{N_1} z_i X_i$ are horizontal vector fields we can define the scalar product by $W \cdot Z = \sum_{i=1}^{N_1} w_i z_i$, so that $|W|^2 = W \cdot W = \sum_{i=1}^{N_1} w_i^2$ and in particular

$$|\nabla_0 u|^2 = \sum_{i=1}^{N_1} |X_i u|^2 .$$

*Example 1.1* A first example of Carnot group is the Heisenberg group $\mathbb{H}^n = \mathbb{R}^{2n} \times \mathbb{R}$ with the group law

$$(x, y, t) \circ (x', y', t') = \left(x + x', y + y', t + t' + 2\left(y \cdot x' - x \cdot y'\right)\right)$$

and the family of dilations

$$\delta_\lambda (x, y, t) = \left(\lambda x, \lambda y, \lambda^2 t\right).$$

In this case the horizontal vector fields are

$$X_i = \frac{\partial}{\partial x_i} + 2y_i \frac{\partial}{\partial t} \quad \text{and} \quad Y_i = \frac{\partial}{\partial y_i} - 2x_i \frac{\partial}{\partial t}$$

and, since

$$[X_i, Y_i] = X_i Y_i - Y_i X_i = -4 \frac{\partial}{\partial t} = -4T,$$

they generate the whole $(2n + 1)$-dimensional Lie algebra of left invariant vector fields. In this case the sub-Laplacian is given by

$$\Delta_{\mathbb{H}^n} = \sum_{i=1}^{n} \left(X_i^2 + Y_i^2\right)$$

and its homogeneous fundamental solution has the closed-form expression

$$\Gamma(x) = \frac{c_n}{\left((x^2+y^2)^2+t^2\right)^{n/2}}$$

where $c_n$ is a suitable constant. In this case the homogeneous dimension is $Q = 2n+2$ so that $\Gamma$ is homogeneous of degree $-2n$ and the homogeneous norm is given by

$$d(x) = \Gamma(x)^{-\frac{1}{2n}} = c_n^{-1/2n}\left((x^2+y^2)^2+t^2\right)^{1/4}.$$

In what follows, we shall consider a non-linear generalization of the sub-Laplacian called $\varphi$-Laplacian defined by

$$\Delta_\varphi u = \mathrm{div}_0\left(\varphi(|\nabla_0 u|)\frac{\nabla_0 u}{|\nabla_0 u|}\right)$$

where the function $\varphi$ satisfies the structural assumptions

$$\begin{cases} \varphi \in C^0\left(\mathbb{R}_0^+\right) \cap C^1\left(\mathbb{R}^+\right), \\ \varphi(0) = 0, \quad \varphi' > 0 \text{ on } \mathbb{R}^+. \end{cases}$$

A meaningful example of $\varphi$-Laplacian is the $p$-Laplacian

$$\Delta_p u = \mathrm{div}_0\left(|\nabla_0 u|^{p-2}\nabla_0 u\right)$$

that can be obtained with the choice $\varphi(t) = t^{p-1}$. Another interesting example comes from the choice $\varphi(t) = \frac{t}{\sqrt{1+t^2}}$ that provides an analog of the mean curvature operator on Carnot groups

$$\Delta_{MC} u = \mathrm{div}_0\left(\frac{\nabla_0 u}{\sqrt{1+|\nabla_0 u|^2}}\right).$$

Operators of this kind, or even more general, have been studied in $\mathbb{R}^n$, in the context of Riemannian geometry and on Carnot groups by several authors (see [8, 9, 19, 21, 23] and references therein). In particular in [19] the authors considered the existence of weak classical solutions to the differential inequality

$$\Delta_\varphi u \geqslant f(u)\,\ell(|\nabla_0 u|)$$

on the Heisenberg group and on $\mathbb{R}^n$. Under the *generalized Keller-Osserman condition*

$$\int_{t_0}^{+\infty} \frac{1}{K^{-1}(F(t))} dt < +\infty$$

where $F(t) = \int_0^t f(s)\,ds$ and $K(t) = \int_0^t \frac{s\varphi'(t)}{\ell(t)} dt$, they proved that such differential inequality admits only constant non-negative entire solutions.

Observe that for the differential inequality

$$\Delta_p u \geqslant u^\gamma |\nabla_0 u|^\theta$$

the generalized Keller-Osserman condition takes the form $\gamma + \theta > p - 1$. One can interpret this condition by saying that the non linearity in the RHS is bigger than the non linearity in the LHS.

## 2 A Brief History of the Keller-Osserman Condition

Around 1957 Joseph Keller [17, 18] and Robert Osserman [22] independently studied non-linear equations of the form

$$\Delta u = f(u) \tag{1}$$

or more generally

$$\Delta u \geqslant f(u). \tag{2}$$

In [17] Keller derived Eq. (1) from the study of the equilibrium of a charged gas inside a container. Let $p$ denote the pressure, $\rho$ the mass density, $a\rho$ the charge density and $E$ the electric field vector. Keller wrote the following equations

$$\begin{cases} \nabla p = a\rho E & \text{equilibrium equation,} \\ \operatorname{div} E = 4\pi a\rho & \text{the electric field is generated by the gas,} \\ p = g(\rho) & \text{equation of state of the gas.} \end{cases}$$

The function $g(\rho)$ is non-negative and increasing. Eliminating $E$ from the first and the second equation gives

$$\operatorname{div}\left(\rho^{-1}\nabla p\right) = 4\pi a^2 \rho.$$

It is not difficult to see that by a change of variable this equation can be reduced to

$$\Delta u = 4\pi a^2 \rho(p(u)) = f(u).$$

For example if $p = c\,\rho$ the change of variable

$$u = \log(\rho)$$

gives

$$c\Delta u = \frac{4\pi a^2}{c} e^u.$$

**Theorem 2.1 (Keller)** *Let $f : \mathbb{R} \to [0, +\infty)$ be increasing and assume*

$$\int_1^{+\infty} \left[ \int_0^x f(z)\, dz \right]^{-1/2} dx < +\infty. \tag{3}$$

*There exists a decreasing function $g : \mathbb{R} \to \mathbb{R}$ such that if $u$ is a solution of $\Delta u = f(u)$ in a domain $D \subset \mathbb{R}^n$ then*

$$u(x) \leqslant g(d(x, \partial D)).$$

*Moreover $g(R) \to -\infty$ as $R \to +\infty$.*

Observe that for $f(u) = u^\gamma$ condition (3) is equivalent to $\gamma > 1$.

The conclusion drawn by Keller using his theorem is the following:

**Conclusion 2.2** *For a certain class of equations of state, both ρ and p are bounded above at every inner point of D, independently of the total mass of fluid within the container D.*

He also found this interesting corollary:

**Corollary 2.1** *Let $f$ be as in the previous theorem, then $\Delta v = f(v)$ has no entire solutions.*

At the same time Osserman studied the differential inequality (2). In [22] he proved the following:

**Theorem 2.3 (Osserman)** *Let $f(z)$ be positive, continuous, and monotone increasing for $z \geqslant z_0$, and suppose*

$$\int^{\infty} \left( \int_0^t f(z)\, dz \right)^{-1/2} dt < +\infty.$$

*Then $u$ cannot satisfy $\Delta v > 0$ on $\mathbb{R}^n$ and $\Delta v \geqslant f(v)$ outside some sphere $S$.*

Osserman's motivation was mainly geometric and as a consequence of his theorem, he obtained the following geometric result.

**Corollary 2.2** *If a simply-connected surface $S$ has a Riemannian metric whose Gauss curvature $K$ satisfies $K \leqslant -\varepsilon < 0$ everywhere, then $S$ is conformally equivalent to the interior of the unit circle.*

Let us briefly sketch out the technique used by Keller and Osserman. Under the condition $\int^{\infty} \left(\int_0^t f(z)\,dz\right)^{-1/2} dt < +\infty$, the initial value problem

$$\begin{cases} \varphi''(r) + \frac{n-1}{r}\varphi'(r) = f(\varphi(r)) \\ \varphi(0) = \varepsilon, \quad \varphi'(0) = 0 \end{cases}$$

has a solution defined on $[0, R_\varepsilon)$ that satisfies $\lim_{r \to R_\varepsilon^-} \varphi(r) = +\infty$. Let now $u$ be a solution of $\Delta u \geqslant f(u)$ defined in a disk of radius $R_\varepsilon$. Without loss of generality we can assume that the disk is centered at the origin and consider $v(x) = u(x) - \varphi(|x|)$. Note that $v(x) \to -\infty$ as $|x| \to R_\varepsilon$. Suppose $v(x)$ has a positive maximum at $x_0$. In a neighborhood of $x_0$

$$\Delta v = \Delta u - \Delta \varphi \geqslant f(u) - f(\varphi) > 0,$$

so that $v$ would be sub-harmonic, contradicting that it has a maximum. It follows that $u(x) \leqslant \varphi(|x|)$.

After the seminal work of Keller and Osserman their technique has been extended to a variety of equations, inequalities and operators. For example Redheffer [24] considered a differential inequality with a gradient term

$$\Delta u \geqslant f(u)\,\ell(|\nabla u|),$$

Naito and Usami [21] and Bandle et al. [2] considered respectively

$$\Delta_\varphi u \geqslant f(u)$$

and

$$\Delta_\varphi u = f(u)\,\ell(|\nabla u|)$$

in $\mathbb{R}^n$. More recently Filippucci et al. [11, 12] gave sufficient conditions for the non-existence of entire positive solutions of differential inequalities of the kind

$$\operatorname{div}\left\{g(|x|)\,|\nabla u|^{p-2}\,\nabla u\right\} \geqslant h(|x|) f(u)$$

and

$$\operatorname{div}\left\{g(|x|)\,|\nabla u|^{p-2}\,\nabla u\right\} \geqslant h(|x|) f(u) \pm \tilde{h}(|x|)\,\ell(|\nabla u|).$$

Mari et al. [20] studied the differential inequality

$$\frac{1}{D}\operatorname{div}\left(D\frac{\varphi\left(|\nabla u|\right)}{|\nabla u|}\nabla u\right) \geqslant b\left(x\right)f\left(u\right)\ell\left(|\nabla u|\right)$$

and

$$\frac{1}{D}\operatorname{div}\left(D\frac{\varphi\left(|\nabla u|\right)}{|\nabla u|}\nabla u\right) \geqslant b\left(x\right)f\left(u\right)\ell\left(|\nabla u|\right) - g\left(u\right)h\left(|\nabla u|\right)$$

on weighted Riemannian manifolds.

Farina and Serrin [9] studied differential inequalities of the kind

$$\operatorname{div}\left(\mathscr{A}\left(x,u,\nabla u\right)\right) \geqslant \mathscr{B}(x,u,\nabla u)$$

with

$$\begin{aligned} |\mathscr{A}\left(x,z,\xi\right)| &\leqslant c\,|x|^{s}\,|z|^{r}\,|\mathscr{A}\left(x,z,\xi\right)\cdot\xi|^{p-1} \\ B(x,z,\xi) &\geqslant C|x|^{-t}|z|^{q} \end{aligned}$$

obtaining non existence results under various assumptions on $p, s, r, t, q$.

Despite the large literature on the subject, the sub-elliptic setting has been considered only in a few papers. D'Ambrosio [7] and D'Ambrosio and Mitidieri [8] considered inequalities of the kind

$$Lu \geqslant f\left(u\right)$$

where

$$Lu = \operatorname{div}_L\left(\mathscr{A}\left(x,u,\nabla_L u\right)\right).$$

In this case the gradient $\nabla_L$ and the divergence $\operatorname{div}_L$ are generated by suitable homogeneous vector fields.

Magliaro et al. [19] considered differential inequalities of the kind

$$\Delta_{\varphi}u \geqslant f\left(u\right)\ell\left(|\nabla_0 u|\right)$$

on the Heisenberg group $\mathbb{H}^n$.

## 3 Coercive Differential Inequalities on Carnot Groups

Our starting point is the following theorem of Magliaro et al. [19, Theorem 1.4].

**Theorem 3.1** *Let $\varphi$, $\ell$ and $f$ satisfy the following* structural assumptions

$$\begin{cases} \varphi \in C^0\left(\mathbb{R}_0^+\right) \cap C^1\left(\mathbb{R}^+\right), \\ \varphi(0) = 0, \quad \varphi' > 0 \text{ on } \mathbb{R}^+, \end{cases} \tag{4}$$

$$\begin{cases} f \in C^0\left(\mathbb{R}_0^+\right), \\ f(0) \geqslant 0, \quad f \text{ increasing}, \end{cases} \tag{5}$$

$$\begin{cases} \ell \in C^0([0,+\infty)), \\ \sup\limits_{[0,t]} \ell(s) \leqslant C\ell(t), \quad \ell(0) > 0, \end{cases} \tag{6}$$

$$\frac{t\varphi'(t)}{\ell(t)} \in L^1\left(0^+\right) \setminus L^1(+\infty), \tag{7}$$

*and the* relaxed homogeneity *conditions*

$$\begin{cases} s\varphi'(st) \leqslant C_1 s^\tau \varphi'(t) \quad \forall s \in [0,1], \\ s^{1+\tau}\ell(t) \leqslant C_2 \ell(st) \quad \forall s \in [0,1] \end{cases} \tag{8}$$

*for positive constants $C$, $C_1$, $C_2$ and $\tau$. Define*

$$K(t) = \int_0^t \frac{s\varphi'(s)}{\ell(s)} ds \text{ and } F(t) = \int_0^t f(s)\, ds.$$

*If the generalized Keller-Osserman condition*

$$\frac{1}{K^{-1}(F(t))} \in L^1(+\infty) \tag{9}$$

*holds and $0 \leqslant u \in C_H^1(\mathbb{H}^n)$ satisfies*

$$\Delta_\varphi u \geqslant f(u)\, \ell(|\nabla_0 u|) \quad \text{on } \mathbb{H}^n$$

*then $u \equiv 0$.*

The steps of their proof are more or less the following.

1. They assume there exists a non-negative, entire solution $u \not\equiv 0$ of $\Delta_\varphi u \geqslant f(u)\, \ell(|\nabla_0 u|)$.

2. They construct a radial function $v$ defined in an annulus of the kind

$$R_1 \leqslant d(x) \leqslant R_2$$

satisfying

$$\Delta_\varphi v \leqslant f(v)\,\ell(|\nabla_0 v|),$$

such that $v(x) \to +\infty$ as $d(x) \to R_2^-$ and such that $v(x)$ is small when $d(x)$ is close to $R_1$. To do this, they implicitly define $\alpha$ by

$$R_1 - t = \int_{\alpha(t)}^{\alpha(R_1)} \frac{ds}{K^{-1}(\sigma F(s))}$$

and observe that $\alpha$ satisfies $\varphi'(\alpha')\alpha'' = \sigma f(\alpha)\,\ell(\alpha')$. Setting $v(x) = \alpha(d(x))$ they obtain

$$\Delta_\varphi v = |\nabla_0 d|^2 \left(\varphi'\left(\alpha' |\nabla_0 d|\right)\alpha'' + \frac{\varphi(\alpha' |\nabla_0 d|)}{|\nabla_0 d|}\frac{Q-1}{d}\right)$$

and using $\varphi'(\alpha')\alpha'' = \sigma f(\alpha)\,\ell(\alpha')$ and the structural assumptions, they are able to show that for small $\sigma$

$$\Delta_\varphi v \leqslant f(v)\,\ell(|\nabla_0 v|).$$

3. A suitable choice of $v$ implies that $u - v$ must have a positive maximum $q$ inside the annulus. At such a point $u > v$ and $\nabla_0 u = \nabla_0 v$. Also, since $\ell > 0$

$$\Delta_\varphi u \geqslant f(u)\,\ell(|\nabla_0 u|) > f(v)\,\ell(|\nabla_0 v|) \geqslant \Delta_\varphi v.$$

It follows that there is a neighborhood of $q$ where

$$\Delta_\varphi u \geqslant \Delta_\varphi v.$$

4. A comparison argument shows that this is not possible (if $\Delta_\varphi$ were linear then $u - v$ would be sub-harmonic in a neighborhood of $q$).

Let now $\mathbb{G}$ be a Carnot group, let $Q$ be its homogeneous dimension, $\Gamma$ the fundamental solution of the sub-Laplacian on $\mathbb{G}$ and $d(x) = \Gamma^{\frac{1}{2-Q}}(x)$ the homogeneous norm. In order to extend the results of Magliaro, Mari, Mastrolia and Rigoli to Carnot groups the main problem to solve is the construction of the radial supersolution. Let $\alpha$ be defined by

$$t_0 - t = \int_{\alpha(t)}^{\alpha(t_0)} \frac{ds}{K^{-1}(\sigma F(s))}$$

and let $v(x) = \alpha(d(x))$. For a *radial* function a lengthy computation gives

$$\Delta_\varphi v = |\nabla_0 d|^2 \left( \varphi'(\alpha' |\nabla_0 d|) \alpha'' + \frac{\varphi(\alpha' |\nabla_0 d|)}{|\nabla_0 d|} \frac{Q-1}{d} \right) + \left[ \varphi'(\alpha' |\nabla_0 d|) \alpha' - \frac{\varphi(\alpha' |\nabla_0 d|)}{|\nabla_0 d|} \right] \nabla_0 |\nabla_0 d| \cdot \frac{\nabla_0 d}{|\nabla_0 d|}.$$

Since on the Heisenberg group the homogeneous norm is explicit, it is possible to check that in this case

$$\nabla_0 |\nabla_0 d| \cdot \frac{\nabla_0 d}{|\nabla_0 d|} = 0$$

giving a much simpler expression for the $\varphi$–Laplacian of a radial function. The Carnot groups where such quantity vanishes have been studied by Balogh and Tyson in [1] and have been called polarizable since they admit polar coordinates. Unfortunately the only known examples of polarizable groups are the Heisenberg group and more generally Kaplan H-type groups (see [16]).

In [4] the authors show that the techniques of Magliaro, Mari, Mastrolia and Rigoli can be extended straightforwardly to polarizable groups. The case of non-polarizable groups turns out to be more complicated and requires extra assumptions. In [4, Theorem 4.1] the following result is proved.

**Theorem 3.2** *Assume the validity of (4), (5), (6) and (7). Also assume that*

$$\begin{cases} s\varphi'(st) \leqslant C_1 s^\tau \varphi'(t), & \forall s \in [0,1], \\ s^{\tau-1} \ell(t) \leqslant C_2 \ell(st), & \forall s \in [0,1], \\ t\varphi'(t) \leqslant C_3 \varphi(t) \end{cases} \tag{10}$$

*for positive constants $C_1, C_2, C_3$ and $\tau$. If the generalized Keller-Osserman condition (9) holds, then every solution $0 \leqslant u \in C^1_H(\mathbb{G})$ of*

$$\Delta_\varphi u \geqslant f(u)\, \ell(|\nabla_0 u|) \quad on\ \mathbb{G}$$

*is identically zero.*

We point out that conditions (10) are stronger than the original conditions (8). Indeed, (10) imply that $at^p \leqslant \varphi(t) \leqslant bt^p$ for some positive constants $a, b$ and $p$.

When condition (9) is not satisfied, the inequality $\Delta_\varphi u \geqslant f(u)\, \ell(|\nabla_0 u|)$ admits entire solutions. Indeed in [4, Theorem 5.1] the following is proved.

**Theorem 3.3** *Assume the validity of (4), (5), (6) and (7). Then, if the generalized Keller-Osserman condition (9) is not satisfied, there exists a non-negative, non-constant solution $u \in C^1_H(\mathbb{G})$ of inequality $\Delta_\varphi u \geqslant f(u)\, \ell(|\nabla_0 u|)$.*

## 4 A Sharper Result for the Heisenberg Group

An unpleasant feature of Theorem 3.2 is the assumption $\ell(0) > 0$ since it excludes some interesting model cases such as

$$\Delta_p u \geqslant f(u) |\nabla_0 u|^\theta .$$

In [5] it is proved that on the Heisenberg group it is possible to relax this hypothesis and assume only $\ell(t) \geqslant 0$.

The condition $\ell(0) > 0$ was necessary in the comparison argument between $u$ and the radial supersolution. As shown in step 4 of the previous section, there exists a point $q$ where $u > v$ and $\nabla_0 u = \nabla_0 v$. Assuming $\ell(0) > 0$ gives

$$\Delta_\varphi u \geqslant f(u) \ell(|\nabla_0 u|) > f(v) \ell(|\nabla_0 v|) \geqslant \Delta_\varphi v,$$

and therefore $\Delta_\varphi u > \Delta_\varphi v$ in a neighborhood of $q$,which is not possible. If $\ell$ is allowed to vanish at 0 we only obtain

$$\Delta_\varphi u \geqslant \Delta_\varphi v$$

at the point $q$ and unfortunately this is not enough to get a contradiction. However since $f(u) > f(v)$ the equality case

$$\Delta_\varphi u \geqslant f(u) \ell(|\nabla_0 u|) = f(v) \ell(|\nabla_0 v|) \geqslant \Delta_\varphi v$$

only happens if $|\nabla_0 u| = |\nabla_0 v| = 0$. Observe that this may actually happen. Indeed, since $v(x) = \alpha(d(x))$ we have

$$\nabla_0 v(q) = \alpha'(d(q)) \nabla_0 d(q)$$

and on the Heisenberg group for $q = (z, t)$ we have

$$|\nabla_0 d(q)| = \frac{|z|^2}{d(z,t)}$$

so that $\nabla_0 v(q)$ vanishes on the vertical line $z = 0$. However, note that, if $\alpha' \neq 0$, for the Euclidean gradient we have

$$|\nabla v| \neq 0.$$

**Idea 1** *Construct $v$ in such a way that $u - v$ cannot have a maximum at a point where $|\nabla_0 u| = |\nabla_0 v| = 0$.*

Since at a point of maximum we have $\nabla u = \nabla v \neq 0$ this will be achieved by showing that the set

$$\mathscr{C} = \{p \in \mathbb{H}^n : \nabla_0 u(p) = 0 \text{ and } \nabla u(p) \neq 0\}$$

cannot be too big and choosing $v$ in such a way that $u - v$ cannot have a maximum in $\mathscr{C}$.

The next theorem (see [5, Proposition 4.1]) shows that, in a suitable sense, $\mathscr{C}$ is "small".

**Theorem 4.1** *Let $u \in C^2_H\left(\overline{\Omega}\right)$, $\Omega$ open set in $\mathbb{H}^m$. Set*

$$\mathscr{C} = \left\{p \in \Omega : X_j u\,(p) = Y_j u\,(p) = 0,\ j = 1, \dots, m,\ Tu\,(p) \neq 0\right\}.$$

*If $m > 1$, for a.e. $z_0 \in \mathbb{R}^{2m}$*

$$\mathscr{C} \cap \{(z_0, t) : t \in \mathbb{R}\}$$

*is countable and discrete.*

*If $m = 1$, for a.e. $z_0 \in \mathbb{R}^{2m}$*

$$\mathscr{C} \cap \{(z_0, t) : t \in \mathbb{R}\}$$

*has Hausdorff dimension $\leqslant \frac{1}{2}$.*

The idea of the proof of this theorem is the following. If $p \in \mathscr{C}$ then

$$0 \neq Tu\,(p) = -\frac{1}{4}\left[X_j Y_j u\,(p) - Y_j X_j u\,(p)\right].$$

It follows that $X_j Y_j u\,(p)$ and $Y_j X_j u\,(p)$ cannot both vanish at $p$. Hence either

$$p \in \left\{q \in \Omega : X_j u\,(q) = 0,\ Y_j X_j u\,(q) \neq 0\right\}$$

or

$$p \in \left\{q \in \Omega : Y_j u\,(q) = 0, X_j Y_j u\,(q) \neq 0\right\}.$$

Since this holds for every $j = 1, \dots, m$ it follows that $p$ is in the intersection of $m$ of the above set.

Such intersections are $m$-codimensional $\mathbb{H}$-regular surfaces by a theorem of Franchi et al. [14]. In particular they have Hausdorff dimension equal to $m + 2$. Thus

$$\dim\left(\mathscr{C}\right) \leqslant m + 2.$$

We now use Eilenberg's inequality (see [10, Theorem 2.10.25] or [6, Theorem 13.3.1]). If $f : X \to Y$ is a Lipschitz map between separable metric spaces, $A \subset X$ and $0 \leqslant k \leqslant d$ we have

$$\int_Y^* \mathscr{H}_{d-k}\left(A \cap f^{-1}\left(y\right)\right) d\mathscr{H}_k\left(y\right) \leqslant c\left(\mathrm{Lip} f\right)^k \mathscr{H}_d\left(A\right).$$

Here $\mathscr{H}_d$ denotes the $d$-dimensional Hausdorff measure and ${}^*\!\int$ the upper Lebesgue integral.

Taking $f : \mathbb{H}^m \to \mathbb{R}^{2m}$ such that $f(z,t) = z$, $d = k = 2m$ and $A = \mathscr{C}$, we obtain

$$ {}^*\!\!\int_{\mathbb{R}^{2m}} \mathscr{H}_0\left(\mathscr{C} \cap f^{-1}(z)\right) d\mathscr{H}_{2m}(z) \leqslant c\,(\mathrm{Lip} f)^k\, \mathscr{H}_{2m}(\mathscr{C}). $$

Since $\dim(\mathscr{C}) \leqslant m + 2 \leqslant 2m$ (if $m > 1$), we have

$$ {}^*\!\!\int_{\mathbb{R}^{2m}} \mathscr{H}_0\left(\mathscr{C} \cap f^{-1}(z)\right) d\mathscr{H}_{2m}(z) < +\infty, $$

which means that

$$ \mathscr{H}_0\left(\mathscr{C} \cap \{(z,t) : t \in \mathbb{R}\}\right) $$

is finite for a.e. $z \in \mathbb{R}^{2m}$.

The fact that it is possible to construct $v$ in such a way that $u - v$ does not have a maximum in $\mathscr{C}$ is a consequence of the following Lemma. See Proposition 3.4 in [5] for a proof.

**Lemma 4.1** *Assume that the* structural assumptions *(4), (5), (6) and (7) and the Keller-Osserman condition (9) hold. Let* $0 < t_0 < t_1$, $0 < \varepsilon < \eta$, $h_1, h_2 : [t_0, +\infty) \to \mathbb{R}$ *and let* $E \subset \mathbb{R}$ *be at most countable. Then there exist* $T > t_1$ *and a strictly increasing convex function* $\alpha \in C^2([t_0, T))$ *such that, for every* $q \in \mathbb{H}^m$, *the radial function* $v(p) = \alpha(d(p,q))$ *satisfies*

$$ \begin{cases} \Delta_\varphi v \leqslant f(v)\,\ell(|\nabla_0 v|) & \text{on } B_T(q) \setminus B_{t_0}(q) \\ v = \varepsilon & \text{on } \partial B_{t_0}(q) \\ v = +\infty & \text{on } \partial B_T(q) \\ \varepsilon \leqslant v \leqslant \eta & \text{on } B_{t_1} \setminus B_{t_0}. \end{cases} \tag{11} $$

*Moreover, for every* $t \in E \cap [t_0, T]$,

$$ \alpha'\left(t^{\frac{1}{2}}\right) \neq h_1(t) \ \ \text{and}\ \alpha'\left(t^{\frac{1}{2}}\right) \neq h_2(t). $$

The idea of the proof of this lemma is to construct a family of supersolutions $v_\sigma(p) = \alpha_\sigma(d(p,q))$ that satisfies (11), prove that for every $t \in E$ there exists at most one value of $\sigma$ such that $\alpha'_\sigma\left(t^{\frac{1}{2}}\right) = h_1(t)$ and then note that there are uncountably many values of $\sigma$ that work.

We are now ready to state the already-advertised result for the Heisenberg group which is Theorem 5.1 in [5].

**Theorem 4.2** *Let $\varphi, f, \ell$ satisfy the usual structural conditions except that we allow $\ell(0) = 0$. Assume the "relaxed homogeneity condition"*

$$\frac{s^2\varphi'(st)}{\ell(st)} \leqslant c\frac{\varphi'(t)}{\ell(t)} \forall s \in [0,1],\, t \in [0,+\infty)$$

*and the Keller-Osserman condition. Let $D = \{(z_0, t) : t \in \mathbb{R}\}$ for some $z_0 \in \mathbb{R}^{2m}$. Let $u$ be a non-negative solution of*

$$\Delta_\varphi u \geqslant f(u)\,\ell(|\nabla_0 u|)$$

*such that $u \in C^1_H(\mathbb{H}^m) \cap C^2_H(\mathbb{H}^m \setminus D)$. Also, in the case $m = 1$ assume that for almost every $z \in \mathbb{R}^2$, $Tu(z,\cdot) \in C^\beta(\mathbb{R})$ for some $\beta > \frac{1}{2}$. Then $u$ is constant.*

When $m > 1$ this theorem follows from Theorem 4.1 and Lemma 4.1 in the following way. Following the method of Magliaro, Mari, Mastrolia and Rigoli one can construct a super-solution defined in an annulus that blows up on the exterior boundary of the annulus. By choosing carefully the center $(z_0, t_0)$ of the annulus it is possible to ensure that

$$E = \mathscr{C} \cap \{(z_0, t) : t \in \mathbb{R}\}$$

is countable. Applying the above lemma with $h_1(t) = 2t^{\frac{1}{2}}Tu(z_0, t)$ and $h_2(t) = -2t^{\frac{1}{2}}Tu(z_0, -t)$ ensures that at the points of $E$

$$Tu \neq Tv$$

and therefore $u - v$ cannot attain a maximum where $|\nabla_0 u| = |\nabla_0 v| = 0$.

For $m = 1$, the proof requires a more refined version of Lemma 4.1 which takes into account the Hausdorff dimension of the set $E$. See Lemma 3.2 in [5].

The next theorem shows that Theorem 4.2 is sharp (see Theorem 6.1 in [5]).

**Theorem 4.3** *Under the usual structural assumptions, if the generalized Keller-Osserman condition is not satisfied, there exists a non-negative, non-constant solution $u \in C^1_H(\mathbb{H}^m) \cap C^2_H(\mathbb{H}^m \setminus \{z = 0\})$ of the inequality*

$$\Delta_\varphi u \geqslant f(u)\,\ell(|\nabla_0 u|).$$

## References

1. Balogh, Z.M., Tyson, J.T.: Polar coordinates in Carnot groups. Math. Z. **241**(4), 697–730 (2002)
2. Bandle, C., Greco, A., Porru, G.: Large solutions of quasilinear elliptic equations: existence and qualitative properties. Boll. Unione Mat. Ital. B (7) **11**(1), 227–252 (1997)

3. Bonfiglioli, A., Lanconelli, E., Uguzzoni, F.: Stratified Lie Groups and Potential Theory for Their Sub-Laplacians. Springer Monographs in Mathematics. Springer, Berlin (2007)
4. Brandolini, L., Magliaro, M.: A note on Keller-Osserman conditions on Carnot groups. Nonlinear Anal. **75**(4), 2326–2337 (2012)
5. Brandolini, L., Magliaro, M.: Liouville type results and a maximum principle for non-linear differential operators on the heisenberg group. Preprint (2013)
6. Burago, Yu.D., Zalgaller, V.A.: Geometric Inequalities. Grundlehren der Mathematischen Wissenschaften [Fundamental Principles of Mathematical Sciences], vol. 285. Springer, Berlin (1988) [Translated from the Russian by A.B. Sosinskiĭ, Springer Series in Soviet Mathematics]
7. D'Ambrosio, L.: Liouville theorems for anisotropic quasilinear inequalities. Nonlinear Anal. **70**(8), 2855–2869 (2009)
8. D'Ambrosio, L., Mitidieri, E.: A priori estimates, positivity results, and nonexistence theorems for quasilinear degenerate elliptic inequalities. Adv. Math. **224**(3), 967–1020 (2010)
9. Farina, A., Serrin, J.: Entire solutions of completely coercive quasilinear elliptic equations, II. J. Differ. Equ. **250**(12), 4409–4436 (2011)
10. Federer, H.: Geometric Measure Theory. Die Grundlehren der Mathematischen Wissenschaften, Band 153. Springer, New York (1969)
11. Filippucci, R., Pucci, P., Rigoli, M.: On entire solutions of degenerate elliptic differential inequalities with nonlinear gradient terms. J. Math. Anal. Appl. **356**(2), 689–697 (2009)
12. Filippucci, R., Pucci, P., Rigoli, M.: On weak solutions of nonlinear weighted $p$-Laplacian elliptic inequalities. Nonlinear Anal. **70**(8), 3008–3019 (2009)
13. Folland, G.B.: Subelliptic estimates and function spaces on nilpotent Lie groups. Ark. Mat. **13**(2), 161–207 (1975)
14. Franchi, B., Serapioni, R., Serra Cassano, F.: Regular submanifolds, graphs and area formula in Heisenberg groups. Adv. Math. **211**(1), 152–203 (2007)
15. Hörmander, L.: Hypoelliptic second order differential equations. Acta Math. **119**, 147–171 (1967)
16. Kaplan, A.: Fundamental solutions for a class of hypoelliptic PDE generated by composition of quadratic forms. Trans. Am. Math. Soc. **258**(1), 147–153 (1980)
17. Keller, J.B.: Electrohydrodynamics, I. The equilibrium of a charged gas in a container. J. Ration. Mech. Anal. **5**, 715–724 (1956)
18. Keller, J.B.: On solutions of $\Delta u = f(u)$. Commun. Pure Appl. Math. **10**, 503–510 (1957)
19. Magliaro, M., Mari, L., Mastrolia, P., Rigoli, M.: Keller-Osserman type conditions for differential inequalities with gradient terms on the Heisenberg group. J. Differ. Equ. **250**(6), 2643–2670 (2011)
20. Mari, L., Rigoli, M., Setti, A.G.: Keller-Osserman conditions for diffusion-type operators on Riemannian manifolds. J. Funct. Anal. **258**(2), 665–712 (2010)
21. Naito, Y., Usami, H.: Entire solutions of the inequality $\mathrm{div}(A(|Du|)Du) \geq f(u)$. Math. Z. **225**(1), 167–175 (1997)
22. Osserman, R.: On the inequality $\Delta u \geq f(u)$. Pac. J. Math. **7**, 1641–1647 (1957)
23. Pigola, S., Rigoli, M., Setti, A.G.: Maximum principles on Riemannian manifolds and applications. Mem. Am. Math. Soc. **174**(822), x+99 (2005)
24. Redheffer, R.: On the inequality $\Delta u \geqq f(u, |\operatorname{grad} u|)$. J. Math. Anal. Appl. **1**, 277–299 (1960)

# Modica Type Gradient Estimates for Reaction-Diffusion Equations

**Agnid Banerjee and Nicola Garofalo**

*Dedicated to Ermanno Lanconelli on the occasion of his 70th birthday*

**Abstract** We continue the study of Modica type gradient estimates for inhomogeneous parabolic equations initiated in Banerjee and Garofalo (Nonlinear Anal. Theory Appl., to appear). First, we show that for the parabolic minimal surface equation with a semilinear force term if a certain gradient estimate is satisfied at $t = 0$, then it holds for all later times $t > 0$. We then establish analogous results for reaction-diffusion equations such as (5) below in $\Omega \times [0, T]$, where $\Omega$ is an epigraph such that the mean curvature of $\partial\Omega$ is nonnegative. We then turn our attention to settings where such gradient estimates are valid without any a priori information on whether the estimate holds at some earlier time. Quite remarkably (see Theorems 4.1, 4.2 and 5.1), this is true for $\mathbb{R}^n \times (-\infty, 0]$ and $\Omega \times (-\infty, 0]$, where $\Omega$ is an epigraph satisfying the geometric assumption mentioned above, and for $M \times (-\infty, 0]$, where $M$ is a connected, compact Riemannian manifold with nonnegative Ricci tensor. As a consequence of the gradient estimate (7), we establish a rigidity result (see Theorem 6.1 below) for solutions to (5) which is the analogue of Theorem 5.1 in Caffarelli et al. (Commun. Pure Appl. Math. **47**, 1457–1473, 1994). Finally, motivated by Theorem 6.1, we close the paper by proposing a parabolic version of the famous conjecture of De Giorgi also known as the $\varepsilon$-version of the Bernstein theorem.

**Keywords** Parabolic Modica type gradient estimates • De Giorgi's conjecture for the heat equation

**Mathematical Subject Classification:** 35K55, 35B45, 35B65

A. Banerjee
University of California, Irvine, CA 92697, USA
e-mail: agnidban@gmail.com

N. Garofalo (✉)
Università di Padova, Padova, Italy
e-mail: rembrandt54@gmail.com



G. Citti et al. (eds.), *Geometric Methods in PDE's*, Springer INdAM Series 13,
DOI 10.1007/978-3-319-02666-4_12

# 1 Introduction

In his pioneering paper [23] L. Modica proved that if $u$ is a (smooth) bounded entire solution of the semilinear Poisson equation $\Delta u = F'(u)$ in $\mathbb{R}^n$, with nonlinearity $F \geq 0$, then $u$ satisfies the a priori gradient bound

$$|Du|^2 \leq 2F(u). \tag{1}$$

With a completely different approach from Modica's original one, this estimate was subsequently extended in [3] to nonlinear equations in which the leading operator is modeled either on the $p$-Laplacian $\operatorname{div}(|Du|^{p-2}Du)$, or on the minimal surface operator $\operatorname{div}((1 + |Du|^2)^{-1/2}Du)$, and later to more general integrands of the calculus of variations in [7]. More recently, in their very interesting paper [12] Farina and Valdinoci have extended the Modica estimate (1) to domains in $\mathbb{R}^n$ which are epigraphs whose boundary has nonnegative mean curvature, and to compact manifolds having nonnegative Ricci tensor, see [13], and also the sequel paper with Sire [15].

It is by now well-known, see [1, 3, 7, 23], that, besides its independent interest, an estimate such as (1) implies Liouville type results, monotonicity properties of the relevant energy and it is also connected to a famous conjecture of De Giorgi (known as the $\varepsilon$-version of the Bernstein theorem) which we discuss at the end of this introduction and in Sect. 6 below, and which nowadays still constitutes a largely unsolved problem.

In the present paper we study Modica type gradient estimates for solutions of some nonlinear parabolic equations in $\mathbb{R}^n$ and, more in general, in complete Riemannian manifolds with nonnegative Ricci tensor, and in unbounded domains satisfying the above mentioned geometric assumptions in [12]. In the first part of the paper we continue the study initiated in the recent work [2], where we considered the following inhomogeneous variant of the normalized $p$-Laplacian evolution in $\mathbb{R}^n \times [0, T]$,

$$|Du|^{2-p}\left\{\operatorname{div}(|Du|^{p-2}Du) - F'(u)\right\} = u_t, \qquad 1 < p \leq 2. \tag{2}$$

In [2] we proved that if a bounded solution $u$ of (2) belonging to a certain class $H$ (see [2] for the relevant definition) satisfies the following gradient estimate at $t = 0$ for a.e. $x \in \mathbb{R}^n$,

$$|Du(x,t)|^p \leq \frac{p}{p-1}F(u(x,t)), \tag{3}$$

then such estimate continues to hold at any given time $t > 0$. On the function $F$ we assumed that $F \in C^{2,\beta}_{loc}(\mathbb{R})$ and $F \geq 0$. These same assumptions will be assumed throughout this whole paper.

In Sect. 2 we show that a similar result is true for the following inhomogeneous variant of the minimal surface parabolic equation

$$(1+|Du|^2)^{1/2}\left\{\operatorname{div}\left(\frac{Du}{(1+|Du|^2)^{1/2}}\right)-F'(u)\right\}=u_t, \tag{4}$$

see Theorem 2.1 below. Equation (4) encompasses two types of equations: when $F(u) = 0$ it represents the equation of motion by mean curvature studied in [10], whereas when $u(x,t) = v(x)$, then (4) corresponds to the steady state which is prescribed mean curvature equation.

In Sect. 3 we establish similar results for the reaction diffusion equation in $\Omega \times [0,T]$

$$\Delta u = u_t + F'(u), \tag{5}$$

where now $\Omega$ is an epigraph and the mean curvature of $\partial\Omega$ is nonnegative. Theorem 3.1 below constitutes the parabolic counterpart of the cited result in [12] for the following problem

$$\begin{cases} \Delta u = F'(u), & \text{in } \Omega, \\ u = 0 \text{ on } \partial\Omega, & u \geq 0 \text{ on } \Omega. \end{cases} \tag{6}$$

In that paper the authors proved that a bounded solution $u$ to (6) satisfies the Modica estimate (1), provided that the mean curvature of $\partial\Omega$ is nonnegative.

In Sect. 4 we turn our attention to settings where global versions of such estimates for solutions to (5) can be established, i.e., when there is no a priori information on whether such an estimate hold at some earlier time $t_0$. In Theorems 4.1 and 4.2 we show that, quite remarkably, respectively in the case $\mathbb{R}^n \times (-\infty, 0]$ and $\Omega \times (-\infty, 0]$, where $\Omega$ is an epigraph that satisfies the geometric assumption mentioned above, the a priori gradient estimate

$$|Du(x,t)|^2 \leq 2F(u(x,t)) \tag{7}$$

holds globally on a bounded solution $u$ of (5).

In Sect. 5 we establish a parabolic generalization of the result in [13], but in the vein of our global results in Sect. 4. In Theorem 5.1 we prove that if $M$ is a compact Riemannian manifold with Ric $\geq 0$, with Laplace-Beltrami $\Delta$, then any bounded entire solution $u$ to (5) in $M \times (-\infty, 0]$ satisfies (7). It remains to be seen whether our result, or for that matter the elliptic result in [13], remain valid when $M$ is only assumed to be complete, but not compact.

Finally in Sect. 6, as a consequence of the a priori estimate (7) in Sect. 4, we establish an analogue of Theorem 5.1 in [3] for solutions to (5) in $\mathbb{R}^n \times (-\infty, 0]$. More precisely, in Theorem 6.1 below we show that if the equality in (7) holds at some $(x_0, t_0)$, then there exists a function $g \in C^2(\mathbb{R})$, $a \in \mathbb{R}^n$, and $\alpha \in \mathbb{R}$, such that

$$u(x,t) = g(< a, x > +\alpha). \tag{8}$$

In particular, $u$ is independent of time and the level sets of $u$ are vertical hyperplanes in $\mathbb{R}^n \times (-\infty, 0]$. This result suggests a parabolic version of the famous conjecture of De Giorgi (also known as the $\varepsilon$-version of the Bernstein theorem for minimal graphs) which asserts that entire solutions to

$$\Delta u = u^3 - u, \tag{9}$$

such that $|u| \leq 1$ and $\frac{\partial u}{\partial x_n} > 0$, must be one-dimensional, i.e., must have level sets which are hyperplanes, at least in dimension $n \leq 8$, see [8]. We recall that the conjecture of De Giorgi has been fully solved for $n = 2$ in [16] and $n = 3$ in [1], and it is known to fail for $n \geq 9$, see [9]. Remarkably, it is still an open question for $4 \leq n \leq 8$. Additional fundamental progress on De Giorgi's conjecture is contained in the papers [17, 24]. For results concerning the $p$-Laplacian version of De Giorgi's conjecture, we refer the reader to the interesting paper [25]. For further results, the state of art and recent progress on De Giorgi's conjecture, we refer to [4, 11, 14] and the references therein.

In Sect. 7 motivated by our Theorem 6.1 below, we close the paper by proposing a parabolic version of De Giorgi's conjecture. It is our hope that it will stimulate interesting further research.

## 2 Forward Modica Type Estimates in $\mathbb{R}^n \times [0, T]$ for the Generalized Motion by Mean Curvature Equation

In [3] it was proved that if $u \in C^2(\mathbb{R}^n) \cap L^\infty(\mathbb{R}^n)$ is a solution to

$$\operatorname{div}\left(\frac{Du}{(1 + |Du|^2)^{1/2}}\right) = F'(u), \tag{10}$$

such that $|Du| \leq C$, then the following Modica type gradient estimate holds

$$\frac{(1 + |Du|^2)^{1/2} - 1}{(1 + |Du|^2)^{1/2}} \leq F(u). \tag{11}$$

In Theorem 2.1 below we generalize this result to the parabolic minimal surface equation (4). Such result also provides the counterpart of the above cited main result (2) in [2] for the normalized parabolic $p$-Laplacian (3). Henceforth, by $v \in C^{2,1}_{loc}$, we mean that $v$ has continuous derivatives of up to order two in the $x$ variable and up to order one in the $t$ variable. We would also like to mention that unlike the case when $F = 0$, further requirements on $F$ need to be imposed to ensure that a bounded solution to (4) has bounded gradient, see e.g. Theorem 4 in [19]. This is why an $L^\infty$ gradient bound is assumed in the hypothesis of the next theorem.

We recall that throughout the whole paper we assume that $F \in C^{2,\beta}_{loc}(\mathbb{R})$ for some $\beta > 0$, and that $F \geq 0$.

**Theorem 2.1** *For a given $\varepsilon > 0$, let $u \in C^{2,1}_{loc}(\mathbb{R}^n \times [0, T]) \cap L^\infty(\mathbb{R}^n \times (-\varepsilon, T])$ be a classical solution to* (4) *in $\mathbb{R}^n \times [0, T]$ such that $|Du| \leq C$. If $u$ satisfies the following gradient estimate*

$$\frac{(1 + |Du|^2)^{1/2} - 1}{(1 + |Du|^2)^{1/2}} \leq F(u) \tag{12}$$

*at $t = 0$, then $u$ satisfies* (12) *for all $t > 0$.*

*Proof* Since $|Du| \leq C$ and $F \in C^{2,\beta}_{loc}$, it follows from the Schauder regularity theory of uniformly parabolic non-divergence equations (see Chaps. 4 and 5 in [22]), that $u \in H_{3+\alpha}(\mathbb{R}^n \times [0, T])$ for some $\alpha > 0$ which depends on $\beta$ and the bounds on $u$ and $Du$ (see Chap. 4 in [22] for the relevant notion). Now we let

$$\phi(s) = (s^2 + 1)^{1/2}, \qquad s \in \mathbb{R}. \tag{13}$$

With this notation we have that $u$ is a classical solution to

$$\operatorname{div}(\phi'(|Du|^2)Du) = \phi'(|Du|^2)u_t + F'(u). \tag{14}$$

Now given that $u \in H_{3+\alpha}(\mathbb{R}^n \times [0, T])$, one can repeat the arguments as in the proof of Theorem 5.1 in [2] with $\phi$ as in (13). We nevertheless provide the details for the sake of completeness and also because the corresponding growth of $\phi$ in $s$ is quite different from the one in Theorem 5.1 in [2]. Let

$$\xi(s) = 2s\phi'(s) - \phi(s), \tag{15}$$

and define $\Lambda = \xi'$. We also define $P$ as follows

$$P(u, x, t) = \xi(|Du(x, t)|^2) - 2F(u(x, t)). \tag{16}$$

With $\phi$ as in (13) above, we have that

$$P = 2\frac{(1 + |Du|^2)^{1/2} - 1}{(1 + |Du|^2)^{1/2}} - 2F(u). \tag{17}$$

We note that the hypothesis that (12) be valid at $t = 0$ can be reformulated by saying that $P(\cdot, 0) \leq 0$. We next write (14) in the following manner

$$a_{ij}(Du)\, u_{ij} = f(u) + \phi'\, u_t,$$

where for $\sigma \in \mathbb{R}^n$ we have let

$$a_{ij}(\sigma) = 2\phi''\sigma_i\,\sigma_j + \phi'\delta_{ij}. \tag{18}$$

Therefore, $u$ satisfies

$$d_{ij}\,u_{ij} = \frac{f}{\Lambda} + \frac{\phi'}{\Lambda}u_t, \tag{19}$$

where $d_{ij} = \frac{a_{ij}}{\Lambda}$. By differentiating (18) with respect to $x_k$, we obtain

$$(a_{ij}\,(u_k)_i)_j = f'\,u_k + \phi'\,u_{tk} + 2\phi''\,u_{hk}\,u_h\,u_t. \tag{20}$$

From the definition of $P$ in (16) we have,

$$P_i = 2\Lambda u_{ki}\,u_k - 2f\,u_i, \quad P_t = 2\Lambda u_{kt}\,u_k - 2f\,u_t. \tag{21}$$

We now consider the following auxiliary function

$$w = w_R = P - \frac{M}{R}\sqrt{|x|^2+1} - \frac{ct}{R^{1/2}},$$

where $R > 1$ and $M$, $c$ are to be determined subsequently. Note that $P \geq w$ for $t \geq 0$. Consider the cylinder $Q_R = B(0,R) \times [0,T]$. One can see that if $M$ is chosen large enough, depending on the $L^\infty$ norm of $u$ and its first derivatives, then $w < 0$ on the lateral boundary of $Q_R$. In this situation we see that if $w$ has a strictly positive maximum at a point $(x_0,t_0)$, then such point cannot be on the parabolic boundary of $Q_R$. In fact, since $w < 0$ on the lateral boundary, the point cannot be on such set. But it cannot be on the bottom of the cylinder either since, in view of (12), at $t = 0$ we have $w(\cdot,0) \leq P(u(\cdot,0)) \leq 0$.

Our objective is to prove the following claim:

$$w \leq K \overset{def}{=} R^{-\frac{1}{2}}, \quad \text{in } Q_R, \tag{22}$$

provided that $M$ and $c$ are chosen appropriately. This claim will be established in (43) below. We first fix a point $(y,s)$ in $\mathbb{R}^n$. Now for all $R$ sufficiently large enough, we have that $(y,s) \in Q_R$. We would like to emphasize over here that finally we let $R \to \infty$. Therefore, once (22) is established, we obtain from it and the definition of $w$ that

$$P(u,y,s) \leq \frac{K'}{R^{1/2}}, \tag{23}$$

where $K'$ depends on $\varepsilon$, $(y,s)$ and the bounds of the derivatives of $u$ of order three. By letting $R \to \infty$ in (23), we find that

$$P(u,y,s) \leq 0. \tag{24}$$

The sought for conclusion thus follows from the arbitrariness of the point $(y, s)$.

In order to prove the claim (22) we argue by contradiction and suppose that there exist $(x_0, t_0) \in \overline{Q}_R$ at which $w$ attains it maximum and for which

$$w(x_0, t_0) > K.$$

It follows that at $(x_0, t_0)$ we must have

$$(\varepsilon^2 + |Du(x_0, t_0)|^2)^{-1/2}|Du(x_0, t_0)|^2 \geq \frac{1}{2}P(x_0, t_0) \geq \frac{1}{2}w(x_0, t_0) > \frac{1}{2}K, \tag{25}$$

which implies, in particular, that $Du(x_0, t_0) \neq 0$. Therefore, we obtain from (25)

$$|Du(x_0, t_0)| \geq (1 + |Du(x_0, t_0)|^2)^{-1/2}|Du(x_0, t_0)|^2 \geq \frac{1}{2}P(x_0, t_0) > \frac{1}{2}K. \tag{26}$$

On the other hand, since $(x_0, t_0)$ does not belong to the parabolic boundary, from the hypothesis that $w$ has its maximum at such point, we conclude that $w_t(x_0, t_0) \geq 0$ and $Dw(x_0, t_0) = 0$. These conditions translate into

$$P_t \geq \frac{c}{R^{1/2}}, \tag{27}$$

and

$$P_i = \frac{M}{R}\frac{x_{0,i}}{(|x_0|^2 + 1)^{1/2}}. \tag{28}$$

Now

$$(d_{ij}w_i)_j = (d_{ij}P_i)_j - \frac{M}{R}(d_{ij}\frac{x_i}{(|x|^2 + 1)^{1/2}})_j,$$

where

$$(d_{ij}P_i)_j = 2(\frac{a_{ij}}{\Lambda}(\Lambda u_{ki}\, u_k - f\, u_i))_j = 2(a_{ij}\,(u_k)_i\, u_k)_j - 2(f\, d_{ij}\, u_i)_j. \tag{29}$$

After a simplification, (29) equals

$$2a_{ij}\,(u_{ki})_j\, u_k + 2a_{ij}\, u_{ki}\, u_{kj} - 2f'\, d_{ij}\, u_i\, u_j - 2f\, d_{ij}\, u_{ij} - 2f\,(d_{ij})_j\, u_i.$$

We notice that

$$d_{ij}u_iu_j = \frac{2\phi''\, u_i\, u_j\, u_i\, u_j + \phi'\,\delta_{ij}\, u_i\, u_j}{\Lambda} = |Du|^2.$$

Now by using (20) and by cancelling the term $2f'|Du|^2$, we get that the right-hand side in (29) equals

$$2\phi' u_{tk}\, u_k + 4\phi''\, u_{hk}\, u_h\, u_k u_t + 2a_{ij}\, u_{ki}\, u_{kj} - 2fd_{ij}\, u_{ij} - 2fd_{ij,j}\, u_i.$$

Therefore by using the Eq. (19), we obtain

$$\begin{aligned}(d_{ij}P_i)_j = {} & 2a_{ij}\, u_{ki}\, u_{kj} + 2\phi'\, u_{tk}\, u_k + 4\phi''\, u_{hk}\, u_h\, u_k\, u_t \\ & - 2\frac{f^2}{\Lambda} - 2\frac{f\,\phi'\, u_t}{\Lambda} - 2fd_{ij,j}\, u_i.\end{aligned} \tag{30}$$

By using the extrema conditions (27) and (28), we have the following two conditions at $(x_0, t_0)$

$$u_{kh}\, u_k\, u_h = \frac{f}{\Lambda}|Du|^2 + \frac{M}{2R\Lambda}\frac{x_h\, u_h}{(|x|^2+1)^{1/2}}, \tag{31}$$

$$2\Lambda\, u_{kt}\, u_k \geq 2fu_t + \frac{c}{R^{1/2}}. \tag{32}$$

Using the extrema conditions and by canceling $2\phi' u_{tk} u_k$ we obtain,

$$\begin{aligned}(d_{ij}w_i)_j \geq {} & 2a_{ij}\, u_{ki}\, u_{kj} + \frac{4\phi'' f}{\Lambda}|Du|^2 u_t - \frac{2f^2}{\Lambda} - 2fd_{ij,j}\, u_i \\ & + \frac{2\phi''\, M\, x_h\, u_h\, u_t}{R\,\Lambda\,(|x|^2+1)^{1/2}} + \frac{c\,\phi'}{R^{1/2}\Lambda} - \frac{M}{R}(d_{ij}\frac{x_i}{(|x|^2+1)^{1/2}})_j.\end{aligned} \tag{33}$$

Now we have the following structure equation, whose proof is lengthy but straightforward,

$$d_{ij,j} u_i = \frac{2\phi''}{\Lambda}(|Du|^2 \Delta u - u_{hk}\, u_h\, u_k). \tag{34}$$

Using (32) in (34), we find

$$d_{ij,i}\, u_i = \frac{2\phi''|Du|^2}{\Lambda}(\Delta u - \frac{f}{\Lambda} - \frac{M\, x_h\, u_h}{2R\,|Du|^2\,\Lambda(|x|^2+1)^{1/2}}).$$

Using the Eq. (14), we have

$$2\phi''\, u_{hk}\, u_h\, u_k + \phi'\,\Delta u = f + \phi'\, u_t.$$

Therefore,

$$\Delta u = \frac{f + \phi' u_t - 2\phi'' u_{hk} u_h u_k}{\phi'}. \tag{35}$$

Substituting the value for $\Delta u$ in (35) and by using the extrema condition (32), we have the following equality at $(x_0, t_0)$,

$$d_{ij,j} u_i = \frac{2\phi'' |Du|^2}{\Lambda \phi'} \Big[ f + u_t \phi' - 2\phi'' \frac{|Du|^2}{\Lambda} f - f \frac{\phi'}{\Lambda} \tag{36}$$

$$- \frac{\phi'' M x_h u_h}{R\Lambda(|x|^2+1)^{1/2}} - \frac{M x_h u_h \phi'}{2R |Du|^2 \Lambda (|x|^2+1)^{1/2}} \Big].$$

Using the definition of $\Lambda$ and cancelling terms in (36), we have that the right-hand side in (36) equals

$$2\phi'' \frac{|Du|^2 u_t}{\Lambda} - \frac{\phi'' M x_h u_h}{\Lambda^2 R (|x|^2+1)^{1/2}} - \frac{2(\phi'')^2 |Du|^2 M x_h u_h}{R \Lambda^2 \phi' (|x|^2+1)^{1/2}}. \tag{37}$$

Therefore, by canceling the terms $4\phi'' f \frac{|Du|^2 u_t}{\Lambda}$ in (33), we obtain the following differential inequality at $(x_0, t_0)$,

$$(d_{ij} w_i)_j \geq \frac{c\,\phi'}{R^{1/2} \Lambda} - \frac{2f^2}{\Lambda} - \frac{M}{R} (d_{ij} \frac{x_i}{(|x|^2+1)^{1/2}})_j + \frac{2\phi'' M x_h u_h u_t}{R \Lambda (|x|^2+1)^{1/2}} \tag{38}$$

$$+ \frac{2f \phi'' M x_h u_h}{\Lambda^2 R (|x|^2+1)^{1/2}} + \frac{4f (\phi'')^2 |Du|^2 M x_h u_h}{R \Lambda^2 \phi' (|x|^2+1)^{1/2}} + 2a_{ij} u_{ki} u_{kj}.$$

Now by using the identity for $DP$ in (21) above, we have

$$u_{ki} u_{kj} u_i u_j = \frac{(P_k + 2fu_k)^2}{4\Lambda^2}. \tag{39}$$

Also,

$$a_{ij} u_{kj} u_{ki} = \phi' u_{ik} u_{ik} + 2\phi'' u_{ik} u_i u_{jk} u_j.$$

Therefore, by Schwarz inequality, we have

$$a_{ij} u_{kj} u_{ki} \geq \phi' \frac{u_{ik} u_{jk} u_i u_j}{|Du|^2} + 2\phi'' u_{ik} u_i u_{jk} u_j = \frac{\Lambda u_{ik} u_i u_{jk} u_j}{|Du|^2}.$$

Then, by using (39) we find

$$a_{ij}\, u_{kj}\, u_{ki} \geq \frac{(P_k + 2fu_k)^2}{4\Lambda |Du|^2} = \frac{|DP|^2 + 4f^2|Du|^2 + 2f < Du, DP >}{4|Du|^2 \Lambda}. \tag{40}$$

At this point, using (40) in (38), we can cancel off $\frac{2f^2}{\Lambda}$ and consequently obtain the following inequality at $(x_0, t_0)$,

$$(d_{ij}w_i)_j \geq \frac{c\phi'}{R^{1/2}\Lambda} + \frac{f < Du, DP >}{|Du|^2 \Lambda} - \frac{M}{R}\,(d_{ij}\, \frac{x_i}{(|x|^2+1)^{1/2}})_j + \frac{2\,\phi''\, M\, x_h\, u_h\, u_t}{R\,\Lambda(|x|^2+1)^{1/2}}$$
$$+ \frac{4f\,(\phi'')^2\,|Du|^2 M\, x_h\, u_h}{R\,\Lambda^2\,\phi'(|x|^2+1)^{1/2}} + \frac{2f\,\phi'' M\, x_h u_h}{\Lambda^2\, R\,(|x|^2+1)^{1/2}}. \tag{41}$$

By assumption, since $w(x_0, t_0) \geq K$, we have that

$$|Du| \geq \frac{1}{2R^{1/2}}.$$

Moreover, since $u$ has bounded derivatives up to order 3, for a fixed $\varepsilon > 0$, we have that $\phi'$ and $\Lambda$ are bounded from below by a positive constant. Therefore by (28), the term $\frac{f<Du,DP>}{|Du|^2\Lambda}$ can be controlled from below by $-\frac{M''}{R^{1/2}}$ where $M''$ depends on $\varepsilon$ and the bounds of the derivatives of $u$. Consequently, from (41), we have at $(x_0, t_0)$,

$$(d_{ij}w_i)_j \geq \frac{C(c)}{R^{1/2}} - \frac{L(M)}{R} - \frac{M''}{R^{1/2}}. \tag{42}$$

Now in the very first place, if $c$ is chosen large enough depending only on $\varepsilon$ and the bounds of the derivatives of $u$ up to order three, we would have the following inequality at $(x_0, t_0)$,

$$(d_{ij}w_i)_j > 0.$$

This contradicts the fact that $w$ has a maximum at $(x_0, t_0)$. Therefore, either $w(x_0, t_0) < K$, or the maximum of $w$ is achieved on the parabolic boundary where $w < 0$. In either case, for an arbitrary point $(y, s)$ such that $|y| \leq R$, we have that

$$w(y, s) \leq \frac{1}{R^{1/2}}. \tag{43}$$

# 3 Forward Gradient Bounds for the Reaction-Diffusion Equation (5) in Epigraphs

In this section we consider Modica type gradient bounds for solutions to the parabolic equation (5) in unbounded generalized cylinders of the type $\Omega \times [0, T]$. On the ground domain $\Omega \subset \mathbb{R}^n$ we assume that it is an epigraph, i.e., that

$$\Omega = \{(x', x_n) \in \mathbb{R}^n \mid x' \in \mathbb{R}^{n-1},\ x_n > h(x')\}. \tag{44}$$

Furthermore, we assume that $h \in C^{2,\alpha}_{loc}(\mathbb{R}^{n-1})$ and that

$$||Dh||_{C^{1,\alpha}(\mathbb{R}^{n-1})} < \infty. \tag{45}$$

Before proving the main result of the section we establish a lemma which will be used throughout the rest of the paper.

**Lemma 3.1** *Let $u$ be a solution to* (5), *and assume that*

$$\inf_G |Du| > 0, \tag{46}$$

*for some open set $G \in \mathbb{R}^n \times \mathbb{R}$. Define*

$$P(x,t) \stackrel{def}{=} P(u,x,t) = |Du(x,t)|^2 - 2F(u(x,t)). \tag{47}$$

*Then, we have in $G$ that*

$$(\Delta - \partial_t)P + < B, DP > \geq \frac{|DP|^2}{2|Du|^2}, \tag{48}$$

*where $B = \frac{2F'(u)Du}{|Du|^2}$.*

*Proof* The proof of the lemma follows from computations similar to that in the proof of Theorem 2.1, but we nevertheless provide details since this lemma will be crucially used in the rest of the paper. We first note that, since $F \in C^{2,\beta}_{loc}$, we have $u \in H_{3+\alpha,loc}$ for some $\alpha$ which also depends on $\beta$. By using (5), it follows from a simple computation that

$$(\Delta - \partial_t)P = 2||D^2u||^2 - 2F'(u)^2. \tag{49}$$

From the definition of $P$, it follows that

$$DP = 2D^2uDu - 2F'(u)Du.$$

This gives

$$4|D^2uDu|^2 = |DP + 2F'(u)Du|^2 = |DP|^2 + 4F'(u)^2|Du|^2 + 4F'(u) < DP, Du > .$$

Therefore, from Cauchy-Schwartz inequality we obtain

$$4||D^2u||^2|Du|^2 \geq |DP|^2 + 4F'(u)^2|Du|^2 + 4F'(u) < DP, Du > .$$

By dividing both sides of this inequality by $2|Du|^2$, and replacing in (49), the desired conclusion follows.

We now state the relevant result which is the parabolic analogue of Theorem 1 in [12].

**Theorem 3.1** *Let $\Omega \subset \mathbb{R}^n$ be as in* (44)*, with $h$ satisfying* (45)*, and assume furthermore that the mean curvature of $\partial\Omega$ is nonnegative. Let $u$ be a nonnegative bounded solution to the following problem*

$$\begin{cases} \Delta u = u_t + F'(u), \\ u = 0 \text{ on } \partial\Omega \times [0, T], \end{cases} \tag{50}$$

*such that*

$$|Du|^2(x, 0) \leq 2F(u)(x, 0). \tag{51}$$

*Furthermore, assume that $||u(\cdot, 0)||_{C^{1,\alpha}(\overline{\Omega})} < \infty$. Then, the following gradient estimate holds for all $t > 0$ and all $x \in \Omega$,*

$$|Du|^2(x, t) \leq 2F(u)(x, t). \tag{52}$$

*Proof* Henceforth in this paper for a given function $g : \mathbb{R}^{n-1} \to \mathbb{R}$ we denote by

$$\Omega_g = \{(x', x_n) \in \mathbb{R}^n \mid x' \in \mathbb{R}^{n-1},\ x_n > g(x')\}$$

its epigraph. With $\alpha$ as in the hypothesis (45) above, we denote

$$\mathscr{F} = \Big\{ g \in C^{2,\alpha}(\mathbb{R}^{n-1}) \mid \partial\Omega_g \text{ has nonnegative mean curvature and } ||Dg||_{C^{1,\alpha}(\mathbb{R}^{n-1})} \leq ||Dh||_{C^{1,\alpha}(\mathbb{R}^{n-1})} \Big\}.$$

We now note that, given a bounded solution $u$ to (50) above, then by Schauder regularity theory (see Chaps. 4, 5 and 12 in [22]) one has

$$||u||_{H_{1+\alpha}(\overline{\Omega}\times[0,T])} \leq C, \tag{53}$$

for some universal $C > 0$ which also depends on $\Omega$ and $||u(\cdot,0)||_{C^{1,\alpha}(\overline{\Omega})}$, and for every $\varepsilon > 0$ there exists $C(\varepsilon) > 0$ such that

$$||u||_{H_{2+\alpha}(\overline{\Omega}\times[\varepsilon,T])} \leq C(\varepsilon). \tag{54}$$

Note that in (54), we cannot take $\varepsilon = 0$, since the compatibility conditions at the corner points need not hold. With $C$ as in (53), we now define

$$\Sigma = \Big\{ v \in C^{2,1}(\overline{\Omega}_g \times [0,T]) \mid \text{there exists } g \in \mathscr{F} \text{ for which } v \text{ solves (50) in } \Omega_g \times [0,T],$$
$$\text{with } 0 \leq v \leq ||u||_{L^\infty},\ ||v||_{H_{1+\alpha}(\overline{\Omega}_g\times[0,T])} \leq C,\ P(v,x,0) \leq 0 \Big\}. \quad .$$

Note that in the definition of $\Sigma$ we have that given any $v \in \Sigma$, there exists a corresponding $g^{(v)}$ in $\mathscr{F}$ such that the assertions in the definition of the class $\Sigma$ hold. Moreover $\Sigma$ is non-empty since $u \in \Sigma$. From now on, with slight abuse of notation, we will denote the corresponding $\Omega_{g^{(v)}}$ by $\Omega_v$.

We now set

$$P_0 = \sup_{v\in\Sigma,(x,t)\in\Omega_v\times[0,T]} P(v;x,t).$$

We note that $P_0$ is finite because by the definition of $\Sigma$, every $v \in \Sigma$ has $H_{1+\alpha}$ norm bounded from above by a constant $C$ which is independent of $v$. Furthermore, by Schauder regularity theory we have that (54) holds uniformly for $v \in \Sigma$ in $\overline{\Omega}_v \times [0,T]$. Our objective is to establish that

$$P_0 \leq 0. \tag{55}$$

Assume on the contrary that $P_0 > 0$. For every $k \in \mathbb{N}$ there exist $v_k \in \Sigma$ and $(x_k,t_k) \in \Omega_{v_k} \times [0,T]$ such that

$$P_0 - \frac{1}{k} < P(v_k,x_k,t_k) \leq P_0.$$

By compactness, possibly passing to a subsequence, we know that there exists $t_0 \in [0,T]$ such that $t_k \to t_0$. We define

$$u_k(x,t) = v_k(x + x_k, t_k).$$

We then have that $u_k \in \Sigma$ and $0 \in \Omega_{u_k}$. Moreover,

$$P(u_k,0,t_k) = P(v_k,x_k,t_k) \to P_0.$$

Now, if we denote by $g_k$ the function corresponding to the graph of $\Omega_{u_k}$, from the fact that $0 \in \Omega_{u_k}$ we infer that

$$g_k(0) \leq 0.$$

We now claim that $g_k(0)$ is bounded. If not, then there exists a subsequence such that

$$g_k(0) \to -\infty.$$

Moreover since $||Dg_k||_{C^{1,\alpha}}$ is bounded uniformly in $k$, we conclude that for every $x' \in \mathbb{R}^{n-1}$

$$g_k(x') \to -\infty, \tag{56}$$

and the same conclusion holds locally uniformly in $x'$. Since the $u_k$'s are uniformly bounded in $H_{1+\alpha}(\Omega_{u_k} \times [0,T])$, we have that $u_k \to w_0$ locally uniformly in $\mathbb{R}^n \times [0,T]$. Note that this can be justified by taking an extension of $u_k$ to $\mathbb{R}^n \times [0,T]$ such that (53) hold in $\mathbb{R}^n \times [0,T]$, uniformly in $k$. Applying (54) to the $u_k$'s we see that the limit function $w_0$ solves (50) in $\mathbb{R}^n \times [0,T]$. Since by the definition of $\Sigma$ we have $P(u_k,0,0) = P(v_k,x_k,0) \leq 0$, we have that $t_0 > 0$, and therefore by (53) we conclude that $P(w_0,0,t_0) = P_0 > 0$. Moreover, again by (53), we have $P(w_0,0,0) \leq 0$. This leads to a contradiction with the case $p = 2$ of Theorem 1.3 established in [2]. Therefore, the sequence $\{g_k(0)\}$ must be bounded.

Now since $g_k$'s are such that $Dg_k$'s have uniformly bounded $C^{1,\alpha}$ norms, we conclude by Ascoli-Arzelà that there exists $g_0 \in \mathscr{F}$ such that $g_k \to g_0$ locally uniformly in $\mathbb{R}^{n-1}$. We denote

$$\Omega_0 = \{(x',x_n) \in \mathbb{R}^n \mid x' \in \mathbb{R}^{n-1},\ x_n > g_0(x')\}.$$

For each $k$, by taking an extension $\tilde{u}_k$ of $u_k$ to $\mathbb{R}^n \times [0,T]$ such that $\tilde{u}_k$ has bounded $H_{1+\alpha}$ norm, we have that (possibly on a subsequence) $\tilde{u}_k \to u_0$ locally uniformly in $\mathbb{R}^n$. Moreover, because of (54) applied to $u_k$'s, the function $u_0$ solves the Eq. (5) in $\Omega_0 \times (0,T]$. We also note that since $Dg_k$'s have uniformly bounded $C^{1,\alpha}$ norms, $\partial\Omega_0$ has nonnegative mean curvature. Moreover, by arguing as in (33) and (34) in [12], we have that $u_0$ vanishes on $\partial\Omega_0 \times [0,T]$. Also, it follows that $P(u_0,x,0) \leq 0$ for $x \in \Omega_0$, and therefore $u_0 \in \Sigma$. Arguing by compactness as previously in this proof, we infer that must be $t_0 > 0$, and since $u_0 \in \Sigma$, that

$$P_0 = P(u_0,0,t_0) = \sup_{(x,t)\in\Omega_0\times[0,T]} P(u_0,x,t) > 0.$$

Since $u_0 \geq 0$ and $u_0$ vanishes on $\partial\Omega_0 \times [0,T]$, indicating by $\nu$ the inward unit normal to $\partial\Omega_0$ at $x$, we have for each $(x,t) \in \partial\Omega_0 \times [0,T]$

$$\partial_\nu u_0(x,t) \geq 0. \tag{57}$$

Given (57) and from the fact that $u_0$ is bounded, by arguing as in (36)–(38) in [12], it follows that for all $t \in [0, T]$

$$\inf_{x \in \Omega_0} |Du_0(x, t)| = 0. \tag{58}$$

Next, we claim that if for a time level $t > 0$ we have $P(u_0, \overline{y}, t) = P_0$, then it must be $\overline{y} \in \partial\Omega_0$. To see this, suppose on the contrary that $\overline{y} \in \Omega_0$. Since $P_0 > 0$, this implies that $|Du_0(\overline{y}, t)| > 0$. Consider now the set

$$U = \{x \in \Omega_0 \mid P(u_0, x, t) = P_0\}.$$

Clearly, $U$ is closed, and since $\overline{y} \in U$ by assumption, we also know that $U \neq \varnothing$. We now prove that $U$ is open. Since $|Du_0(x, t)| > 0$ for every $x \in U$, by Lemma 3.1 and the strong maximum principle (we note that since $F \in C^{2,\beta}_{loc}$, we have that $u_0 \in H_{3+\alpha'}$ in the interior for some $\alpha'$ which also depends on $\beta$. Hence, $P(u_0, \cdot, \cdot)$ is a classical subsolution), we conclude that for every $x \in U$ there exists $\delta_x > 0$ such that $P(u_0, z, t) = P_0$ for $z \in B(x, \delta_x)$. This implies that $U$ is open.

Since $\Omega_0$, being an epigraph, is connected, we conclude that $U = \Omega_0$. Now from (58) we have that for every fixed $t \in [0, T]$ there exists a sequence $x_j \in \Omega_0$ such that $Du_0(x_j, t) \to 0$ as $j \to \infty$. As a consequence, $\liminf_{j \to \infty} P(u_0, x_j, t) \leq 0$. This implies that for large enough $j$ we must have $P(u_0, x_j, t) < P_0$, which contradicts the above conclusion that $U = \Omega_0$. Therefore this establishes the claim that if $P(u_0, \overline{y}, t) = P_0$, then $\overline{y} \in \partial\Omega_0$. Since $P(u_0, 0, t_0) = P_0$, and $P_0$ is assumed to be positive, this implies in particular that $(0, t_0) \in \partial\Omega_0 \times (0, T]$. Again, since $P_0 > 0$ by assumption, we must have that in (57) a strict inequality holds at $(0, t_0)$, i.e.

$$\partial_\nu u_0(0, t_0) > 0. \tag{59}$$

This is because if the normal derivative is zero at $(0, t_0)$, then it must also be $Du_0(0, t_0) = 0$ (since $u_0$ vanishes on the lateral boundary of $\Omega_0 \times [0, T]$), and this contradicts the fact that $P_0 > 0$.

From (59) we infer that $Du_0(0, t_0) \neq 0$, and therefore Lemma 3.1 implies that, near $(0, t_0)$, the function $P(u_0, \cdot, \cdot)$ is a subsolution to a uniformly parabolic equation. Now, by an application of the Hopf Lemma (see for instance Theorem 3′ in [21]) we have that

$$\partial_\nu P(u_0, 0, t_0) < 0. \tag{60}$$

Again by noting that $u_0$ vanishes on the lateral boundary, we have that $\partial_t u_0 = 0$ at $(0, t_0)$. Therefore, at $(0, t_0)$, the function $u_0$ satisfies the elliptic equation

$$\Delta u_0 = F'(u_0). \tag{61}$$

At this point, by using the fact that the mean curvature of $\partial\Omega_0$ is nonnegative and the Eq. (61) satisfied by $u_0$ at $(0, t_0)$, one can argue as in (50)–(55) in [12] to reach a contradiction with (60) above. Such contradiction being generated from having assumed that $P_0 > 0$, we conclude that (55) must hold, and this implies the sought for conclusion of the theorem.

*Remark 3.1* Note that in the hypothesis of Theorem 3.1 instead of $u \geq 0$ we could have assumed that $\partial_\nu u \geq 0$ on $\partial\Omega \times [0, T]$.

*Remark 3.2* We also note that the conclusion in Theorem 3.1 holds if $\Omega$ is of the form $\Omega = \Omega_0 \times \mathbb{R}^{n-n_0}$ for $1 \leq n_0 \leq n$, where $\Omega_0$ is a bounded $C^{2,\alpha}$ domain with nonnegative mean curvature. The corresponding modifications in the proof would be as follows. Let $D$ be the set of domains which are all translates of $\Omega$. The classes $H$ and $\Sigma$ would be defined corresponding to $D$ as in [12]. Then, by arguing as in the proof of Theorem 3.1, we can assume that sets $\Omega_{u_k} \in D$ are such that $\Omega_{u_k} = p_k + \Omega$ where $p_k = (p'_k, 0) \in \mathbb{R}^{n_0} \times \mathbb{R}^{n-n_0}$. Since $0 \in \Omega_{u_k}$ and $\Omega_0$ is bounded, this implies that $p'_k$ is bounded independent of $k$. Therefore, up to a subsequence, $p_k \to p_0$ and $u_k \to u_0$ such that $u_0$ solves (50) in $\Omega_1 = p_0 + \Omega$. The rest of the proof remains the same as that of Theorem 3.1.

*Remark 3.3* It remains an interesting open question whether Theorem 3.1 holds for the inhomogeneous variant of the normalized $p$-Laplacian evolution studied in [2]. Note that unlike the case of $\mathbb{R}^n$, the Hopf lemma applied to $P$ is a crucial step in the proof of Theorem 3.1 for which Lemma 3.1 is the key ingredient. As far as we are aware of, an appropriate analogue of Lemma 3.1 is not known to be valid for $p \neq 2$, even in the case when $F = 0$. Therefore, to be able to generalize Theorem 3.1 to the case of inhomogeneous normalized $p$-Laplacian evolution as studied in [2], lack of an appropriate subsolution-type argument (i.e., Lemma 3.1), and a priori $H_{1+\alpha}$ estimates seem to be the two major obstructions at this point.

## 4 Gradient Estimates for the Reaction-Diffusion Equation (5) in $\mathbb{R}^n \times (-\infty, 0]$ and $\Omega \times (-\infty, 0]$

In this section we turn our attention to the settings $\mathbb{R}^n \times (-\infty, 0]$ and $\Omega \times (-\infty, 0]$, where $\Omega$ is an epigraph satisfying the geometric assumptions as in the previous section. We investigate the validity of Modica type gradient estimates in a different situation with respect to that of Sect. 3, where such estimates were established under the crucial hypothesis that the initial datum satisfies a similar inequality. We first note that such unconstrained global estimates cannot be expected in $\mathbb{R}^n \times [0, T]$ without any assumption on the initial datum. This depends of the fact that, if at time $t = 0$ the initial datum is such that the function defined in (47) above satisfies $P(u, x, 0) > 0$ at some $x \in \mathbb{R}^n$, then by continuity $P(u, x, t) > 0$ for all $t \in [0, \varepsilon]$, for some $\varepsilon > 0$. This justifies our choice of the setting in this section. We now state our first main result.

**Theorem 4.1** *Let $u$ be a bounded solution to* (5) *in* $\mathbb{R}^n \times (-\infty, 0]$. *Then, with* $P(u, \cdot, \cdot)$ *as in* (47) *we have*

$$P(u, x, t) \leq 0, \qquad \textit{for all } (x, t) \in \mathbb{R}^n \times (-\infty, 0]. \tag{62}$$

*Remark 4.1* An explicit example of a bounded solution to (5) is an eternal travelling wave as in Sect. 7.

*Proof* The proof is inspired to that of Theorem 1.6 in [3]. We define the class $\Sigma$ as follows.

$$\Sigma = \{v \mid v \text{ solves (5) in } \mathbb{R}^n \times (-\infty, 0],\ \ ||v||_{L^\infty} \leq ||u||_{L^\infty}\}. \tag{63}$$

Note that $u \in \Sigma$. Set

$$P_0 = \sup_{v \in \Sigma, (x,t) \in \mathbb{R}^n \times (-\infty, 0]} P(v, x, t). \tag{64}$$

Since $F \in C^{2,\beta}_{loc}(\mathbb{R})$ and the $L^\infty$ norm of $v \in \Sigma$ is uniformly bounded by that of $u$, from the Schauder theory we infer all elements $v \in \Sigma$ have uniformly bounded $H_{3+\alpha}$ norms in $\mathbb{R}^n \times (-\infty, 0]$, for some $\alpha$ depending also on $\beta$. Therefore, $P_0$ is bounded.

We claim that $P_0 \leq 0$. Suppose, on the contrary, that $P_0 > 0$. Then, there exists $v_k \in \Sigma$ and corresponding points $(x_k, t_k) \in \mathbb{R}^n \times (-\infty, 0]$ such that $P(v_k, x_k, t_k) \to P_0$. Define now

$$u_k(x, t) = v_k(x + x_k, t + t_k). \tag{65}$$

Note that since $t_k \leq 0$, we have that $u_k \in \Sigma$ and $P(u_k, 0, 0) = P(v_k, x_k, t_k) \to P_0$. Moreover, since $u_k$'s have uniformly bounded $H_{3+\alpha}$ norms, for a subsequence, $u_k \to u_0$ which belongs to $\Sigma$. Moreover,

$$P(u_0, 0, 0) = \sup_{(x,t) \in \mathbb{R}^n \times (-\infty, 0]} P(u_0, x, t) = P_0 > 0.$$

As before, this implies that $Du_0(0, 0) \neq 0$. Now, by an application of Lemma 3.1, the strong maximum principle and the connectedness of $\mathbb{R}^n$, we have that $P(u_0, x, 0) = P_0$ for all $x \in \mathbb{R}^n$. On the other hand, since $u_0$ is bounded, it follows that

$$\inf_{x \in \mathbb{R}^n} |Du_0(x, 0)| = 0.$$

Then, there exists $x_j \in \mathbb{R}^n$ such that $|Du_0(x_j, 0)| \to 0$. However, we have that $P(u_0, x_j, 0) = P_0 > 0$ by assumption which is a contradiction for large enough $j$. Therefore, $P_0 \leq 0$ and the conclusion follows.

As an application of Theorem 4.1 one has the following result on the propagation of zeros whose proof is identical to that of Theorem 1.8 in [3] (see also Theorem 1.6 in [2]).

**Corollary 4.1** *Let $u$ be a bounded solution to* (5) *in* $\mathbb{R}^n \times (-\infty, 0]$. *If* $F(u(x_0, t_0)) = 0$ *for some point* $(x_0, t_0) \in \mathbb{R}^n \times (-\infty, 0]$, *then* $u(x, t_0) = u(x_0, t_0)$ *for all* $x \in \mathbb{R}^n$.

We also have the following counterpart of Theorem 4.1 in an infinite cylinder of the type $\Omega \times (-\infty, 0]$ where $\Omega$ satisfies the hypothesis in Theorem 3.1.

**Theorem 4.2** *Let* $\Omega \subset \mathbb{R}^n$ *be as in* (44) *above, with $h$ satisfying* (45). *Furthermore, assume that the mean curvature of* $\partial\Omega$ *is nonnegative. Let $u$ be a nonnegative bounded solution to the following problem*

$$\begin{cases} \Delta u = u_t + F'(u) \\ u = 0 \text{ on } \partial\Omega \times (-\infty, 0]. \end{cases} \tag{66}$$

*Then, we have that* $P(u, x, t) \leq 0$ *for all* $(x, t) \in \Omega \times (-\infty, 0]$.

*Proof* By Schauder theory we have that

$$||u||_{H_{3+\alpha}(\Omega \times (-\infty, 0])} \leq C, \tag{67}$$

for some $C$ which also depends on $\beta$. We let $\mathscr{F}$ be as in the proof of Theorem 3.1 and define

$$\begin{aligned} \Sigma = &\{v \in C^{2,1}(\overline{\Omega}_g \times [0, T]) \mid \text{there exists } g \in \mathscr{F} \text{ for which } v \text{ solves (5) in } \Omega_g \times [0, T], \\ &\text{with } 0 \leq v \leq ||u||_{L^\infty},\ v = 0 \text{ on } \partial\Omega_g \times (-\infty, 0]\}. \end{aligned}$$

As before, note that in the definition of $\Sigma$ we have that, given any $v \in \Sigma$, there exists a corresponding $g^{(v)}$ in $\mathscr{F}$ such that the assertions in the definition of the class $\Sigma$ hold. With slight abuse of notation, we will denote the corresponding $\Omega_{g^{(v)}}$ by $\Omega_v$. Again by Schauder theory, we have that any $v \in \Sigma$ satisfies (67) in $\Omega_v \times (-\infty, 0]$, where the constant $C$ is independent of $v$. We now set,

$$P_0 = \sup_{v \in \Sigma, (x,t) \in \Omega_v \times (-\infty, 0]} P(v, x, t).$$

As before, we claim that $P_0 \leq 0$. This claim would of course imply the sought for conclusion. From the definition of $\mathscr{H}$, we note that $P_0$ is bounded. Suppose, on the contrary, that $P_0 > 0$. Then, there exists $v_k$'s and corresponding points $(x_k, t_k)$ such that $P(v_k, x_k, t_k) \to P_0$. We define,

$$u_k(x, t) = v_k(x + x_k, t + t_k). \tag{68}$$

Since $t_k \le 0$, we note that $u_k \in \Sigma$. Now, by an application of Theorem 4.1 and a compactness type argument as in the proof of Theorem 3.1, we conclude that if $g_k$ is function corresponding to $\Omega_{g_k} = \Omega_{u_k}$ for each $k$, then $g_k(0)$'s are bounded and since $Dg_k$'s have uniformly bounded $C^{1,\alpha}$ norms, then $g_k$'s are bounded locally uniformly in $\mathbb{R}^{n-1}$. From this point on the proof follows step by step the lines of that of Theorem 3.1 and we thus skip pointless repetitions. There exists a $g_0 \in \mathscr{F}$ for which $g_k \to g_0$ locally uniformly in $\mathbb{R}^{n-1}$ as in that proof and we call $\Omega_0$ the epigraph of $g_0$. From the uniform Schauder type estimates, possibly passing to a subsequence, we conclude the existence of a solution $u_0 \ge 0$ of (5) in $\Omega_0 \times (-\infty, 0]$ such that $\Omega_{u_k} \to \Omega_0$, and $u_k \to u_0$ which solves such that $\partial\Omega_0$ has nonnegative mean curvature. Moreover, $u_0$ vanishes on the lateral boundary and $P(u_0, 0, 0) = \sup P(u_0, 0, 0) = P_0$. The rest of the proof remains the same as that of Theorem 3.1, but with $(0, 0)$ in place of $(0, t_0)$.

*Remark 4.2* As indicated in Remark 3.3, the conclusion of Theorem 4.2 remains valid with minor modifications in the proof when $\Omega = \Omega_0 \times \mathbb{R}^{n-n_0}$, $1 \le n_0 \le n$, where $\Omega_0$ is a bounded smooth domain with boundary having nonnegative mean curvature.

## 5 Modica Type Estimates for Reaction-Diffusion Equations on Compact Manifolds with Nonnegative Ricci Tensor

Let $(M, g)$ be a connected, compact Riemannian manifold with Laplace-Beltrami $\Delta_g$, and suppose that the Ricci tensor be nonnegative. In the paper [13] the authors established a Modica type estimate for bounded solutions in $M$ of the semilinear Poisson equation

$$\Delta_g u = F'(u), \tag{69}$$

under the assumption that $F \in C^2(\mathbb{R})$, and $F \ge 0$. Precisely, they proved that following inequality holds

$$|\nabla_g u(x)|^2 \le 2F(u), \tag{70}$$

where $\nabla_g$ is the Riemannian gradient on $M$.

In this section, we prove a parabolic analogue of (70). Our main result can be stated as follows.

**Theorem 5.1** *Let $M$ be a connected compact Riemannian manifold with* Ric $\ge 0$, *and let $u$ be a bounded solution to*

$$\Delta_g u = u_t + F'(u) \tag{71}$$

*on* $M \times (-\infty, 0]$ *where* $F \in C^{2,\beta}(\mathbb{R})$ *and* $F \geq 0$. *Then, the following estimate holds in* $M \times (-\infty, 0]$

$$|\nabla_g u(x,t)|^2 \leq 2F(u(x,t)). \tag{72}$$

*Proof* By Schauder theory, we have that $u \in H_{3+\alpha}(M \times (-\infty, 0])$ for some $\alpha$ which additionally depends on $\beta$. This follows from writing the equation in local coordinates and by using the compactness of $M$. We next recall the Bochner-Weitzenbock formula, which holds for any $\phi \in C^3(M)$

$$\frac{1}{2}\Delta_g|\nabla_g\phi|^2 = |H_\phi|^2 + < \nabla_g\phi, \nabla_g\Delta_g\phi > + \mathrm{Ric}_g < \nabla_g\phi, \nabla_g\phi > . \tag{73}$$

Here, $H_\phi$ is the Hessian of $\phi$ and the square of the Hilbert-Schmidt norm of $H_\phi$ is given by

$$|H_\phi|^2 = \Sigma_i < \nabla_{X_i}\nabla_g, \nabla_{X_i}, \nabla_g >,$$

where $\{X_i\}$ is a local orthonormal frame. Moreover, Cauchy-Schwarz inequality gives

$$|H_\phi|^2 \geq |\nabla_g|\nabla_g\phi||^2. \tag{74}$$

See for instance [15] for a proof of this fact. Now we define the class

$$\mathscr{F} = \left\{v \mid v \text{ is a classical solution to (71) in } M\times(-\infty,0],\ ||v||_{L^\infty(M)} \leq ||u||_{L^\infty(M)}\right\}.$$

By the Schauder theory we see as before that for every $v \in \mathscr{F}$ the norm of $v$ in $H_{3+\alpha}(M \times (-\infty, 0])$ is bounded independent of $v$ for some $\alpha$ which additionally depends on $\beta$. In particular, without loss of generality, one may assume that the choice of the exponent $\alpha$ is the same as for $u$. Now, given any $v \in \mathscr{F}$, we let

$$P(v,x,t) = |\nabla_g v(x,t)|^2 - 2F(v(x,t)). \tag{75}$$

Applying (73) we find

$$\begin{aligned}(\Delta_g - \partial_t)P(v,x,t) &= 2|H_v|^2 + 2(< \nabla_g v, \nabla_g\Delta_g v > + \mathrm{Ric}_g < \nabla_g v, \nabla_g v >) - \\ &\quad 2 < \nabla_g v, \nabla_g v_t > -2F'(v)(\Delta_g v - v_t) - 2 < \nabla_g v, \nabla_g F'(v) > .\end{aligned}$$

Using the fact that $v$ solves (71), we obtain

$$\begin{aligned}(\Delta_g - \partial_t)P(v,x,t) &= 2|H_v|^2 + 2 < \nabla_g v, \nabla_g F'(v) > \\ &\quad + 2\,\mathrm{Ric}_g < \nabla_g v, \nabla_g v > -2F'(v)^2 - 2 < \nabla_g v, \nabla_g F'(v) > .\end{aligned}$$

After cancelling off the term $2 < \nabla_g v \nabla_g F'(v) >$, and by using (74) and the fact that the Ricci tensor is nonnegative, we find

$$(\Delta_g - \partial_t)P(v,x,t) = 2|H_v|^2 + 2\,\mathrm{Ric}_g < \nabla_g v, \nabla_g v > -2F'(v)^2 \geq 2|\nabla_g(|\nabla_g v|)|^2 - 2F'(v)^2. \tag{76}$$

Now from the definition of $P$,

$$\nabla_g P - 2F'(v)\nabla_g v = \nabla_g(|\nabla_g v|^2).$$

Therefore,

$$|\nabla_g P|^2 + 4F'(v)^2|\nabla_g v|^2 - 4F'(v) < \nabla_g v, \nabla_g P >= |\nabla_g(|\nabla_g v|^2)|^2 = 4|\nabla_g v|^2|\nabla_g(|\nabla_g v|)|^2.$$

By dividing by $2|\nabla_g v|^2$ in the latter equation, we find

$$\frac{|\nabla_g P|^2}{2|\nabla_g v|^2} = 2|\nabla_g(|\nabla_g v|)|^2 - 2F'(v)^2 + 2\frac{F'(v)}{|\nabla_g v|^2} < \nabla_g v, \nabla_g P > . \tag{77}$$

Combining (76) and (77), we finally obtain

$$(\Delta_g - \partial_t)P + 2\frac{F'(v)}{|\nabla_g v|^2} < \nabla_g v, \nabla_g P > \geq \frac{|\nabla_g P|^2}{2|\nabla_g v|^2}. \tag{78}$$

The inequality (78) shows that $P(v,x,t)$ is a subsolution to a uniformly parabolic equation in any open set where $|\nabla_g v| > 0$. Now we define

$$P_0 = \sup_{v \in \mathscr{F}, (x,t) \in M \times (-\infty, 0])} P(v,x,t).$$

Our goal as before is to show that $P_0 \leq 0$, from the which the conclusion of the theorem would follow. Suppose on the contrary that $P_0 > 0$. Then, there exists $v_k \in \mathscr{F}$ and $(x_k, t_k) \in M \times (-\infty, 0]$ such that $P(v_k, x_k, t_k) \to P_0$. We define

$$u_k(x,t) = v_k(x_k, t + t_k).$$

Since $t_k \leq 0$ we have that $u_k \in \mathscr{F}$, and since $M$ is compact, $x_k \to x_0$ after possibly passing to a subsequence. Moreover, $P(u_k, x_0, 0) \to P_0$. By compactness, we have that $u_k \to u_0$ in $H_{3+\alpha}$, where $u_0$ is a solution to (71), and $P(u_0, x_0, 0) = P_0 > 0$. Since since $F \geq 0$ this implies that $\nabla_g u_0(x_0, 0) \neq 0$. By continuity, we see that $\nabla_g u_0 \neq 0$ in a parabolic neighborhood of $(x_0, 0)$. By (78) and by the strong maximum principle we infer that $P(u_0, x, 0) = P_0$ in a neighborhood of $x_0$, and since $M$ is connected, we conclude that for all $x \in M$

$$P(u_0, x, 0) = P_0 > 0. \tag{79}$$

Since $u_0(\cdot, 0) \in C^1(M)$ and $M$ is compact, there exists $y_0 \in M$ at which $u_0(\cdot, 0)$ attains its absolute minimum. At such point one has

$$\nabla_g u_0(y_0, 0) = 0.$$

Since $F \geq 0$, this implies that

$$P(u_0, y_0, 0) \leq 0, \tag{80}$$

which is a contradiction to (79). Therefore $P_0 \leq 0$ and the theorem is proved.

*Remark 5.1* It remains an interesting question whether the conclusion of Theorem 5.1 (and for that matter even the corresponding elliptic result in [13]) continue to hold when $M$ is only assumed to be complete and not compact. In such a case, one would need to bypass the compactness argument which uses translation in a crucial way (see for instance (65) as in the proof of Theorem 4.1). We intend to come back to this question in a future study.

## 6 On a Conjecture of De Giorgi and Level Sets of Solutions to (5)

In 1978 Ennio De Giorgi formulated the following conjecture, also known as $\varepsilon$-version of the Bernstein theorem: *let $u$ be an entire solution to*

$$\Delta u = u^3 - u, \tag{81}$$

*such that $|u| \leq 1$ and $\frac{\partial u}{\partial x_n} > 0$. Then, $u$ must be one-dimensional, i.e., must have level sets which are hyperplanes, at least in dimension $n \leq 8$.*

As mentioned in the introduction, the conjecture of De Giorgi has been fully solved for $n = 2$ in [16] and $n = 3$ in [1], and it is known to fail for $n \geq 9$, see [9]. For $4 \leq n \leq 8$ it is still an open question. Additional fundamental progress on De Giorgi's conjecture is contained in the papers [17, 24]. Besides these developments, in [3] it was established that for entire bounded solutions to

$$\operatorname{div}(|Du|^{p-2}Du) = F'(u), \tag{82}$$

if the equality holds at some point $x_0 \in \mathbb{R}^n$ for the corresponding gradient estimate

$$|Du|^p \leq \frac{p}{p-1} F(u), \tag{83}$$

then $u$ must be one dimensional. The result in [3] actually regarded a more general class of equations than (82), and in [7] some further generalizations were presented. We now establish a parabolic analogue of that result in the case $p = 2$.

**Theorem 6.1** *Let $u$ be a bounded solution to* (5) *in $\mathbb{R}^n \times (-\infty, 0]$. Furthermore, assume that the zero set of $F$ is discrete. With $P$ as in* (47) *above, if $P(u, x_0, t_0) = 0$ for some point $(x_0, t_0) \in \mathbb{R}^n \times (-\infty, 0]$, then there exists $g \in C^2(\mathbb{R})$ such that $u(x,t) = g(< a, x > +\alpha)$ for some $a \in \mathbb{R}^n$ and $\alpha \in \mathbb{R}$. In particular, $u$ is independent of time, and the level sets of $u$ are vertical hyperplanes in $\mathbb{R}^n \times (-\infty, 0]$.*

*Proof* We begin by observing that it suffices to prove the theorem under the hypothesis that $t_0 = 0$. In fact, once that is done, then if $t_0 < 0$ we consider the function $v(x,t) = u(x, t + t_0)$. For such function we have $P(v, x, 0) = P(u, x, t_0)$ and therefore $v$ satisfies the same hypothesis as $u$, except that $P(v, x_0, 0) = 0$. But then we conclude that $v(x,t) = u(x, t + t_0) = g(< a, x > +\alpha)$, which implies the desired conclusion for $u$ as well.

We thus assume without restriction that $P(u, x_0, 0) = 0$, and consider the set

$$A = \{x \in \mathbb{R}^n \mid P(u, x, 0) = 0\}.$$

By the continuity of $P$ we have that $A$ is closed, and since $(x_0, 0) \in A$, this set is also non-empty. We distinguish two cases:

*Case 1* There exists $x_1 \in A$ such that $Du(x_1, 0) = 0$.

*Case 2* $Du(x, 0) \neq 0$ for every $x \in A$.

If Case 1 occurs, then from the fact that $P(u, x_1, 0) = 0$ we obtain that $F(u(x_1, 0)) = 0$. By Corollary 4.1 we thus conclude that must be $u(\cdot, 0) \equiv u_0 = u(x_1, 0)$. At this point we observe that, since by assumption $F \geq 0$, and $F(u_0) = 0$, we must also have $F'(u_0) = 0$. Therefore, if we set $v = u - u_0$, then by the continuity of $F''$ and the fact that $u \in L^\infty(\mathbb{R}^n)$, we have

$$|F'(u)| = |F'(v + u_0)| = |F'(v + u_0) - F'(u_0)| \leq \int_{u_0}^{v+u_0} |F''(s)| ds \leq C|v|.$$

Since by (5) we have $\Delta v - \partial_t v = \Delta u - \partial_t u = F'(u)$, we see that $v$ is thus a solution of the following inequality

$$|\Delta v - \partial_t v| \leq C|v|.$$

Since $v(\cdot, 0) = 0$, by the backward uniqueness result in Theorem 2.2 in [5], we have that $u \equiv u_0$ in $\mathbb{R}^n \times (-\infty, 0]$, from which the desired conclusion follows in this case.

If instead Case 2 occurs, we prove that $A$ is also open. But then, by connectedness, we conclude in such case that $A = \mathbb{R}^n$. To see that $A$ is open fix $x_1 \in A$. Since $Du(x_1, 0) \neq 0$, by the continuity of $Du$ we conclude the existence of $r > 0$ such that $Du(x, t) \neq 0$ for every $(x, t) \in G = B(x_1, r) \times (-r^2, 0]$. By Lemma 3.1 above we conclude that $P(u,\cdot,\cdot)$ is a sub-caloric function in $G$. Since by Theorem 4.1 we know that $P(u,\cdot,\cdot) \leq 0$, by the strong maximum principle we conclude that $P(u,\cdot,\cdot) \equiv 0$ in $G$. In particular, $P(u,x,0) = 0$ for every $x \in B(x_1,r)$, which implies that $A$ is open.

Since as we have seen the desired conclusion of the theorem does hold in Case 1, we can without loss of generality assume that we are in Case 2, and therefore $Du(x,0) \neq 0$ for every $x \in A = \mathbb{R}^n$. Furthermore, since for $x \in A$ we have $P(u,x,0) = 0$, we also have

$$|Du(x,0)|^2 = 2F(u(x,0)), \quad \text{for every } x \in \mathbb{R}^n. \tag{84}$$

Next, we consider the set

$$K = \big\{(x,t) \in \mathbb{R}^n \times (-\infty,0] \mid P(u,x,t) = 0\big\}.$$

We note that $K$ is closed and non-empty since by assumption we know that $(x_0,0) \in A$ (in fact, by (84) we now know that $\mathbb{R}^n \times \{0\} \subset K$). Let $(y_1,t_1) \in K$. If $Du(y_1,t_1) = 0$, we can argue as above (i.e., as if it were $t_1 = 0$) and conclude by backward uniqueness that $u \equiv u(y_1,t_1)$ in $\mathbb{R}^n \times (-\infty,t_1]$. Then, by the forward uniqueness of bounded solutions, see Theorem 2.5 in [20], we can infer that $u \equiv u(y_1,t_1)$ in $\mathbb{R}^n \times (t_1,0]$. All together, we would have proved that $u \equiv u(y_1,t_1)$ in $\mathbb{R}^n \times (-\infty,0]$ and therefore the conclusion of the theorem would follow.

Therefore from now on, without loss of generality, we may assume that $Du$ never vanishes in $K$. With this assumption in place, if $(y_1,t_1) \in K$, then since $Du(y_1,t_1) \neq 0$, by continuity there exists $r > 0$ such that $Du$ does not vanish in $G = B_r(y_1) \times (t_1 - r^2, t_1)$. But then, again by Lemma 3.1, the function $P(u,\cdot,\cdot)$ is sub-caloric in $G$. Since $P(u,\cdot,\cdot) \leq 0$ in $G$ (Theorem 4.2) and $P(u,y_1,t_1) = 0$ ($(y_1,t_1) \in K$), we can apply the strong maximum principle to conclude that $P \equiv 0$ in $G$. Then, again by connectedness, as in the case when $t_1 = 0$, we conclude that $\mathbb{R}^n \times \{t_1\} \subset K$. In particular, we have that $P(u,y_1,t) = 0$ when $t \in (t_1 - r^2, t_1]$. Therefore, we can now repeat the arguments above with $(y_1,t)$ in place of $(y_1,t_1)$ for each such $t$ and conclude that $P \equiv 0$ in $\mathbb{R}^n \times (t_1 - r^2, t_1]$.

We now claim that:

$$K = \mathbb{R}^n \times (-\infty,0], \quad \text{or equivalently} \quad P(u,x,t) = 0, \text{ for every } (x,t) \in \mathbb{R}^n \times (-\infty,0]. \tag{85}$$

Suppose the claim not true, hence $P \not\equiv 0$ in $\mathbb{R}^n \times (-\infty,0]$. From the above arguments it follows that if for $t_2 < 0$ there exists $y_2 \in \mathbb{R}^n$ such that $P(u,y_2,t_2) \neq 0$, then it must be $P(u,x,t_2) \neq 0$ for all $x \in \mathbb{R}^n$. We define

$$T_0 = \sup\{t < 0 \mid P(u,\cdot,t) \neq 0\}.$$

Since we are assuming the claim not true, we must have $\{t < 0 \mid P(u,\cdot,t) \neq 0\} \neq \varnothing$, hence $T_0 \leq 0$ is well-defined. We first observe that $T_0 < 0$. In fact, since by the hypothesis $(x_0, 0) \in K$ and we are assuming that we are in Case 2, we have already proved above the existence of $r > 0$ such that $\mathbb{R}^n \times (-r^2, 0] \subset K$. This fact shows that $T_0 \leq -r^2 < 0$. Next, we see that it must be $P(u, \cdot, T_0) = 0$. In fact, if this were not the case there would exist $y_2 \in \mathbb{R}^n$ such that $P(u, y_2, T_0) < 0$. Since $T_0 < 0$, by continuity we would have that $P(u, y_2, t) < 0$, for all $t \in [T_0, T_0 + \delta_1)$ for some $\delta_1 > 0$. By the arguments above, this would imply that $P$ never vanishes in $\mathbb{R}^n \times [T_0, T_0 + \delta_1)$, in contradiction with the definition of $T_0$. Since, as we have just seen, $P(u, \cdot, T_0) = 0$, arguing again as above we conclude that $P \equiv 0$ in $\mathbb{R}^n \times (T_0 - r^2, T_0]$ for some $r > 0$. But this contradicts the definition of $T_0$.

This contradiction shows that $\{t < 0 \mid P(u, \cdot, t) \neq 0\} = \varnothing$, hence the claim (85) must be true. We also recall that we are assuming that $Du$ never vanishes in $K = \mathbb{R}^n \times (-\infty, 0]$.

In conclusion, we have that

$$|Du|^2 = 2F(u) \quad \text{in } \mathbb{R}^n \times (-\infty, 0], \quad \text{and } Du \neq 0. \tag{86}$$

At this point we argue as in the proof of Theorem 5.1 in [3], and we let $v = H(u)$, where $H$ is a function to be suitably chosen subsequently. Then, we have that

$$\Delta v - v_t = H''(u)|Du|^2 + H'(u)\Delta u - H'(u)u_t.$$

By using (5) and (86), we conclude that

$$\Delta v - v_t = 2H''(u)F(u) + H'(u)F'(u). \tag{87}$$

Let $u_0 = u(0,0)$ and define

$$H(u) = \int_{u_0}^{u} (2F(s))^{-1/2} ds.$$

Since $|Du|(x,t) > 0$ for all $(x,t) \in \mathbb{R}^n \times (-\infty, 0]$, we have from (86) that $F(u(x,t)) > 0$. Therefore, if the zero set of $F$ is ordered in the following manner, $a_0 < a_1 < a_2 < a_3 < a_4 < \ldots$, then by connectedness, we have that $F(u(\mathbb{R}^n \times (-\infty, 0])) \subset (a_i, a_{i+1})$ for some $i$. We infer that $H$ is well defined and is $C^{2,\beta}$, and with this $H$ it is easy to check that the right-hand side in (87) is zero, i.e., $v$ is a solution to the heat equation in $\mathbb{R}^n \times (-\infty, 0]$. Moreover, by the definition of $H$ and (86),

$$|Dv|^2 = H'(u)^2|Du|^2 = 1,$$

i.e., $Dv$ is bounded in $\mathbb{R}^n \times (-\infty, 0]$. Since $v_i = D_{x_i}v$ is a solution to the heat equation for each $i \in 1, \ldots n$, by Liouville's theorem in $\mathbb{R}^n \times (-\infty, 0]$ applied to $v_i$, we conclude that $Dv$ is constant, hence $\Delta v = 0$. This implies $v_t = 0$, hence $v$ is time-independent. Hence, there exist $a \in \mathbb{R}^n$ and $\alpha \in \mathbb{R}$ such that $v =< a, x > +\alpha$. The desired conclusion now follows by taking $g = H^{-1}$. This completes the proof of the theorem.

# 7 A Parabolic Version of the Conjecture of De Giorgi

Motivated by the result in Theorem 6.1, the fact that $\mathbb{R}^n \times (-\infty, 0]$ is the appropriate setting for the parabolic Liouville type theorems, and the crucial role played by them in the proof of the original conjecture of De Giorgi, at least for $n \leq 3$ (see [1, 16, 17]), it is tempting to propose the following parabolic version of De Giorgi's conjecture:

*Conjecture 1* Let $u$ be a solution in $\mathbb{R}^n \times (-\infty, 0]$ to

$$\Delta u - u_t = u^3 - u,$$

such that $|u| \leq 1$, and $\partial_{x_n} u(x,t) > 0$ for all $(x,t) \in \mathbb{R}^n \times (-\infty, 0]$. Then, $u$ must be one dimensional and independent of $t$, at least for $n \leq 8$. In other words, for $n \leq 8$ the level sets of $u$ must be vertical hyperplanes, parallel to the $t$ axis.

However, Matteo Novaga has kindly brought to our attention that, stated this way, the conjecture is not true. There exist in fact eternal traveling wave solutions of the form

$$v(x', x_n, t) = u(x', x_n - ct), \qquad c \geq 0, \tag{88}$$

for which $\partial_{x_n} u(x) > 0$. This suggests that one should amend the above in the following way.

*Conjecture 2* Let $u$ be a solution in $\mathbb{R}^n \times (-\infty, 0]$ to

$$\Delta u - u_t = u^3 - u,$$

such that $|u| \leq 1$, and $\partial_{x_n} u(x,t) > 0$ for all $(x,t) \in \mathbb{R}^n \times (-\infty, 0]$. Then, $u$ must be an eternal traveling wave, i.e. after a change of coordinates, $u$ must be of the form (88). For interesting accounts of traveling waves solutions we refer the reader to the papers [6, 18].

It also remains to be seen what additional assumption needs to be imposed in the hypothesis of Conjecture 1 so that the corresponding conclusion holds. We hope that these questions will stimulate interesting further research.

**Acknowledgements** First author was supported in part by the second author's NSF Grant DMS-1001317 and by a postdoctoral grant of the Institute Mittag-Leffler. Second author was supported in part by NSF Grant DMS-1001317 and by a grant of the University of Padova, "Progetti d'Ateneo 2013". The paper was finalized during the first author's stay at the Institut Mittag-Leffler during the semester-long program *Homogenization and Random Phenomenon*. The first author would like to thank the Institute and the organizers of the program for the kind hospitality and the excellent working conditions. We would like to thank Matteo Novaga for kindly bringing to our attention that Conjecture 1 at the end of this paper is violated by the traveling wave solutions in [6, 18] and for suggesting the amended Conjecture 2.

## References


1. Ambrosio, L., Cabré, X.: Entire solutions of semilinear elliptic equations in $\mathbb{R}^3$ and a conjecture of De Giorgi. J. Am. Math. Soc. **13**(4), 725–739 (2000)
2. Banerjee, A., Garofalo, N.: Modica type gradient estimates for an inhomogeneous variant of the normalized $p$-Laplacian evolution. Nonlinear Anal. Theory Appl. **121**, 458–468 (2015)
3. Caffarelli, L., Garofalo, N., Segala, F.: A gradient bound for entire solutions of quasilinear equations and its consequences. Commun. Pure Appl. Math. **47**, 1457–1473 (1994)
4. Cesaroni, A., Novaga, M., Valdinoci, E.: A symmetry result for the Ornstein-Uhlenbeck operator. Discrete Contin. Dyn. Syst. **34**(6), 2451–2467 (2014)
5. Chen, X.-Y.: A strong unique continuation theorem for parabolic equations. Math. Ann. **311**(4), 603–630 (1998)
6. Chen, X., Guo, J.-S., Hamel, F., Ninomiya, H., Roquejoffre, J.-M.: Traveling waves with paraboloid like interfaces for balanced bistable dynamics. Ann. Inst. H. Poincaré Anal. Non Linéaire **24**(3), 369–393 (2007)
7. Danielli, D., Garofalo, N.: Properties of entire solutions of non-uniformly elliptic equations arising in geometry and in phase transitions. Calc. Var. Partial Differ. Equ. **15**(4), 451–491 (2002)
8. De Giorgi, E.: Convergence problems for functionals and operators. In: de Giorgi, E., Magenes, E., Mosco, U. (eds.) Proceedings of International Meeting on Recent Methods in Nonlinear Analysis, Rome 1978, Pitagora, Bologna, pp. 131–188 (1979)
9. del Pino, M., Kowalczyk, M., Wei, J.: On De Giorgi's conjecture in dimension $N \geq 9$. Ann. Math. (2) **174**(3), 1485–1569 (2011)
10. Ecker, K., Huisken, G.: Interior estimates for hypersurfaces moving by mean curvature. Invent. Math. **105**(3), 547–569 (1991)
11. Farina, A., Valdinoci, E.: The state of the art for a conjecture of De Giorgi and related problems. In: Recent Progress on Reaction-Diffusion Systems and Viscosity Solutions, pp. 74–96. World Scientific Publishing, Hackensack (2009) [35J60 (35B05)]
12. Farina, A., Valdinoci, E.: A pointwise gradient estimate in possibly unbounded domains with nonnegative mean curvature. Adv. Math. **225**(5), 2808–2827 (2010)
13. Farina, A., Valdinoci, E.: A pointwise gradient bound for elliptic equations on compact manifolds with nonnegative Ricci curvature. Discrete Contin. Dyn. Syst. **30**(4), 1139–1144 (2011)
14. Farina, A., Sciunzi, B., Valdinoci, E.: Bernstein and De Giorgi type problems: new results via a geometric approach (English summary) . Ann. Sc. Norm. Super. Pisa Cl. Sci. (5) **7**(4), 741–791 (2008) [58E12 (35B05 35J60 35J70 49J45)]
15. Farina, A., Sire, Y., Valdinoci, E.: Stable solutions of elliptic equations on Riemannian manifolds. J. Geom. Anal. **23**, 1158–1172 (2013)
16. Ghoussoub, N., Gui, C.: On a conjecture of De Giorgi and some related problems. Math. Ann. **311**(3), 481–491 (1998)
17. Ghoussoub, N., Gui, C.: On De Giorgi's conjecture in dimensions 4 and 5. Ann. Math. (2) **157**(1), 313–334 (2003)
18. Gui, C.: Symmetry of traveling wave solutions to the Allen-Cahn equation in $\mathbb{R}^2$. Arch. Ration. Mech. Anal. **203**(3), 1037–1065 (2012)
19. Ladyzhenskaya, O., Uratlseva, N.: Local estimtes for gradients of solutions to non-uniformly elliptic and parabolic equations. Commun. Pure. Appl. Math. **23**, 677–703 (1970)
20. Ladyzhenskaja, O., Solonnikov, V.A., Uraltseva, N.: Linear and Quasilinear Equations of Parabolic Type. Translations of Mathematical Monographs, vol. 23, xi+648 pp. American Mathematical Society, Providence (1967)
21. Li, Y., Nirenberg, L.: On the Hopf lemma. arXiv:0709.3531v1
22. Lieberman, G.: Second Order Parabolic Differential Equations, xii+439 pp. World Scientific, River Edge (1996)

23. Modica, L.: A gradient bound and a Liouville theorem for nonlinear Poisson equations. Commun. Pure Appl. Math. **38**, 679–684 (1985)
24. Savin, O.: Regularity of flat level sets in phase transitions. Ann. Math. (2) **169**(1), 41–78 (2009)
25. Savin, O., Sciunzi, B., Valdinoci, E.: Flat level set regularity of p-Laplace phase transitions. Mem. Am. Math. Soc. **182**(858), vi+144 pp. (2006)

# A Few Recent Results on Fully Nonlinear PDE's

**Italo Capuzzo Dolcetta**

*Dedicated to Ermanno Lanconelli on the occasion of his 70th birthday, with sympathy and esteem*

**Abstract** This note dedicated to Ermanno Lanconelli reports on some research in collaboration with Fabiana Leoni (Sapienza Università di Roma) and Antonio Vitolo (Università di Salerno) on viscosity solutions of elliptic partial differential equations of the form

$$F(D^2u) = f(u) - h(x). \tag{1}$$

In the first part I will discuss local gradient estimates for non-negative solutions of (1) in the spirit of a 2005 paper by Yan Yan Li and Louis Nirenberg.

The second part of the note focuses on entire solutions of (1) with semilinear term $f$ satisfying a Keller-Osserman type integrability condition.

**Keywords** Fully nonlinear PDEs • Keller-Osserman condition

**AMS Classification:** Primary: 35J60, Secondary: 35J70

## 1 Introduction

I wish to contribute to the present volume with a short overview of some recent results about second order fully non linear elliptic partial differential equations of the form

$$F(D^2u) = g(u) + f(x). \tag{2}$$

I. Capuzzo Dolcetta (✉)
Dipartimento di Matematica, Sapienza Università di Roma, Roma, Italy

Istituto Nazionale di Alta Matematica "F. Severi", Roma, Italy
e-mail: capuzzo@mat.uniroma1.it



G. Citti et al. (eds.), *Geometric Methods in PDE's*, Springer INdAM Series 13,
DOI 10.1007/978-3-319-02666-4_13

The results presented here are part of an ongoing program in collaboration with F. Leoni (Sapienza Università di Roma) and A. Vitolo (Università di Salerno).

Here, and in the whole paper, $f$ is a bounded continuous function and $F$ denotes a continuous scalar mapping defined on $S^n$, the set of symmetric $n \times n$ matrices with the standard partial order $Y \geq O$ if and only if $Y$ is non-negative definite. We will assume also, for simplicity, that $F(O) = 0$.

In the first part we will discuss some $L^\infty$-gradient estimates for positive viscosity solutions of

$$F(D^2u) = g(u) + f(x)$$

in a ball $B_R$ under the main assumptions of uniform ellipticity of $F$, that is

$$0 < \lambda||Y|| \leq F(X+Y) - F(X) \leq \Lambda||Y|| \;\; \forall X, Y \in S^n, \quad Y \geq 0,$$

and sublinearity of $g$. The results in this section of the paper are inspired by and generalize those obtained in the linear case by Li and Nirenberg [23].

In the second part we address the issue of necessary and sufficient conditions for existence of entire viscosity solutions of the differential inequality

$$F(D^2u) \geq g(u) + f(x), \; x \in R^n.$$

In this context, the nonlinear term $g$ is typically superlinear and fulfills a Keller-Osserman type integrability condition [18, 28]. Most of the existing literature on both of the above type of problems deals with $F = \Delta$ or more general operators under the assumption of uniform ellipticity, see for example [4, 11, 13, 23].

Let us point out explicitly that our results apply to a class of degenerate elliptic operators satisfying some form of comparison principle. The model examples considered in this paper are the *partial Laplacian* $\mathscr{P}_k^+$ defined for real symmetric matrices $X \in \mathscr{S}^n$ and a positive integer $1 \leq k \leq n$ as

$$\mathscr{P}_k^+(X) = \mu_{n-k+1}(X) + \ldots + \mu_n(X)$$

where $\mu_1(X) \leq \mu_2(X) \leq \ldots \leq \mu_n(X)$ are the eigenvalues of the matrix $X$, see [17] and *degenerate maximal Pucci* operator defined by

$$\mathscr{M}_{0,1}^+(X) = \sum_{\mu_i > 0} \mu_i(X).$$

# 2 Glaeser's Type Inequalities

In this section we discuss some interpolation inequalities for *non-negative functions* which are classical for smooth functions and can be generalized to some non-smooth case.

In order to introduce the reader to this topic, let us consider first the case of a non-negative single variable function defined on $\mathbb{R}$. A classical interpolation inequality states that if $u : \mathbb{R} \to \mathbb{R}$ is a bounded $C^2$ function with bounded second derivative, then

$$||u'||_{L^\infty} \le \sqrt{2\,||u||_{L^\infty}\,||u''||_{L^\infty}}\ . \tag{3}$$

This well-known interpolation inequality due to Landau [20], see also Kolmogorov [19], holds also for bounded functions of $n$ variables having bounded Hessian matrix $D^2u$. Many authors contributed later several extensions and variants, for quite recent results see e.g. [24, 25] and the material presented in this section.

## 2.1 An n-Dimensional Non-Smooth Glaeser Type Inequality

A similar but perhaps less known interpolation inequality holds under just unilateral bounds on $u$ and $D^2u$. More precisely, assume that $u \in C^2(\mathbb{R}^n)$ is non-negative and that its Hessian matrix $D^2u$ satisfies, for some $M \ge 0$,

$$D^2u(x)\,h \cdot h \le M\,|h|^2 \quad \text{for all } x, h \in \mathbb{R}^n\,. \tag{4}$$

For such functions the following inequality holds:

$$|Du(x)| \le \sqrt{2Mu(x)} \quad \text{for all } x \in \mathbb{R}^n\,. \tag{5}$$

Note the pointwise character of the above estimate and also that boundedness from above of $u$ is not required a priori. For bounded above $u$ the above gives back the Landau inequality (3) in any dimension $n$. Observe also that for $M = 0$, (5) implies the well-known fact that concave non-negative functions defined on the whole $\mathbb{R}^n$ are constants.

The above inequality (5), in the particular case of a strictly positive one variable functions $u$ with bounded second derivative, can be found in the paper [16] (and attributed there to B. Malgrange) in the equivalent form

$$\left|\left(\sqrt{u}\right)'(x)\right| \le \sqrt{\frac{\sup|u''|}{2}}$$

and employed later in [26].

The elementary proof of the validity of (5) is as follows: the Taylor's expansion around a point $x$ gives, since $u \geq 0$,

$$0 \leq u(x+h) \leq u(x) + Du(x) \cdot h + \frac{M}{2}|h|^2. \tag{6}$$

For any fixed $x$, the convex quadratic polynomial

$$Q(h) := u(x) + Du(x) \cdot h + \frac{M}{2}|h|^2$$

attains its global minimum at $h^* = -\frac{1}{M}\, Du(x)$. Thanks to (6), one deduces that

$$0 \leq Q(h^*) = u(x) - \frac{1}{2M}\, |Du(x)|^2$$

and inequality (5) follows.

A more general non smooth version of the weak Landau inequality (5) is valid for semiconcave, non necessarily differentiable, functions $u : \mathbb{R}^n \to \mathbb{R}$. By this we mean that $u$ is continuous and there exists $M \geq 0$ such that $x \to u(x) - \frac{M}{2}|x|^2$ is concave on $\mathbb{R}^n$.

It is well-known that for semiconcave functions, at any point $x$ the superdifferential of $u$ at $x$, that is the set

$$D^+u(x) = \left\{p \in \mathbb{R}^n \,:\, \limsup_{y \to x} \frac{u(y) - u(x) - p \cdot (y-x)}{|y-x|} \leq 0\right\}$$

is non empty, closed and convex. and that semiconcave functions are locally Lipschitz continuous and twice differentiable almost everywhere as sums of a $C^2$ function and a concave one. Of course, if $u \in C^2(\mathbb{R}^n)$ and the previously considered condition

$$D^2u(x)\, h \cdot h \leq M\, |h|^2 \quad \text{for all } x, h \in \mathbb{R}^n$$

holds, then $u$ semiconcave with constant $M$. It is also easy to check for semiconcave functions a vector $p \in D^+u(x)$ if and only if

$$u(y) \leq u(x) + p \cdot (y-x) + \frac{M}{2}|y-x|^2 \quad \text{for all} \quad y \in \mathbb{R}^n.$$

It is therefore immediate to derive the simple generalization of estimate (5) stated in the next

**Proposition 2.1** *Assume that $u \in C(\mathbb{R}^n)$ is semiconcave with semiconcavity constant $M$ and non-negative. Then,*

$$D^+u(x) \subseteq B_r(0) \quad \textit{with} \quad r = \sqrt{2Mu(x)}\,.$$

As an application of this estimate, consider the Hamilton-Jacobi equation

$$u + H(Du) = f \text{ in } \mathbb{R}^n \tag{7}$$

where $H$ is convex and coercive and $f$ semiconcave. Equations of this type arise in the Dynamic Programming approach to deterministic optimal control problems.

Assuming that $H(0) = 0$ and $f \geq 0$, then the unique bounded viscosity solution of (7) is Lipschitz continuous, non-negative and semiconcave for some semiconcavity constant $M$ depending on $H$ and $f$.

Therefore, by the preceding Proposition and the Rademacher's theorem,

$$|Du(x)| \leq \sqrt{2Mu(x)} \quad \text{almost everywhere in } \mathbb{R}^n.$$

We refer to [2, 6] for properties of viscosity solutions of Hamilton-Jacobi equations.

## 2.2 Non-negative Functions in Bounded Sets

Let us consider first $C^2$ non-negative functions $u$ defined on a finite interval $(-R, R)$ with $u'' \leq M$. In this setting, the following form of the Glaeser inequality holds,:

$$(\star) \quad |u'(0)| \leq \sqrt{2u(0)M} \text{ if } M \geq \frac{2u(0)}{R^2}$$

$$(\star\star) \quad |u'(0)| \leq \left(\frac{u(0)}{R} + \frac{R}{2}M\right) \text{ if } M < \frac{2u(0)}{R^2}.$$

Similar estimates holding at a generic $x \in (-R, R)$ can be easily deduced from the above. The constant $\sqrt{2}$ is optimal in the first inequality as shown by $u(x) = (x-R)^2$ The elementary proof is as follows:

$$0 \leq u(x) = u(0) + u'(0)x + \int_0^x (x-y)u''(y)\,dy.$$

Hence,

$$|u'(0)| \leq \frac{u(0)}{|x|} + \frac{|x|}{2}M.$$

The conclusion is then obtained by constrained optimization: if $M \geq \frac{2u(0)}{R^2}$, minimize the right hand side of above with respect to $|x| \in [0, R]$ to obtain $(\star)$. In the other case, the choice $|x| = R$ optimizes the right hand side, yielding $(\star\star)$.

Similar Glaeser's type inequalities hold for non-negative functions defined on a higher dimensional ball, see [23]. A model result from that paper is as follows: if $u \in C^2(B_R(0))$ is such that $|\Delta u| \leq M$ and $u \geq 0$, then

$$|Du(x)| \leq C\sqrt{u(0)M} \text{ if } 2|x| \leq \sqrt{\frac{u(0)}{M}} \leq R$$

$$|Du(x)| \leq C\left(\frac{u(0)}{R} + MR\right) \quad \text{if} \quad 2|x| \leq R \leq \sqrt{\frac{u(0)}{M}}$$

for some constant $C$ depending only on the space dimension.

Observe in this respect that condition (4) implies of course just the unilateral bound $\Delta u \leq nM$.

Consider now the following setting, generalizing that of [23]:

$$|F(D^2u) - g(u)| \leq M \quad \text{in} \quad B_R. \tag{8}$$

Examples of functions satisfying the above are of course, continuous viscosity solutions of the second order partial differential equation

$$F(D^2u) - g(u) = f(x) \quad \text{in} \quad B_R$$

with bounded right-hand side $f$.

Concerning the zero-order nonlinear term $g$ we shall assume that, for some positive constant $G$,

$$g : \mathbb{R}_+ \to \mathbb{R} \quad \text{is continuous and} \quad |g(s)| \leq Gs \quad \text{for all } s > 0\,. \tag{9}$$

For the principal part, we assume uniform ellipticity:

$$0 < \lambda||Y|| \leq F(X+Y) - F(X) \leq \Lambda||Y|| \;\; \forall\, X, Y \in S^n, \quad Y \geq 0\,.$$

For $R > 0$, let $u \in C(B_R)$ be a non-negative viscosity solution of

$$F(D^2u) - g(u) = f(x), \quad x \in B_R$$

with $f \in C(B_R)$, $\|f\|_{L^\infty(B_R)} \leq M$ and set

$$R_* = \sqrt{u(0)/M} \quad , \quad R_G = \sqrt{\sigma/G} \quad , \quad R_0 = \min(R_*, R_G, R).$$

Our result in this setting is as follows:

**Theorem 2.1** *For $F$, $g$ and $u$ as above, there exist positive constants $\chi$ and $\sigma$ depending only on $n$, $\lambda$ and $\Lambda$ such that*

$$\chi \sup_{B_{R_0/2}} |Du| \leq \begin{cases} \sqrt{u(0)M} & \text{if } R_* \leq \min(R_G, R) \\ \frac{u(0)}{R} + MR & \text{if } R \leq \min(R_*, R_G)\,. \\ u(0)\sqrt{G} + \frac{M}{\sqrt{G}} & \text{if } R_G \leq \min(R, R_*) \end{cases} \tag{10}$$

The main tools used in our proof of Theorem 2.1, which is notably different from the one in [23] for the linear case, are comparison principles and, in a crucial way, the weak Harnack inequality and the $C^{1,\alpha}$ regularity theory for viscosity solutions as developed by Caffarelli [5]. Here below is a sketch of the proof, for simplicity in the case $F(D^2u) = \Delta u$. We refer to [7] for full details and to [8] for complementary result in this direction.

Let us assume then that

$$|\Delta u - g(u)| \leq M$$
$$u \in C(B_R(0))\,, \quad u \geq 0\,, \quad g(u(x)) \leq Gu(x).$$

The first step is to show that for $0 < r < \frac{2(n+2)}{G}$, the following form of weak Harnack inequality holds:

$$\sup_{d\in(0,r)} \fint_{B_d} u\,dx \leq \frac{1}{1 - \frac{Gr^2}{2(n+2)}} \left( u(0) + \frac{Mr^2}{2(n+2)} \right). \tag{11}$$

The proof uses the upper bound $\Delta u - g(u) \leq M$, the divergence theorem and the coarea formula.

In the second step we use the lower bound $\Delta u - g(u) \geq M$ in order to show that for $0 < r < \frac{2(n+2)}{G}$, the following inequality holds:

$$u(0) \leq \frac{1 - \frac{Gr^2}{2(n+2)}}{1 - \frac{Gr^2}{n+2}} \fint_{B_r} u\,dx + \frac{1}{1 - \frac{Gr^2}{n+2}} \frac{Mr^2}{2(n+2)}. \tag{12}$$

Combining the two inequalities above we obtain that the following form of the Harnack inequality holds

$$\sup_{B_{r/2}} u \leq \frac{3 \cdot 2^{n-1}}{1 - \frac{1}{2}\frac{Gr^2}{n+2}} u(0) + \left( \frac{3 \cdot 2^n}{1 - \frac{1}{2}\frac{Gr^2}{n+2}} + 1 \right) \frac{M}{4} \frac{r^2}{n+2} \tag{13}$$

for $r < \min(\sqrt{2(n+2)/G}; R)$.

The final step of the proof makes use of the classical gradient estimate for solutions of the Poisson equation:

$$|Du(0)| \leq \frac{1}{\sqrt{2}} \left[ \frac{1}{r} \left( 4n + \frac{G}{4} r^2 \right) \sup_{B_{r/2}} u + \frac{M}{4} r \right] \tag{14}$$

which, together with the above inequality (13), allows to derive the claim after some further computations. The Glaeser's inequality easily produces a (perhaps) unusual proof of the classical Liouville theorem for harmonic functions:

$$u \in C^2(\mathbb{R}^n)\,,\ \Delta u = 0\,, u \geq 0 \ \text{ imply } \ u \equiv \text{constant}. \tag{15}$$

Indeed, if $u(0) = 0$ then, by the Maximum Principle, $u \equiv 0$. The other possible case is $u(0) > 0$; since $u$ is harmonic,$-\varepsilon \leq \Delta u \leq \varepsilon$ for any arbitrarily small $\varepsilon > 0$. Thanks to the above Theorem,

$$\sup_{B_{R_\varepsilon}} |Du(x)| \leq C\sqrt{\varepsilon\, u(0)}\ ,\ R_\varepsilon = \frac{1}{2}\sqrt{\frac{u(0)}{\varepsilon}} > 0$$

for some constant $C$ depending only on $n$.

Since $R_\varepsilon \to +\infty$ as $\varepsilon \to 0^+$, one can pass to the limit by monotonicity in the above to conclude that $\sup_{R^n} |Du(x)| = 0$.

## 3 Entire Subsolutions

Consider the semilinear equation

$$\Delta u = |u|^{\gamma-1}u + f(x) \tag{16}$$

with $\gamma > 1$ and $f(x) \geq \varepsilon > 0$ bounded and continuous. We know from Brezis [4] that this equation has a unique classical entire solution $u < 0$ in $\mathbb{R}^n$.

Then $v = -u$ solves

$$\Delta v = |v|^\gamma - f(x)\,.$$

Consider now the equation

$$\Delta u = |u|^\gamma + f(x)\,, \tag{17}$$

and observe that if $u$ is a solution of the above then $u$ solves also

$$\Delta u \geq g(u) \tag{18}$$

where

$$g(t) = t^\gamma + \varepsilon \quad \text{for} \quad t \geq 0 \quad , \quad g(t) \equiv \varepsilon \quad \text{for} \quad t < 0$$

is a positive, non decreasing and continuously differentiable function such that

$$\int_0^{+\infty} \left(\int_0^t g(s)ds\right)^{-\frac{1}{2}} dt = \int_0^{+\infty} \left((\gamma+1)^{-1}t^{\gamma+1} + \varepsilon t\right)^{-\frac{1}{2}} dt < +\infty.$$

Therefore, the so called Keller-Ossermann condition

$$\int_0^{+\infty} \left(\int_0^t g(s)ds\right)^{-\frac{1}{2}} dt = +\infty$$

is not verified by $g$. hence, well-known results by Keller [18] and Osserman [28], imply that inequality (18), and therefore also Eq. (17), does not have entire solutions.

We are interested here in investigating the validity of this type of results for inequalities where the Laplace operator is replaced by somefully nonlinear, possibly degenerate elliptic operator. More precisely, we consider viscosity solutions of the partial differential inequality

$$F(D^2u) \geq g(u) \quad \text{in } \mathbb{R}^n \tag{19}$$

where $F$ is a second order degenerate elliptic operator in the sense of Crandall–Ishii–Lions [10], that is

$$0 \leq F(x, X+Y) - F(x, X) \leq \text{Trace}\,(Y)$$

for all $x \in \mathbb{R}^n$ and $X,\ Y \in \mathcal{S}^n$ with $Y \geq O$ and $g(u)$ is a positive, non decreasing zero order term.

In our presentation here, $F$ will be either the "partial" Laplacian operator $\mathcal{P}_k^+$, see [17, 23] and also [1, 27], defined for real symmetric matrices $X \in \mathcal{S}^n$ and a positive integer $1 \leq k \leq n$ as

$$\mathcal{P}_k^+(X) = \mu_{n-k+1}(X) + \ldots + \mu_n(X)$$

where $\mu_1(X) \leq \mu_2(X) \leq \ldots \leq \mu_n(X)$ are the eigenvalues of the matrix $X$ or the degenerate maximal Pucci operator defined by

$$\mathcal{M}_{0,1}^+(X) = \sum_{\mu_i > 0} \mu_i(X)\,. \tag{20}$$

Observe that

$$\mathcal{P}_k^+(X) \leq \mathcal{M}_{0,1}^+(X)$$

for any $1 \leq k \leq n$ and for all $X \in \mathcal{S}^n$.

Our main results in this context are, for the "partial" Laplacian,

**Theorem 3.1** *Let $1 \leq k \leq n$ and $g : \mathbb{R} \to \mathbb{R}$ be positive, continuous and non decreasing. Then the inequality*

$$\mathcal{P}_k^+(D^2u) \geq g(u) \tag{21}$$

*has an entire viscosity solution $u \in C(\mathbb{R}^n)$ if and only if $g$ satisfies the Keller-Osserman condition*

$$\int_0^{+\infty} \left( \int_0^t g(s)ds \right)^{-\frac{1}{2}} dt = +\infty \tag{22}$$

while for the degenerate Pucci operator we have

**Theorem 3.2** *Let $g : \mathbb{R} \to \mathbb{R}$ be positive, continuous and strictly increasing. Then the inequality*

$$\mathcal{M}_{0,1}^{+}(D^2u) \geq g(u) \tag{23}$$

*has an entire viscosity solution $u \in C(\mathbb{R}^n)$ if and only if $g$ satisfies the Keller-Osserman condition (22).*

In order to appreciate the novelty of the above results, let us mention that after the classical results of Keller, Osserman and Brezis for $F = \Delta$ several extensions have been established, see [3, 11, 21, 22] for operators in divergence form and [12, 13, 15] for fully nonlinear equations. In these papers, existence, uniqueness and comparison results are given for the equation $F(D^2u) = g(u)$ assuming that the zero order term $g : \mathbb{R} \to \mathbb{R}$ is odd, continuous, increasing, convex in $[0, +\infty)$ and satisfying the growth condition

$$\int^{+\infty} \left( \int_0^t g(s)ds \right)^{-\frac{1}{2}} dt < +\infty.$$

In all these papers with the exception of [12], the principal part $F$ of the operator is assumed to be uniformly elliptic.

Let us emphasize that our results cover in fact the somewhat complementary cases in which $g$ is bounded below, say positive, and non decreasing. As far as we know, no previous results in this spirit were known before for degenerate elliptic equations.

The proof of both theorems is based on a comparison argument with radial symmetric functions obtained as solutions of an associated ODE.

It is worth to point out that the comparison principle works also in the present cases where a strong degeneracy may occur in the principal part and possibly in the zero order term as well.

Let us indicate the main steps of the proof of Theorem 3.1. The proof of Theorem 3.2 is completely analogous; observe however that the stronger assumption required there, namely $g$ strictly increasing, is used to prove the validity of the comparison principle for this strongly degenerate elliptic operator.

We refer for details to the forthcoming paper [9].

The first step is to consider as in [28] an auxiliary Cauchy problem, namely

$$\varphi''(r) + \frac{c-1}{r}\varphi'(r) = g(\varphi(r)), \quad \varphi'(0) = 0 \tag{24}$$

with $g : \mathbb{R} \to \mathbb{R}$ continuous, non negative and non decreasing. For this ODE one can prove the following facts:

- If $c > 0$ then (24) has solutions $\varphi \in C^2([0, R))$, continuous in $[0, R)$, twice differentiable in $(0, R)$ and such that

$$0 = \varphi'(0) = \lim_{r\to 0+} \varphi'(r)\,,\ \varphi''(0) = \lim_{r\to 0+} \varphi''(r) = \lim_{r\to 0+} \frac{\varphi'(r)}{r} \neq \infty\,.$$

- If $c > 0$ then every solution $\varphi$ of (24) in some interval $[0, R)$ is non decreasing and convex.
- If $c \geq 1$ then every maximal solution of (24) is globally defined in $[0, +\infty)$ if and only if $g$ satisfies the Keller-Osserman condition

$$\int_0^{+\infty} \left( \int_0^t g(s)ds \right)^{-\frac{1}{2}} dt = +\infty.$$

The second step amounts to establish a comparison principle between viscosity sub solutions and classical super solutions:

- assume $g : \mathbb{R} \to \mathbb{R}$ continuous and nondecreasing, let $u \in C(B_R)$ and $\Phi \in C^2(B_R)$ be, respectively, a viscosity subsolution and a classical supersolution of $\mathscr{P}_k^+(D^2\Phi) = g(\Phi) \quad \text{in } B_R$.
  Or, alternatively,
- assume $g : \mathbb{R} \to \mathbb{R}$ continuous and strictly increasing, let $u \in C(B_R)$ and $\Phi \in C^2(B_R)$ be, respectively, a viscosity subsolution and a classical supersolution of $\mathscr{M}_{0,1}^+(X) = \sum_{\mu_i > 0} \mu_i(X) = g(\Phi) \quad \text{in } B_R$.

Then the boundary sign condition $\limsup_{|x|\to R^-} (u(x) - \Phi(x)) \leq 0$ propagates to the interior, that is $u(x) \leq \Phi(x)$ for all $x \in B_R$.

The proof of this property, see also [14], is by contradiction: suppose there is a point $x \in B_R$ where $u(x) > \Phi(x)$. As easy to check the set $\Omega :\equiv \{x \in B_R,\ u(x) - \Phi(x) > \varepsilon\}$ is non-empty and $\overline{\Omega} \subset B_R$ for $\varepsilon > 0$ small enough. Since $u$ is a viscosity subsolution and $\Phi$ is a classical supersolution, then by viscosity calculus it turns out that the function $v = u - \Phi$ satisfies

$$\mathscr{P}_k^+(D^2v) \geq \mathscr{P}_k^+(D^2u) - \mathscr{P}_k^+(D^2\Phi) \geq g(u) - g(\Phi)$$

in the viscosity sense in $B_R$. Since $g$ is non decreasing, it follows then that $\mathscr{P}_k^+(D^2v) \geq 0$ in $\Omega$.

Moreover, $v > \varepsilon$ in $\Omega$ $v = \varepsilon$ on $\partial\Omega$. Hence, there exists a concave paraboloid $\Psi(x)$ touching $v$ from above at some point $x_0 \in \Omega$, a contradiction to the inequality $\mathscr{P}_k^+(D^2\Psi(x_0)) \geq 0$ which holds since $\Phi$ is a super solution.

The following step is to use the ODE to build a local radial solutions. To this end, let $1 \leq k \leq n$, $g$ non negative, non decreasing and continuous and $\varphi \in C^2([0, R))$ be a solution of the auxiliary Cauchy problem with $c = k$.

Then, $\Phi(x) = \varphi(|x|)$ is a classical solution of $\mathscr{P}_k^+(D^2\Phi) = g(\Phi)$ in $B_R$.
To check this, note that

$$D^2\Phi(x) = \begin{cases} \varphi''(0)\, I_n, & \text{if } x = 0 \\ \frac{\varphi'(|x|)}{|x|} I_n + \left(\varphi''(|x|) - \frac{\varphi'(|x|)}{|x|}\right) \frac{x}{|x|} \otimes \frac{x}{|x|} & \text{if } x \neq 0. \end{cases}$$

Hence, it is easy to check that $\Phi \in C^2(B_R)$ and that the eigenvalues of $D^2\Phi(x)$ are:

- $\varphi''(0)$ with multiplicity $n$, if $x = 0$.
- $\varphi''(|x|)$, which is simple, and $\frac{\varphi'(|x|)}{|x|}$ with multiplicity $n - 1$ for $x \neq 0$.

Therefore,

$$\mathscr{P}_k^+(D^2\Phi(0)) = k\,\varphi''(0) = g(\varphi(0)) = g(\Phi(0))$$

so that, by step 1,

$$\mathscr{P}_k^+(D^2\Phi(x)) = \varphi''(|x|) + \frac{k-1}{|x|}\varphi'(|x|) = g(\varphi(|x|)) = g(\Phi(x)) \qquad \text{for } x \neq 0.$$

Hence $\Phi$ is a classical solution of $\mathscr{P}_k^+(D^2\Phi) = g(\Phi)$ in $B_R$.

Suppose now that the Keller-Osserman condition (22) holds and let $\varphi$ be a maximal solution of the ODE . By the first step we know that $\varphi$ is globally defined on $[0, +\infty)$ which implies that $u(x) = \varphi(|x|)$ is an entire solution of $\mathscr{P}_k^+(D^2u) \geq g(u)$.

Conversely, assume that $u \in C(\mathbb{R}^n)$ solves $\mathscr{P}_k^+(D^2u) \geq g(u)$ and let $\varphi \in C^2([0, R))$ be a maximal solution of the ODE problem such that $u(0) > \varphi(0)$. The claim is that $R = +\infty$. If, on the contrary, $R < +\infty$, then $\varphi(r) \to +\infty$ as $r \to R^-$ and $\Phi(x) = \varphi(|x|)$ blows up on the boundary $\partial B_R$.

Hence, by comparison, $u(x) \leq \Phi(x)$ in $B_R$, a contradiction to $u(0) > \varphi(0)$.

Therefore, the maximal interval of existence of $\varphi$ is $[0, +\infty)$ and it follows that (22) is satisfied.

## References


1. Ambrosio, L., Soner, H.M.: Level set approach to mean curvature flow in arbitrary codimension. J. Differ. Geom. **43**(4), 693–737 (1996)
2. Bardi, M., Capuzzo Dolcetta, I.: Optimal Control and Viscosity Solutions of Hamilton-Jacobi-Bellman Equations. Systems and Control: Foundations and Applications. Birkhauser, Basel (1997)
3. Boccardo, L., Gallouet, T., Vazquez, J.L.: Nonlinear elliptic equations in $\mathbb{R}^N$ without growth restriction on the data. J. Differ. Equ. **105**(2), 334–363 (1993)

4. Brezis, H.: Semilinear equations in $\mathbb{R}^n$ without conditions at infinity. Appl. Math. Optim. **12**, 271–282 (1984)
5. Caffarelli, L.A.: Interior a priori estimates for solutions of fully nonlinear equations. Ann. Math. **130**, 189–213 (1989)
6. Cannarsa, P., Sinestrari, C.: Semiconcave Functions, Hamilton-Jacobi Equations and Optimal Control. Progress in Nonlinear Differential Equations and Their Applications. Birkhauser, Basel (2004)
7. Capuzzo Dolcetta, I., Vitolo, A.: Pointwise gradient estimates of Glaeser's type. Boll. Unione Mat. Ital. **5**, 211–224 (2012)
8. Capuzzo Dolcetta, I., Vitolo, A.: Glaeser's type gradient estimates for non-negative solutions of fully nonlinear elliptic equations. Discrete Contin. Dyn. Syst. A **28**, 539–557 (2010) [Special Issue Dedicated to Louis Nirenberg on the Occasion of his 85th Birthday]
9. Capuzzo Dolcetta, I., Leoni, F., Vitolo, A.: (in preparation)
10. Crandall, M.G., Ishii, H., Lions, P.L., User's guide to viscosity solutions of second order partial differential equations. Bull. Am. Math. Soc. (N.S.) **27**, 1–67 (1992)
11. D'Ambrosio, L., Mitidieri, E.: A priori estimates, positivity results, and nonexistence theorems for quasilinear degenerate elliptic inequalities. Adv. Math. **224**, 967–1020 (2010)
12. Diaz, G.: A note on the Liouville method applied to elliptic eventually degenerate fully nonlinear equations governed by the Pucci operators and the Keller–Osserman condition. Math. Ann. **353**, 145–159 (2012)
13. Esteban, M.J., Felmer, P.L., Quaas, A.: Super-linear elliptic equations for fully nonlinear operators without growth restrictions for the data. Proc. R. Soc. Edinb. **53**, 125–141 (2010)
14. Galise, G.: Maximum principles, entire solutions and removable singularities of fully nonlinear second order equations. Ph.D. Thesis, Università di Salerno a.a. (2011/2012)
15. Galise, G., Vitolo, A.: Viscosity solutions of uniformly elliptic equations without boundary and growth conditions at infinity. Int. J. Differ. Equ. **2011**, 18 pp. (2011)
16. Glaeser, G.: Racine carrée d'une fonction différentiable. Ann. Inst. Fourier **13**, 203–210 (1963)
17. Harvey, R., Lawson, B., Jr.: Dirichlet duality and the nonlinear Dirichlet problem. Commun. Pure Appl. Math. **62**, 396–443 (2009)
18. Keller, J.B.: On solutions of $\Delta u = f(u)$. Commun. Pure Appl. Math. **10**, 503–510 (1957)
19. Kolmogorov, A.N.: Une géneralization de l'inégalité de M.J. Hadamard entre les bornes supériores des dérivées successives d'une fonction. C. R. Acad. Sci. Paris **207**, 764–765 (1938)
20. Landau, E.: Einige Ungleichungen für zweimal differenzierbare Funktionen. Proc. Lond. Math. Soc. **13**, 43–49 (1913)
21. Leoni, F.: Nonlinear elliptic equations in $\mathbb{R}^N$ with "absorbing" zero order terms. Adv. Differ. Equ. **5**, 681–722 (2000)
22. Leoni, F., Pellacci, B.: Local estimates and global existence for strongly nonlinear parabolic equations with locally integrable data. J. Evol. Equ. **6**, 113–144 (2006)
23. Li, Y.Y., Nirenberg, L.: Generalization of a well-known inequality. Prog. Nonlinear Differ. Equ. Appl. **66**, 365–370 (2005)
24. Maz'ya, V.G., Kufner, A.: Variations on the theme of the inequality $(f')^2 \leq 2f \sup |f''|$. Manuscripta Math. **56**, 89–104 (1986)
25. Maz'ya, V.G., Shaposhnikova, T.O.: Sharp pointwise interpolation inequalities for derivatives. Funct. Anal. Appl. **36**, 30–48 (2002)
26. Nirenberg, L., Treves, F.: Solvability of the first order linear partial differential equation. Commun. Pure Appl. Math. **16**, 331–351 (1963)
27. Oberman, A., Silvestre, L.: The Dirichlet problem for the convex envelope. Trans. Am. Math. Soc. **363**(11), 5871–5886 (2011)
28. Osserman, R.: On the inequality $\Delta u \geq f(u)$. Pac. J. Math. **7**, 1641–1647 (1957)

# Hölder Regularity of the Gradient for Solutions of Fully Nonlinear Equations with Sub Linear First Order Term

**Isabeau Birindelli and Francoise Demengel**

*Dedicated to Ermanno Lanconelli and his imperishable mathematical enthusiasm on the occasion of his 70th birthday*

**Abstract** Using an improvement of flatness Lemma, we prove Hölder regularity of the gradient of solutions with higher order term a uniformly elliptic fully nonlinear operator and with Hamiltonian which is sub-linear. The result is based on some general compactness results.

**Keywords** Holder regularity • Fully nonlinear

**Mathematical Subject Classification:** 35J25, 35J60, 35P30

## 1 Introduction

In this paper we shall establish some regularity results of solutions of a class of fully nonlinear equations, with a first order term which is sub-linear; it is a natural continuation of [5, 12]. Precisely we shall consider the following family of equations

$$F(D^2u) + b(x)|\nabla u|^\beta = f(x) \text{ in } \Omega \subset \mathbb{R}^N. \tag{1}$$

See also [1] for related recent results.

**Theorem 1.1** *Suppose that $F$ is uniformly elliptic, that $\beta \in (0,1)$, $f$ and $b$ are in $\mathscr{C}(\overline{\Omega})$. For any $u$, bounded viscosity solution of (1) and for any $r < 1$, there exist $\gamma \in (0,1)$ depending on ellipticity constants of $F$, $\|b\|_\infty$, $\omega(b)$ and $\beta$ and $C = C(\gamma)$*

I. Birindelli (✉)
Dipartimento di Matematica G. Castelnuovo, Sapienza Università di Roma, Roma, Italy
e-mail: isabeau@mat.uniroma1.it

F. Demengel
Laboratoire d'Analyse et Géométrie, Université de Cergy Pontoise, Paris, France
e-mail: Francoise.Demengel@u-cergy.fr



G. Citti et al. (eds.), *Geometric Methods in PDE's*, Springer INdAM Series 13,
DOI 10.1007/978-3-319-02666-4_14

*such that*

$$\|u\|_{\mathscr{C}^{1,\gamma}(B_r(x_o))} \leq C\left(\|u\|_\infty + \|b\|_\infty^{\frac{1}{1-\beta}} + \|f\|_\infty\right),$$

*as long as* $B_1(x_o) \subset \Omega$.

Answering a question that we raised in [4], Imbert and Silvestre in [12] proved an interior Hölder regularity for the gradient of the solutions of

$$|\nabla u|^\alpha F(D^2u) = f(x)$$

when $\alpha \geq 0$. Their proof relies on a priori Lipschitz bounds, rescaling and an improvement of flatness Lemma, in this way they are lead to use the classical regularity results of Caffarelli, and Evans [7, 8, 11] for uniformly elliptic equations.

Following their breakthrough, in [5], we proved the same interior regularity when $\alpha \geq 0$ in the presence of lower order terms. We also proved $C^{1,\gamma}$ regularity up to the boundary if the boundary datum is sufficiently smooth. Our main motivation to investigate the regularity of these solutions i.e. the simplicity of the first eigenvalue associated to the Dirichlet problem for $|\nabla u|^\alpha F(D^2u)$, required continuity of the gradient up to the boundary.

When $\alpha \in (-1, 0)$, in [4] we proved $\mathscr{C}^{1,\gamma}$ regularity for solutions of the Dirichlet problem, using a fixed point argument which required global Dirichlet conditions on the whole boundary. So one of the question left open was: is the local regularity valid for $\alpha < 0$?

Theorem 1.1 answers to this question since the following holds:

**Proposition 1.1** *Suppose that, for* $\alpha \in (-1, 0)$, *u is a viscosity solution of*

$$|\nabla u|^\alpha F(D^2u) = f(x) \text{ in } \Omega$$

*then u is a viscosity solution of*

$$F(D^2u) - f(x)|\nabla u|^{-\alpha} = 0 \text{ in } \Omega.$$

The proof is postponed to the appendix, but recall that singular equations require a special definition of viscosity solutions.

Theorem 1.1 concerns continuous viscosity solutions of (1); we should point out that in the case of $L^p$ viscosity solutions (see [9]) it is possible to use a different strategy. Indeed one could prove first, using the argument below, that the solutions are Lipschitz continuous. By Rademacher theorem they are almost everywhere differentiable and hence they will be an $L^p$ viscosity solution of

$$F(D^2u) = g(x)$$

with $g \in L^\infty$. The classical result of Caffarelli [7] implies that the solution are $C^{1,\alpha}$. But this is a different result from ours, since continuous viscosity solutions are $L^p$ viscosity solutions only when $g$ is continuous, which somehow is what we want to prove.

In turn the $C^{1,\alpha}$ regularity implies that $g$ is Hölder continuous, so further regularities can be obtained (see e.g. [6, 14]).

Even for $F(D^2u) = \Delta u$ it would be impossible to mention all the work that has been done on equation of the form

$$F(D^2u) + |\nabla u|^p = f(x).$$

Interestingly most of the literature is concerned with the case $p > 1$. In particular the so called natural growth i.e. $p = 2$ has been much studied in variational contexts and the behaviours are quite different when $p \geq 2$ or $1 < p < 2$. We will just mention the fundamental papers of Lasry and Lions [13] and Trudinger [15]. And more recently the papers of Capuzzo Dolcetta et al. [10] and Barles et al. [2]. In the latter the Hölder regularity of the solution is proved for non local uniformly elliptic operators, and with lower order terms that may be sublinear.

*Remark 1.1* Observe that the operator is not Lipschitz continuous with respect to $\nabla u$. This implies that in general uniqueness of the Dirichlet problem does not hold. For example, when $\Omega$ is the ball of radius 1, then $u \equiv 0$ and $u(x) = C(1 - |x|^\gamma)$ with $\gamma = \frac{2-\beta}{1-\beta}$ and $C = \gamma^{-1}(\gamma + N - 2)^{\frac{1}{\beta-1}}$ are both solutions of equation

$$\begin{cases} \Delta u + |\nabla u|^\beta = 0 & \text{in } \Omega, \\ u = 0 & \text{on } \partial\Omega. \end{cases}$$

## 2 Interior Regularity Results

Let $S^N$ denote the symmetric $N \times N$ matrices. In the whole paper $F$ indicates a uniformly elliptic operator i.e. $F$ satisfies $F(0) = 0$ and, for some $0 < \lambda \leq \Lambda$,

$$\lambda \text{tr} N \leq F(M + N) - F(M) \leq \Lambda \text{tr} N$$

for any $M \in S^N$ and any $N \in S^N$ such that $N \geq 0$. The constants appearing in the estimates below often depend on $\lambda$ and $\Lambda$, but we will not specify them explicitly when it happens.

We recall that we want to prove

**Theorem 2.1** *Let $f$ and $b$ continuous in $\overline{B}_1 \subset \Omega$. For any $u$, bounded viscosity solution of (1) in $B_1$, and for any $r < 1$ there exist*

$$\gamma = \gamma(\|f\|_\infty, \|b\|_\infty, \beta, \omega_b(\delta)) \text{ and } C = C(\gamma)$$

*such that*

$$\|u\|_{\mathscr{C}^{1,\gamma}(B_r)} \le C\left(\|u\|_\infty + \|b\|_\infty^{\frac{1}{1-\beta}} + \|f\|_\infty\right).$$

Before proving Theorem 2.1, we shall prove a local Lipschitz continuity result.

**Lemma 2.1** *Suppose that* $H : B_1 \times \mathbb{R}^N \to \mathbb{R}$ *is such that*
$H(.,0)$ *is bounded in* $B_1$ *and there exist* $C > 0$ *such that for all* $q \in \mathbb{R}^N$,

$$|H(x,q) - H(x,0)| \le C(|q|^\beta + |q|).$$

*Then there exists* $C_o$ *such that if* $C < C_o$, *any bounded solution* $u$ *of*

$$F(D^2u) + H(x,\nabla u) = f(x) \text{ in } B_1$$

*is Lipschitz continuous in* $B_r$, *for* $r < 1$ *with some Lipschitz constant depending on* $r$, $\|f\|_\infty$, $C_o$ *and* $\|H(.0)\|_\infty$.

*Proof of Lemma 2.1* The proof proceeds as in [5, 12]. We outline it here, in order to indicate the changes that need to be done.

Let $r < r' < 1$ and $x_o \in B_r$, we consider on $B_{r'} \times B_{r'}$ the function

$$\Phi(x,y) = u(x) - u(y) - L^2\omega(|x-y|) - L|x-x_o|^2 - L|y-x_o|^2$$

where the continuous function $\omega$ is given by $\omega(s) = s - w_o s^{\frac{3}{2}}$ for $s \le (2/3w_o)^2$ and constant elsewhere; here $w_o$ is chosen in order that $(2/3w_o)^2 > 1$.

The scope is to prove that, for $L$ independent of $x_o$, chosen large enough,

$$\Phi(x,y) \le 0 \text{ on } B_r^2. \tag{2}$$

This will imply that $u$ is Lipschitz continuous on $B_r$ by taking $x = x_o$, and letting $x_o$ vary.

So we begin to choose $L > \frac{8\sup u}{(r'-r)^2}$. Suppose by contradiction that $\Phi(\bar{x},\bar{y}) = \sup \Phi(x,y) > 0$. By the hypothesis on $L$, $(\bar{x},\bar{y})$ is in the interior of $B_{r'}^2$. Proceeding in the calculations as in [2] (see also [3, 12]) we get that if (2) is not true then there exist $X$ and $Y$ such that

$$(q_x, X) \in J^{2,+}u(\bar{x}), (q_y, -Y) \in J^{2,-}u(\bar{y})$$

where $\overline{J}^{2,+}, \overline{J}^{2,-}$ are the standard semi-jets, while $q_x = L^2\omega'(|x-y|)\frac{x-y}{|x-y|} + 2L(x-x_o)$ and $q_y = L^2\omega'(|x-y|)\frac{x-y}{|x-y|} - 2L(y-x_o)$.

Then, there exist constant $\kappa_1, \kappa_2$ depending only on $\lambda, \Lambda, \omega_o$ such that

$$\mathscr{M}^+(X+Y) \le -\kappa_1 L^2$$

and $|q_x|, |q_y| \le \kappa_2 L^2$.

Using the equation,

$$\begin{aligned} f(\bar{x}) &\leq H(\bar{x}, q_x) + F(X) \\ &\leq H(\bar{x}, q_x) + F(-Y) + \mathscr{M}^+(X+Y) \\ &\leq f(\bar{y}) - \kappa_1 L^2 \\ &\quad + \|H(.,0)\|_\infty + C(|q_x|^\beta + |q_y|^\beta + |q_x| + |q_y|). \end{aligned}$$

The term $\|H(.,0)\|_\infty$ is $o(L^2)$, while for $C_o \leq \frac{\kappa_1}{16\kappa_2}$

$$\begin{aligned} C(|q_x|^\beta + |q_y|^\beta + |q_x| + |q_y|) &\leq \frac{\kappa_1 L^2}{2} + 4C_o(1 + \kappa_2 L^2) \\ &\leq \frac{3\kappa_1 L^2}{4} + 4C_o. \end{aligned}$$

In conclusion we have obtained that $f(\bar{x}) \leq f(\bar{y}) - \frac{\kappa_1 L^2}{4} + o(L^2)$. This is a contradiction for $L$ large.

**Corollary 2.1** *Suppose that $(f_n)_n$ and $(H_n(\cdot,0))_n$ are sequences converging uniformly respectively to $f_\infty$ and $H_\infty$ on any compact subset of $B_1$, such that for all $q \in \mathbb{R}^N$,*

$$|H_n(x,q) - H_n(x,0)| \leq \epsilon_n(|q|^\beta + |q|) \tag{3}$$

*with $\epsilon_n \to 0$. Let $u_n$ be a sequence of solutions of*

$$F(D^2 u_n) + H_n(x, \nabla u_n) = f_n(x) \text{ in } B_1.$$

*If $\|u_n\|_\infty$ is a bounded sequence, then up to subsequences, $u_n$ converges, in any compact subset of $B_1$, to $u_\infty$ a solution of the limit equation*

$$F(D^2 u_\infty) + H_\infty(x) = f_\infty(x) \text{ in } B_1.$$

## 2.1 Holder Regularity of the Gradient: Main Ingredients

We will follow the line of proof in [5, 12]. The modulus of continuity of a function $g$ is defined by $\omega_g(\delta) = \sup_{[x-y|\leq\delta} |g(x) - g(y)|$. In the following, $\omega$ will denote some continuous increasing function on $[0, \delta_o]$ such that $\omega(0) = 0$.

**Lemma 2.2 (Improvement of Flatness)** *There exist $\epsilon_o \in (0,1)$ and there exists $\rho \in (0,1)$ depending on $(\beta, N, \lambda, \Lambda, \omega)$ such that : for any $\epsilon < \epsilon_o$, for any $p \in \mathbb{R}^N$ and for any $f$ and $b$ such that $\|f\|_\infty \leq \epsilon$, $\|b\|_\infty \leq \epsilon$ and such that*

$\omega_b(\delta) \leq \|b\|_\infty \omega(\delta)$, *if u is a solution of*

$$F(D^2u) + b(x)|\nabla u + p|^\beta = f(x) \text{ in } B_1$$

*with* $\operatorname{osc}_{B_1} u \leq 1$ , *then there exists* $q^\star \in \mathbb{R}^N$ *such that*

$$\operatorname*{osc}_{B_\rho}(u - q^\star \cdot x) \leq \frac{1}{2}\rho.$$

*Proof of Lemma 2.2* We argue by contradiction i.e. we suppose that, for any $n \in \mathbb{N}$, there exist $p_n \in \mathbb{R}^N$, and $u_n$ a solution of

$$F(D^2u_n) + b_n(x)|\nabla u_n + p_n|^\beta = f_n(x) \text{ in } B_1$$

with $\operatorname{osc}_{B_1} u_n \leq 1$ and such that, for any $\rho \in (0,1)$ and any $q^\star \in \mathbb{R}^N$,

$$\operatorname*{osc}_{B_\rho}(u_n - q^\star \cdot x) \geq \frac{1}{2}\rho.$$

Observe that $u_n - u_n(0)$ satisfies the same equation as $u_n$, it has oscillation 1 and it is bounded, we can then suppose that the sequence $(u_n)$ is bounded. Suppose first that $|p_n|$ is bounded, so it converges, up to subsequences. Let $v_n(x) = u_n(x) + p_n \cdot x$, which is a solution of

$$F(D^2v_n) + b_n(x)|\nabla v_n|^\beta = f_n(x).$$

We can apply Corollary 2.1 with $H_n(x,q) = b_n(x)|q|^\beta$, since (3) holds.

Hence $v_n$ converges uniformly to $v_\infty$, a solution of the limit equation

$$F(D^2v_\infty) = 0 \text{ in } B_1.$$

Furthermore $v_\infty$ satisfies, for any $\rho \in (0,1)$ and any $q^\star \in \mathbb{R}^N$,

$$\operatorname*{osc}_{B_\rho}(v_\infty - q^\star \cdot x) \geq \frac{1}{2}\rho. \tag{4}$$

This contradicts the classical $\mathscr{C}^{1,\alpha}$ regularity results, see Evans [11] and Caffarelli [7].

We suppose now that $|p_n|$ goes to infinity. There are two cases, suppose first that $|p_n|^\beta \|b_n\|_\infty$ is bounded. Let $H_n(x,q) = b_n(x)|q + p_n|^\beta$. Since $\omega_{|p_n|^\beta b_n}(\delta) \leq |p_n|^\beta \|b_n\|_\infty \omega(\delta)$, $H_n(x,0)$ is equicontinuous and up to a subsequence, it converges uniformly to some function $H_\infty(x)$, while $(u_n)_n$ is a uniformly bounded sequence of solutions of

$$F(D^2u_n) + H_n(x, \nabla u_n) = f_n(x).$$

We can apply Corollary 2.1 and up to a subsequence, $u_n$ converges to $u_\infty$ which is a solution of

$$F(D^2u_\infty) + H_\infty(x) = 0.$$

Furthermore $u_\infty$ satisfies (4), for any $\rho \in (0,1)$ and any $q^\star \in \mathbb{R}^N$. As in the case $p_n$ bounded, this contradicts the classical $\mathscr{C}^{1,\gamma}$ regularity results cited above.

We are left to treat the case where $a_n = |p_n|^\beta \|b_n\|_\infty$ is unbounded. Hence, up to a subsequence, it goes to $+\infty$. We divide the equation by $a_n$, so $v_n := \frac{u_n}{a_n}$ satisfies

$$F(D^2v_n) + \frac{b_n(x)}{a_n}|a_n\nabla v_n + p_n|^\beta = \frac{f_n(x)}{a_n}.$$

We can apply Corollary 2.1 with

$$H_n(x,q) = b_n(x)a_n^{\beta-1}|q + a_n^{-1}p_n|^\beta.$$

Observe that, $H_n(x,0) = b_n(x)a_n^{-1}|p_n|^\beta$ is equicontinuous, of $L^\infty$ norm 1 and up to a subsequence, it converges uniformly to some function $H_\infty(x)$.

Passing to the limit one gets that the limit equation is

$$F(0) + H_\infty(x) = 0.$$

This yields a contradiction, since $H_\infty$ has norm 1 and it ends the proof of Lemma 2.2.

The next step is an iteration process which is needed in order to prove Theorem 2.1.

**Lemma 2.3** *Given $\epsilon_o$, $\omega$ and $\rho$ as in Lemma 2.2. Let $b$ and $f$ be such that $\|f\|_\infty, \|b\|_\infty \le \epsilon_o$ and such that $\omega_b(\delta) \le \|b\|_\infty\omega(\delta)$. Suppose that $u$ is a viscosity solution of*

$$F(D^2u) + b(x)|\nabla u|^\beta = f(x) \text{ in } B_1 \tag{5}$$

*and, $\operatorname{osc}_{B_1} u \le 1$. Then, there exists $\gamma \in (0,1)$, such that for all $k > 1$, $k \in \mathbb{N}$ there exists $p_k \in \mathbb{R}^N$ such that*

$$\operatorname*{osc}_{B_{r_k}}(u(x) - p_k \cdot x) \le r_k^{1+\gamma} \tag{6}$$

*where $r_k := \rho^k$.*

The proof is by induction and rescaling. For $k = 0$ just take $p_k = 0$. Suppose now that, for a fixe $k$, (6) holds with some $p_k$. Choose $\gamma \in (0,1)$ such that $\rho^\gamma > \frac{1}{2}$.

Define the function $u_k(x) = r_k^{-1-\gamma}(u(r_k x) - p_k \cdot (r_k x))$. By the induction hypothesis, $p_k$ is such that $\operatorname{osc}_{B_1} u_k \le 1$ and $u_k$ is a solution of

$$F(D^2 u_k) + r_k^{1-\gamma} b(r_k x)|r_k^{\gamma}(\nabla u_k + p_k r_k^{-\gamma})|^{\beta} = r_k^{1-\gamma} f(r_k x).$$

Denoting by $b_k$ the function $b_k(x) = r_k^{1-\gamma(1-\beta)} b(r_k x)$ which satisfies $\omega_{b_k}(\delta) = r_k^{1-\gamma(1-\beta)} \omega_b(r_k\delta) \le r_k^{1-\gamma(1-\beta)} \|b\|_\infty \omega(r_k \delta) \le \|b_k\|_\infty \omega(\delta)$, the equation above can be written as

$$F(D^2 u_k) + b_k(x)|\nabla u_k + p_k r_k^{-\gamma}|^{\beta} = r_k^{1-\gamma} f(r_k x).$$

Since the $L^\infty$ norm of $f_k = r_k^{1-\gamma} f(r_k \cdot)$ is less than $\epsilon$, we can conclude that there exists $q_k$ such that

$$\operatorname*{osc}_{B_\rho}(u_k(x) - q_k \cdot x) \le \frac{1}{2}\rho.$$

So that, for $p_{k+1} = p_k + q_k r_k^{\gamma+1}$,

$$\operatorname*{osc}_{B_{r_{k+1}}}(u(x) - p_{k+1} \cdot x) \le \frac{\rho}{2} r_k^{1+\gamma} \le r_{k+1}^{1+\gamma}.$$

This ends the proof of Lemma 2.3.

## 2.2 Holder Regularity of the Gradient: Conclusion

**Lemma 2.4** *Suppose that for any $r$, there exists $p_r$ such that*

$$\operatorname*{osc}_{B_r}(u(x) - p_r \cdot x) \le C r^{1+\gamma}$$

*then $u$ is $\mathscr{C}^{1,\gamma}$ in* 0.

*Proof* It is clear that it is sufficient to prove that $p_r$ converges when $r$ goes to 0.

We will prove that the sequence $p_{2^{-k}}$ converges and then conclude for the whole sequence. Let $r_k = \frac{1}{2^k}$, since $r_{k+1} < r_k$ for $x, y$ in $\overline{B_{r_{k+1}}}$

$$|u(x) - u(y) - p_{k+1} \cdot (x - y)| \le C r_{k+1}^{1+\gamma}$$

and

$$|u(x) - u(y) - p_k \cdot (x - y)| \le C r_k^{1+\gamma}.$$

Subtracting

$$|(p_{k+1} - p_k \cdot x - y))| \le C(r_{k+1}^{1+\gamma} + r_k^{1+\gamma}).$$

Then, choosing $x = \frac{p_{k+1}-p_k}{|p_{k+1}-p_k|} r_{k+1} = -y$, one gets

$$2|p_{k+1} - p_k| r_{k+1} \le C(r_{k+1}^{1+\gamma} + r_k^{1+\gamma})$$

which implies

$$|p_{k+1} - p_k| \le C2r_k^{\gamma}.$$

This proves that the series of general term $(p_{k+1} - p_k)$ converges; hence so does the sequence $p_k$.

We deduce the convergence of the whole sequence $p_\rho$ when $\rho$ goes to zero. Let $k$ be such that $r^{k+1} \le \rho \le r^k$. Then for all $x \in B_\rho$

$$(u(x) - p_\rho \cdot x) \le C\rho^{1+\gamma} \le Cr_k^{1+\gamma}$$

and also, since $x \in B_{r^k}$,

$$(u(x) - p_{r^k} \cdot x) \le Cr_k^{1+\gamma}.$$

Hence, by subtracting, $(p_\rho - p_{r^k}) \cdot x \le 2Cr_k^{1+\gamma}$. Then, taking $x = \frac{p_\rho - p_{r^k}}{|p_\rho - p_{r^k}|}\rho$, we get $|p_\rho - p_{r^k}| \le C\frac{r_k^{1+\gamma}}{\rho} \le C\frac{r_k^{1+\gamma}}{r_{k+1}} = 2Cr_k^{\gamma}$. This implies that $p_\rho$ has the same limit as $p_k$. This ends the proof of Lemma 2.4.

Suppose now that $u$ is a bounded solution of (5), for general $f$ bounded in $L^\infty$, and $b$ continuous. The function $v(x) = \epsilon u(x)$ with $\epsilon^{-1} = \operatorname{osc} u + \frac{1}{\epsilon_o}(\|f\|_\infty + \|b\|_\infty^{\frac{1}{1-\beta}})$ satisfies the equation

$$F(D^2 v) + b(x)\epsilon^{1-\beta}|\nabla v|^\beta = \epsilon f(x).$$

Our choice of $\epsilon$ implies that we are under the conditions of Lemma 2.3, so $v$ is in $\mathscr{C}^{1,\gamma}$, by Lemma 2.4, and so is $u$.

## Appendix

*Proof of Proposition 1.1* We assume that $\alpha \in (-1, 0)$ and that $u$ is a supersolution of

$$|\nabla u|^\alpha F(D^2 u) = f(x) \text{ in } \Omega \tag{7}$$

i.e. we suppose that for any $x_o \in \Omega$ either $u$ is locally constant in a neighbourhood of $x_o$ and then $0 \leq f$ in that neighbourhood, or, if it is not constant, for any $\varphi$ test function that touches $u$ by below at $x_o$ and such that $\nabla\varphi(x_o) \neq 0$, we require that

$$|\nabla\varphi(x_o)|^\alpha F(D^2\varphi(x_o)) \leq f(x_o).$$

We need to prove that this implies that $u$ is a supersolution of

$$F(D^2u) - f(x)|\nabla u|^{-\alpha} = 0 \text{ in } \Omega, \tag{8}$$

in the usual viscosity sense. Without loss of generality we let $x_o = 0$. If $u$ is constant around 0, $D^2u(0) = 0$ and $Du(0) = 0$, so the conclusion is immediate. If $\varphi$ is some test function by below at zero such that $\nabla\varphi(0) \neq 0$, the conclusion is also immediate. We then suppose that there exists $M \in S$ such that

$$u(x) \geq u(0) + \frac{1}{2}\langle Mx, x\rangle + o(|x|^2). \tag{9}$$

We want to prove that

$$F(M) \leq 0.$$

Let us observe first that one can suppose that $M$ is invertible, since if it is not, it can be replaced by $M_n = M - \frac{1}{n}I$ which satisfies (9) and tends to $M$.

Let $k > 2$ and $R > 0$ such that

$$\inf_{|x|<R}\left(u(x) - \frac{1}{2}\langle Mx, x\rangle + |x|^k\right) = u(0)$$

where the infimum is strict. We choose $\delta < R$ such that $k(2\delta)^{k-2} < \frac{1}{2}\inf_i |\lambda_i(M)|$. Let $\epsilon$ be such that

$$\inf_{\delta<|x|<R}\left(u(x) - \frac{1}{2}\langle Mx, x\rangle + |x|^k\right) = u(0) + \epsilon$$

and let $\delta_2 < \delta$ and such that $k(2\delta)^{k-1}\delta_2 + \|M\|_\infty(\delta_2^2 + 2\delta_2\delta) < \frac{\epsilon}{4}$. Then, for $x$ such that $|x| < \delta_2$,

$$\begin{aligned}\inf_{|y|\leq\delta}\{u(y) - \frac{1}{2}\langle M(y-x), y-x\rangle + |y-x|^k\} &\leq \inf_{|y|\leq\delta}\{u(y) - \frac{1}{2}\langle My, y\rangle + |y|^k\} + \frac{\epsilon}{4} \\ &= u(0) + \frac{\epsilon}{4}\end{aligned}$$

and on the opposite

$$\inf_{R>|y|>\delta}\{u(y) - \frac{1}{2}\langle M(y-x), y-x\rangle + |y-x|^k\}$$

$$\geq \inf_{|y|>\delta}\{u(y) - \frac{1}{2}\langle My, y\rangle + |y|^k\} - \frac{\epsilon}{4} > u(0) + 3\frac{\epsilon}{4}.$$

Since the function $u$ is supposed to be non locally constant, there exist $x_\delta$ and $y_\delta$ in $B(0,\delta_2)$ such that

$$u(x_\delta) > u(y_\delta) - \frac{1}{2}\langle M(x_\delta - y_\delta), x_\delta - y_\delta\rangle + |x_\delta - y_\delta|^k$$

and then the infimum $\inf_{y,|y|\leq\delta}\{u(y) - \frac{1}{2}\langle M(x_\delta - y), x_\delta - y\rangle + |x_\delta - y|^k\}$ is achieved on some point $z_\delta$ different from $x_\delta$. This implies that the function

$$\varphi(z) := u(z_\delta) + \frac{1}{2}\langle M(x_\delta - z), x_\delta - z\rangle - |x_\delta - z|^k + \frac{1}{2}\langle M(x_\delta - z_\delta), x_\delta - z_\delta\rangle + |x_\delta - z_\delta|^k$$

touches $u$ by below at the point $z_\delta$. But

$$\nabla\varphi(z_\delta) = M(z_\delta - x_\delta) - k|x_\delta - z_\delta|^{k-2}(z_\delta - x_\delta) \neq 0,$$

indeed, if it was equal to zero, $z_\delta - x_\delta$ would be an eigenvector corresponding to the eigenvalue $k|x_\delta - z_\delta|^{k-2}$ which is supposed to be strictly less than any eigenvalue of $M$.

Since $u$ is a super-solution of (7), multiplying by $|\nabla\varphi(z_\delta)|^{-\alpha}$, we get

$$F\left(M - \frac{d^2}{dz^2}(|x_\delta - z|^k)(z_\delta)\right) \leq f(z_\delta)|\nabla\varphi(z_\delta)|^{-\alpha}.$$

By passing to the limit for $\delta \to 0$ we obtain the desired conclusion i.e. $F(M) \leq 0$.

We would argue in the same manner for sub-solutions.

## References


1. Araujo, D., Ricarte, G., Teixeira, E.: Optimal gradient continuity for degenerate elliptic equations. (2013). Preprint [arXiv:1206.4089]
2. Barles, G., Chasseigne, E., Imbert, C.: Hölder continuity of solutions of second-order non-linear elliptic integro-differential equations. J. Eur. Math. Soc. **13**, 1–26 (2011)
3. Birindelli, I., Demengel, F.: Eigenvalue, maximum principle and regularity for fully non linear homogeneous operators. Commun. Pure Appl. Anal. **6**, 335–366 (2007)
4. Birindelli, I., Demengel, F.: Regularity and uniqueness of the first eigenfunction for singular fully non linear operators. J. Differ. Equ. **249**, 1089–1110 (2010)

5. Birindelli, I., Demengel, F.: $\mathscr{C}^{1,\beta}$ regularity for Dirichlet problems associated to fully nonlinear degenerate elliptic equations. Control Optim. Calc. Var. **20**, 1009–1024 (2014)
6. Cabré, X., Caffarelli, L.: Regularity for viscosity solutions of fully nonlinear equations $F(D^2u) = 0$. Topol. Meth. Nonlinear Anal. **6**, 31–48 (1995)
7. Caffarelli, L.: Interior a priori estimates for solutions of fully nonlinear equations. Ann. Math. 2nd Ser. **130**, 189–213 (1989)
8. Caffarelli, L., Cabré, X.: Fully-Nonlinear Equations Colloquium Publications, vol. 43. American Mathematical Society, Providence (1995)
9. Caffarelli, L., Crandall, M.G., , Kocan, M., Swięch, A.: On viscosity solutions of fully nonlinear equations with measurable ingredients. Commun. Pure Appl. Math. **49**(4), 365–397 (1996)
10. Capuzzo Dolcetta, I., Leoni, F., Porretta, A.: Hölder estimates for degenerate elliptic equations with coercive Hamiltonians. Trans. Am. Math. Soc. **362**(9), 4511–4536 (2010)
11. Evans, L.C.: Classical solutions of fully nonlinear, convex, second-order elliptic equations. Commun. Pure Appl. Math. **25**, 333–363 (1982)
12. Imbert, C., Silvestre, L.: $C^{1,\alpha}$ regularity of solutions of degenerate fully non-linear elliptic equations. Adv. Math. **233**, 196–206 (2013)
13. Lasry, J.M., Lions, P.L.: Nonlinear elliptic equations with singular boundary conditions and stochastic control with state constraints, I. The model problem. Math. Ann. **283**(4), 583–630 (1989)
14. Teixeira, E.V.: Universal moduli of continuity for solutions to fully nonlinear elliptic equations. Arch. Ration. Mech. Anal. **211**(3), 911–927 (2014)
15. Trudinger, N.S.: Fully nonlinear, uniformly elliptic equations under natural structure conditions. Trans. Am. Math. Soc. **278**, 751–769 (1983)

# The Reflector Problem and the Inverse Square Law

**Cristian E. Gutiérrez and Ahmad Sabra**

*Al diletto amico Ermanno Lanconelli in occasione del suo settantesimo compleanno*[1]

**Abstract** We introduce a model to design reflectors that take into account the inverse square law for radiation. We prove existence of solutions in the near field case when the input and output energies are prescribed.

**Keywords** geometric optics • Monge-Ampere type equations • radiometry

**AMS Subject Classification:** 78A05, 35J96, 35Q60

## 1 Introduction

It is known that the intensity of radiation is inversely proportional to the square of the distance from the source. In particular, at large distances from the source, the radiation intensity is distributed over larger surfaces and therefore the intensity per unit area decreases as the distance from the surface to the source increases. The purpose in this paper is to describe and solve a problem in radiometry involving the inverse square law, see e.g., [2, Sect. 4.8.1, formula (10)] and [8, Chap. 4]. We will present here only the essentials for the solution; details and further results can be found in [6].

We begin explaining the concepts needed to pose the problem. Let $\Omega \subseteq S^2$, and suppose that radiation is emanating from the origin $O$ for each direction $x \in \Omega$; $f(x)$ denotes the radiant intensity in the direction $x$, measured in Watts per steradian.

[1]Dedicated to Ermanno Lanconelli on the occasion of his 70th birthday.

C.E. Gutiérrez (✉) • A. Sabra
Department of Mathematics, Temple University, Philadelphia, PA, USA
e-mail: gutierre@temple.edu; ahmad.sabra@temple.edu





G. Citti et al. (eds.), *Geometric Methods in PDE's*, Springer INdAM Series 13,
DOI 10.1007/978-3-319-02666-4_15

Then the total amount of energy received at $\Omega$, or in other words the radiant flux through $\Omega$, is given by the surface integral

$$\Phi = \int_{\Omega} f(x)\, dx.^{2}$$

The *irradiance E* is the amount of energy or radiant flux incident on a surface per unit area; it is measured in $W/m^2$. If radiation is received on a surface $\sigma$, given parametrically by $\rho(x)x$ for $x \in \Omega$, then by [2, Sect. 4.8.1, formula (10)] the irradiance over an infinitesimal patch around $\rho(x)x$ is given by

$$E(x) = \frac{f(x)\, x \cdot \nu(x)}{\rho(x)^2},$$

where $\nu(x)$ is the outer unit normal to the surface $\sigma$ at the point $\rho(x)x$. Based on the above considerations we introduce the following quantity, measuring the amount of irradiance at the patch of surface $\sigma(\Omega)$, and given by the surface integral

$$\int_{\Omega} f(x)\, \frac{x \cdot \nu(x)}{\rho(x)^2}\, dx. \tag{1}$$

The problem we propose and solve in this paper is the following. Suppose $f$ is a positive function defined in $\Omega \subseteq S^2$, and $\eta$ is a Radon measure on a bounded set $D$ contained on a surface in $R^3$ with $\text{dist}(0, D) > 0$. We want to find a reflector surface $\sigma$ parameterized by $\rho(x)x$ for $x \in \Omega$ such that the radiation emanated from $O$ is reflected off by $\sigma$ into the set $D$ and such that the irradiance received on each patch of $D$ has at least measure $\eta$. In other words, we propose to find $\sigma$ such that

$$\int_{\tau_\sigma(E)} f(x)\, \frac{x \cdot \nu(x)}{\rho(x)^2}\, dx \geq \eta(E), \tag{2}$$

for each $E \subseteq D$, where the set $\tau_\sigma(E)$ is the collection of directions $x \in \Omega$ that $\sigma$ reflects off in $E$. We ask the reflector to cover all the target $D$, that is, $\tau_\sigma(D) = \Omega$. In particular, from (2) we need to have

$$\int_{\Omega} f(x)\, \frac{x \cdot \nu(x)}{\rho(x)^2}\, dx \geq \eta(D); \tag{3}$$

we say in this case that the reflector $\sigma$ is admissible. Since $f$ and $\eta$ are given but we do not know the reflector $\sigma$, we do not know a priori if (3) holds. However, assuming

[2]The units for this quantity are Watts because the units for $\Omega \subseteq S^2$ are considered non dimensional units, i.e., $\Omega$ is measured in steradians.

that the input and output energies satisfy

$$\int_{\Omega} f(x)\,dx \geq \frac{1}{C}\eta(D),$$

we will show that there exists a reflector $\sigma$ satisfying (2); see condition (12) and Theorems 4.1 and 5.1. Here $C$ is an appropriate constant depending only on the distance between the farthest point on the target and the source, and from how close to the source we want to place the reflector. In particular, we will see that if the target $D$ has a point very far away from the source, then the constant $1/C$ will be very large and therefore, for a given $\eta$ we will need more energy $f$ at the outset to prove the existence of a reflector satisfying (2). We will also see that, in general, for each fixed point $P_0$ in the support of the measure $\eta$, a reflector can be constructed so that it overshoots energy only at $P_0$, that is, for each set $E \subseteq \Omega$ such that $P_0 \notin E$ we have equality in (2); see Theorems 4.1 [parts (2) and (3)] and Theorem 5.1. In Sects. 4.1 and 5.1, we show that it is possible to construct a reflector that minimizes the energy overshot at $P_0$, that is unique in the discrete case, see Theorem 4.2.

To solve our problem, we introduce the notion of reflector and reflector measure with supporting ellipsoids of revolution, and show that (2) makes sense in terms of Lebesgue measurability, Sect. 3. With this definition, reflectors are concave functions and therefore differentiable a.e., so the normal $\nu(x)$ exists a.e.. To obtain the $\sigma$-additivity of the reflector measure given in Proposition 3.5, we need to assume that the target $D$ is contained in a hyperplane or $D$ is countable. This is needed in the proof of Lemma 3.2 and Remark 3.1, those results might fail otherwise, see Remark 3.2.

With this definition of the reflector, the reflected rays might cross the reflector to reach the target, in other words, the reflector might obstruct the target in certain directions. This is physically undesirable and it can be avoided by assuming that the supporting ellipsoids used in the definition of reflector have the target contained in their interiors. Another kind of physical obstruction might happen when the target obstructs the incident rays in their way to the reflector. All of these are discussed and illustrated in Sect. 3.1.

When the reflector is smooth, it satisfies a Monge-Ampère type pde that is indicated in Sect. 6.

We finish this introduction mentioning results in the literature that are relevant for this work and place our results in perspective. The reflector problem in the far field case has been considered by L. Caffarelli, V. Oliker and X-J. Wang, see [3, 9, 10]. The near field case is in [7] where the notion of reflector defined with supporting ellipsoids is introduced. In all these papers it is assumed that $\int_{\Omega} f(x)\,dx = \eta(\text{target})$, and the model does not take into account the inverse square law. For the far field refraction, models taking into account the loss of energy due to internal reflection are considered in [5].

We believe that this work is the first contribution to the problem of constructing a reflector that takes into account how far it is from the source.

## 2 Ellipsoids

Let $O$ be the origin in $\mathbb{R}^3$, $P \neq O$ be an arbitrary point, and $c > OP$. The ellipsoid of revolution $E_d(P)$ with foci $O$ and $P$ is given by $\{X : |X| + |X - P| = c\}$ and has polar radius

$$\rho_d(x) = \frac{d}{1 - \varepsilon x \cdot m}, \tag{4}$$

where $d = \dfrac{c^2 - OP^2}{2c}$ is the so called semi-latus rectum, $\varepsilon = \dfrac{OP}{c}$ is the eccentricity, and $m = \dfrac{\overrightarrow{OP}}{OP}$, $x \in S^2$. From (4), we obtain that the outer unit normal to the ellipsoid $E_d(P)$ at the point $\rho_d(x)x$ is given by

$$\nu_d(x) = \frac{x - \varepsilon m}{|x - \varepsilon m|}. \tag{5}$$

Using the formula for $\varepsilon$, and $d$ we get that $OP = \dfrac{2\varepsilon d}{1 - \varepsilon^2}$. Solving for $\varepsilon$ we obtain the following simple proposition that will be used frequently in the paper.

**Proposition 2.1** *Let $O$ be the origin in $\mathbb{R}^3$ and $P \neq O$. Fix $\delta > 0$ and consider an ellipsoid $E_d(P)$ with $d \geq \delta\, OP$. Then there exists a constant $0 < c_\delta < 1$, independent of $P$, such that $E_d(P)$ has eccentricity $\varepsilon \leq c_\delta$ and we have*

$$\frac{d}{1 + c_\delta} \leq \min_{x \in S^2} \rho_d(x) \leq \max_{x \in S^2} \rho_d(x) \leq \frac{d}{1 - c_\delta}. \tag{6}$$

*In fact,*

$$\varepsilon \leq -\delta + \sqrt{1 + \delta^2} := c_\delta < 1. \tag{7}$$

To conclude this section, we recall the following proposition borrowed from [7, Lemma 6].

**Proposition 2.2** *For ellipsoids of fixed foci $O$ and $P$, the eccentricity $\varepsilon$ is a strictly decreasing function of $d$, and for each fixed $x$, $\rho_d(x)$ is strictly increasing function of $d$.*

## 3 Reflectors and Reflector Measures

In this section we introduce the definition of reflector and reflector measure, and prove some properties that will be used later in the paper.

**Definition 3.1** Let $\Omega \subseteq S^2$ such that $|\partial\Omega| = 0$. The surface $\sigma = \{\rho(x)x\}_{x\in\bar{\Omega}}$ is a reflector from $\bar{\Omega}$ to $D$ if for each $x_0 \in \bar{\Omega}$ there exists an ellipsoid $E_d(P)$ with $P \in D$ that supports $\sigma$ at $\rho(x_0)\,x_0$. That is, $E_d(P)$ is given by $\rho_d(x) = \dfrac{d}{1-\varepsilon\, x\cdot m}$ with $m = \dfrac{\overrightarrow{OP}}{OP}$ and satisfies $\rho(x) \leq \rho_d(x)$ for all $x \in \bar{\Omega}$ with equality at $x = x_0$.

Notice that reflectors are concave and therefore continuous.

The reflector mapping associated with a reflector $\sigma$ is given by

$$\mathcal{N}_\sigma(x_0) = \{P \in D : \text{there exists } E_d(P) \text{ supporting } \sigma \text{ at } \rho(x_0)x_0\};$$

and the tracing mapping is

$$\tau_\sigma(P) = \{x \in \bar{\Omega} : P \in \mathcal{N}_\sigma(x)\}.$$

For the proof of the properties of the reflector measure, we will need the following measure theory result.

**Lemma 3.1** *Let $S \subseteq \mathbb{R}^n$ be a set, not necessarily Lebesgue measurable, and consider*

$$f(x) := \limsup_{r\to 0} \frac{|S \cap B_r(x)|_*}{|B_r(x)|},$$

*where $|\cdot|_*$ and $|\cdot|$ denote the Lebesgue outer measure and Lebesgue measure respectively. If*

$$M = \{x \in S : f(x) < 1\},$$

*then $|M| = 0$. Here $B_r(x)$ is the Euclidean ball centered at $x$ with radius $r$.*

*Moreover, if $B_r(x)$ is a ball in a metric space $X$ and $\mu^*$ is a Carathéodory outer measure on $X$,[3] then a similar result holds true for all $S \subseteq X$.*

The proof of Lemma 3.1 uses Vitali's covering theorem as stated in [11, Corollaries (7.18) and (7.19)]. From the book of Ambrosio and Tilli [1, Theorem 2.2.2.], we conclude that the same result applies to the case of a general metric space.

The following lemma is essential for the reflector measure.

**Lemma 3.2** *Suppose $D$ is contained in a plane $\Pi$ in $\mathbb{R}^3$, and let $\sigma$ be a reflector from $\bar{\Omega}$ to $D$. Then the set*

$$S = \{x \in \bar{\Omega} : \textit{there exist } P_1 \neq P_2 \textit{ in } D \textit{ such that } x \in \tau_\sigma(P_1) \cap \tau_\sigma(P_2)\}$$

*has measure zero in $S^2$.*

[3]From [11, Theorem (11.5)] every Borel subset of $X$ is Carathéodory measurable.

*Proof* Let $N$ be the set of points where $\sigma$ is not differentiable. Since $\rho$ is concave, it is locally Lipschitz and so the measure of $N$ in $S^2$ is zero. Let us write $S = (S \cap N) \cup (S \cap N^c)$. We shall prove that the measure, in $S^2$, of $F := S \cap N^c$ is zero.

Let $x_0 \in S \cap N^c$, then there exist $E_{d_1}(P_1)$ and $E_{d_2}(P_2)$ supporting ellipsoids to $\sigma$ at $x_0$ with $P_1 \neq P_2$, and $x_0$ is not a singular point of $\sigma$. Then there is a unique normal $\nu_0$ to $\sigma$ at $x_0$ and $\nu_0$ is also normal to both ellipsoids $E_{d_1}(P_1)$ and $E_{d_2}(P_2)$ at $x_0$. Hence from Snell's law $\rho(x_0)x_0$ is on the line joining $P_1$ and $P_2$. Since $D \subseteq \Pi$ we get that $E := \{\rho(x_0)\,x_0 : x_0 \in S \cap N^c\} \subseteq \Pi$. That is, the graph of $\sigma$, for $x \in S \cap N^c$, is contained in the plane $\Pi$.

We will prove that the set $S \cap N^c$ has measure zero in $S^2$.

*Case 1* $O \in \Pi$. Since $\rho(z)\,z \in \Pi$, for each $z \in F$, then the incident ray is contained in $\Pi$. Therefore $F$ is contained in a great circle of $S^2$ and hence has surface measure zero.

*Case 2* $O \notin \Pi$. In this case, it can be shown, see [6], that for every $z_0 \in F$, $|B_r(z_0) \cap F|_* \leq \frac{1}{2}|B_r(z_0)|$ for all $r > 0$, and hence by Lemma 3.1 the measure of $F$ is zero in $S^2$ and the proof of the lemma is complete.

*Remark 3.1* If $D$ is a finite or countable set, then the conclusion of Lemma 3.2 holds regardless if $D$ is on a plane. In fact, let $D = \{P_i\}_{i=1}^{\infty}$, with $O \notin D$. Let $\Pi_{ij}$ be the plane (or line) generated by the points $O, P_i, P_j$. Following the proof of Lemma 3.2 we have that

$$S \cap N^c \subseteq \cup_{i \neq j} \Pi_{ij},$$

and since the surface measure of $S \cap N^c \cap \Pi_{ij}$ is zero we are done.

*Remark 3.2* We present an example of a target $D$ that is not contained in a plane, a set $\Omega \subseteq S^2$, and a reflector $\sigma$ from $\bar{\Omega}$ to $D$ such that the set $S$ in Lemma 3.2 has positive measure. Consider the origin $O$, the point $P_0 = (0,2,0)$, and the half sphere $S_- = \{X = (x_1,x_2,x_3) : |X - P_0| = 1, x_3 \leq 0\}$. Let $D = S_- \cup \{P_0\}$, and consider the ellipsoid $E_d(P_0) = \{\rho_d(x)x\}_{x \in S^2}$ with $d$ large enough such that it contains $D$. Let $\Omega = \{x = (a,b,c) \in S^2 : c > 0\}$ and the reflector $\sigma = \{\rho_d(x)x\}_{x \in \bar{\Omega}}$.

Each point $P \in D$ is reached by reflection, because the ray from $P_0$ passing through $P$ intersects $\sigma$ at some point $P'$ and since $\sigma$ is an ellipsoid, the ray emanating from $O$ with direction $P'/|P'|$ is reflected off to $P$. We can see in this case that the set $S$ in Lemma 3.2 has positive measure, see [6].

Similarly, Lemma 3.2 does not hold when $D$ is contained in a finite union of planes. For example, let $P = (0,2,0)$, $\mathscr{C}$ be the closed disk centered at $(0,2,-1)$ and radius 1 contained on the plane $z = -1$, and let the target be $D = P \cup \mathscr{C}$. If $\sigma$ is a sufficiently large ellipsoid with foci $O$, $P$, and containing $D$, then the set $\tau_\sigma(P) \cap \tau_\sigma(\mathscr{C})$ has positive measure.

As a consequence of Lemma 3.2 we obtain the following.

**Proposition 3.1** *Suppose $\sigma$ is a reflector from $\bar{\Omega}$ to D and the target D is contained in a plane or D is countable. If A and B are disjoint subsets of D, then $\tau_\sigma(A) \cap \tau_\sigma(B)$ has Lebesgue measure zero.*

**Definition 3.2** Suppose $D$ is compact not containing $O$, and $M = \max_{P \in D} OP$. For each $\delta > 0$, we let $\mathscr{A}(\delta)$ be the collection of all reflectors $\sigma = \{\rho(x)x\}_{x \in \bar{\Omega}}$ from $\bar{\Omega}$ to $D$, such that for each $x_0 \in \bar{\Omega}$ there exist a supporting ellipsoid $E_d(P)$ to $\sigma$ at $\rho(x_0)\,x_0$ with $P \in D$ and $d \geq \delta M$.

The following proposition will be used in Sect. 5.

**Proposition 3.2** *If $\sigma = \{\rho(x)x\}_{x \in \bar{\Omega}}$ is a reflector in $\mathscr{A}(\delta)$, then $\rho$ is globally Lipschitz in $\bar{\Omega}$, the Lipschitz constant of $\rho$ is bounded uniformly by a constant depending only on $\delta$ and M.*

*Proof* Let $x, y \in \bar{\Omega}$. Then there exist $P \in D$ with $d \geq \delta M$ such that $E_d(P)$ supports $\sigma$ at $\rho(x)x$, i.e., $\rho(z) \leq \rho_d(z)$ for all $z \in \bar{\Omega}$ with equality at $z = x$. Therefore since $OP = \dfrac{2\varepsilon d}{1-\varepsilon^2}$, using (7) we get that

$$\rho(y) - \rho(x) = \rho(y) - \rho_d(x) \leq \rho_d(y) - \rho_d(x) \leq \frac{OP}{2}\,\frac{1+\varepsilon}{1-\varepsilon}\,|x-y| \leq \frac{M}{2}\,\frac{1+c_\delta}{1-c_\delta}\,|x-y|.$$

Interchanging the roles of $x, y$, we conclude the proposition.

**Proposition 3.3** *Assume $O \notin D$ with D compact such that either D is contained on a plane or D is countable. Let $\sigma \in \mathscr{A}(\delta)$ and let S be the set from Lemma 3.2. Suppose $\{x_n\}_{n=1}^{\infty}, x_0$ are in $\bar{\Omega} \setminus S$ and $x_n \to x_0$. If $E_{d_n}(P_n)$ and $E_{d_0}(P_0)$ are the corresponding supporting ellipsoids to $\sigma$ at $x_n$ and $x_0$, and $\nu(x_n), \nu(x_0)$ are the corresponding unit normal vectors, then we have*

1. $\lim_{n\to\infty} d_n = d_0$.
2. $\lim_{n\to\infty} P_n = P_0$.
3. $\lim_{n\to\infty} \nu(x_n) = \nu(x_0)$.

**Definition 3.3** Let $\delta > 0$ and let $D$ be compact with $O \notin D$. Given $\sigma \in \mathscr{A}(\delta)$, we define $\mathscr{S} = \{\, E \subseteq D : \tau_\sigma(E)$ is Lebesgue measurable$\}$.

**Proposition 3.4** *If $O \notin D$, with D a compact set contained either on a plane or D is countable, then $\mathscr{S}$ is a sigma-algebra on D containing all Borel sets.*

We are now ready to define the notion of reflector measure.

**Proposition 3.5** *Assume the target D is a compact set with $O \notin D$ such that D is contained on a plane or D is countable. Let $f \in L^1(\bar{\Omega})$ be non-negative, and let $\sigma$ be a reflector in $\mathscr{A}(\delta)$ for some $\delta > 0$. We define*

$$\mu(E) = \int_{\tau_\sigma(E)} f(x)\,\frac{x \cdot \nu(x)}{\rho^2(x)}\,dx$$

*for each Borel set E. Then $\mu$ is a finite Borel measure on D.*

*Proof* By Proposition 3.4, $\tau_\sigma(E)$ is Lebesgue measurable for each Borel set $E$. By Proposition 3.3 the function $x \cdot \nu(x)$ is continuous relative to $\bar{\Omega} \setminus S$. Since $\sigma \in A(\delta)$, $\rho$ is continuous and bounded below. It then follows that the function $\dfrac{x \cdot \nu(x)}{\rho^2(x)}$ is continuous relative to $\bar{\Omega} \setminus S$.

Let $x \in \bar{\Omega} \setminus S$, by (5) and Proposition 2.1, we have $x \cdot \nu(x) = \dfrac{1 - \varepsilon\, x \cdot m}{|x - \varepsilon m|} \geq \dfrac{1 - c_\delta}{1 + c_\delta} > 0$, where $\varepsilon$ is the eccentricity of the supporting ellipsoid to $\sigma$ at $\rho(x)x$. From Proposition 3.1, $|S| = 0$ and therefore $\mu(E)$ is well defined for each $E$ Borel set and is non negative. To prove the sigma additivity of $\mu$, let $E_1, E_2, \dots$ be countable mutually disjoint sequence of Borel sets. Then by Proposition 3.1, $\mu(E_i \cap E_j) = 0$ for all $i \neq j$, and hence $\mu\left(\cup_{i=1}^{+\infty} E_i\right) = \sum_{i=1}^{+\infty} \mu(E_i)$.

To complete the list of properties of the reflector measure, we state the following stability property.

**Proposition 3.6** *Suppose $D$ is a compact set with $O \notin D$ and such that $D$ is contained on a plane or $D$ is countable, and let $f \in L^1(\bar{\Omega})$ be non-negative. Let $\sigma_n$ be a sequence of reflectors in $\mathscr{A}(\delta)$ for some fixed $\delta > 0$, where $\sigma_n = \{\rho_n(x)\, x\}_{x \in \bar{\Omega}}$ are such that $\rho_n(x) \leq b$ for all $x \in \bar{\Omega}$, for all $n$ and for some $b > 0$, and $\rho_n$ converges point-wise to $\rho$ in $\bar{\Omega}$. Let $\sigma = \{\rho(x)x\}_{x \in \bar{\Omega}}$. Then we have*

1. *$\sigma \in \mathscr{A}(\delta)$, i.e., for all $x \in \bar{\Omega}$ there exist $P$ in $D$ and $d \geq \delta M$ such that $E_d(P)$ supports $\sigma$ at $\rho(x)x$.*
2. *If $\mu$ is the reflector measure corresponding to $\sigma$, then $\mu_n$ converges weakly to $\mu$.*

**Corollary 3.1** *If $D = \{P_1, P_2, \dots, P_N\}$ and $\sigma_n, \sigma, \mu_n, \mu$ are as in Proposition 3.6, then*

$$\lim_{n \to +\infty} \mu_n(P_i) = \mu(P_i).$$

*Proof* Define $h_i(P_j) = \delta_i^j$, $1 \leq i, j \leq N$. Since $D$ is discrete, $h$ is continuous on $D$, then by the previous proposition

$$\lim_{n \to +\infty} \mu_n(P_i) = \lim_{n \to +\infty} \int_D h_i(y)\, d\mu_n(y) = \int_D h_i(y)\, d\mu(y) = \mu(P_i).$$

## 3.1 Physical Visibility Issues

With the Definition 3.1 of reflector, the reflected rays might cross the reflector to reach the target, in other words, the reflector might obstruct the target in certain directions. This is illustrated in Fig. 1a. In this section, we show by convexity that if the ellipsoids used in the definition of reflector are chosen such that they contain

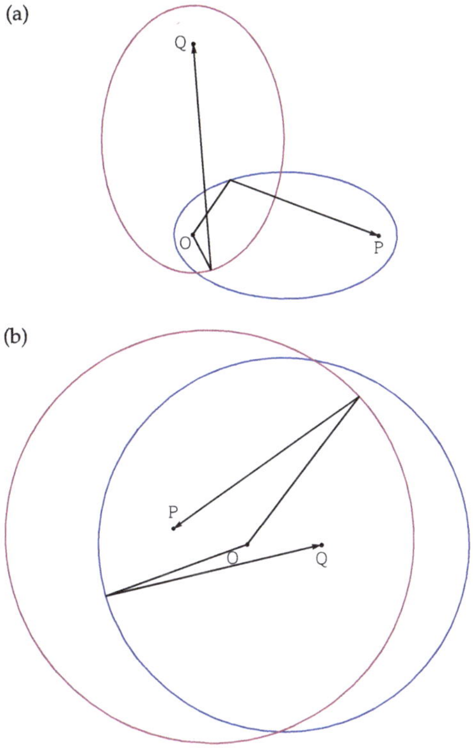


**Fig. 1** Reflectors illustrating obstruction. (**a**) $P = (1, 0)$, $Q = (0, 1)$, $E_{1.2}(P)$ and $E_{1.4}(Q)$; (**b**) $P = (-1, 0.2)$, $Q = (1, 0)$, $E_{5.5}(P)$ and $E_5(Q)$

all points of $D$ in their interiors, this obstruction can be avoided, that is, the reflector will not obstruct the target in any direction. For example, in Fig. 1b each reflected ray will not cross the reflector to reach the target.

Indeed, let $\{E_{d_i}\}_{i \in I}$ be a family of ellipsoids with foci $O$ and $P_i$, such that the convex body $B$ enclosed by all $\{E_{d_i}\}_{i \in I}$ is a reflector. Let us assume that all $P_i$'s are in the interior of $B$, and $D = \{P_i\}_{i \in I}$ is compact. We shall prove that under this condition any ray emanating from $O$ is reflected into a ray that does not cross the boundary of $B$ to reach the target. Suppose by contradiction that there is ray $r$ emanating from $O$ so that the reflected ray $r'$ crosses the boundary of $B$ to reach the target. Then $r$ hits $\partial B$ at some point $P$, and so $P$ is on the boundary of some ellipsoid $E_{d_i}$, and the reflected ray $r'$ crosses the boundary of $B$ at a point $Q$ to reach the target at say $P_i$. Since $P_i$ is in the interior of $B$, $P \in \partial B$, and $B$ is convex, the segment $(1 - t)P + tP_i \in \text{Int}\,(B)$ for all $0 < t \leq 1$. Since for some $t$, $Q$ is on this segment, then $Q$ belongs to the interior of $B$, a contradiction.

To assure that each supporting ellipsoid in the definition of reflector contain all points in $D$, we proceed as follows. Take $m = \min_{P\in D} OP, M = \max_{P\in D} OP$. Let $P_i, P_j \in D$, $P_i$ is inside the body of the ellipsoid with focus $O$ and $P_j$ if and only if

$$OP_i + P_iP_j < c_j = \frac{OP_j}{\varepsilon_j},$$

that is, if and only if $\varepsilon_j < \dfrac{OP_j}{OP_i + P_iP_j}$ for all $P_i, P_j \in D$. Therefore, in the definition of reflector it is enough to choose ellipsoids with eccentricity $\varepsilon$ satisfying

$$\varepsilon < \frac{m}{M + \operatorname{diam}(D)}. \tag{8}$$

Since $d_j = \dfrac{(1-\varepsilon_j^2)\, OP_j}{2\,\varepsilon_j}$ then by monotonicity of $d$ we have that (8) is then equivalent to

$$d_j > \frac{1 - \left(\dfrac{m}{M + \operatorname{diam}(D)}\right)^2}{2\dfrac{m}{M + \operatorname{diam}(D)}} OP_j \qquad \text{for all } P_j \in D,$$

and so it is enough to choose $d > \dfrac{1 - \left(\dfrac{m}{M + \operatorname{diam}(D)}\right)^2}{2\dfrac{m}{M + \operatorname{diam}(D)}} M.$

Therefore, to avoid obstruction of the target by the reflector, we can consider reflectors in the class $\mathscr{A}(\delta)$ with

$$\delta > \frac{1 - \left(\dfrac{m}{M + \operatorname{diam}(D)}\right)^2}{2\dfrac{m}{M + \operatorname{diam}(D)}} := \delta_D. \tag{9}$$

To complete this section, we mention the case when the target is on the way to the reflector, that is, the incident rays cross the target before reaching the reflector. Clearly, this can be avoided if we assume that $\bar{\Omega} \cap D^* = \emptyset$, where $D^*$ is the projection of $D$ on $S^2$.

# 4 Solution of the Problem in the Discrete Case

**Definition 4.1** Let $\Omega \subseteq S^2$ with $|\partial\Omega| = 0$, $D = \{P_1, P_2, \ldots, P_N\}$ is such that $O \notin D$, and $M = \max_{P\in D} OP$. Let $d_1, \cdots, d_N$ be positive numbers and $w = (d_1, d_2, \cdots, d_n)$. Define the reflector $\sigma = \{\rho_w(x)x\}$, $x \in \bar{\Omega}$, where

$$\rho_w(x) = \min_{1\le i\le N} \rho_{d_i}(x),$$

with $\rho_{d_i}(x) = \dfrac{d_i}{1-\varepsilon_i x\cdot m_i}$, $\varepsilon_i = \sqrt{1+\dfrac{d_i^2}{OP_i^2}} - \dfrac{d_i}{OP_i}$ and $m_i = \dfrac{\overrightarrow{OP_i}}{OP_i}$, $OP_i = |\overrightarrow{OP_i}|$.

**Lemma 4.1** *Let $0 < \delta \le \delta'$, and let $\{\rho_w(x)x\}$ be the reflector with $w = (d_1, \cdots, d_N)$, where $d_1 \le \delta' M$ and $d_i \ge \delta M$ for $1 \le i \le N$. If $f \in L^1(\bar{\Omega})$ and $f > 0$ a.e. then*

$$\mu_w(D) = \int_{\bar{\Omega}} f(x)\,\frac{x\cdot\nu_w(x)}{\rho_w^2(x)}\,dx > C(\delta,\delta',M)\int_{\bar{\Omega}} f(x)\,dx \tag{10}$$

*where $C(\delta,\delta',M)$ is a constant depending only on $\delta$, $\delta'$ and $M$.*

*Proof* From Proposition 3.2, the set of singular points of the reflector $\rho_w$ has measure zero. For each $x \in \Omega$ not a singular point the normal $\nu_w(x)$ exists and by (5) and Proposition 2.1, $\dfrac{x\cdot\nu_w(x)}{\rho_w^2(x)} \ge \dfrac{(1-c_\delta)^3}{(1+c_\delta)\,(\delta' M)^2} := C(\delta,\delta',M)$ and

$$\mu_w(D) = \int_{\bar{\Omega}} f(x)\,\frac{x\cdot\nu_w(x)}{\rho_w^2(x)}\,dx \ge C(\delta,\delta',M)\int_{\bar{\Omega}} f(x)\,dx.$$

To prove the strict inequality, suppose by contradiction that we have equality. Since $f > 0$ a.e., we then would get that

$$\frac{x\cdot\nu_w(x)}{\rho_w^2(x)} = C(\delta,\delta',M), \quad \text{for a.e. } x \in \Omega, \tag{11}$$

then $\rho_w$ is constant a.e. and hence by continuity $\rho_w$ is constant on $\Omega$, a contradiction.

The following lemma is similar to [7, Lemma 9].

**Lemma 4.2** *Consider the reflectors $\sigma = \{\rho_w(x)x\}_{x\in\bar{\Omega}}$ and $\tilde{\sigma} = \{\rho_{\tilde{w}}(x)x\}_{x\in\bar{\Omega}}$, with $w = (d_1, d_2, \cdots, d_l, \cdots, d_N)$ and $\tilde{w} = (d_1, d_2, \cdots, \tilde{d}_l, \cdots, d_N)$, such that $\tilde{d}_l \le d_l$. We write in this case $w \ge_l \tilde{w}$.*

*If $\mu$ and $\tilde{\mu}$ are the corresponding reflector measures, then $\tilde{\mu}(P_i) \le \mu(P_i)$ for $i \ne l$.*

As in [7] we obtain the following corollary.

**Corollary 4.1** *Let* $w_1 = (d_1^1, d_2^1, \cdots, d_N^1)$ *and* $w_2 = (d_1^2, d_2^2, \cdots, d_N^2)$. *Define* $w = (d_1, d_2, \cdots, d_N)$ *where* $d_i = \min(d_i^1, d_i^2)$, *we write* $w = \min(w_1, w_2)$. *Let* $\mu_1, \mu_2, \mu$ *be their corresponding reflector measures. Then*

$$\mu(P_i) \le \max(\mu_1(P_i), \mu_2(P_i)) \text{ for all } 1 \le i \le N.$$

We now prove existence of solutions in the discrete case.

**Theorem 4.1** *Let* $\Omega \subseteq S^2$ *with* $|\partial\Omega| = 0$, $f \in L^1(\bar{\Omega})$ *such that* $f > 0$ *a.e,* $g_1, g_2, \cdots, g_N$ *positive numbers with* $N > 1$. *Let* $D = \{P_1, P_2, \ldots., P_N\}$ *be such that* $O \notin D$, *and let* $M = \max_{1 \le i \le N} OP_i$. *Define the measure* $\eta$ *on* $D$ *by* $\eta = \sum_{i=1}^N g_i \delta_{P_i}$. *Fix* $\delta > 0$, *let* $k \ge \dfrac{1 + c_\delta}{1 - c_\delta}$, *where* $c_\delta$ *is from (7), and suppose that*

$$\int_{\bar{\Omega}} f(x)\, dx \ge \frac{1}{C(\delta, k\delta, M)} \eta(D), \tag{12}$$

*where* $C(\delta, k\delta, M) = \dfrac{(1 - c_\delta)^3}{(1 + c_\delta)(k\delta M)^2}$.

*Then there exists a reflector* $\bar{w} = (\bar{d}_1, \cdots, \bar{d}_N)$ *in* $\mathscr{A}(\delta)$, *i.e., with* $\bar{d}_i \ge \delta M$ *for* $1 \le i \le N$, *satisfying:*

1. $\bar{\Omega} = \bigcup_{i=1}^N \tau_{\bar{\sigma}}(P_i)$.
2. $\bar{\mu}(P_i) = g_i$ *for* $2 \le i \le N$, *where* $\bar{\mu}$ *is the reflector measure corresponding to* $\bar{w}$; *and*
3. $\bar{\mu}(P_1) > g_1$.

*Proof* Consider the set:

$$W = \{w = (d_1, \cdots, d_N) : d_1 = k\delta M, d_i \ge \delta M,$$
$$\mu_w(P_i) = \int_{\tau_{\sigma_w}(P_i)} f(x) \frac{x \cdot \nu_w(x)}{\rho_w^2(x)}\, dx \le g_i, i = 2, \cdots, N\}.$$

We first show that $W \ne \emptyset$. In fact, since $d_1$ is fixed we can choose $d_i$ large enough for $i \ne 1$ so that $\rho_w(x) = \rho_{d_1}(x)$, in that case $\mu_w(P_i) = 0 < g_i$ for $i = 2, \cdots, N$ and $w \in W$.

$W$ is closed. In fact, let $w_n = (d_1^n, \cdots, d_N^n) \in W$ converging to $w = (d_1, \cdots, d_N)$, and let $\mu_n$ and $\mu$ be their corresponding reflector measures. By Proposition 2.1 we have $\rho_{w_n}(x) = \rho_{d_1}(x) \le \dfrac{k\delta M}{1 - c_\delta}$. Then by Corollary 3.1 $\mu(P_i) = \lim_{n\to\infty} \mu_n(P_i) \le g_i$ for all $i = 2, \cdots, N$. Therefore $w \in W$.

Note that $\mu_w(P_1) > g_1$ for every $w \in W$. In fact, by Lemma 4.1 and condition (12), we have that

$$\begin{aligned}\mu_w(P_1) - g_1 =& \mu_w(D) - (g_1 + \mu_w(P_2) + \cdots + \mu_w(P_N))\\ \geq& \mu_w(D) - (g_1 + g_2 + \cdots + g_N) > 0.\end{aligned}$$

Let $\bar{d}_1 = k\delta M$, and $\bar{d}_i = \inf_{w \in W} d_i$ for $2 \leq i \leq N$. Take the reflector $\bar{\sigma} = \{\rho_{\bar{w}}(x)x\}$ and it's corresponding measure $\bar{\mu}$, with $\bar{w} = (\bar{d}_1, \cdots, \bar{d}_N)$. We have that $\bar{d}_i \geq \delta M$ for $2 \leq i \leq N$. Since $W$ is closed and the $d_i'$s are bounded below, the infimum is attained at some reflector $\bar{w}_i = (k\delta M, \bar{d}_2^i, \cdots, \bar{d}_{i-1}^i, \bar{d}_i, \bar{d}_{i+1}^i, \cdots, \bar{d}_N^i) \in W$ for $2 \leq i \leq N$. Let $\bar{\mu}_i$ be the reflector measure corresponding to $\bar{w}_i$. Since $\bar{w} = \min_{2 \leq i \leq N} \bar{w}_i$, it follows from Corollary 4.1 that $\bar{\mu}(P_i) \leq \max(\bar{\mu}_2(P_i), \bar{\mu}_3(P_i), \cdots, \bar{\mu}_N(P_i)) \leq g_i$ for $2 \leq i \leq N$, and so $\bar{w} \in W$.

It remains to prove that in fact we have $\bar{\mu}(P_i) = g_i$ for all $i \geq 2$. Without loss of generality, suppose that the inequality is strict for $i = 2$, that is, $\bar{\mu}(P_2) < g_2$. Take $0 < \lambda < 1$, $w_\lambda = (k\delta M, \lambda \bar{d}_2, \bar{d}_3, \ldots \bar{d}_N)$, and let $\mu_\lambda$ be the corresponding reflector measure. We claim that $\bar{d}_2 > \delta M$. Suppose by contradiction that $\bar{d}_2 = \delta M$. Then by Proposition 2.1, $\rho_{\bar{d}_2} \leq \dfrac{\delta M}{1 - c_\delta}$ and $\rho_{\bar{d}_1} \geq \dfrac{k\delta M}{1 + c_\delta}$, but since $k \geq \dfrac{1 + c_\delta}{1 - c_\delta}$, we have $\rho_{\bar{d}_1} \geq \rho_{\bar{d}_2}$. Therefore $\tau_{\bar{\sigma}}(P_1) \subseteq \tau_{\bar{\sigma}}(P_2)$ and hence by Proposition 3.1 $\bar{\mu}(P_1) = 0$, a contradiction. This proves the claim, and therefore $\lambda \bar{d}_2 > \delta M$ for all $\lambda$ sufficiently close to one. Moreover, by Lemma 4.2, $\mu_\lambda(P_i) \leq \bar{\mu}(P_i) \leq g_i$ for $i \geq 3$, and by Corollary 3.1 $\lim_{\lambda \to 1} \mu_\lambda(P_2) = \bar{\mu}(P_2) < g_2$ Then there exist $\lambda_0$ close to one such that $\mu_\lambda(P_2) < g_2$ and $\lambda \bar{d}_2 \geq \delta M$, for $\lambda_0 \leq \lambda < 1$. Hence $w_\lambda \in W$ contradicting the definition of $\bar{d}_2$. We conclude that $\bar{w}$ satisfies conditions (1)–(3).

## 4.1 Discussion About Overshooting in the Discrete Case

Theorem 4.1 shows the existence of a solution that overshoots energy at $P_1$.

**Definition 4.2** With the notation of Theorem 4.1 we define the following

1. $W = \{w = (d_1, \cdots, d_N) : d_1 = k\delta M, d_i \geq \delta M, \mu_w(P_i) \leq g_i \textit{ for } 2 \leq i \leq N\}$.
2. The reflector $\bar{\sigma} = \{\rho_{\bar{w}}(x)x\}$, and its corresponding reflector measure $\bar{\mu}$, where $\bar{w} = (\bar{d}_1, \cdots, \bar{d}_N)$ with $\bar{d}_1 = k\delta M$, and $\bar{d}_i = \inf_{w \in W} d_i$ for $2 \leq i \leq N$.

**Theorem 4.2** *Let $w = (d_1, d_2, \ldots, d_N) \in \mathscr{C}$ and $\mu$ its corresponding reflector measure. Then*

$$\bar{\mu}(D) \leq \mu(D),$$

*where $\bar{\mu}$ is the reflector measure corresponding to $\bar{w}$. Moreover, if $\bar{\Omega}$ is connected and $\bar{\mu}(D) = \mu(D)$ then:*

$$\bar{d}_i = d_i \quad \textit{for all } 1 \leq i \leq N.$$

*Proof* Since $w$ and $\bar{w}$ are in $\mathscr{C}$, then $\mu(P_i) = \bar{\mu}(P_i)$ for all $2 \leq i \leq N$. By definition of $\bar{w} = (\bar{d}_1, \cdots, \bar{d}_N)$ we have $\bar{d}_1 = k\delta M = d_1$ and $\bar{d}_i \leq d_i$ for all $2 \leq i \leq N$, since $\mathscr{C} \subseteq W$. Let $\sigma$ and $\bar{\sigma}$ be the reflectors corresponding to $w$ and $\bar{w}$, respectively, then by Proposition 2.2 $\tau_{\bar{\sigma}}(P_1) \subseteq \tau_\sigma(P_1)$ and $\bar{\mu}(P_1) \leq \mu(P_1)$. We conclude that $\bar{\mu}(D) \leq \mu(D)$.

Suppose now that $\bar{\Omega}$ is connected and we have equality, i.e., $\bar{\mu}(P_i) = \mu(P_i)$ for all $1 \leq i \leq N$. Let $I = \{1 \leq i \leq N : \bar{d}_i = d_i\}$ and $J = \{1 \leq i \leq N : \bar{d}_i < d_i\}$. Our goal is to prove that $J$ is empty. First notice that $I \neq \emptyset$, since $1 \in I$. Similarly as before $\tau_{\bar{\sigma}}(P_i) \subseteq \tau_\sigma(P_i)$ for all $i \in I$, and therefore

$$\begin{aligned}
\mu(P_i) &= \int_{\tau_\sigma(P_i)} f(x) \frac{x \cdot \nu_{d_i}(x)}{\rho_{d_i}^2(x)} \, dx = \int_{\tau_\sigma(P_i)} f(x) \frac{x \cdot \nu_{\bar{d}_i}(x)}{\rho_{\bar{d}_i}^2(x)} \, dx, \\
&= \int_{\tau_{\bar{\sigma}}(P_i)} f(x) \frac{x \cdot \nu_{\bar{d}_i}(x)}{\rho_{\bar{d}_i}{}^2(x)} \, dx + \int_{\tau_\sigma(P_i) \setminus \tau_{\bar{\sigma}}(P_i)} f(x) \frac{x \cdot \nu_{\bar{d}_i}(x)}{\rho_{\bar{d}_i}^2(x)} \, dx \\
&= \bar{\mu}(P_i) + \int_{\tau_\sigma(P_i) \setminus \tau_{\bar{\sigma}}(P_i)} f(x) \frac{x \cdot \nu_{\bar{d}_i}(x)}{\rho_{\bar{d}_i}^2(x)} \, dx.
\end{aligned}$$

Since $\mu(P_i) = \bar{\mu}(P_i)$ and $\dfrac{x \cdot \nu_{\bar{d}_i}(x)}{\rho_{\bar{d}_i}^2(x)} f(x) > 0$ a.e., we get $|\tau_\sigma(P_i) \setminus \tau_{\bar{\sigma}}(P_i)| = 0$, and so $|\tau_\sigma(P_i)| = |\tau_{\bar{\sigma}}(P_i)|$ for $i \in I$.

Suppose now that $J \neq \emptyset$ and let $x \in \bigcup_{j \in J} \tau_\sigma(P_j)$, then $x \in \tau_\sigma(P_{j_0})$ for some $j_0 \in J$. By Proposition 2.2 we have:

$$\rho_{\bar{w}}(x) = \rho_{\bar{d}_{j_0}}(x) < \rho_{d_{j_0}}(x) \leq \rho_{d_i}(x) = \rho_{\bar{d}_i}(x) \qquad \text{for all } i \in I.$$

Then by continuity of $\rho_{\bar{w}}$,

$$x \in \operatorname{Int}\left(\bigcup_{j \in J} \tau_{\bar{\sigma}}(P_j)\right) \qquad \text{and so} \quad \bigcup_{j \in J} \tau_\sigma(P_j) \subseteq \operatorname{Int}\left(\bigcup_{j \in J} \tau_{\bar{\sigma}}(P_j)\right).$$

Since $\bar{\Omega}$ is connected and $\bigcup_{j \in J} \tau_\sigma(P_j)$ is closed, we get that the set $A = \bigcup_{j \in J} \tau_{\bar{\sigma}}(P_j) \setminus \bigcup_{j \in J} \tau_\sigma(P_j)$ contains the non empty open set then $\bigcup_{j \in J} \tau_{\bar{\sigma}}(P_j) = \left(\bigcup_{j \in J} \tau_\sigma(P_j)\right) \cup A$ with $|A| > 0$, a contradiction.

## 5 Solution for a General Measure $\mu$

**Theorem 5.1** *Suppose the target D is compact, $O \notin D$, and either D is contained on a plane, or D is countable, and let $M = \max_{P \in D} OP$. Let $\Omega \subseteq S^2$, with $|\partial\Omega| = 0$, $f \in L^1(\bar{\Omega})$ with $f > 0$ a.e, and let $\eta$ be a Radon measure on D.*

*Given $\delta > 0$, $k \geq \dfrac{1+c_\delta}{1-c_\delta}$, with $c_\delta$ from* (7)*, we assume that*

$$\int_{\bar{\Omega}} f(x)\,dx \geq \frac{1}{C(\delta, k\delta, M)}\eta(D), \tag{13}$$

*where $C(\delta, k\delta, M) = \dfrac{(1-c_\delta)^3}{(1+c_\delta)(k\delta M)^2}$.*

*Given $P_0 \in \operatorname{supp}(\eta)$, the support of the measure $\eta$, there exists a reflector $\sigma = \{\rho(x)x\}_{x\in\bar{\Omega}}$ from $\bar{\Omega}$ to D in $\mathscr{A}(\delta)$ such that*

$$\eta(E) \leq \int_{\tau_\sigma(E)} f(x)\,\frac{x\cdot\nu_\rho(x)}{\rho^2(x)}\,dx,$$

*for each Borel set $E \subseteq D$, and*

$$\eta(E) = \int_{\tau_\sigma(E)} f(x)\,\frac{x\cdot\nu_\rho(x)}{\rho^2(x)}\,dx,$$

*for each Borel set $E \subseteq D$ with $P_0 \notin E$; and $\dfrac{\delta M}{1+c_\delta} \leq \rho(x) \leq \dfrac{k\delta M}{1-c_\delta}$ for all $x \in \bar{\Omega}$.*

*Proof* Partition the domain $D$ into a disjoint finite union of Borel sets with small diameter, say less than $\varepsilon$, so that $P_0$ is in the interior of one of them ($D$ has the relative topology inherited from $\mathbb{R}^3$). Notice that the $\eta$-measure of such a set is positive since $P_0 \in \operatorname{supp}(\eta)$. Of all these sets discard the ones that have $\eta$-measure zero. We then label the remaining sets $D_1^1, \dots, D_{N_1}^1$, and we may assume $P_0 \in (D_1^1)^\circ$ and $\eta(D_j^1) > 0$ for $1 \leq j \leq N_1$. Next pick $P_i^1 \in D_i^1$, so that $P_1^1 = P_0$, and define a measure on $D$ by $\eta_1 = \sum_{i=1}^{N_1} \eta(D_i^1)\delta_{P_i^1}$. Then from (13), $\eta_1(D) = \eta(D) \leq C(\delta, k\delta, M)\int_{\bar{\Omega}} f(x)\,dx$. Thus by Theorem 4.1, there exists a reflector $\sigma_1 = \left\{\rho_1(x)x : \rho_1(x) = \min_{1\leq i\leq N_1} \dfrac{d_i^1}{1-\varepsilon_i^1\, x\cdot m_i^1}\right\}$ with $d_1^1 = k\delta M$, $d_i^1 \geq \delta M$ for $2 \leq i \leq N_1$, $m_i^1 = \dfrac{\overrightarrow{OP_i^1}}{OP_i^1}$ for $1 \leq i \leq N_1$, and satisfying $\eta_1(E) \leq \displaystyle\int_{\tau_{\sigma_1}(E)} f(x)\,\frac{x\cdot\nu_{\rho_1}(x)}{\rho_1^2(x)}\,dx$, with equality if $P_0 \notin E$, for each $E$ Borel subset of $D$.

By this way for each $\ell = 1, 2, \cdots$, we obtain a finite disjoint sequence of Borel sets $D_j^\ell$, $1 \leq j \leq N_\ell$, with diameters less that $\varepsilon/2^\ell$ and $\eta(D_j^\ell) > 0$ such that $P_0 \in (D_1^\ell)^\circ$, $D_1^{\ell+1} \subseteq D_1^\ell$, and pick $P_j^\ell \in D_j^\ell$ with $P_1^\ell = P_0$, for all $\ell$ and $j$. The corresponding measures on $D$ are given by $\eta_\ell = \sum_{i=1}^{N_\ell} \eta(D_i^\ell)\delta_{P_i^\ell}$ satisfying $\eta_\ell(D) = \eta(D) \leq C(\delta, k\delta, M) \int_{\bar{\Omega}} f(x)\,dx$. We then have a corresponding sequence of reflectors given by $\sigma_\ell = \left\{\rho_\ell(x)x : \rho_\ell(x) = \min_{1\leq i\leq N_\ell} \dfrac{d_i^\ell}{1 - \varepsilon_i^\ell x \cdot m_i^\ell}\right\}$ with $d_1^\ell = k\delta M$, $d_i^\ell \geq \delta M$ for $2 \leq i \leq N_\ell$, $m_i^\ell = \dfrac{\overrightarrow{OP_i^\ell}}{OP_i^\ell}$ for $1 \leq i \leq N_\ell$, and satisfying $\eta_\ell(E) \leq \displaystyle\int_{\tau_{\sigma_\ell}(E)} f(x) \frac{x \cdot \nu_{\rho_\ell}(x)}{\rho_\ell^2(x)} dx$, with equality if $P_0 \notin E$, for each $E$ Borel subset of $D$.

Since $\sigma_\ell \in \mathscr{A}(\delta)$ for all $\ell$, it follows by Proposition 3.2 that $\rho_\ell$ are Lipschitz continuous in $\bar{\Omega}$ with a constant depending only on $\delta$ and $M$. In addition, from Proposition 2.1, and since $d_1^\ell = k\delta M$, we have $\dfrac{\delta M}{1 + c_\delta} \leq \rho_\ell(x) \leq \dfrac{k\delta M}{1 - c_\delta}$ $\forall \ell, x$. By Arzelá-Ascoli theorem, there is a subsequence, denoted also by $\rho_\ell$, converging to $\rho$ uniformly in $\bar{\Omega}$. From Proposition 3.6, $\sigma = \{\rho(x)x\}$ is a reflector in $\mathscr{A}(\delta)$ and the reflector measures $\mu_\ell$, corresponding to $\sigma_\ell$, converge weakly to $\mu$, the reflector measure corresponding to $\sigma$. We also have that $\eta_\ell$ converges weakly to $\eta$, and $\eta_\ell(E) = \mu_\ell(E)$ for every Borel set $E \subseteq D$ with $P_0 \notin E$, and each $\ell$. Then we obtain that $\eta(E) = \mu(E)$ for every Borel set $E \subseteq D$ with $P_0 \notin E$. Since $\eta_\ell(E) \leq \mu_\ell(E)$ for any Borel set $E \subseteq D$, we also conclude that $\eta(E) \leq \mu(E)$.

## *5.1 Discussion About Overshooting*

In this section, we will discuss the issue of overshooting energy to the point $P_0 \in \operatorname{supp}(\eta)$ and show that there is a reflector that minimizes the overshooting. Indeed, let $P_0 \in \operatorname{supp}(\eta)$ and

$$I = \inf\left\{\int_{\tau_\sigma(P_0)} f(x) \frac{x \cdot \nu_\rho(x)}{\rho^2(x)} dx : \sigma \text{ is a reflector as in Theorem 5.1}\right\}. \tag{14}$$

There exists a sequence of reflectors $\sigma_k = \{\rho_k(x)x\}$ such that

$$I = \lim_{k\to\infty} \int_{\tau_{\sigma_k}(P_0)} f(x) \frac{x \cdot \nu_{\rho_k}(x)}{\rho_k^2(x)} dx.$$

Therefore from Proposition 3.2, $\rho_k$ are uniformly Lipschitz in $\bar{\Omega}$, and by Theorem 5.1 uniformly bounded. Then by Arzelá-Ascoli there exists a subsequence, also denoted $\rho_k$, converging uniformly to $\rho$. By Proposition 3.6, $\sigma = \{\rho(x)x\} \in \mathscr{A}(\delta)$, and the corresponding reflector measures $\mu_k$ and $\mu$ satisfy $\mu_k \to \mu$ weakly. In particular, $I = \int_{\tau_\sigma(P_0)} f(x) \frac{x \cdot \nu_\rho(x)}{\rho^2(x)} dx$, and we are done.

We now compare $\mu(P_0)$ with $\eta(P_0)$.

*Case 1* $\mu(P_0) = \int_{\tau_\sigma(P_0)} f(x) \frac{x \cdot \nu_\rho(x)}{\rho^2(x)} dx > 0.$

In this case, we shall prove that for each open set $G \subseteq D$, with $P_0 \in G$, we have:

$$\int_{\tau_\sigma(G)} f(x) \frac{x \cdot \nu_\rho(x)}{\rho^2(x)} dx > \eta(G), \tag{15}$$

in other words, the reflector overshoots on each open set containing $P_0$. Notice that from Theorem 5.1 we have equality in (15) for each Borel set not containing $P_0$. Suppose by contradiction there exists an open set $G$, with $P_0 \in G$, such that $\int_{\tau_\sigma(G)} f(x) \frac{x \cdot \nu_\rho(x)}{\rho^2(x)} dx = \eta(G)$. Then

$$\begin{aligned}
\int_{\bar{\Omega}} f(x) \frac{x \cdot \nu_\rho(x)}{\rho^2(x)} dx &= \int_{\tau_\sigma(D)} f(x) \frac{x \cdot \nu_\rho(x)}{\rho^2(x)} dx \\
&= \int_{\tau_\sigma(D \setminus G)} f(x) \frac{x \cdot \nu_\rho(x)}{\rho^2(x)} dx + \int_{\tau_\sigma(G)} f(x) \frac{x \cdot \nu_\rho(x)}{\rho^2(x)} dx \\
&= \eta(D \setminus G) + \eta(G) = \eta(D) \leq C(\delta, k\delta, M) \int_{\bar{\Omega}} f(x)\, dx
\end{aligned}$$

from (13). But $f > 0$ a.e. and by Proposition 2.1, $\frac{x \cdot \nu_\rho(x)}{\rho^2(x)} - C(\delta, k\delta, D) \geq 0$, thus $\frac{x \cdot \nu_\rho(x)}{\rho^2(x)} = C(\delta, k\delta, M)$ for a.e. $x \in \bar{\Omega}$. Hence, again by Proposition 2.1, $\rho(x)$ is constant a.e., and since $\rho$ is continuous, then is constant in $\bar{\Omega}$ . This is a contradiction. Notice that if $\eta(P_0) > 0$, then $\mu(P_0) > 0$ and so the reflector overshoots.

*Case 2* $\mu(P_0) = \int_{\tau_\sigma(P_0)} f(x) \frac{x \cdot \nu_\rho(x)}{\rho^2(x)} dx = 0.$

This implies that $|\tau_\sigma(P_0)| = 0$ and $\eta(P_0) = 0$. Then for each $G$ open neighborhood of $P_0$ we have

$$\mu(G) = \int_{\tau_\sigma(G \setminus P_0)} f(x) \frac{x \cdot \nu_\rho(x)}{\rho^2(x)} dx + \int_{\tau_\sigma(P_0)} f(x) \frac{x \cdot \nu_\rho(x)}{\rho^2(x)} dx = \eta(G \setminus P_0) = \eta(G).$$

This identity also holds for every open set not containing $P_0$, and so for any open set in $D$. Since both measures $\mu$ and $\eta$ are outer regular then they are equal. Therefore in this case the reflector doesn't overshoot.

## 6 The Differential Equation for the Problem

Proceeding similarly as in [4, Appendix] we conclude that $\rho$ satisfies the following Monge-Ampère type equation

$$\left|\det\left(D^2\rho + \mathscr{A}(x,\rho(x),D\rho(x))\right)\right| \tag{16}$$
$$\le \frac{f(x)}{4\,g(T(x))\,\sqrt{1-|x|^2}\,\left|F(F+D\rho\cdot D_pF)\right|\,\rho^3\sqrt{\rho^2+|D\rho|^2-(x\cdot D\rho)^2}}$$

with $T$ the map from $\bar{\Omega}$ to $D$ and $F := F(x,\rho(x),D\rho(x))$,where

$$F(x,u,p) = \frac{\dfrac{u}{\sqrt{u^2+|p|^2-(p\cdot x)^2}}}{-\sqrt{u^2+|p|^2-(p\cdot x)^2}+2(u+p\cdot x)\dfrac{u}{\sqrt{u^2+|p|^2-(p\cdot x)^2}}},$$

and

$$\mathscr{A}(x,\rho,D\rho) = \frac{1}{\rho\,(F+D\rho\cdot D_pF)}\left[(F+\rho F_u)D\rho\otimes D\rho+\rho D\rho\otimes D_xF\right].$$ [4]

## References


1. Ambrosio, L., Tilli, P.: Topics on Analysis in Metric Spaces. Oxford Lecture Series in Mathematics and Its Applications, vol. 25. Oxford University Press, Oxford (2004)
2. Born, M., Wolf, E.: Principles of Optics, Electromagnetic Theory, Propagation, Interference and Diffraction of Light, 7th (expanded), 2006 edn. Cambridge University Press, Cambridge (1959)
3. Caffarelli, L.A., Oliker, V.: Weak solutions of one inverse problem in geometric optics. J. Math. Sci. **154**(1), 37–46 (2008)
4. Gutiérrez, C.E., Huang, Q.: The near field refractor. Ann. Inst. Henri Poincaré (C) Anal. Non Linéaire. **31**(4), 655–684 (2014) https://www.math.temple.edu/~gutierre/papers/nearfield.final.version.pdf
5. Gutiérrez, C.E., Mawi, H.: The far field refractor with loss of energy. Nonlinear Anal. Theory Methods Appl. **82**, 12–46 (2013)
6. Gutiérrez, C.E., Sabra, A.: The reflector problem and the inverse square law. Nonlinear Anal. Theory Methods Appl. **96**, 109–133 (2014)
7. Kochengin, S., Oliker, V.: Determination of reflector surfaces from near-field scattering data. Inverse Prob. **13**, 363–373 (1997)
8. McCluney, W.R.: Introduction to Radiometry and Photometry. Artech House, Boston (1994)
9. Wang, X.-J.: On the design of a reflector antenna. Inverse Prob. **12**, 351–375 (1996)
10. Wang, X.-J.: On the design of a reflector antenna II. Calc. Var. Partial Differ. Equ. **20**(3), 329–341 (2004)
11. Wheeden, R.L., Zygmund, A.: Measure and Integral. Marcel Dekker, New York (1977)


[4] For $a, b$ vectors in $\mathbb{R}^3$, $a\otimes b$ is the matrix $a^t b$.

# Gagliardo-Nirenberg Inequalities for Horizontal Vector Fields in the Engel Group and in the Seven-Dimensional Quaternionic Heisenberg Group

**Annalisa Baldi, Bruno Franchi, and Francesca Tripaldi**

*Dedicated to Ermanno Lanconelli on the occasion of his 70th birthday*

**Abstract** Recently, Bourgain and Brezis and Lanzani and Stein considered a class of div-curl inequalities in de Rham's complex. In this note we prove the natural counterpart of these inequalities for horizontal vector fields in the Engel group and in the seven-dimensional quaternionic Heisenberg group.

**Keywords** Differential forms • Div-curl systems • Engel group • Gagliardo-Nirenberg inequalities • Quaternionic Heisenberg groups

**2010 MSC:** 58A10, 35R03, 26D15, 43A80, 46E35, 35F35

## 1 Introduction

Let $Z = Z(x) = (Z_1(x), Z_2(x), Z_3(x))$ be a compactly supported smooth vector field in $\mathbb{R}^3$, and consider the system

$$\begin{cases} \operatorname{curl} Z = f \\ \operatorname{div} Z = 0 . \end{cases} \tag{1}$$

A. Baldi • B. Franchi (✉)
Dipartimento di Matematica, Università di Bologna, Bologna, Italy
e-mail: annalisa.baldi2@unibo.it; bruno.franchi@unibo.it

F. Tripaldi
Department of Mathematics, King's College London, Strand, London, UK
e-mail: francesca.tripaldi@kcl.ac.uk

 
G. Citti et al. (eds.), *Geometric Methods in PDE's*, Springer INdAM Series 13,
DOI 10.1007/978-3-319-02666-4_16

It is well known that $Z = (-\Delta)^{-1}\mathrm{curl} f$ is a solution of (1). Then, by the Calderón-Zygmund theory we can say that

$$\|\nabla Z\|_{L^p(\mathbb{R}^3)} \le C_p \|f\|_{L^p(\mathbb{R}^3)}, \quad \text{for} \quad 1 < p < \infty.$$

Then, by Sobolev inequality, if $1 < p < 3$ we have:

$$\|\mathbb{Z}\|_{L^{p*}(\mathbb{R}^3)} \le \|f\|_{L^p(\mathbb{R}^3)},$$

where $\frac{1}{p*} = \frac{1}{p} - \frac{1}{3}$. When we turn to the case $p = 1$ the first inequality fails. The second remains true. This is exactly the result proved in [11] by Bourgain and Brezis.

More precisely, in [11], Bourgain and Brezis establish new estimates for the Laplacian, the divi-curl system, and more general Hodge systems in $\mathbb{R}^n$ and they show in particular that if Z is a compactly supported smooth vector field in $\mathbb{R}^n$, with $n \ge 3$, and if curl $Z = f$ and div $Z = 0$, then there exists a constant $C > 0$ so that

$$\|\mathbb{Z}\|_{L^{n/(n-1)}(\mathbb{R}^n)} \le \|f\|_{L^1(\mathbb{R}^n)}. \tag{2}$$

This result does not follow straightforwardly from Calderòn-Zygmund theory and Sobolev inequality. The inequality (2) is a generalization of the classical sharp Sobolev inequality (the so-called Gagliardo-Nirenberg inequality) valid for all $n \ge 1$: let $u$ be a compactly supported scalar smooth function in $\mathbb{R}^n$ then

$$\|u\|_{L^{n/(n-1)}(\mathbb{R}^n)} \le c\|\nabla u\|_{L^1(\mathbb{R}^n)}. \tag{3}$$

In [28] Lanzani and Stein proved that the classical Gagliardo–Niremberg inequality (3) is the first link of a chain of analogous inequalities for compactly supported smooth differential $h$-forms in $\mathbb{R}^n$. In particular, their result for one-forms read as

$$\|u\|_{L^{n/(n-1)}(\mathbb{R}^n)} \le C\left(\|du\|_{L^1(\mathbb{R}^n)} + \|\delta u\|_{\mathcal{H}^1(\mathbb{R}^n)}\right) \tag{4}$$

where $d$ is the exterior differential, and $\delta$ (the exterior codifferential) is its formal $L^2$-adjoint.

Here $\mathcal{H}^1(\mathbb{R}^n)$ is the real Hardy space (see e.g. [34]). In other words, the main result of [28] provides a priori estimates for the div-curl system

$$\begin{cases} du = f \\ \delta u = g, \end{cases}$$

when the data $f, g$ belong to $L^1(\mathbb{R}^n)$. This result contains in particular Burgain–Brezis inequality (2); see also [36] for divergence-free vector fields in $\mathbb{R}^n$ [10, 11]. Related results have been obtained again by Burgain–Brezis in [12].

We refer the reader to all previous references for an extensive discussion about the presence of the Hardy space in (4). We stress explicitly that (4) holds for $n = 2$ (see [11]).

Recently, in [14], Chanillo and Van Schaftingen extended Burgain–Brezis inequality to a class of vector fields in Carnot groups. Some of the results of [14] are presented in Theorem 2.1 below.

We recall that a connected and simply connected Lie group $(\mathbb{G}, \cdot)$ (in general non-commutative) is said a *Carnot group of step* $\kappa$ if its Lie algebra $\mathfrak{g}$ admits a *step* $\kappa$ *stratification*, i.e. there exist linear subspaces $V_1, \ldots, V_\kappa$ such that

$$\mathfrak{g} = V_1 \oplus \ldots \oplus V_\kappa, \quad [V_1, V_i] = V_{i+1}, \quad V_\kappa \neq \{0\}, \quad V_i = \{0\} \text{ if } i > \kappa,$$

where $[V_1, V_i]$ is the subspace of $\mathfrak{g}$ generated by the commutators $[X, Y]$ with $X \in V_1$ and $Y \in V_i$. The first layer $V_1$ is called the *horizontal layer* and plays a key role in the theory, since it generates the whole of $\mathfrak{g}$ by commutation.

A Carnot group $\mathbb{G}$ can always be identified, through exponential coordinates, with the Euclidean space $\mathbb{R}^n$, where $n$ is the dimension of $\mathfrak{g}$, endowed with a suitable group operation.

In addition, the stratification of the Lie algebra induces a family of anisotropic dilations $\delta_\lambda$ ($\lambda > 0$) on $\mathfrak{g}$ and therefore, through the exponential map, on $\mathbb{G}$. We refer to [18] or [9] for an exhaustive introduction.

We denote by $Q$ the *homogeneous dimension* of $\mathbb{G}$, i.e. we set

$$Q := \sum_{i=1}^{\kappa} i \dim(V_i).$$

It is well known that $Q$ is the Hausdorff dimension of the metric space $\mathbb{G}$ endowed with any left invariant distance that is homogeneous with respect to group dilations. In general, $Q > N$.

The Lie algebra $\mathfrak{g}$ of $\mathbb{G}$ can be identified with the tangent space at the origin $e$ of $\mathbb{G}$, and hence the horizontal layer of $\mathfrak{g}$ can be identified with a subspace $H\mathbb{G}_e$ of $T\mathbb{G}_e$. By left translation, $H\mathbb{G}_e$ generates a subbundle $H\mathbb{G}$ of the tangent bundle $T\mathbb{G}$, called the horizontal bundle. A section of $H\mathbb{G}$ is called a horizontal vector field. Since, as usual, vector fields are identified with differential operators, we refer to the elements of $V_1$ as the *horizontal derivatives*.

The scalar Gagliardo-Nirenberg inequality is already well known in the setting of Carnot groups, as well as its geometric counterpart, the isoperimetric inequality (see [13, 20, 21, 23, 29, 30]) but, in spite of the extensive study of differential equations in Carnot groups (and, more generally, in sub-Riemannian spaces) carried out during the last few decades, very few results are known for pde's involving differential forms in groups (see, e.g., [1, 2, 6, 8, 19, 22, 31, 33]).

A natural setting for div-curl type systems in Carnot groups is provided by the so-called Rumin's complex $(E_0^*, d_c)$ of differential forms in $\mathbb{G}$. In fact, De Rham's complex $(\Omega^*, d)$ of differential forms, endowed with the usual exterior differential, does not fit the very structure of the group, since it is not invariant under group dilations: the differential $d$ mixes derivatives along all the layers of the stratification. Rumin's complex is meant precisely to overcome this difficulty.

As a matter of fact, the construction of the complex $(E_0^*, d_c)$ is rather technical and will be illustrated in Sect. 2. However, it is important to stress here that Rumin's differential $d_c$ may be a differential operator of higher order in the horizontal derivatives. This property affects crucially our results, that are therefore a distinct counterpart of those of Lanzani and Stein.

Among Carnot groups, the simplest but, at the same time, non-trivial instance is provided by Heisenberg groups $\mathbb{H}^n$, with $n \geq 1$, and, in particular, by the first Heisenberg group $\mathbb{H}^1$ which is in some sense the "model" of all topologically three-dimensional contact structures. These are step 2 Carnot groups and Lanzani-Stein inequalities for $\mathbb{H}^n$ are studied in [3–5].

The aim of the present note is to attack the study of inequality (4) for differential forms and their related div-curl type system in some distinguished Carnot groups of higher step: the first Engel group and the seven-dimensional quaternionic Heisenberg group. The Engel group is the model of the class of the so-called *filiform groups*; the quaternionic Heisenberg groups represent in some way an extension of Heisenberg groups: they are defined by replacing the complex field $\mathbb{C}$ by the field of quaternions $\mathbb{H}$ in t he definition of $\mathbb{H}^n$. This generates a two-step Carnot group whose centre is three-dimensional (while the centre in $\mathbb{H}^n$ is one-dimensional instead). The study of quaternions has received a boost in recent years, especially from their application to computer graphics. Quaternion multiplication can be used to rotate vectors in $\mathbb{R}^3$ and it is much better suited than the usual multiplication by $3 \times 3$ rotation matrices: data storage is reduced to speed up calculations and distortions of lengths and angles due to numerical inaccuracies can be avoided since quaternions can be easily renormalised without floating point computations.

Sections 3 and 4 contain a detailed presentation about these groups.

This note is organized as follows: in Sect. 2 we fix our notations and we collect some known results about Carnot groups and in particular we recall a crucial estimate proved by Chanillo and Van Schaftingen [14] for "divergence free" horizontal vector fields in Carnot groups. Moreover, we sketch the construction of Rumin's complex of differential forms in Carnot groups, and we remind some properties of the fundamental solution for a suitable Laplace operator on Rumin's forms [6, 7]. In Sect. 3 we present some basic facts about differential forms in the first Engel group and we collect our main result in Theorem 3.1 In Sect. 4, we extend the results seen in the Engel group to the seven-dimensional quaternionic Heisenberg group.

Finally, we recall that different generalizations of the global inequalities proved by Lanzani and Stein and Bourgain and Brezis have been proved in [27] (for the differential complex associated with an involutive elliptic structure), in [38] (for pseudoconvex CR manifolds) and in [37] (where, in particular, a Gagliardo-Nirenberg inequality for the subelliptic $\overline{\partial}$-operator in $\mathbb{H}^n$ is obtained).

## 2 Preliminary Results and Differential Forms in Carnot Groups

With the same notations used above, let us denote by $(\mathbb{G}, \cdot)$ a *Carnot group of step* $\kappa$ identified to $\mathbb{R}^n$ through exponential coordinates, and with $\mathfrak{g}$ its Lie algebra.

Let $X = \{X_1, \dots, X_n\}$ be the family of left invariant vector fields where the subset $\{X_1, \dots, X_m\}$ spans $V_1$, and hence generates by commutations all the other vector fields; we will refer to $X_1, \dots, X_m$ as the *generating vector fields* of the algebra, or as the *horizontal derivatives* of the group.

The Lie algebra $\mathfrak{g}$ can be endowed with a scalar product $\langle \cdot, \cdot \rangle$, making $\{X_1, \dots, X_n\}$ an orthonormal basis.

We can write the elements of $\mathbb{G}$ in *exponential coordinates*, identifying $p$ with the n-tuple $(p_1, \dots, p_n) \in \mathbb{R}^n$ and we identify $\mathbb{G}$ with $(\mathbb{R}^n, \cdot)$, where the explicit expression of the group operation $\cdot$ is determined by the Campbell-Hausdorff formula.

For any $x \in \mathbb{G}$, the *(left) translation* $\tau_x : \mathbb{G} \to \mathbb{G}$ is defined as

$$z \mapsto \tau_x z := x \cdot z.$$

For any $\lambda > 0$, the *dilation* $\delta_\lambda : \mathbb{G} \to \mathbb{G}$, is defined as

$$\delta_\lambda(x_1, \dots, x_n) = (\lambda^{d_1} x_1, \dots, \lambda^{d_n} x_n),$$

where $d_i \in \mathbb{N}$ is called the *homogeneity of the variable* $x_i$ in $\mathbb{G}$ (see [18], Chap. 1).

The Haar measure of $\mathbb{G} = (\mathbb{R}^n, \cdot)$ is the Lebesgue measure $\mathcal{L}^n$ in $\mathbb{R}^n$.

The dual space of $\mathfrak{g}$ is denoted by $\bigwedge^1 \mathfrak{g}$. The basis of $\bigwedge^1 \mathfrak{g}$, dual of the basis $X_1, \cdots, X_n$, is the family of covectors $\{\theta_1, \cdots, \theta_n\}$. We still indicate by $\langle \cdot, \cdot \rangle$ the inner product in $\bigwedge^1 \mathfrak{g}$ that makes $\theta_1, \cdots, \theta_n$ an orthonormal basis. We point out that, except for the trivial case of the commutative group $\mathbb{R}^n$, the forms $\theta_1, \cdots, \theta_n$ may have polynomial (hence variable) coefficients.

Following Federer (see [16], 1.3), the exterior algebras of $\mathfrak{g}$ and of $\bigwedge^1 \mathfrak{g}$ are the graded algebras indicated as $\bigwedge_* \mathfrak{g} = \bigoplus_{h=0}^{n} \bigwedge_h \mathfrak{g}$ and $\bigwedge^* \mathfrak{g} = \bigoplus_{h=0}^{n} \bigwedge^h \mathfrak{g}$ where $\bigwedge_0 \mathfrak{g} = \bigwedge^0 \mathfrak{g} = \mathbb{R}$ and, for $1 \le h \le n$,

$$\bigwedge_h \mathfrak{g} := \operatorname{span}\{X_{i_1} \wedge \cdots \wedge X_{i_h} : 1 \le i_1 < \cdots < i_h \le n\},$$

$$\bigwedge^h \mathfrak{g} := \operatorname{span}\{\theta_{i_1} \wedge \cdots \wedge \theta_{i_h} : 1 \le i_1 < \cdots < i_h \le n\}.$$

The elements of $\bigwedge_h \mathfrak{g}$ and $\bigwedge^h \mathfrak{g}$ are called *h-vectors* and *h-covectors*.

We denote by $\Theta^h$ the basis $\{\theta_{i_1} \wedge \cdots \wedge \theta_{i_h} : 1 \le i_1 < \cdots < i_h \le n\}$ of $\bigwedge^h \mathfrak{g}$. We remind that $\dim \bigwedge^h \mathfrak{g} = \dim \bigwedge_h \mathfrak{g} = \binom{n}{h}$.

The dual space $\bigwedge^1(\bigwedge_h \mathfrak{g})$ of $\bigwedge_h \mathfrak{g}$ can be naturally identified with $\bigwedge^h \mathfrak{g}$. The action of a $h$-covector $\varphi$ on an $h$-vector $v$ is denoted as $\langle \varphi | v \rangle$.

The inner product $\langle \cdot, \cdot \rangle$ extends canonically to $\bigwedge_h \mathfrak{g}$ and to $\bigwedge^h \mathfrak{g}$ making the bases $X_{i_1} \wedge \cdots \wedge X_{i_h}$ and $\theta_{i_1} \wedge \cdots \wedge \theta_{i_h}$ orthonormal.

We also set$X_{\{1,\cdots,n\}} := X_1 \wedge \cdots \wedge X_n$ and $\theta_{\{1,\cdots,n\}} := \theta_1 \wedge \cdots \wedge \theta_n$.

Starting from $\bigwedge_* \mathfrak{g}$ and $\bigwedge^* \mathfrak{g}$, by left translation, we can now define two families of vector bundles (still denoted by $\bigwedge_* \mathfrak{g}$ and $\bigwedge^* \mathfrak{g}$) over $\mathbb{G}$ (see [8] for details). Sections of these vector bundles are said respectively vector fields and differential forms.

If $f : \mathbb{G} \to \mathbb{R}$, we denote by $\nabla_{\mathbb{G}} f$ the horizontal vector field

$$\nabla_{\mathbb{G}} f := \sum_{i=1}^{m} (X_i f) X_i,$$

whose coordinates are $(X_1 f, \ldots, X_m f)$. Moreover, if $\Phi = (\phi_1, \ldots, \phi_m)$ is a horizontal vector field, we define $\mathrm{div}_{\mathbb{G}}\, \phi$ as the real valued function

$$\mathrm{div}_{\mathbb{G}}\,(\Phi) := \sum_{j=1}^{m} X_j \phi_j.$$

As costumary, we set

$$\Delta_{\mathbb{G}} f := \mathrm{div}_{\mathbb{G}}\,(\nabla_{\mathbb{G}} f).$$

Following e.g. [18], we can define a group convolution on $\mathbb{G}$: if, for instance, $f \in \mathcal{D}(\mathbb{G})$ and $g \in L^1_{\mathrm{loc}}(\mathbb{G})$, we set

$$f * g(p) := \int f(q) g(q^{-1}p)\, dq \quad \text{for } p \in \mathbb{G}.$$

We remind that, if (say) $g$ is a smooth function and $L$ is a left invariant differential operator, then

$$L(f * g) = f * Lg. \tag{5}$$

In addition

$$\langle f * g | \varphi \rangle = \langle g | {}^{\mathrm{v}}\! f * \varphi \rangle \quad \text{and} \quad \langle f * g | \varphi \rangle = \langle f | \varphi * {}^{\mathrm{v}} g \rangle$$

for any test function $\varphi$. Suppose now $f \in \mathscr{E}'(\mathbb{G})$ and $g \in \mathscr{D}'(\mathbb{G})$. Then, if $\psi \in \mathscr{D}(\mathbb{G})$, we have (all convolutions being well defined)

$$\begin{aligned} \langle (Lf) * g|\psi\rangle &= \langle Lf|\psi * {}^{\mathrm{v}}g\rangle = (-1)^{|I|}\langle f|\psi * (L^{*}\,{}^{\mathrm{v}}g)\rangle \\ &= (-1)^{|I|}\langle f * {}^{\mathrm{v}}L^{*}\,{}^{\mathrm{v}}g|\psi\rangle. \end{aligned} \tag{6}$$

We should also remind the notion of *kernel of order* $\alpha$. Following [17], a kernel of order $\alpha$ is a homogeneous distribution of degree $\alpha - Q$ (with respect to group dilations), that is smooth outside of the origin.

**Proposition 2.1** *Let $K \in \mathscr{D}'(\Omega)$ be a kernel of order $\alpha$.*

*(i)* ${}^{\mathrm{v}}K$ *is again a kernel of order* $\alpha$.
*(ii)* $X_\ell K$ *is a kernel of order* $\alpha - 1$ *for any horizontal derivative* $X_\ell K$, $\ell = 1, \dots, m$.
*(iii) If* $\alpha > 0$, *then* $K \in L^1_{\mathrm{loc}}(\mathbb{H}^n)$.
*(iv) If* $\alpha = 0$, *then the map* $f \to f * K$ *is* $L^p$ *continuous for* $1 < p < \infty$.

We now recall a remarkable estimate proved by Chanillo and Van Schaftingen in the spirit of Bourgain–Bresis's inequality which is crucial to our proof.

Let $k \geq 1$ be fixed, and let $F \in L^1(\mathbb{G}, \otimes^k \bigwedge_1 \mathfrak{h}_1)$ belong to the space of horizontal $k$-tensors. We can write

$$F = \sum_{i_1,\dots,i_k} F_{i_1,\dots,i_k} X_{i_1} \otimes \cdots \otimes X_{i_k}.$$

The tensor $F$ can be identified with the differential operator

$$u \to Fu := \sum_{i_1,\dots,i_k} F_{i_1,\dots,i_k} X_{i_1} \cdots X_{i_k} u.$$

Moreover, we denote by $\mathscr{S}(\mathbb{G}, \mathrm{Sym}(\otimes^k \bigwedge_1 \mathfrak{h}_1))$ the subspace of smooth symmetric horizontal $k$-tensors with coefficients in $\mathscr{S}(\mathbb{G})$.

**Theorem 2.1 ([14], Theorem 5)** *Let* $k \geq 1$, $F \in L^1(\mathbb{G}, \otimes^k \bigwedge_1 \mathfrak{h}_1)$, $\Phi \in \mathscr{S}(\mathbb{G}, \mathrm{Sym}(\otimes^k \bigwedge_1 \mathfrak{h}_1))$.

*Suppose*

$$\int_{\mathbb{G}} F\psi \, dV = 0 \quad \textit{for all } \psi \in \mathscr{D}(\mathbb{G}),$$

*i.e. suppose that*

$$\sum_{i_1,\dots,i_k} W_{i_k} \cdots W_{i_1} F_{i_1,\dots,i_k} = 0 \quad \textit{in } \mathscr{D}'(\mathbb{G}).$$

*Then*

$$\left|\int_{\mathbb{G}} \langle \Phi, F\rangle \, dV\right| \le C_k \|F\|_{L^1(\mathbb{G},\otimes^k \bigwedge_1 \mathfrak{h}_1)} \|\nabla_{\mathbb{G}} \Phi\|_{L^Q(\mathbb{G},\otimes^k \bigwedge_1 \mathfrak{h}_1)}.$$

Let us turn to differential forms in Carnot groups. The notion of intrinsic forms in Carnot groups is due to Rumin [32, 33]. A more extended presentation of the results of this section can be found in [8, 19, 35].

The following notion of weight of a differential form plays a key role.

**Definition 2.1** If $\alpha \in \bigwedge^1 \mathfrak{g}$, $\alpha \neq 0$, we say that $\alpha$ has *pure weight* $p$, and we write $w(\alpha) = p$, if $\alpha^\natural \in V_p$. More generally, if $\alpha \in \bigwedge^h \mathfrak{g}$, we say that $\alpha$ has pure weight $p$ if $\alpha$ is a linear combination of covectors $\theta_{i_1} \wedge \cdots \wedge \theta_{i_h}$ with $w(\theta_{i_1}) + \cdots + w(\theta_{i_h}) = p$.

In particular, the canonical volume form $dV$ has weight $Q$ (the homogeneous dimension of the group).

We have ([8], formula (16))

$$\bigwedge^h \mathfrak{g} = \bigoplus_{p=M_h^{\min}}^{M_h^{\max}} \bigwedge^{h,p} \mathfrak{g}, \tag{7}$$

where $\bigwedge^{h,p} \mathfrak{g}$ is the linear span of the $h$-covectors of weight $p$ and $M_h^{\min}$, $M_h^{\max}$ are respectively the smallest and the largest weights of left-invariant $h$-covectors.

Since the elements of the basis $\Theta^h$ have pure weights, a basis of $\bigwedge^{h,p} \mathfrak{g}$ is given by $\Theta^{h,p} := \Theta^h \cap \bigwedge^{h,p} \mathfrak{g}$. In other words, the basis $\Theta^h = \cup_p \Theta^{h,p}$ is a basis adapted to the filtration of $\bigwedge^h \mathfrak{g}$ associated with (7).

We denote by $\Omega^{h,p}$ the vector space of all smooth $h$-forms in $\mathbb{G}$ of pure weight $p$, i.e. the space of all smooth sections of $\bigwedge^{h,p} \mathfrak{g}$. We have

$$\Omega^h = \bigoplus_{p=M_h^{\min}}^{M_h^{\max}} \Omega^{h,p}.$$

**Definition 2.2** Let now $\alpha = \sum_{\theta_i^h \in \Theta^{h,p}} \alpha_i \, \theta_i^h \in \Omega^{h,p}$ be a (say) smooth form of pure weight $p$. Then we can write $d\alpha = d_0\alpha + d_1\alpha + \cdots + d_\kappa\alpha$, where $d_i$ increases the weight by $i$ with $i = 1, \ldots, \kappa$.

**Definition 2.3 (M. Rumin)** If $0 \le h \le n$ we set

$$E_0^h := \ker d_0 \cap (\mathrm{Im}\, d_0)^\perp \subset \Omega^h.$$

In the sequel, we refer to the elements of $E_0^h$ as the *intrinsic h-forms on* $\mathbb{G}$. Since the construction of $E_0^h$ is left invariant, this space of forms can be seen as the space of sections of a fiber subbundle of $\bigwedge^h \mathfrak{g}$, generated by left translations and still denoted by $E_0^h$. In particular $E_0^h$ inherits from $\bigwedge^h \mathfrak{g}$ the scalar product on the fibers.

As a consequence, we can obtain a left invariant orthonormal basis $\Xi_0^h = \{\xi_j\}$ of $E_0^h$ such that

$$\Xi_0^h = \bigcup_{p=N_h^{\min}}^{N_h^{\max}} \Xi_0^{h,p},$$

where $\Xi_0^{h,p} := \Xi^h \cap \bigwedge^{h,p} \mathfrak{g}$ is a left invariant orthonormal basis of $E_0^{h,p}$. All the elements of $\Xi_0^{h,p}$ have pure weight $p$. Without loss of generality, the indices $j$ of $\Xi_0^h = \{\xi_j^h\}$ are ordered once and for all in an increasing way according to the weight of the respective element of the basis.

Correspondingly, the set of indices $\{1, 2, \ldots, \dim E_0^h\}$ can be written as the union of finite (possibly empty) sets of indices

$$\{1, 2, \ldots, \dim E_0^h\} = \bigcup_{p=N_h^{\min}}^{N_h^{\max}} I_{0,p}^h,$$

where

$$j \in I_{0,p}^h \quad \text{if and only if} \quad \xi_j^h \in \Xi_0^{h,p}.$$

Without loss of generality, if $m := \dim V_1$, we can take

$$\Xi_0^1 = \Xi_0^{1,1} = \{dx_1, \ldots, dx_m\}.$$

Once the basis $\Theta_0^h$ is chosen, the spaces $\mathscr{E}(\Omega, E_0^h)$, $\mathscr{D}(\Omega, E_0^h)$, $\mathscr{S}(\mathbb{G}, E_0^h)$ can be identified with $\mathscr{E}(\Omega)^{\dim E_0^h}$, $\mathscr{D}(\Omega)^{\dim E_0^h}$, $S(\mathbb{G})^{\dim E_0^h}$, respectively.

The differential $d_c$ acting on $h$-forms can be identified with respect to the bases $\Xi_0^h$ and $\Xi_0^{h+1}$ with a matrix-valued differential operator $L^h := \big(L_{i,j}^h\big)$. If $j \in I_{0,p}^h$ and $i \in I_{0,q}^{h+1}$, then the $L_{i,j}^h$'s are homogeneous left invariant differential operators of order $q - p \geq 1$ in the horizontal derivatives, and $L_{i,j}^h = 0$ if $j \in I_{0,p}^h$ and $i \in I_{0,q}^{h+1}$, with $q - p < 1$ (see, e.g., [33] and [8], Sect. 2).

Analogously, $\delta_c$ can be identified, with a matrix-valued differential operator $P^h := \big(P_{i,j}^h\big)$.

We have:

$$P_{i,j}^h = \big(L_{j,i}^{h-1}\big)^*.$$

Let now $\mathscr{L} := \big(\mathscr{L}_{ji}\big)_{j,i=1,\ldots,N}$ be a differential operator on $\mathscr{E}(\mathbb{G}, \mathbb{R}^N)$ defined by

$$\mathscr{L}(\alpha_1, \ldots, \alpha_N) = \Big(\sum_i \mathscr{L}_{i1}\alpha_i, \ldots, \sum_i \mathscr{L}_{iN}\alpha_i\Big),$$

where the $\mathcal{L}_{ij}$'s are constant coefficient homogeneous polynomials of degree $a$ in $X_1, \ldots, X_m$. Due to the left invariance and the homogeneity (with respect to the group dilations) of the vector fields $X_1, \ldots, X_m$, the operator $\mathcal{L}$ is left invariant and homogeneous of degree $a$. We notice that the formal adjoint ${}^t\mathcal{L}$ of $\mathcal{L}$ is given by

$$ {}^t\mathcal{L}(\alpha_1, \ldots, \alpha_N) = \Big(\sum_i {}^t\mathcal{L}_{1i}\alpha_i, \ldots, \sum_i {}^t\mathcal{L}_{Ni}\alpha_i\Big). $$

The following result is contained in [7], Theorem 3.1.

**Theorem 2.2** *Suppose $\mathcal{L}$ is a left-invariant hypoelliptic differential operator on $\mathcal{E}(\mathbb{G}, \mathbb{R}^N)$ such that ${}^t\mathcal{L} = \mathcal{L}$. Suppose also that $\mathcal{L}$ is homogeneous of degree $a < Q$. Then for $j = 1, \ldots, N$ there exists*

$$ K_j = \big(K_{1j}, \ldots, K_{Nj}\big) $$

*with $K_{ij} \in \mathcal{D}'(\mathbb{G}) \cap \mathcal{E}(\mathbb{G} \setminus \{0\})$, $i, j = 1, \ldots, N$ such that*

*(i) We have*

$$ \sum_i \mathcal{L}_{i\ell} K_{ij} = \begin{cases} \delta \text{ if } \ell = j \\ 0 \text{ if } \ell \neq j. \end{cases} $$

*(ii) The $K_{ij}$'s are kernels of type $a$ in the sense of [17], for $i, j = 1, \ldots, N$ (i.e. they are smooth functions outside of the origin, homogeneous of degree $a - Q$, and hence belonging to $L^1_{\mathrm{loc}}(\mathbb{G})$, by Corollary 1.7 of [17]).*

*(iii) When $\alpha \in \mathcal{D}(\mathbb{G}, \mathbb{R}^N)$, if we set*

$$ \mathcal{K}\alpha := \Big(\sum_j \alpha_j * K_{1j}, \ldots, \sum_j \alpha_j * K_{Nj}\Big), $$

*then $\mathcal{L}\mathcal{K}\alpha = \alpha$ and $\mathcal{K}\mathcal{L}\alpha = \alpha$.*

## 3 Engel Group

The first Engel group is a three-step Carnot group whose Lie algebra is given by $\mathfrak{g} = V_1 \oplus V_2 \oplus V_3$, where

$$ V_1 = \operatorname{span}\{X_1, X_2\}\,,\; V_2 = \operatorname{span}\{T_1 = [X_1, X_2]\}\,,\; V_3 = \operatorname{span}\{T_2 = [X_1, T_1]\} $$

and thus will be realized on $\mathbb{R}^4$. In exponential coordinates an explicit representation of the vector fields is

$$X_1 = \partial_1 - \frac{x_2}{2}\partial_3 - (\frac{x_3}{2} + \frac{x_1 x_2}{12})\partial_4 \ , \ X_2 = \partial_2 + \frac{x_1}{2}\partial_3 + \frac{x_1^2}{12}\partial_4$$
$$T_1 = \partial_3 + \frac{x_1}{2}\partial_4 \ , \ T_2 = \partial_4.$$

The Lie algebra $\mathfrak{g}$ is endowed with a scalar product making $\{X_1, X_2, T_1, T_2\}$ an orthonormal basis. We denote by $\{dx_1, dx_2, \tau_1, \tau_2\}$ the dual basis of $\Lambda^1\mathfrak{g}$. The forms $dx_1$, $dx_2$ have weight 1, whereas $\tau_1$ has weight 2 and $\tau_2$ has weight 3.

One can easily compute the intrinsic classes of forms $E_0 = \ker d_0 \cap (\operatorname{Im} d_0)^{\perp}$:

$E_0^0 = \Omega^0$;
$E_0^1 = \Omega^{1,1} = \operatorname{span}\{dx_1, dx_2\}$;
$E_0^2 = \operatorname{span}\{dx_2 \wedge \tau_1, dx_1 \wedge \tau_2\} \subset \Omega^{2,3} \oplus \Omega^{2,4}$;
$E_0^3 = \Omega^{3,6} = \operatorname{span}\{dx_1 \wedge \tau_1 \wedge \tau_2, dx_2 \wedge \tau_1 \wedge \tau_2\}$;
$E_0^4 = \Omega^{4,7} = \operatorname{span}\{dx_1 \wedge dx_2 \wedge \tau_1 \wedge \tau_2\}$.

We want to compute the action of the differential operator $d_c$ on $E_0^*$ as a matrix-valued operator as follows:

- $d_c : E_0^0 \longrightarrow E_0^1$ can be seen in matrix form as

$$d_c = \begin{pmatrix} L^0_{1,1} \\ L^0_{2,1} \end{pmatrix}$$

  with $L^0_{1,1} = X_1$ and $L^0_{2,1} = X_2$ .
- $d_c : E_0^1 \longrightarrow E_0^2$ can be expressed as

$$d_c = \begin{pmatrix} L^1_{1,1} & L^1_{1,2} \\ L^1_{2,1} & L^1_{2,2} \end{pmatrix}$$

  where

$$L^1_{1,1} = -X_2^2 \, , L^1_{1,2} = T_1 + X_2X_1 \, , \, L^1_{2,1} = T_2 - X_1T_1 + X_1^2X_2 \, , L^1_{2,2} = -X_1^3 \, .$$

- $d_c : E_0^2 \longrightarrow E_0^3$ is given by

$$d_c = \begin{pmatrix} L^2_{1,1} & L^2_{1,2} \\ L^2_{2,1} & L^2_{2,2} \end{pmatrix}$$

  where

$$L^2_{1,1} = X_1^3 \, , L^2_{1,2} = -T_1 - X_1X_2 \, , L^2_{2,1} = X_2X_1^2 + T_2 - T_1X_1 \, , L^2_{2,2} = -X_2^2 \, .$$

- $d_c : E_0^3 \longrightarrow E_0^4$ can be expressed as

$$d_c = \left(L^3_{1,1}\ L^3_{2,1}\right)$$

with

$$L^3_{1,1} = X_2\,,\ L^3_{2,1} = -X_1\,.$$

Analogously

- $\delta_c : E_0^1 \longrightarrow E_0^0$ is given by:

$$\delta_c(f_1 dx_1 + f_2 dx_2) = X_1 f_1 + X_2 f_2$$

so that, in matrix form we obtain

$$\delta_c = \left(P^1_{1,1}\,,\ P^1_{1,2}\right)$$

where $P^1_{1,1} = X_1$ and $P^1_{1,2} = X_2$.

- $\delta_c : E_0^2 \longrightarrow E_0^1$ has the form:

$$\delta_c(f_1 dx_2 \wedge \tau_1 + f_2 dx_1 \wedge \tau_2) = (L^2_{2,2} f_1 + L^2_{2,1} f_2) dx_1 - (L^2_{1,2} f_1 + L^2_{1,1} f_2) dx_2\,.$$

The matrix form will be then:

$$\delta_c = \begin{pmatrix} P^2_{1,1} & P^2_{1,2} \\ P^2_{2,1} & P^2_{2,2} \end{pmatrix}$$

where $P^2_{1,1} = -X_2^2\,, P^2_{1,2} = X_2 X_1^2 + T_2 - T_1 X_1\,, P^2_{2,1} = T_1 + X_1 X_2\,, P^2_{2,2} = -X_1^3\,.$

- $\delta_c : E_0^3 \longrightarrow E_0^2$ has the form

$$\begin{aligned}\delta_c(f_1 dx_1 \wedge \tau_1 \wedge \tau_2 + f_2 dx_2 \wedge \tau_1 \wedge \tau_2) =& (-L^1_{2,2} f_1 + L^1_{2,1} f_2) dx_2 \wedge \tau_1 \\ &+ (-L^1_{1,2} f_1 + L^1_{1,1} f_2) dx_1 \wedge \tau_2\,.\end{aligned}$$

The matrix form will be then:

$$\delta_c = \begin{pmatrix} P^3_{1,1} & P^3_{1,2} \\ P^3_{2,1} & P^3_{2,2} \end{pmatrix}$$

where $P^3_{1,1} = X_1^3\,, P^3_{1,2} = X_1^2 X_2 + T_2 - X_1 T_1\,, P^3_{2,1} = -T_1 - X_2 X_1\,, P^3_{2,2} = -X_2^2\,.$

We must remind now few definitions of the function spaces we need for our results.

If $p, q \in [1, \infty]$, we define the space

$$L^{p,q}(\mathbb{G}) := L^p(\mathbb{G}) \cap L^q(\mathbb{G})$$

endowed with the norm

$$\|u\|_{L^{p,q}(\mathbb{G})} := (\|u\|^2_{L^p(\mathbb{G})} + \|u\|^2_{L^q(\mathbb{G})})^{1/2}.$$

We have:

- $L^{p,q}(\mathbb{G})$ is a Banach space.
- $\mathcal{D}(\mathbb{G})$ is dense in $L^{p,q}(\mathbb{G})$.

Again if $p, q \in [1, \infty]$, we can endow the vector space $L^p(\mathbb{G}) + L^q(\mathbb{G})$ with the norm

$$\|u\|_{L^p(\mathbb{G})+L^q(\mathbb{G})} := \inf\{(\|u_1\|^2_{L^p(\mathbb{G})} + \|u_2\|^2_{L^q(\mathbb{G})})^{1/2}; \\ u_1 \in L^p(\mathbb{G}), u_2 \in L^q(\mathbb{G}), u = u_1 + u_2\}.$$

We stress that $L^p(\mathbb{G}) + L^q(\mathbb{G}) \subset L^1_{\text{loc}}(\mathbb{G})$. Analogous spaces of forms can be defined in the usual way.

The following characterization of $(L^{p,q}(\mathbb{G}))^*$ can be proved by standard arguments of functional analysis.

**Proposition 3.1** *If $p, q \in (1, \infty)$ and $p', q'$ are their conjugate exponents, then*

*(i) If $u = u_1 + u_2 \in L^{p'}(\mathbb{G}) + L^{q'}(\mathbb{G})$, with $u_1 \in L^{p'}(\mathbb{G})$ and $u_2 \in L^{q'}(\mathbb{G})$, then the map*

$$\phi \to \int_{\mathbb{G}} (u_1\phi + u_2\phi)\, dV \quad \textit{for } \phi \in L^{p,q}(\mathbb{G})$$

*belongs to $(L^{p,q}(\mathbb{G}))^*$ and $\|u\|_{L^{p'}(\mathbb{G})+L^{q'}\mathbb{G})} \geq \|F\|$.*

*(ii) If $u \in L^{p'}(\mathbb{G}) + L^{q'}(\mathbb{G})$, then there exist $u_1 \in L^{p'}(\mathbb{G})$ and $u_2 \in L^{q'}(\mathbb{G})$ such that $u = u_1 + u_2$ and $\|u\|_{L^{p'}(\mathbb{G})+L^{q'}\mathbb{G})} = (\|u_1\|^2_{L^p(\mathbb{G})} + \|u_2\|^2_{L^q(\mathbb{G})})^{1/2}$. Then the functional*

$$\phi \to F(\phi) := \int_{\mathbb{G}} (u_1\phi + u_2\phi)\, dV \quad \textit{for } \phi \in L^{p,q}(\mathbb{G})$$

*belongs to $(L^{p,q}(\mathbb{G}))^*$ and $\|F\| \approx \|u\|_{L^{p'}(\mathbb{G})+L^{q'}\mathbb{G})}$.*

*(iii) Reciprocally, if $F \in (L^{p,q}(\mathbb{G}))^*$, then there exist $u = u_1+u_2 \in L^{p'}(\mathbb{G})+L^{q'}(\mathbb{G})$, $u_1 \in L^{p'}(\mathbb{G})$ and $u_2 \in L^{q'}(\mathbb{G})$ such that*

$$F(\phi) = \int_{\mathbb{G}} (u_1\phi + u_2\phi)\, dV \quad \textit{for all } \phi \in L^{p,q}(\mathbb{G}).$$

*Moreover $\|u\|_{L^{p'}(\mathbb{G})+L^{q'}\mathbb{G})} = \|F\|$.*

We can state now our Gagliardo-Nirenberg inequality for horizontal vector fields in Engel group.

**Theorem 3.1** *There exists a constant* $C > 0$ *such that, if* $u = u_1\,dx + u_2\,dy \in \mathscr{D}(\mathbb{G}, E_0^1)$ *and we set* $d_c u = f dx_2 \wedge \tau_1 + g dx_1 \wedge \tau_2$*, then*

$$\|u\|_{L^{Q/(Q-2)}(\mathbb{G},E_0^1)+L^{Q/(Q-3)}(\mathbb{G},E_0^1)} \le C\big(\|f\|_{L^1(\mathbb{G})} + \|g\|_{L^1(\mathbb{G})} + \|\Delta_{\mathbb{G}}\delta_c u\|_{\mathscr{H}^1(\mathbb{G})}\big). \tag{8}$$

*Proof* To prove (8), first we need to define a suitable differential operator $\Delta_{\mathbb{G},1}$ on $E_0^1$ that is provided of a homogeneous fundamental solution. We set

$$\begin{aligned}
\Delta_{\mathbb{G},1} &:= \delta_c \begin{pmatrix} -\Delta_{\mathbb{G}} & 0 \\ 0 & 1 \end{pmatrix} d_c + (d_c\delta_c)^3 \\
&= \delta_c \begin{pmatrix} -\Delta_{\mathbb{G}} L^1_{1,1} & -\Delta_{\mathbb{G}} L^1_{1,2} \\ L^1_{2,1} & L^1_{2,2} \end{pmatrix} \\
&= \begin{pmatrix} -(L^1_{1,1})^*\Delta_{\mathbb{G}} L^1_{1,1} + (L^1_{2,1})^* L^1_{2,1} & -(L^1_{1,1})^*\Delta_{\mathbb{G}} L^1_{1,2} + (L^1_{2,1})^* L^1_{2,2} \\ -(L^1_{1,2})^*\Delta_{\mathbb{G}} L^1_{1,1} + (L^1_{2,2})^* L^1_{2,1} & -(L^1_{1,2})^*\Delta_{\mathbb{G}} L^1_{1,2} + (L^1_{2,2})^* L^1_{2,2} \end{pmatrix} \\
&\quad + (d_c\delta_c)^3.
\end{aligned}$$

It is easy to see that $\Delta_{\mathbb{G},1}$ is a self-adjoint non-negative left invariant differential operator that is homogeneous (with respect to group dilations) of degree $6 < 7 = Q$.

**Lemma 3.1** *The operator* $\Delta_{\mathbb{G},1}$ *is hypoelliptic.*

*Proof* We use Rockland's approach as in [15, 24, 25]. Let $\pi$ be a nontrivial irreducible unitary representation of $\mathbb{G}$. Without loss of generality, if $S_\pi$ is the space of $^\infty$ vectors of the representation, we may assume that

$$S_\pi = \mathscr{S}(\mathbb{R}^k),$$

for a suitable $k \in \mathbb{N}$. By [25] (see also [15], p. 63, Remark 5), the hypoellipticity of $\Delta_{\mathbb{G},1}$ is equivalent to the injectivity of $\pi\big(\Delta_{\mathbb{G},1}\big)$ on $S_\pi^{N_1}$.

Let now $u = (u_1, u_2) \in \big(\mathscr{S}(\mathbb{R}^k)\big)^2$ be such that

$$\pi(\delta_c)\begin{pmatrix} -\pi(\Delta_{\mathbb{G}}) & 0 \\ 0 & 1 \end{pmatrix} \pi(d_c)u + (\pi(d_c)\pi(\delta_c))^3 u = 0. \tag{9}$$

We multiply (9) by $u$ and we integrate the identity on $(\mathbb{R}^k)^{N_1}$.

We consider first the term and we integrate by parts. We obtain

$$\int_{\mathbb{G}} \langle \pi(\delta_c) \begin{pmatrix} -\pi(\Delta_{\mathbb{G}}) & 0 \\ 0 & 1 \end{pmatrix} \pi(d_c)u, u\rangle \, dV$$

$$= \int_{\mathbb{G}} \Big\{ (-\pi(\Delta_{\mathbb{G}})\pi(L^1_{1,1})u_1)(\pi(L^1_{1,1})u_1) + (\pi(L^1_{2,1})u_1)^2$$

$$+ (\pi(-\Delta_{\mathbb{G}})\pi(L^1_{1,2})u_2)(\pi(L^1_{1,1})u_1) + (\pi(L^1_{2,2})u_2)(\pi(L^1_{2,1})u_1)$$

$$+ (\pi(-\Delta_{\mathbb{G}})\pi(L^1_{1,1})u_1)(\pi(L^1_{1,2})u_2) + (\pi(L^1_{2,1})u_1)(\pi(L^1_{2,2})u_2)$$

$$+ (\pi(-\Delta_{\mathbb{G}})\pi(L^1_{1,2})u_2)(\pi(L^1_{1,2})u_2) + (\pi(L^1_{2,2})u_2)^2 \Big\} \, dV := I.$$

Consider now the term $-\pi(\Delta_{\mathbb{G}})$. We have

$$-\pi(\Delta_{\mathbb{G}}) = \pi(-X_1^2) + \pi(-X_2^2) = \pi(X_1)^*\pi(X_1) + \pi(X_2)^*\pi(X_2).$$

Hence

$$I = \int_{\mathbb{G}} \Big\{ (\pi(X_1)\pi(L^1_{1,1})u_1))^2 + (\pi(L^1_{2,1})u_1)^2$$

$$+ 2(\pi(X_1)\pi(L^1_{1,2})u_2)(\pi(X_1)\pi(L^1_{1,1})u_1) + 2(\pi(L^1_{2,2})u_2)(\pi(L^1_{2,1})u_1)$$

$$+ (\pi(X_1)\pi(L^1_{1,2})u_2)^2 + (\pi(L^1_{2,2})u_2)^2 \Big\} \, dV$$

$$+ \int_{\mathbb{G}} \Big\{ (\pi(X_2)\pi(L^1_{1,1})u_1))^2$$

$$+ 2(\pi(X_2)\pi(L^1_{1,2})u_2)(\pi(X_2)\pi(L^1_{1,1})u_1)$$

$$+ (\pi(X_2)\pi(L^1_{1,2})u_2)^2 \Big\} \, dV$$

$$= \int_{\mathbb{G}} \Big\{ (\pi(X_1)\pi(L^1_{1,2})u_2 + \pi(X_1)\pi(L^1_{1,1})u_1)^2$$

$$+ (\pi(L^1_{2,2})u_2 + \pi(L^1_{2,1})u_1)^2 \Big\} \, dV$$

$$+ \int_{\mathbb{G}} \Big\{ (\pi(X_2)\pi(L^1_{1,2})u_2 + \pi(X_2)\pi(L^1_{1,1})u_1)^2 \Big\} \, dV.$$

Therefore (9) yields

$$\pi(X_1)\pi(L^1_{1,2})u_2 + \pi(X_1)\pi(L^1_{1,1})u_1 = 0, \tag{10}$$

$$\pi(X_2)\pi(L^1_{1,2})u_2 + \pi(X_2)\pi(L^1_{1,1})u_1 = 0, \tag{11}$$

$$\pi(L^1_{2,2})u_2 + \pi(L^1_{2,1})u_1 = 0 \tag{12}$$

and

$$\pi(-\Delta_{\mathbb{G}})\pi(d_c^*)u = 0. \tag{13}$$

We apply $\pi(X_1)$ to (10) and $\pi(X_2)$ to (11). Summing up we obtain

$$\pi(\Delta_G)(\pi(L^1_{1,2})u_2 + \pi(L^1_{1,1})u_1) = 0.$$

But $\pi(\Delta_G)$ is injective, since $\Delta_G$ is hypoelliptic, by [26], and hence

$$\pi(L^1_{1,2})u_2 + \pi(L^1_{1,1})u_1 = 0. \tag{14}$$

Combining (12) and (14) we obtain

$$\pi(d_c)u = 0. \tag{15}$$

By [33], proof of Theorem 5.2, there exists $X \in \mathfrak{g}$ such that, for any $v \in (\mathscr{S}(\mathbb{R}^k))^{N_1}$,

$$v = Q_X\pi(d_c)v + \pi(d_c)Q_Xv, \tag{16}$$

where

$$Q_X := \pi(\Pi_{E_0}\Pi_E)P_Xi_X\pi(\Pi_E\Pi_{E_0}).$$

Here $P_X$ is the inverse of $\pi(\mathscr{L}_X)$, $\mathscr{L}_X$ being the Lie derivative along $X$.

Replacing (15) in (16), we get

$$u = \pi(d_c)Q_Xu. \tag{17}$$

Thus, if we replace (17) in (13), we get

$$\pi((-\Delta_{\mathbb{G}})^2)Q_Xu = 0,$$

yielding eventually $u = 0$, since $(-\Delta_{\mathbb{G}})^2$ is hypoelliptic and then $\pi((-\Delta_{\mathbb{G}})^2)$ is injective. Thus $Q_Xu = 0$ and therefore $u = 0$, by (17).

To achieve the proof of Theorem 3.1, we notice that, by Theorem 2.2, if $u \in \mathscr{D}(\mathbb{G}, E_0^1)$ and $\phi = \phi_1 dx + \phi_2 dy \in \mathscr{D}(\mathbb{G}, E_0^1)$, we can write

$$\begin{aligned}
\langle u, \phi \rangle_{L^2(\mathbb{H}^1, E_0^1)} &= \langle u, \Delta_{\mathbb{G},1} \mathscr{K} \phi \rangle_{L^2(\mathbb{H}^1, E_0^1)} \\
&= \langle d_c u, \begin{pmatrix} -\Delta_{\mathbb{G}} L^1_{1,1} & -\Delta_{\mathbb{G}} L^1_{1,2} \\ L^1_{2,1} & L^1_{2,2} \end{pmatrix} \mathscr{K} \phi \rangle_{L^2(\mathbb{G}, E_0^2)} \\
&+ \langle u, (d_c \delta_c)^3 \mathscr{K} \phi \rangle_{L^2(\mathbb{G}, E_0^1)}.
\end{aligned} \tag{18}$$

Since $d_c u$ is a closed form in $E_0^2$, we can write

$$d_c u = f\, dy \wedge \tau_1 + g\, dx \wedge \tau_2,$$

with

$$-T_1 g + X_1^3 f - X_1 X_2 g = 0 \tag{19}$$

and

$$T_2 f + X_2 X_1^2 f - X_2^2 g - T_1 X_1 f = 0. \tag{20}$$

We notice also that, if

$$\mathscr{K} \phi = (\mathscr{K} \phi)_1 dx + (\mathscr{K} \phi)_2 dy,$$

we can write

$$\begin{aligned}
&\begin{pmatrix} -\Delta_{\mathbb{G}} L^1_{1,1} & -\Delta_{\mathbb{G}} L^1_{1,2} \\ L^1_{2,1} & L^1_{2,2} \end{pmatrix} \mathscr{K} \phi \\
&= \big( - \Delta_G L^1_{1,1} (\mathscr{K} \phi)_1 - \Delta_G L^1_{1,2} (\mathscr{K} \phi)_2 \big)\, dy \wedge \tau_1 \\
&\quad + \big( L^1_{2,1} (\mathscr{K} \phi)_1 + L^1_{2,2} (\mathscr{K} \phi)_2 \big)\, dx \wedge \tau_2.
\end{aligned}$$

Thus

$$\begin{aligned}
&\langle d_c u, \begin{pmatrix} -\Delta_{\mathbb{G}} L^1_{1,1} & -\Delta_{\mathbb{G}} L^1_{1,2} \\ L^1_{2,1} & L^1_{2,2} \end{pmatrix} \mathscr{K} \phi \rangle_{L^2(\mathbb{G}, E_0^2)} \\
&= \int_{\mathbb{G}} f \big( - \Delta_G L^1_{1,1} (\mathscr{K} \phi)_1 - \Delta_G L^1_{1,2} (\mathscr{K} \phi)_2 \big)\, dV \\
&\quad + \int_{\mathbb{G}} g \big( L^1_{2,1} (\mathscr{K} \phi)_1 + L^1_{2,2} (\mathscr{K} \phi)_2 \big)\, dV := I_1 + I_2.
\end{aligned}$$

We notice now that (19) can be written as

$$-2X_1X_2g + X_2X_1g + X_1^3f = 0. \tag{21}$$

On the other hand, we can write $g = X_1g_1 + X_2g_2$, with $g_1, g_2 \in \mathscr{S}(\mathbb{G})$.
Therefore, (21) can be written as

$$-2X_1X_2X_1g_1 - 2X_1X_2^2g_2 + X_2X_1^2g_1 + X_2X_1X_2g_2 + X_1^3f = 0.$$

We denote by $F$ the differential operator

$$F := 2g_1X_1X_2X_1 + 2g_2X_1X_2^2 - g_1X_2X_1^2 - g_2X_2X_1X_2g_2 - fX_1^3$$

that, in turn, can be identified with the horizontal three-tensor

$$\begin{aligned} F :=& 2g_1X_1 \otimes X_2 \otimes X_1 + 2g_2X_1 \otimes X_2 \otimes X_2 - g_1X_2 \otimes X_1 \otimes X_1 \\ &- g_2X_2 \otimes X_1 \otimes X_2g_2 - fX_1 \otimes X_1 \otimes X_1. \end{aligned}$$

Identity (19) can be written as

$$\int_{\mathbb{G}} F\psi \, dV = 0 \quad \text{for all } \psi \in \mathscr{D}(\mathbb{G}).$$

Now we can estimate $I_1$. Consider for instance

$$\left| \int_{\mathbb{G}} f \, \Delta_G L^1_{1,1}(\mathscr{K}\phi)_1 \, dV \right|,$$

and let $\Phi$ be the symmetric three-tensor (with coefficients in $\mathscr{S}(\mathbb{G})$)

$$\Phi := \Delta_{\mathbb{G}} L^1_{1,1}(\mathscr{K}\phi)_1 X_1 \otimes X_1 \otimes X_1.$$

By Theorem 2.2,

$$\Delta_{\mathbb{G}} L^1_{1,1}(\mathscr{K}\phi)_1 = \Delta_{\mathbb{G}} L^1_{1,1}\mathscr{K}_{11}\phi_1 + \Delta_{\mathbb{G}} L^1_{1,1}\mathscr{K}_{12}\phi_2,$$

where $\Delta_{\mathbb{G}} L^1_{1,1}\mathscr{K}_{11}$ and $\Delta_{\mathbb{G}} L^1_{1,1}\mathscr{K}_{12}$ are convolution operators associated with kernel of type 2. Thus by Theorem 2.1 and [17], Proposition 1.11,

$$\begin{aligned} \left| \int_{\mathbb{G}} f \, \Delta_{\mathbb{G}} L^1_{1,1}(\mathscr{K}\phi)_1 \, dV \right| &= \int_{\mathbb{G}} \langle F, \Phi \rangle \, dV \\ &\le C\big(\|f\|_{L^1(\mathbb{G})} + \|g\|_{L^1(\mathbb{G})}\big) \cdot \big(\|\nabla_{\mathbb{G}} \Delta_{\mathbb{G}} L^1_{1,1}\mathscr{K}_{11}\phi_1\|_{L^Q(\mathbb{G})} + \|\nabla_{\mathbb{G}} \Delta_{\mathbb{G}} L^1_{1,1}\mathscr{K}_{12}\phi_1\|_{L^Q(\mathbb{G})}\big) \\ &\le C\big(\|f\|_{L^1(\mathbb{G})} + \|g\|_{L^1(\mathbb{G})}\big)\|\phi\|_{L^{Q/2}(\mathbb{G}, E_0^1)}. \end{aligned}$$

The same argument can be repeated for all terms that appear in $I_1$ and we obtain

$$\begin{aligned} I_1 &\le C\big(\|f\|_{L^1(\mathbb{G})} + \|g\|_{L^1(\mathbb{G})}\big)\|\phi\|_{L^{Q/2}(\mathbb{G},E_0^1)} \\ &\le C\big(\|f\|_{L^1(\mathbb{G})} + \|g\|_{L^1(\mathbb{G})}\big)\|\phi\|_{L^{Q/2,Q/3}(\mathbb{G},E_0^1)}. \end{aligned} \tag{22}$$

In order to estimate $I_2$, we note first that (20) can be written as

$$X_1^2X_2f - 3X_1X_2X_1f + 2X_2X_1^2f - X_2^2g = 0.$$

We denote by $G$ the differential operator

$$G := (X_2f)X_1^2 - 3(X_1f)X_2X_1 + 2(X_1f)X_1X_2 - gX_2^2$$

that, in turn, can be identified with the horizontal two-tensor

$$G := (X_2f)X_1 \otimes X_1 - 3(X_1f)X_2 \otimes X_1 + 2(X_1f)X_1 \otimes X_2 - gX_2 \otimes X_2.$$

Identity (20) can be written as

$$\int_{\mathbb{G}} G\psi \, dV = 0 \quad \text{for all } \psi \in \mathscr{D}(\mathbb{G}).$$

Since the function $g$ appears as the coefficient of the symmetric tensor $X_2 \otimes X_2$, we can try to repeat all the arguments yielding to the previous estimate of $I_1$. However, we are facing a crucial difference: when handling, e.g., the first term in $I_2$, the terms $\Delta_{\mathbb{G}}L^1_{1,1}\mathscr{K}_{11}$ and $\Delta_{\mathbb{G}}L^1_{1,1}\mathscr{K}_{12}$ (that are convolution operators associated with kernels of type 2) are now replaced by $L^1_{2,1}\mathscr{K}_{11}$ and $L^1_{2,1}\mathscr{K}_{12}$ that are convolution operators associated with kernels of type 3. Thus, we are led to the estimate

$$\begin{aligned} I_2 &\le C\big(\|f\|_{L^1(\mathbb{G})} + \|g\|_{L^1(\mathbb{G})}\big)\|\phi\|_{L^{Q/3}(\mathbb{G},E_0^1)} \\ &\le C\big(\|f\|_{L^1(\mathbb{G})} + \|g\|_{L^1(\mathbb{G})}\big)\|\phi\|_{L^{Q/2,Q/3}(\mathbb{G},E_0^1)}. \end{aligned} \tag{23}$$

Combining (22) and (23) we obtain an estimate of the first term of the right-hand side of (18).

Let us proceed to consider the second term of the right-hand side of (18). We can write

$$\begin{aligned} \langle u, (d_c\delta_c)^3\mathscr{K}\phi\rangle_{L^2(\mathbb{G},E_0^1)} &= \langle \delta_c d_c\delta_c u, \delta_c d_c\delta_c u\mathscr{K}\phi\rangle_{L^2(\mathbb{G})} \\ &= \langle \delta_c d_c\delta_c u, \delta_c d_c\delta_c u((\phi_1 * K_{1,1})dx + (\phi_1 * K_{2,1})dy)\rangle_{L^2(\mathbb{G})} \\ &+ \langle \delta_c d_c\delta_c u, \delta_c d_c\delta_c u((\phi_2 * K_{1,2})dx + (\phi_1 * K_{2,2})dy)\rangle_{L^2(\mathbb{G})}. \end{aligned}$$

By (5), the last term is a sum of terms of the form

$$\langle \delta_c d_c \delta_c u, \phi_j * LK_{i,j}) \rangle_{L^2(\mathbb{G})},$$

where $L$ is a left invariant differential operator of order 3 in the horizontal derivatives. On the other hand, by (6),

$$\begin{aligned}\Big| \langle \delta_c d_c \delta_c u, \phi_j * LK_{i,j}) \rangle_{L^2(\mathbb{G})} \Big| &= \Big| \langle \delta_c d_c \delta_c u * {}^{\mathrm{v}}LK_{i,j}, \phi_j) \rangle_{L^2(\mathbb{G})} \Big| \\ &\le \| \delta_c d_c \delta_c u * {}^{\mathrm{v}}LK_{i,j} \|_{L^{Q/(Q-3)}(\mathbb{G})} \| \phi_j \|_{L^{Q/3}(\mathbb{G})} \\ &\le C \| \delta_c d_c \delta_c u \|_{\mathscr{H}^1(\mathbb{G})} \| \phi_j \|_{L^{Q/3}(\mathbb{G})},\end{aligned} \tag{24}$$

by Theorem 6.10 in [18], since ${}^{\mathrm{v}}LK_{i,j}$ is a kernel of type 3, by Proposition 2.1.

Combining (18), (22), (23) and (24) we achieve the proof of the theorem by a duality argument.

*Remark 3.1* If $d_c u = g dx_1 \wedge \tau_2$ (in particular if $u$ is closed), it follows from the proof of the previous theorem that (8) may be improved as follows:

$$\|u\|_{L^{Q/(Q-3)}(\mathbb{G},E_0^1)} \le C\big(\|g\|_{L^1(\mathbb{G})} + \|\Delta_{\mathbb{G}} \delta_c u\|_{\mathscr{H}^1(\mathbb{G})}\big).$$

## 4 The Seven-Dimensional Quaternionic Heisenberg Group

Let $\mathbb{H}$ be the space of quaternionic numbers and let $\mathbf{i}, \mathbf{j}, \mathbf{k}$ be three imaginary units such that

$$\mathbf{i}^2 = \mathbf{j}^2 = \mathbf{k}^2 = \mathbf{i}{\cdot}\mathbf{j}{\cdot}\mathbf{k} = -1.$$

The quaternionic Heisenberg group (in dimension 7) is a nilpotent Lie group with underlying manifold $\mathbb{R}^4 \times \mathbb{R}^3$, where the group structure is given by:

$$[x, t] * [y, s] = [x + y, t + s + \frac{1}{2}\mathrm{Im}(\bar{y}x)]$$

where $x, y \in \mathbb{H} \cong \mathbb{R}^4$ and $t, s \in \mathrm{Im}(\mathbb{H}) \cong \mathbb{R}^3$, where one identifies $x = x_1 + x_2\mathbf{i} + x_3\mathbf{j} + x_4\mathbf{k}$, with $x = (x_1, x_2, x_3, x_4)$ and $t = t_1\mathbf{i} + t_2\mathbf{j} + t_3\mathbf{k}$ with $t = (t_1, t_2, t_3)$.

A basis for the Lie algebra of left-invariant vector fields on the group is given by:

$$X_1 = \frac{\partial}{\partial x_1} + \frac{1}{2}x_2\frac{\partial}{\partial t_1} + \frac{1}{2}x_3\frac{\partial}{\partial t_2} + \frac{1}{2}x_4\frac{\partial}{\partial t_3};$$

$$X_2 = \frac{\partial}{\partial x_2} - \frac{1}{2}x_1\frac{\partial}{\partial t_1} + \frac{1}{2}x_4\frac{\partial}{\partial t_2} - \frac{1}{2}x_3\frac{\partial}{\partial t_3};$$

$$X_3 = \frac{\partial}{\partial x_3} - \frac{1}{2}x_4\frac{\partial}{\partial t_1} - \frac{1}{2}x_1\frac{\partial}{\partial t_2} + \frac{1}{2}x_2\frac{\partial}{\partial t_3};$$
$$X_4 = \frac{\partial}{\partial x_4} + \frac{1}{2}x_3\frac{\partial}{\partial t_1} - \frac{1}{2}x_2\frac{\partial}{\partial t_2} - \frac{1}{2}x_1\frac{\partial}{\partial t_3};$$
$$T_k = \frac{\partial}{\partial t_k} \text{ for } k = 1,2,3\,.$$

The non-trivial commutation relations are:

$$[X_1,X_2] = -[X_3,X_4] = -T_1\,;\; [X_1,X_3] = [X_2,X_4] = -T_2;$$
$$[X_1,X_4] = -[X_2,X_3] = -T_3\,.$$

The standard quaternionic contact forms $\tau_1$, $\tau_2$, $\tau_3$ are given by:

$$\tau_1 = \mathrm{d}t_1 - \frac{1}{2}\,x_2\mathrm{d}x_1 + \frac{1}{2}\,x_1\mathrm{d}x_2 - \frac{1}{2}\,x_4\mathrm{d}x_3 + \frac{1}{2}\,x_3\mathrm{d}x_4;$$
$$\tau_2 = \mathrm{d}t_2 - \frac{1}{2}\,x_3\mathrm{d}x_1 + \frac{1}{2}\,x_4\mathrm{d}x_2 + \frac{1}{2}\,x_1\mathrm{d}x_3 - \frac{1}{2}\,x_2\mathrm{d}x_4;$$
$$\tau_1 = \mathrm{d}t_3 - \frac{1}{2}\,x_4\mathrm{d}x_1 - \frac{1}{2}\,x_3\mathrm{d}x_2 + \frac{1}{2}\,x_2\mathrm{d}x_3 + \frac{1}{2}\,x_1\mathrm{d}x_4\,.$$

So that:

$$\mathrm{d}\tau_1 = -\mathrm{d}x_1 \wedge \mathrm{d}x_2 - \mathrm{d}x_3 \wedge \mathrm{d}x_4;$$
$$\mathrm{d}\tau_2 = -\mathrm{d}x_1 \wedge \mathrm{d}x_3 + \mathrm{d}x_2 \wedge \mathrm{d}x_4;$$
$$\mathrm{d}\tau_3 = -\mathrm{d}x_1 \wedge \mathrm{d}x_4 - \mathrm{d}x_2 \wedge \mathrm{d}x_3\,.$$

The space of intrinsic one-forms and two-forms are

$$E_0^1 = \Omega^{1,1} = \mathrm{span}\{dx_1, dx_2, dx_3, dx_4\},$$

and

$$E_0^2 = \mathrm{span}\{\alpha_2,\alpha_4,\alpha_6\} \oplus \mathrm{span}\{\beta_1,\beta_2,\beta_3,\beta_4,\beta_5,\beta_6,\beta_7,\beta_8\}\,,$$

where

$$\alpha_1 := dx_1 \wedge dx_2 + dx_3 \wedge dx_4\;,\; \alpha_2 := dx_1 \wedge dx_2 - dx_3 \wedge dx_4\;,$$
$$\alpha_3 := dx_1 \wedge dx_3 - dx_2 \wedge dx_4\;,\; \alpha_4 := dx_1 \wedge dx_3 + dx_2 \wedge dx_4\;,$$
$$\alpha_5 := dx_1 \wedge dx_4 + dx_2 \wedge dx_3\;,\; \alpha_6 := dx_1 \wedge dx_4 - dx_2 \wedge dx_3\;,$$

and

$$\beta_1 := dx_1 \wedge \tau_2 + dx_4 \wedge \tau_1\,,\ \beta_2 := dx_2 \wedge \tau_3 + dx_4 \wedge \tau_1,\ \beta_3 := dx_1 \wedge \tau_3 + dx_2 \wedge \tau_2\,,$$

$$\beta_4 := dx_3 \wedge \tau_1 + dx_2 \wedge \tau_2,\ \beta_5 := dx_1 \wedge \tau_1 + dx_3 \wedge \tau_3\,,\ \beta_6 := dx_4 \wedge \tau_2 + dx_3 \wedge \tau_3\,,$$

$$\beta_7 := dx_2 \wedge \tau_1 - dx_4 \wedge \tau_3\,,\ \beta_8 := -dx_3 \wedge \tau_2 + dx_4 \wedge \tau_3,\ \beta_9 := dx_1 \wedge \tau_2 - dx_4 \wedge \tau_1\,,$$

$$\beta_{10} := dx_1 \wedge \tau_3 - dx_2 \wedge \tau_2\,,\ \beta_{11} := dx_1 \wedge \tau_1 - dx_3 \wedge \tau_3,\ \beta_{12} := dx_2 \wedge \tau_1 + dx_4 \wedge \tau_3\,,$$

respectively.

We want to compute the action of the differential operator $d_c$ on $E_0^*$ as a matrix-valued operator as follows:

- $d_c : E_0^0 \longrightarrow E_0^1$ can be seen in matrix form as

$$d_c = \begin{pmatrix} L^0_{1,1} \\ L^0_{2,1} \\ L^0_{3,1} \\ L^0_{4,1} \end{pmatrix}$$

with $L^0_{1,1} = X_1\,, L^0_{2,1} = X_2\,, L^0_{3,1} = X_3$ and $L^0_{4,1} = X_4\,.$

- $d_c : E_0^1 \longrightarrow E_0^2$ can be expressed as

$$d_c = \begin{pmatrix} Q^1_{1,1} & Q^1_{1,2} & Q^1_{1,3} & Q^1_{1,4} \\ Q^1_{2,1} & Q^1_{2,2} & Q^1_{2,3} & Q^1_{2,4} \\ Q^1_{3,1} & Q^1_{3,2} & Q^1_{3,3} & Q^1_{3,4} \\ Q^1_{4,1} & Q^1_{4,2} & Q^1_{4,3} & Q^1_{4,4} \\ L^1_{1,1} & L^1_{1,2} & L^1_{1,3} & L^1_{1,4} \\ L^1_{2,1} & L^1_{2,2} & L^1_{2,3} & L^1_{2,4} \\ L^1_{3,1} & L^1_{3,2} & L^1_{3,3} & L^1_{3,4} \\ L^1_{4,1} & L^1_{4,2} & L^1_{4,3} & L^1_{4,4} \\ L^1_{5,1} & L^1_{5,2} & L^1_{5,3} & L^1_{5,4} \\ L^1_{6,1} & L^1_{6,2} & L^1_{6,3} & L^1_{6,4} \\ L^1_{7,1} & L^1_{7,2} & L^1_{7,3} & L^1_{7,4} \\ L^1_{8,1} & L^1_{8,2} & L^1_{8,3} & L^1_{8,4} \end{pmatrix}$$

where $Q^1_{1,1} = \frac{1}{2}X_2$, $Q^1_{1,2} = -\frac{1}{2}X_1$, $Q^1_{1,3} = -\frac{1}{2}X_4$, $Q^1_{1,4} = \frac{1}{2}X_3$, $Q^1_{2,1} = \frac{1}{2}X_3$, $Q^1_{2,2} = \frac{1}{2}X_4$, $Q^1_{2,3} = -\frac{1}{2}X_1$, $Q^1_{2,4} = -\frac{1}{2}X_2$, $Q^1_{3,1} = \frac{1}{2}X_4$, $Q^1_{3,2} = -\frac{1}{2}X_3$, $Q^1_{3,3} = \frac{1}{2}X_2$, $Q^1_{3,4} = -\frac{1}{2}X_1$, $L^1_{1,1} = \frac{1}{4}(T_2 - X_1X_3)$, $L^1_{1,2} = \frac{1}{4}(3X_1X_4 - X_4X_1 + X_2X_3)$, $L^1_{1,3} = \frac{1}{4}(X_1^2 - X_2^2 - X_4^2)$, $L^1_{1,4} = \frac{1}{4}(X_4X_3 - 3X_1X_2 + X_2X_1)$, $L^1_{2,1} = -\frac{1}{2}X_2X_4$, $L^1_{2,2} = \frac{1}{2}(T_3 - X_3X_2)$, $L^1_{2,3} = \frac{1}{2}X_2^2$, $L^1_{2,4} = \frac{1}{2}X_2X_1$, $L^1_{3,1} = \frac{1}{4}(T_3 - X_1X_4)$, $L^1_{3,2} = \frac{1}{4}(X_2X_4 - 3X_1X_3 + X_3X_1)$, $L^1_{3,3} = \frac{1}{4}(X_3X_4 + 3X_1X_2 - X_2X_1)$, $L^1_{3,4} = \frac{1}{4}(X_1^2 - X_2^2 - X_3^2)$, $L^1_{4,1} = -\frac{1}{2}X_3X_2$, $L^1_{4,2} = \frac{1}{2}X_3X_1$, $L^1_{4,3} = \frac{1}{2}(T_1 - X_4X_3)$, $L^1_{4,4} = \frac{1}{2}X_3^2$, $L^1_{5,1} =$

$\frac{1}{4}(T_1 - X_1X_2)$, $L^1_{5,2} = \frac{1}{4}(X_1^2 - X_3^2 - X_4^2)$, $L^1_{5,3} = \frac{1}{4}(X_3X_2 - 3X_1X_4 + X_4X_1)$, $L^1_{5,4} = \frac{1}{4}(X_4X_2 - X_1X_3 + 3X_3X_1)$, $L^1_{6,1} = -\frac{1}{2}X_4X_3$, $L^1_{6,2} = \frac{1}{2}X_4^2$, $L^1_{6,3} = \frac{1}{2}X_4X_1$, $L^1_{6,4} = \frac{1}{2}(T_2 - X_2X_4)$, $L^1_{7,1} = \frac{1}{4}(-X_2^2 + X_3^2 + X_4^2)$, $L^1_{7,2} = \frac{1}{4}(T_1 + X_2X_1)$, $L^1_{7,3} = \frac{1}{4}(X_2X_4 - 3X_4X_2 - X_3X_1)$, $L^1_{7,4} = \frac{1}{4}(3X_3X_2 - X_2X_3 - X_4X_1)$, $L^1_{8,1} = -\frac{1}{2}X_3^2$, $L^1_{8,2} = \frac{1}{2}X_3X_4$, $L^1_{8,3} = \frac{1}{2}(X_1X_3 - T_2)$ and $L_{8,4} = -\frac{1}{2}X_3X_2$.

- $\delta_c : E_0^1 \longrightarrow E_0^0$ is given by:

$$\delta_c(f_1dx_1 + f_2dx_2 + f_3dx_3 + f_4dx_4) = -X_1f_1 - X_2f_2 - X_3f_3 - X_4f_4$$

so that, in matrix form we obtain

$$\delta_c = \left(P^1_{1,1}\,,\, P^1_{1,2}\,,\, P^1_{1,3}\,,\, P^1_{1,4}\right)$$

where $P^1_{1,1} = -X_1$, $P^1_{1,2} = -X_2$, $P^1_{1,3} = -X_3$ and $P^1_{1,4} = -X_4$.

- $\delta_c : E_0^2 \longrightarrow E_0^1$ can be expressed in matrix form as:

$$\delta_c = \left(P^2_1\,,\, P^2_2\right)$$

where

$$P^2_1 = \begin{pmatrix} -Q^5_{4,1} & -Q^5_{4,2} & -Q^5_{4,3} \\ Q^5_{3,1} & Q^5_{3,2} & Q^5_{3,3} \\ -Q^5_{2,1} & -Q^5_{2,2} & -Q^5_{2,3} \\ Q^5_{1,1} & Q^5_{1,2} & Q^5_{1,3} \end{pmatrix}$$

with $Q^5_{1,1} = -X_3$, $Q^5_{1,2} = X_2$, $Q^5_{1,3} = X_1$, $Q^5_{2,1} = -X_4$, $Q^5_{2,2} = -X_1$, $Q^5_{2,3} = X_2$, $Q^5_{3,1} = X_1$, $Q^5_{3,2} = -X_4$, $Q^5_{3,3} = X_3$, $Q^5_{4,1} = X_2$, $Q^5_{4,2} = X_3$, $Q^5_{4,3} = X_4$ and

$$P^2_2 = \begin{pmatrix} L^5_{4,1} & L^5_{4,2} & -L^5_{4,3} & -L^5_{4,4} & -L^5_{4,5} & -L^5_{4,6} & L^5_{4,7} & L^5_{4,8} \\ -L^5_{3,1} & -L^5_{3,2} & L^5_{3,3} & L^5_{3,4} & L^5_{3,5} & L^5_{3,6} & -L^5_{3,7} & -L^5_{3,8} \\ L^5_{2,1} & L^5_{2,2} & -L^5_{2,3} & -L^5_{2,4} & -L^5_{2,5} & -L^5_{2,6} & L^5_{2,7} & L^5_{2,8} \\ -L^5_{1,1} & -L^5_{1,2} & L^5_{1,3} & L^5_{1,4} & L^5_{1,5} & L^5_{1,6} & -L^5_{1,7} & -L^5_{1,8} \end{pmatrix}$$

with $L^5_{1,1} = T_1 + \frac{1}{2}(X_2X_1 - X_3X_4)$, $L^5_{1,2} = T_1 - \frac{1}{2}(X_1X_3 + X_3X_4)$, $L^5_{1,3} = \frac{1}{2}(X_1^2 - X_2^2)$, $L^5_{1,4} = \frac{1}{2}(X_3^2 - X_2^2)$, $L^5_{1,5} = \frac{1}{2}(X_1X_3 + X_3X_1)$, $L^5_{1,6} = -T_2 + \frac{1}{2}(X_1X_3 - X_2X_4)$, $L^5_{1,7} = -T_3 + \frac{1}{2}(X_1X_4 - X_3X_2)$, $L^5_{1,8} = T_3 - \frac{1}{2}(X_1X_4 + X_2X_3)$, $L^5_{2,1} = \frac{1}{2}(X_1^2 - X_4^2)$, $L^5_{2,2} = \frac{1}{2}(X_2^2 - X_4^2)$, $L^5_{2,3} = -\frac{1}{2}(X_1X_2 + X_2X_1)$, $L^5_{2,4} = T_1 + \frac{1}{2}(X_4X_3 - X_1X_2)$, $L^5_{2,5} = T_3 + \frac{1}{2}(X_4X_1 - X_2X_3)$, $L^5_{2,6} = T_3 - \frac{1}{2}(X_1X_4 + X_2X_3)$, $L^5_{2,7} = -\frac{1}{2}(X_2X_4 + X_4X_2)$, $L^5_{2,8} = T_2 + \frac{1}{2}(X_2X_4 - X_1X_3)$, $L^5_{3,1} = -\frac{1}{2}(X_1X_4 + X_4X_1)$, $L^5_{3,2} = T_3 + \frac{1}{2}(X_3X_2 - X_1X_4)$, $L^5_{3,3} = -T_2 + \frac{1}{2}(X_4X_2 - X_3X_1)$, $L^5_{3,4} = -T_2 + \frac{1}{2}(X_4X_2 + X_1X_3)$, $L^5_{3,5} = \frac{1}{2}(X_1^2 - X_3^2)$, $L^5_{3,6} = \frac{1}{2}(X_4^2 - X_3^2)$, $L^5_{3,7} = T_1 - \frac{1}{2}(X_1X_2 + X_3X_4)$, $L^5_{3,8} =$

$\frac{1}{2}(X_3X_4+X_4X_3)$, $L^5_{4,1} = -T_2-\frac{1}{2}(X_2X_4+X_3X_1)$, $L^5_{4,2} = -\frac{1}{2}(X_4X_2+X_2X_4)$, $L^5_{4,3} = T_3 + \frac{1}{2}(X_4X_1 + X_3X_2)$, $L^5_{4,4} = \frac{1}{2}(X_3X_2 + X_2X_3)$, $L^5_{4,5} = T_1 + \frac{1}{2}(X_2X_1 + X_4X_3)$, $L^5_{4,6} = \frac{1}{2}(X_3X_4 + X_4X_3)$, $L^5_{4,7} = \frac{1}{2}(X_4^2 - X_2^2)$, $L^5_{4,8} = \frac{1}{2}(X_3^2 - X_4^2)$.

We can state now our Gagliardo-Nirenberg inequality for horizontal vector fields in the seven-dimensional quaternionic Heisenberg group.

**Theorem 4.1** *There exists a constant $C > 0$ such that, if*

$$u = u_1\,dx_1 + u_2\,dx_2 + u_3\,dx_3 + u_4\,dx_4 \in \mathcal{D}(\mathbb{G}, E_0^1)$$

*and we set*

$$d_c u := \sum_{i=1}^{3} f_i\alpha_i + \sum_{j=1}^{8} g_i\beta_i,$$

*then*

$$\|u\|_{L^{Q/(Q-1)}(\mathbb{G},E_0^1)+L^{Q/(Q-2)}(\mathbb{G},E_0^1)} \le C\Big(\sum_{i=1}^{3}\|f_i\|_{L^1(\mathbb{G})} + \sum_{j=1}^{8}\|g_j\|_{L^1(\mathbb{G})} + \|\Delta_{\mathbb{G}}\delta_c u\|_{H^1(\mathbb{G})}\Big).$$

*Proof* Thanks to the explicit form of the intrinsic differential $d_c$ given above, we can repeat more or less the proof of Theorem 3.1. In particular, it is crucial to provide preliminarily a suitable differential operator $\Delta_{\mathbb{G},1}$ on $E_0^1$ with a homogeneous fundamental solution. Let us set:

$$\Delta_{\mathbb{G},1} := \delta_c \begin{pmatrix} -\Delta_{\mathbb{G}} \cdot Id_{3\times3} & 0 \\ 0 & Id_{8\times8} \end{pmatrix} d_c + (d_c\delta_c)^2 .$$

It is easy to check that $\Delta_{\mathbb{G},1}$ is a hypoelliptic operator and then Theorem 2.2 applies.

**Acknowledgements** Annalisa Baldi and Bruno Franchi are supported by MIUR, Italy and by University of Bologna, Italy, funds for selected research topics, by GNAMPA of INdAM and by EC project CG-DICE.

Francesca Tripaldi is supported by the School of Natural and Mathematical Sciences of King's College London through a GTA scheme.

Francesca Tripaldi would also like to thank Prof. Pierre Pansu for his invaluable support during the preparation of her Master Thesis.

## References

1. Baldi, A., Franchi, B.: Maxwell's equations in anisotropic media and Carnot groups as variational limits. Adv. Nonlinear Stud. **15**, 325–354 (2015)
2. Baldi, A., Franchi, B.: Some remarks on vector potentials for Maxwell's equations in space-time Carnot groups. Boll. Unione Mat. Ital. (9) **5**(2), 337–355 (2012)

3. Baldi, A., Franchi, B.: Sharp a priori estimates for div-curl systems in Heisenberg groups. J. Funct. Anal. **265**(10), 2388–2419 (2013)
4. Baldi, A., Bernabei, M., Franchi, B.: A recursive basis for primitive forms in symplectic spaces and applications to Heisenberg groups. To appear in Acta Math. Sin. (Engl. Ser.)
5. Baldi, A., Franchi, B., Pansu, P. : Gagliardo-Nirenberg inequalities for differential forms in Heisenberg groups. hal - 01140910, arXiv 1506.05904
6. Baldi, A., Franchi, B., Tesi, M.C.: Compensated compactness in the contact complex of Heisenberg groups. Indiana Univ. Math. J. **57**, 133–186 (2008)
7. Baldi, A., Franchi, B., Tesi, M.C.: Hypoellipticity, fundamental solution and Liouville type theorem for matrix-valued differential operators in Carnot groups. J. Eur. Math. Soc. **11**(4), 777–798 (2009)
8. Baldi, A., Franchi, B., Tchou, N., Tesi, M.C.: Compensated compactness for differential forms in Carnot groups and applications. Adv. Math. **223**(5), 1555–1607 (2010)
9. Bonfiglioli, A., Lanconelli, E., Uguzzoni, F: Stratified Lie Groups and Potential Theory for Their Sub-Laplacians. Springer Monographs in Mathematics. Springer, Berlin (2007)
10. Bourgain, J., Brezis, H: On the equation div $Y = f$ and application to control of phases. J. Am. Math. Soc. **16**(2), 393–426 (2003)
11. Bourgain, J., Brezis, H: New estimates for the Laplacian, the div-curl, and related Hodge systems. C. R. Math. Acad. Sci. Paris **338**(7), 539–543 (2004)
12. Bourgain, J., Brezis, H: New estimates for elliptic equations and Hodge type systems. J. Eur. Math. Soc. **9**(2), 277–315 (2007)
13. Capogna, L., Danielli, D., Garofalo, N.: The geometric Sobolev embedding for vector fields and the isoperimetric inequality. Commun. Anal. Geom. **2**(2), 203–215 (1994)
14. Chanillo, S., Van Schaftingen, J.: Subelliptic Bourgain-Brezis estimates on groups. Math. Res. Lett. **16**(3), 487–501 (2009)
15. Christ, M., Geller, D., Głowacki, P., Polin, L.: Pseudodifferential operators on groups with dilations. Duke Math. J. **68**(1), 31–65 (1992)
16. Federer, H.: Geometric Measure Theory. Die Grundlehren der Mathematischen Wissenschaften, Band 153. Springer, New York (1969)
17. Folland, G.B.: Subelliptic estimates and function spaces on nilpotent Lie groups. Ark. Mat. **13**(2), 161–207 (1975)
18. Folland, G.B., Stein, E.M.: Hardy Spaces on Homogeneous Groups. Mathematical Notes, vol. 28. Princeton University Press, Princeton (1982)
19. Franchi, B., Tesi, M.C.: Wave and Maxwell's equations in Carnot groups. Commun. Contemp. Math. **14**(5), 1250032, 62 (2012)
20. Franchi, B., Gallot, S., Wheeden, R.L.: Sobolev and isoperimetric inequalities for degenerate metrics. Math. Ann. **300**(4), 557–571 (1994)
21. Franchi, B., Lu, G., Wheeden, R.L.: Representation formulas and weighted Poincaré inequalities for Hörmander vector fields. Ann. Inst. Fourier (Grenoble) **45**(2), 577–604 (1995)
22. Franchi, B., Obrecht, E., Vecchi, E.: On a class of semilinear evolution equations for vector potentials associated with Maxwell's equations in Carnot groups. Nonlinear Anal. **90**, 56–69 (2013)
23. Garofalo, N., Nhieu, D.-M.: Isoperimetric and Sobolev inequalities for Carnot-Carathéodory spaces and the existence of minimal surfaces. Commun. Pure Appl. Math. **49**(10), 1081–1144 (1996)
24. Głowacki, P.: The Rockland condition for nondifferential convolution operators, II. Stud. Math. **98**(2), 99–114 (1991)
25. Helffer, B., Nourrigat, J: Hypoellipticité Maximale Pour des Opérateurs Polynômes de Camps de Vecteurs. Progress in Mathematics, vol. 58. Birkhäuser, Boston (1985)
26. Hörmander, L.: Hypoelliptic second order differential equations. Acta Math. **119**, 147–171 (1967)
27. Hounie, J., Picon, T.: Local Gagliardo-Nirenberg estimates for elliptic systems of vector fields. Math. Res. Lett. **18**(4), 791–804 (2011)
28. Lanzani, L., Stein, E.M.: A note on div curl inequalities. Math. Res. Lett. **12**(1), 57–61 (2005)

29. Maheux, P., Saloff-Coste, L.: Analyse sur les boules d'un opérateur sous-elliptique. Math. Ann. **303**(4), 713–740 (1995)
30. Pansu, P.: Géomedu group de Heisenberg. Ph.D. thesis, Université Paris VII (1982)
31. Rumin, M.: Formes différentielles sur les variétés de contact. J. Differ. Geom. **39**(2), 281–330 (1994)
32. Rumin, M.: Differential geometry on C-C spaces and application to the Novikov-Shubin numbers of nilpotent Lie groups. C. R. Acad. Sci. Paris Sér. I Math. **329**(11), 985–990 (1999)
33. Rumin, M.: Around heat decay on forms and relations of nilpotent Lie groups. In: Séminaire de Théorie Spectrale et Géométrie, Année 2000–2001, vol. 19, pp. 123–164. University of Grenoble I (2001)
34. Stein, E.M.: Harmonic Analysis: Real-Variable Methods, Orthogonality, and Oscillatory Integrals. Princeton Mathematical Series, vol. 43. Princeton University Press, Princeton (1993) [With the assistance of Timothy S. Murphy, Monographs in Harmonic Analysis, III]
35. Tripaldi, F.: Differential forms on Carnot groups. Master's thesis, School of Sciences, University of Bologna (2013)
36. Van Schaftingen, J.: Estimates for $L^1$-vector fields. C. R. Math. Acad. Sci. Paris **339**(3), 181–186 (2004)
37. Wang, Y., Yung, P.-L.: A subelliptic Bourgain-Brezis inequality. J. Eur. Math. Soc. **16**(4), 649–693 (2014)
38. Yung, P.-L.: Sobolev inequalities for $(0, q)$ forms on CR manifolds of finite type. Math. Res. Lett. **17**(1), 177–196 (2010)

# Regularity of the Free Boundary in Problems with Distributed Sources

**Daniela De Silva, Fausto Ferrari, and Sandro Salsa**

**Abstract** In this survey paper we describe some recent progress on the analysis of two phase free boundary problems governed by elliptic inhomogeneous equations. We also discuss several open questions.

**Keywords** Free boundary problems • Nonhomogeneous elliptic equations • Regularity of the free boundary

**Mathematics Subject Classification:** 35R35, 35D40

## 1 Main Definitions and Results

In this brief survey we describe new regularity results concerning two phase problems governed by elliptic equations with forcing terms. In absence of distributed sources, this theory has been developed by a number of authors (see for example [2, 13–16]) along the ideas of Caffarelli in the seminal papers [3, 5]. Here we present a new approach introduced in [8] by the first author and subsequently refined in [10–12] to cover a broad spectrum of applications.

D. De Silva
Department of Mathematics, Barnard College, Columbia University, New York, NY 10027, USA
e-mail: desilva@math.columbia.edu

F. Ferrari
Dipartimento di Matematica dell' Università, Piazza di Porta S. Donato, 5, 40126 Bologna, Italy
e-mail: fausto.ferrari@unibo.it

S. Salsa (✉)
Dipartimento di Matematica del Politecnico, Piazza Leonardo da Vinci, 32, 20133 Milano, Italy
e-mail: sandro.salsa@polimi.it



G. Citti et al. (eds.), *Geometric Methods in PDE's*, Springer INdAM Series 13,
DOI 10.1007/978-3-319-02666-4_17

Precisely we are interested in the regularity properties of the free boundary for the following kind of problems. In a bounded domain $\Omega \subset \mathbf{R}^n$, consider the problem

$$\begin{cases} \Delta u = f & \text{in } \Omega^+(u) \cup \Omega^-(u), \\ u_\nu^+ = G(u_\nu^-, x) & \text{on } F(u) = \partial\Omega^+(u) \cap \Omega, \end{cases} \tag{1}$$

where

$$\Omega^+(u) = \{x \in \Omega : u(x) > 0\}, \quad \Omega^-(u) = \{x \in \Omega : u(x) \leq 0\}^\circ.$$

Here $f$ is bounded on $\Omega$ and continuous in $\Omega^+(u) \cup \Omega^-(u)$, while $u_\nu^+$ and $u_\nu^-$ denote the normal derivatives in the inward direction to $\Omega^+(u)$ and $\Omega^-(u)$ respectively. $F(u)$ is called the *free boundary*.

The function

$$G(\eta, x) : [0, \infty) \times \Omega \to (0, \infty)$$

satisfies the following assumptions.

(H1) $G(\eta, \cdot) \in C^{0,\bar{\gamma}}(\Omega)$ uniformly in $\eta$; $G(\cdot, x) \in C^{1,\bar{\gamma}}([0, L])$ for every $x \in \Omega$.
(H2) $G_\eta(\cdot, x) > 0$ with $G(0, x) \geq \gamma_0 > 0$ uniformly in $x$.
(H3) There exists $N > 0$ such that $\eta^{-N} G(\eta, x)$ is strictly decreasing in $\eta$, uniformly in $x$.

We describe two typical model problems of this type arising in classical fluid-dynamics. A traveling two-dimensional gravity wave moves with constant speed on the surface of an incompressible, inviscid, heavy fluid. The bottom is horizontal. With respect to a reference domain moving with the wave speed, the motion is steady and occupies a fixed region $\Omega$, delimited from above by an unknown free line $S$, representing the wave profile.

Since the flow is incompressible, the velocity can be expressed by the gradient of a *stream function* $\psi$. If some suitable hypotheses on the flow speed are satisfied, then $\psi$ and the *vorticity*, $\omega$ are functionally dependent i.e. $\omega = \Delta\psi$.

Assuming furthermore that the bottom and $S$ are *streamlines*, from Bernoulli law on $S$ we derive the following model:

$$\begin{cases} \Delta\psi = -\gamma(\psi) & \text{in } \Omega = \{0 < \psi < B\} \\ 0 \leq \psi \leq B & \text{in } \bar{\Omega} \\ \psi = B & \text{on } y = 0 \\ |\nabla\psi|^2 + 2gy = Q, \ \ \psi = 0 \text{ on } S. \end{cases}$$

Here $Q$ is constant, $B, g$ are positive constants and $\gamma : [0, B] \to \mathbf{R}$ is called *vorticity function*.

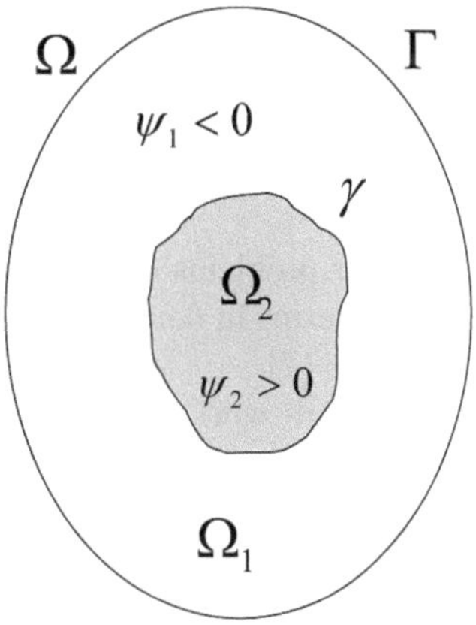


**Fig. 1** Prandtl-Batchelor flow configuration

The problem is to find $S$ such that there exists a function $\psi$ satisfying the above system.

Since $\psi^- \equiv 0$ this is a *one phase* problem and several papers have been recently devoted to it. Of particular interest is the proof of the so-called *Stokes conjecture,* according to which at points where the gradient vanishes (*stagnation points*) the wave profile presents a 120° corner. Away from stagnation points the free boundary is Lipschitz and moreover $Q - 2gy > 0$.

We refer to [21, 22] and the reference therein for more details and known results. Among the various problems left open there was the regularity of $S$ away from stagnation points. The answer is given in [8], where the author shows that in this regions $S$ is a smooth curve.

The second model is a two phase problem called Prantl-Batchelor flow. A bounded 2-dimensional domain is delimited by two simple closed curves $\gamma, \Gamma$. Let $\Omega_1, \Omega_2$ be as in the Fig. 1.

For given constants $\mu < 0, \omega > 0$, consider functions $\psi_1, \psi_2$ satisfying

$$\Delta\psi_1 = 0 \text{ in } \Omega_1, \psi_1 = 0 \text{ on } \gamma, \psi_1 = \mu \text{ on } \Gamma,$$

$$\Delta\psi_2 = \omega \text{ in } \Omega_2, \psi_2 = 0 \text{ on } \gamma.$$

The two functions $\psi_1, \psi_2$ are interpreted as stream functions of an irrotational flow in $\Omega_1$ and of a constant vorticity flow in $\Omega_2$. In the model proposed by Batchelor, coming from the limit of large Reynold number in the steady Navier-Stokes equation, a flow of this type is hypothesized in which there is a jump in the tangential velocity along $\gamma$, namely

$$|\nabla\psi_2|^2 - |\nabla\psi_1|^2 = \sigma$$

for some positive constant $\sigma$. In this problem $\gamma$ is to be determined and plays the role of a free boundary.

There is no satisfactory theory for this problem. Viscosity solutions are Lipschitz across $\gamma$ as shown in [6], but neither existence nor regularity is known (uniqueness fails already in the radial case, where two explicit solutions can be found).

As a consequence of the results in [11], *flat* or Lipschitz free boundaries are smooth.

Other problems of the type (1) arise from singular perturbation problems with forcing terms in flame propagation theory (see [18]) or from magnetohydrodynamics as in [17].

We shall work in the context of viscosity solutions which we introduce below. First classical comparison sub/super solutions are defined as follows.

**Definition 1.1** We say that $v \in C(\Omega)$ is a $C^2$ strict (comparison) subsolution (resp. supersolution) to our f.b.p. in $\Omega$, if $v \in C^2(\overline{\Omega^+(v)}) \cap C^2(\overline{\Omega^-(v)})$ and the following conditions are satisfied:

1. $\Delta v > f$ (resp. $< f$) in $\Omega^+(v) \cup \Omega^-(v)$.
2. If $x_0 \in F(v)$, then

$$v_\nu^+(x_0) > G(v_\nu^-(x_0), x_0) \quad (\text{resp. } v_\nu^+(x_0) < G(v_\nu^-(x_0), x_0), \;\; v_\nu^+(x_0) \neq 0).$$

Observe that the free boundary of a strict comparison sub/supersolution is $C^2$.

Given $u, \varphi \in C(\Omega)$, we say that $\varphi$ *touches u by below* (resp. *above*) at $x_0 \in \Omega$ if $u(x_0) = \varphi(x_0)$, and

$$u(x) \geq \varphi(x) \quad (\text{resp. } u(x) \leq \varphi(x)) \quad \text{in a neighborhood } O \text{ of } x_0.$$

**Definition 1.2** Let $u$ be a continuous function in $\Omega$. We say that $u$ is a viscosity solution (resp. supesolution) to our f.b.p. in $\Omega$, if the following conditions are satisfied:

1. $\Delta u = f$ in $\Omega^+(u) \cup \Omega^-(u)$ in the viscosity sense.
2. Let $x_0 \in F(u)$ and $v \in C^2(\overline{B^+(v)}) \cap C^2(\overline{B^-(v)})$ $(B = B_\delta(x_0))$ with $F(v) \in C^2$. If $v$ touches $u$ by below (resp.above) at $x_0 \in F(v)$, then

$$v_\nu^+(x_0)) \leq G(v_\nu^-(x_0)) \;\; (\text{resp. } \;\; \geq).$$

When $f = 0$ the existence of Lipschitz viscosity solutions has been settled by Caffarelli in [4]. In particular, the positivity set of $u$ has finite perimeter and, with respect to the $n-1$ Hausdorff measure $H^{n-1}$, a.e. point on $F(u)$ has a normal in the measure theoretical sense.

Under the assumption $G(\eta, x) \to \infty$, as $\eta \to \infty$, the Lipschitz continuity of the solution in the nonhomogeneous case has been proven in [6], Theorem 4.5, as a consequence of the following monotonicity formula:

**Theorem 1.1** *Let $u, v$ be nonnegative, continuous functions in $B_1$, with*

$$\Delta u \geq -1, \Delta v \geq -1 \quad \textit{in the sense of distributions}$$

*and* $u(0) = v(0) = 0$, $u(x)\,v(x) = 0$ *in* $B_1$. *Then, for* $r \leq 1/2$,

$$\Phi(r) = \frac{1}{r^4}\int_{B_r}\frac{|\nabla u|^2}{|x|^{n-2}}\int_{B_r}\frac{|\nabla v|^2}{|x|^{n-2}} \leq c(n)\left(1 + \|u\|^2_{L^2(B_1)}\right)\left(1 + \|v\|^2_{L^2(B_1)}\right).$$

Observe that if the supports of $u$ and $v$ were separated by a smooth surface with normal $\nu$ at $x = 0$ then, by taking the limit as $r \to 0$, we could deduce that

$$(u_\nu(0))^2 (v_\nu(0))^2 \leq \Phi(1/2).$$

Hence $\Phi(r)$ "morally" gives a control in average of the product of the normal derivatives of $u$ at the origin.

As we have said, we are mainly interested in the regularity properties of the free boundary, in particular in proving that *flat* or *Lipschitz* free boundaries are smooth ($C^{1,\gamma}$).

A way to express the flatness of the free boundary is to assume that $F(u)$ is trapped between two parallel hyperplanes at $\delta$-distance from each other, for a small $\delta$ ($\delta$-flatness). While this looks like a somewhat strong assumption, it is indeed a natural one since it is satisfied for example by rescaling a solution around a point of the free boundary where there is a normal in some weak sense (*regular points*), for instance in the measure theoretical one. We have seen that in the homogeneous case $H^{n-1}$-a.e. points on $F(u)$ are of this kind. Moreover, starting form a Lipschitz free boundary, $H^{n-1}$-a.e. points on $F(u)$ are regular, by Rademacher Theorem.

The following results are proved in [11]. A constant depending only on (some of) the parameters $n$, Lip$(u)$, $\gamma_0$ and $N$ is called universal. The $C^{1,\bar{\gamma}}$ norm of $G(\cdot, x)$ may depend on $x$ and enters in a qualitative way only. We will always assume that

$$0 \in F(u).$$

**Theorem 1.2** *(Flatness implies* $C^{1,\gamma}$*) Let* $u$ *be a viscosity solution to* (1) *in* $B_1$*, with Lip*$(u) \leq L$*. Assume that* $f$ *is continuous in* $B_1^+(u) \cup B_1^-(u)$, $\|f\|_{L^\infty(B_1)} \leq L$ *and* $G$ *satisfies* $(H1)$*-*$(H3)$*.*

*There exists a universal constant* $\bar{\delta} > 0$ *such that, if*

$$\{x_n \leq -\delta\} \subset B_1 \cap \{u^+(x) = 0\} \subset \{x_n \leq \delta\}, \qquad (\delta - \textit{flatness}) \tag{2}$$

*with* $0 \leq \delta \leq \bar{\delta}$*, then* $F(u)$ *is* $C^{1,\gamma}$ *in* $B_{1/2}$*.*

We also have:

**Theorem 1.3** *(Lipschitz implies* $C^{1,\gamma}$*) Let* $u$ *be a viscosity solution to* (1) *in* $B_1$*, with Lip*$(u) \leq L$*. Assume that* $f$ *is continuous in* $B_1^+(u) \cup B_1^-(u)$, $\|f\|_{L^\infty(B_1)} \leq L$ *and* $G$ *satisfies* $(H1)$–$(H3)$*. If* $F(u)$ *is a Lipschitz graph in a neighborhood of* 0*, then* $F(u)$ *is* $C^{1,\gamma}$ *in a (smaller) neighborhood of* 0*.*

As we shall see later, Theorem 1.3 follows from Theorem 1.2 and the main result in [3] via a blow-up argument.

The flatness conditions present in the literature (see, for instance [5]), are often stated in terms of "$\varepsilon$-monotonicity" along a large cone of directions $\Gamma(\theta_0, e)$ of axis $e$ and opening $\theta_0$. Precisely, a function $u$ is said to be $\varepsilon$-monotone ($\varepsilon > 0$ small) along the direction $\tau$ in the cone $\Gamma(\theta_0, e)$ if for every $\varepsilon' \geq \varepsilon$,

$$u(x + \varepsilon'\tau) \geq u(x).$$

A variant of Theorem 1.2, found in [11] states the following.

**Theorem 1.4** *Let $u$ be a solution to our f.b.p in $B_1$, $0 \in F(u)$. Suppose that $u^+$ is non-degenerate. Then there exist $\theta_0 < \pi/2$ and $\varepsilon_0 > 0$ such that if $u^+$ is $\varepsilon$-monotone along every direction in $\Gamma(\theta_0, e_n)$ for some $\varepsilon \leq \varepsilon_0$, then $u^+$ is fully monotone in $B_{1/2}$ along any direction in $\Gamma(\theta_1, e_n)$ for some $\theta_1$ depending on $\theta_0, \varepsilon_0$. In particular $F(u)$ is Lipschitz and therefore $C^{1,\gamma}$.*

Geometrically, the $\varepsilon$-monotonicity of $u^+$ can be interpreted as $\varepsilon$-closeness of $F(u)$ to the graph of a Lipschitz function (Fig. 2). Our flatness assumption requires $\varepsilon$-closeness of $F(u)$ to a hyperplane. If $\|f\|_\infty$ is small enough, depending on $\varepsilon$, it is not hard to check that $\varepsilon$-flatness of $F(u)$ implies $c\varepsilon$-monotonicity of $u^+$ along the directions of a flat cone, for a $c$ depending on its opening.

The proof of Theorem 1.4 follows immediately from the fact that if $u^+$ is non-degenerate and $\varepsilon$-monotone along every direction in $\Gamma(\theta_0, e_n)$ for some $\varepsilon \leq \varepsilon_0$, then there exist a radius $r_0 > 0$ and $\delta_0 > 0$ depending on $\varepsilon_0, \theta_0$ such that $u^+$ is $\delta_0$-flat in $B_{r_0}$.

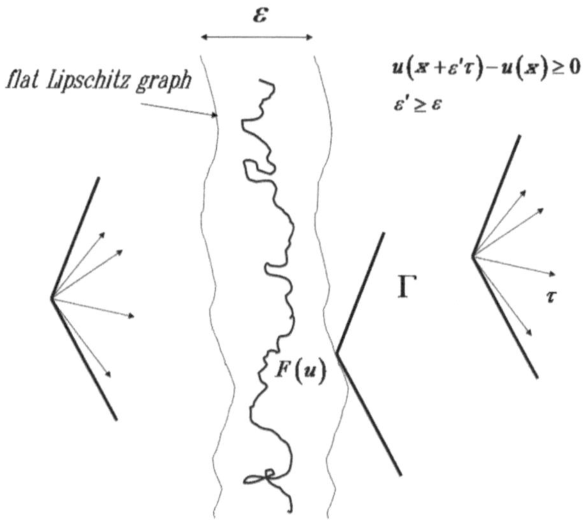


**Fig. 2** $\varepsilon$-monotonicity along a cone of directions

## 2 Reduction of Theorem 1.2 to a Localized Form

The proof of Theorem 1.2 is based on an iterative procedure that "squeezes" our solution around an optimal configuration $U_\beta(x \cdot \nu)$ at a geometric rate in dyadically decreasing balls. Here $U_\beta = U_\beta(t)$ is given by

$$U_\beta(t) = \alpha t^+ - \beta t^-, \ \ \beta \geq 0, \ \ \alpha = G_0(\beta) \equiv G(\beta, 0)$$

and $\nu$ is a unit vector which plays the role of the normal vector at the origin. $U_\beta(x \cdot \nu)$ is a so-called *two plane solution.*

This strategy of flatness improvement works nicely in the one phase case ($\beta = 0$) or as long as the two phases $u^+, u^-$ are, say, comparable (*nondegenerate case*). The difficulties arise when the negative phase becomes very small but at the same time not negligible (*degenerate case.*) In this case the flatness assumption in Theorem 1.2 gives a control of the positive phase only, through the closeness to a *one plane solution* $U_0(x_n) = x_n^+$.

As we shall see, this requires to face a dychotomy in the final iteration. A similar situation is already present in the homogeneous case $f = 0$ (see e.g. [5]).

The first step is to check that the flatness condition (2) implies that $u$ is close to $U_\beta$ for some $\beta$. Indeed we prove the following lemma.

**Lemma 2.1** *Let u satisfy* (2). *Given any* $\eta > 0$ *there exist* $\bar{\delta}, \bar{\rho} > 0$ *depending only on* $\eta, n,$ *and* $L$ *such that if* $\delta \leq \bar{\delta}$, *then*

$$\|u - U_\beta\|_{L^\infty(B_{\bar{\rho}})} \leq \eta \bar{\rho} \tag{3}$$

*for some* $0 \leq \beta \leq L$.

The proof, by contradiction, follows from the following compactness result.

**Lemma 2.2** *Let* $u_k$ *be a sequence of (Lipschitz) viscosity solutions to*

$$\begin{cases} |\Delta u_k| \leq M, & \text{in } \Omega^+(u_k) \cup \Omega^-(u_k), \\ (u_k^+)_\nu = G_k((u_k^-)_\nu, x), & \text{on } F(u_k). \end{cases}$$

*Assume that:*

1. $u_k \to u^*$ *uniformly on compact sets of* $\Omega$.
2. $G_k(\eta, \cdot) \to G(\eta, \cdot)$ *on compact sets of* $\Omega$, *uniformly on* $0 \leq \eta \leq L = Lip(u_k)$.
3. $\{u_k^+ = 0\} \to \{(u^*)^+ = 0\}$ *in the Hausdorff distance.*

*Then*

$$|\Delta u^*| \leq M, \quad \text{in } \Omega^+(u^*) \cup \Omega^-(u^*)$$

*and $u^*$ satisfies the free boundary condition*

$$(u^*)^+_\nu = G((u^*)^-_\nu, x) \quad on \ F(u^*),$$

*both in the viscosity sense.*

In view of Lemma 2.1, after proper rescaling, Theorem 1.2 follows from the following result.

**Lemma 2.3** *Let $u$ be a viscosity solution to our f.b.p. in $B_1$ with $Lip\,(u) \leq L$. There exists a universal constant $\bar{\eta} > 0$ such that, if*

$$\|u - U_\beta\|_{L^\infty(B_1)} \leq \bar{\eta} \quad for\ some\ 0 \leq \beta \leq L, \tag{4}$$

$$\{x_n \leq -\bar{\eta}\} \subset B_1 \cap \{u^+(x) = 0\} \subset \{x_n \leq \bar{\eta}\}, \tag{5}$$

$$\|f\|_{L^\infty(B_1)} \leq \bar{\eta}, [G(\eta, \cdot)]_{C^{0,\bar{\gamma}}(B_1)} \leq \bar{\eta}, \quad \forall 0 \leq \eta \leq L,$$

*then $F(u)$ is $C^{1,\gamma}$ in $B_{1/2}$.*

We are almost ready to start the *improvement of flatness procedure.* This means that from (4) and (5) we should be able to squeeze more the graph of $u$ (and therefore $F(u)$) around a possibly rotated new two plane solution in a neighborhood of the origin. A closer look to (4) reveals that, when $\alpha$ and $\beta$ are comparable, a nice control on the location of $F(u)$ is available but when $\beta \ll \alpha$ only a one side control of $F(u)$ is possible. This dichotomy is well reflected in the following elementary lemma that we state for a general continuous function. In particular, it translates the "vertical" closeness between the graphs of $u$ and $U_\beta$ given by (4) into "horizontal" closeness.

**Lemma 2.4** *Let $u$ be a continuous function. If, for a small $\eta > 0$,*

$$\|u - U_\beta\|_{L^\infty(B_1)} \leq \eta$$

*and*

$$\{x_n \leq -\eta\} \subset B_1 \cap \{u^+(x) = 0\} \subset \{x_n \leq \eta\},$$

*then:*

*If $\beta \geq \eta^{1/3}$,*

$$U_\beta\left(x_n - \eta^{1/3}\right) \leq u(x) \leq U_\beta\left(x_n + \eta^{1/3}\right) \quad in \ B_{3/4}.$$

*If $\beta < \eta^{1/3}$,*

$$U_0\left(x_n - \eta^{1/3}\right) \leq u^+(x) \leq U_0\left(x_n + \eta^{1/3}\right) \quad in \ B_{3/4}.$$

Set $\bar{\eta} = \tilde{\varepsilon}^3$ in Lemma 2.3. Then, according to Lemma 2.4, the dichotomy *nondegenerate versus degenerate* will translate quantitatively into the two cases:

$$\beta \geq \tilde{\varepsilon} : \ \textbf{nondegenerate}, \ \beta < \tilde{\varepsilon} : \ \textbf{degenerate}.$$

The parameter $\tilde{\varepsilon}$ will be chosen later in the final iteration.

## 3 Lipschitz Implies $C^{1,\gamma}$

In this section we show how Theorem 1.3 follows from Theorem 1.2. For simplicity of exposition we consider the model case

$$G(\eta, x) = \sqrt{1+\eta^2}.$$

We use the following Liouville type result for global viscosity solutions to a homogeneous free boundary problem, that could be of independent interest. Note that no growth at infinity is needed.

**Lemma 3.1** *Let $U$ be a global viscosity solution to*

$$\begin{cases} \Delta U = 0, & \text{in } \{U > 0\} \cup \{U \leq 0\}^0, \\ (U_\nu^+)^2 - (U_\nu^-)^2 = 1, & \text{on } F(U) := \partial\{U > 0\}. \end{cases} \tag{6}$$

*Assume that $F(U) = \{x_n = g(x'), x' \in \mathbf{R}^{n-1}\}$ with $Lip(g) \leq M$. Then $g$ is linear and (in a proper system of coordinates) $U(x) = U_\beta(x)$ for some $\beta \geq 0$.*

*Proof* Balls of radius $\rho$ and centered at 0 in $\mathbf{R}^{n-1}$ are denoted by $B'_\rho$.

By the regularity theory in [3] , since $U$ is a solution in $B_2$, the free boundary $F(U)$ is $C^{1,\gamma}$ in $B_1$ with a bound depending only on $n$ and on $M$. Thus,

$$|g(x') - g(0) - \nabla g(0) \cdot x'| \leq C|x'|^{1+\alpha}, \quad x' \in B'_1$$

with $C$ depending only on $n, M$. Moreover, since $U$ is a global solution, the rescaling

$$g_R(x') = \frac{1}{R} g(Rx'), \quad x' \in B'_2,$$

which preserves the same Lipschitz constant as $g$, satisfies the same inequality as above i.e.

$$|g_R(x') - g_R(0) - \nabla g_R(0) \cdot x'| \leq C|x'|^{1+\alpha}, \quad x' \in B'_1.$$

This reads,

$$|g(Rx') - g(0) - \nabla g(0) \cdot Rx'| \leq CR|x'|^{1+\alpha}, \quad x' \in B_1'.$$

Thus,

$$|g(y') - g(0) - \nabla g(0) \cdot y'| \leq C\frac{1}{R^\alpha}|y'|^{1+\alpha}, \quad y' \in B_R'.$$

Passing to the limit as $R \to \infty$ we obtain that $g$ is linear.

After a change of coordinates, the free boundary reduces to $x_n = 0$. Since $u_{x_n}$ is harmonic and (it can be shown) positive on $x_n > 0$, by Liouville Theorem we conclude the proof.

We need another Lemma stating that if the free boundary $F(u)$ is trapped in a $\delta$-neighborhood of a Lipschitz graph, then our solution grows linearly $\delta$-away from the free boundary.

**Lemma 3.2** *Let u be a solution to* (1) *in* $B_2$ *with Lip*$(u) \leq L$ *and* $\|f\|_{L^\infty} \leq L$*. If*

$$\{x_n \leq g(x') - \delta\} \subset \{u^+ = 0\} \subset \{x_n \leq g(x') + \delta\},$$

*with g a Lipschitz function, Lip*$(g) \leq L, g(0) = 0$*, then*

$$u(x) \geq c_0(x_n - g(x')), \quad x \in \{x_n \geq g(x') + 2\delta\} \cap B_{c_0},$$

*for some* $c_0 > 0$ *depending on* $n, L$ *as long as* $\delta \leq c_0$.

*Proof* All constants in this proof will depend on $n, L$. It suffices to show that our statement holds for $\{x_n \geq g(x') + C\delta\}$ for some large constant $C$. Then one can apply Harnack inequality to obtain the full statement. To this aim, we want to show that

$$u(de_n) \geq c_0 d, \quad d \geq C\delta.$$

After rescaling, we are reduced to proving that

$$u(e_n) \geq c_0$$

as long as $\delta \leq 1/C$ and $\|f\|_\infty$ is sufficiently small. Let

$$w(x) = \frac{1}{\gamma}(1 - |x|^{-\gamma})$$

be defined on the closure of the annulus $B_2 \setminus \overline{B}_1$ with $\|f\|_\infty$ small enough so that

$$\Delta w > \|f\| \quad \text{on } B_2 \setminus \overline{B}_1.$$

Let

$$w_t(x) = w(x + te_n),$$

$t \in \mathbf{R}$. Notice that

$$|\nabla w_0| < 1 \quad \text{on } \partial B_1.$$

From our flatness assumption for $|t|$ sufficiently large (depending on the Lipschitz constant of $g$), $w_t$ is strictly above $u$. We increase $t$ and let $\bar{t}$ be the first $t$ such that $w_t$ touches $u$ by above. Since $w_{\bar{t}}$ is a strict supersolution to our free boundary problem, the touching point $z$ can occur only on the $\eta := \frac{1}{\gamma}(1 - 2^{-\gamma})$ level set in the positive phase of $u$, and $|z| \leq C = C(L)$ . Since $u$ is Lipschitz continuous, $0 < u(z) = \eta \leq Ld(z, F(u))$, that is a full ball around $z$ of radius $\eta/L$ is contained in the positive phase of $u$. Thus, for $\bar{\delta}$ small depending on $\eta, L$ we have that $B_{\eta/2L}(z) \subset \{x_n \geq g(x') + 2\bar{\delta}\}$. Since $x_n = g(x') + 2\bar{\delta}$ is Lipschitz we can connect $e_n$ and $z$ with a chain of intersecting balls included in the positive side of $u$ with radii comparable to $\eta/2L$. The number of balls depends on $L$ . Then we can apply Harnack inequality and obtain

$$u(e_n) \geq cu(z) = c_0,$$

as desired.

We can now provide the proof of Theorem 1.3.

*Proof* Let $\bar{\eta}$ be the universal constant in Lemma 2.3. Consider the blow-up sequence

$$u_k(x) = \frac{u(\delta_k x)}{\delta_k}$$

with $\delta_k \to 0$ as $k \to \infty$. Each $u_k$ solves (1) with right hand side

$$f_k(x) = \delta_k f(\delta_k x)$$

and

$$\|f_k(x)\| \leq \delta_k \|f\|_{L^\infty} \leq \bar{\eta}$$

for $k$ large enough. Standard arguments (see for example [1]), using the uniform Lipschitz continuity of the $u_k$'s and the non-degeneracy of their positive part $u_k^+$ (see Lemma 3.2), imply that (up to a subsequence)

$$u_k \to \tilde{u} \quad \text{uniformly on compacts}$$

and

$$\{u_k^+ = 0\} \to \{\tilde{u} = 0\} \quad \text{in the Hausdorff distance,}$$

where the blow-up limit $\tilde{u}$ solves the global homogeneous two-phase free boundary problem

$$\begin{cases} \Delta \tilde{u} = 0, & \text{in } \{\tilde{u} > 0\} \cup \{\tilde{u} \le 0\}^0, \\ (\tilde{u}_\nu^+)^2 - (\tilde{u}_\nu^-)^2 = 1, & \text{on } F(\tilde{u}) := \partial\{\tilde{u} > 0\}. \end{cases}$$

Since $F(u)$ is a Lipschitz graph in a neighborhood of 0, it follows from Lemma 3.1 that $F(\tilde{u})$ is a two-plane solutions, $\tilde{u} = U_\beta$ for some $\beta \ge 0$. Thus, for $k$ large enough

$$\|u_k - U_\beta\|_{L^\infty} \le \bar{\eta}$$

and

$$\{x_n \le -\bar{\eta}\} \subset B_1 \cap \{u_k^+(x) = 0\} \subset \{x_n \le \bar{\eta}\}.$$

Therefore, we can apply our flatness Theorem 1.2 and conclude that $F(u_k)$ and hence $F(u)$ is smooth.

# 4 The Nondegenerate Case

## 4.1 *Improvement of Flatness*

Assume that for some $\varepsilon > 0$ small, we have

$$U_\beta(x_n - \varepsilon) \le u(x) \le U_\beta(x_n + \varepsilon) \quad \text{in } B_1, \tag{7}$$

with $0 < \beta \le L$, $\alpha = G(\beta, 0) \equiv G_0(\beta)$. One would like to get in a smaller ball a geometric improvement of (7).We assume that (this will be achieved at the end by rescaling)

$$\|f\|_{L^\infty(B_1)} \le \varepsilon^2 \min\{\alpha, \beta\}, \tag{8}$$

and

$$\|G(\eta, \cdot) - G_0(\eta)\|_{L^\infty(B_1)} \le \varepsilon^2, \quad \forall 0 \le \eta \le L.$$

Then the basic step in the improvement of flatness reads as follows.

**Lemma 4.1** *If* $0 < r \leq r_0$ *for* $r_0$ *universal, and* $0 < \varepsilon \leq \varepsilon_0$ *for some* $\varepsilon_0$ *depending on* $r$*, then*

$$U_{\beta'}(x \cdot \nu_1 - r\frac{\varepsilon}{2}) \leq u(x) \leq U_{\beta'}(x \cdot \nu_1 + r\frac{\varepsilon}{2}) \quad in\ B_r, \tag{9}$$

*with* $|\nu_1| = 1$, $|\nu_1 - e_n| \leq \tilde{C}\varepsilon$, *and* $|\beta - \beta'| \leq \tilde{C}\beta\varepsilon$ *for a universal constant* $\tilde{C}$.

Assume the lemma above holds. To prove Lemma 2.3 we rescale considering a blow up sequence

$$u_k(x) = \frac{u(\rho_k x)}{\rho_k} \qquad \rho_k = \bar{r}^k, \quad x \in B_1 \tag{10}$$

for suitable $\bar{r} \leq \min\left\{r_0, \frac{1}{16}\right\}$, $\tilde{\varepsilon} \leq \varepsilon_0(\bar{r})$, as in Lemma 4.1, and iterate to get, at the $k$th step,

$$U_{\beta_k}(x \cdot \nu_k - \rho_k\varepsilon_k) \leq u_k(x) \leq U_{\beta_k}(x \cdot \nu_k + \rho_k\varepsilon_k) \quad \text{in } B_{\rho_k},$$

with $\varepsilon_k = 2^{-k}\tilde{\varepsilon}$, $|\nu_k| = 1$, $|\nu_k - \nu_{k-1}| \leq \tilde{C}\varepsilon_{k-1}$,

$$|\beta_k - \beta_{k-1}| \leq \tilde{C}\beta_{k-1}\varepsilon_{k-1}, \quad \varepsilon_k \leq \beta_k \leq L.$$

Note that **in the non-degenerate case**, $\beta \geq \tilde{\varepsilon}$, at each step we have the correct inductive hypotheses. For simplicity, say we are in the model case $G(\eta, x) = \sqrt{1 + \eta^2}$. Starting with $\beta = \beta_0 \geq \varepsilon_0 = \tilde{\varepsilon}$, if $k \geq 1$ and $\beta_{k-1} \geq \varepsilon_{k-1}$, then

$$\begin{aligned}\beta_k &\geq \beta_{k-1}(1 - \tilde{C}\varepsilon_{k-1}) \geq 2^{-k+1}\tilde{\varepsilon}\left(1 - \tilde{C}2^{-k+1}\tilde{\varepsilon}\right) \\ &\geq 2^{-k}\tilde{\varepsilon} = \varepsilon_k.\end{aligned}$$

Thus, since

$$f_k(x) = \rho_k f(\rho_k x), \quad x \in B_1$$

(recall that $\bar{\eta} = \tilde{\varepsilon}^3$)

$$\|f_k\|_{L^\infty(B_1)} \leq \rho_k\tilde{\varepsilon}^3 \leq \tilde{\varepsilon}_k^2\beta_k = \tilde{\varepsilon}_k^2 \min\{\alpha_k, \beta_k\}.$$

The Fig. 3 describes the step from $k$ to $k + 1$.

This implies that $F(u)$ is $C^{1,\gamma}$ at the origin. Repeating the procedure for points in a neighborhood of $x = 0$, since all estimates are universal, we conclude that there exists a unit vector $\nu_\infty = \lim \nu_k$ and $C > 0$, $\gamma \in (0, 1]$, both universal, such that, in the coordinate system $e_1, \ldots, e_{n-1}, \nu_\infty$, $\nu_\infty \perp e_j$, $e_j \cdot e_k = \delta_{jk}$, $F(u)$ is $C^{1,\gamma}$ graph,

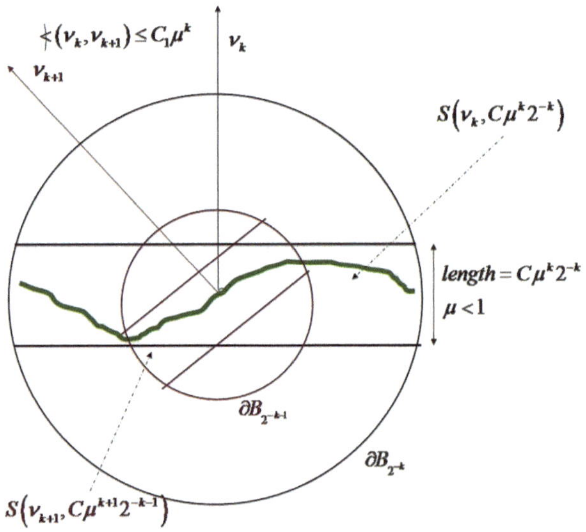


**Fig. 3** Improvement of flatness (here $\rho_k = 2^{-k}$)

say $x_n = g(x')$, with $g(0') = 0$ and

$$|g(x') - \nu_\infty \cdot x'| \leq C|x'|^{1+\gamma}$$

in a neighborhood of $x = 0$.

The main question is: *where is it hidden the information allowing one to realize the step from (7) to (9)?*

Here a *linearized problem* comes into play.

## 4.2 The Linearized Problem

Let us first consider the one-phase case (see [8]) where $u \geq 0$ in $B_1$,

$$\Delta u = f \quad \text{in } B_1^+(u)$$

and $u_\nu^+ = |\nabla u^+| = g(x)$ on the free boundary. Assume that

$$|f| \leq \varepsilon^2, \ |g(x) - 1| \leq \varepsilon^2.$$

The flatness condition writes $(U_0(x) = x_n^+)$

$$(x_n - \varepsilon)^+ \leq u(x) \leq (x_n + \varepsilon)^+ \quad \text{in } B_1. \tag{11}$$

Renormalize by setting

$$\tilde{u}_\varepsilon(x) = \frac{u(x) - x_n}{\varepsilon} \quad \text{in } B_1^+(u) \cup F(u)$$

or

$$u(x) = x_n + \varepsilon \tilde{u}_\varepsilon(x) \quad \text{in } B_1^+(u) \cup F(u). \tag{12}$$

In (12), $u$ appears as a first order perturbation of the hyperplane $x_n$.

The idea is that the key information we are looking for is **stored precisely in the "corrector"** $\tilde{u}_\varepsilon$. To extract it, we look at what happens to $\tilde{u}_\varepsilon$, asymptotically as $\varepsilon \to 0$. Note that, as $\varepsilon \to 0$, $B_1^+(u) \to \{x_n > 0\}$ and $F(u)$ goes to $\{x_n = 0\}$, both in Hausdorff distance.

We have

$$\Delta \tilde{u}_\varepsilon = \frac{f}{\varepsilon} \sim \varepsilon \quad \text{in } B_1^+(u)$$

and on $F(u)$,

$$|\nabla u|^2 = |e_n + \varepsilon \nabla \tilde{u}_\varepsilon|^2 = g^2 \sim 1 + \varepsilon^2$$

that is, after simplifying by $\varepsilon$,

$$2\tilde{u}_{x_n} + \varepsilon |\nabla \tilde{u}_\varepsilon|^2 \sim \varepsilon.$$

Thus, formally, letting $\varepsilon \to 0$, we get "for the limit" $\tilde{u} = \tilde{u}_0$ the following problem:

$$\Delta \tilde{u} = 0, \quad \text{in } B_{1/2}^+ = B_{1/2} \cap \{x_n > 0\} \tag{13}$$

and the Neumann condition (linearization of the free boundary condition)

$$\tilde{u}_{x_n} = 0 \quad \text{on } B_{1/2} \cap \{x_n = 0\}. \tag{14}$$

We call (13), (14) the *linearized problem.*

Let us see how the general condition

$$|\nabla u^+| = G(|\nabla u^-|, x)$$

linearizes in the *nondegenerate* two phase problem.

First let

$$\Delta u = f \quad \text{in } B_1$$

with

$$|f| \leq \varepsilon^2 \min\{\alpha, \beta\}$$

and

$$|G(\eta, \cdot) - G_0(\eta)| \leq \varepsilon^2 \quad \forall \eta \in [0, L].$$

The flatness condition

$$\alpha(x_n - \varepsilon)^+ - \beta(x_n - \varepsilon)^- \leq u(x) \leq \alpha(x_n + \varepsilon)^+ - \beta(x_n + \varepsilon)^- \quad \text{in } B_1, \tag{15}$$

with $0 < \beta \leq L, \alpha = G_0(\beta)$, suggests the renormalization

$$\tilde{u}_\varepsilon(x) = \begin{cases} \frac{u(x) - \alpha x_n}{\alpha\varepsilon}, \ x \in B_1^+(u) \cup F(u) \\ \\ \frac{u(x) - \beta x_n}{\beta\varepsilon}, \ x \in B_1^-(u) \end{cases}$$

or

$$u(x) = \begin{cases} \alpha x_n + \varepsilon\alpha\tilde{u}_\varepsilon(x), \ x \in B_1^+(u) \cup F(u) \\ \beta x_n + \varepsilon\beta\tilde{u}_\varepsilon(x), \ x \in B_1^-(u). \end{cases} \tag{16}$$

We have

$$\Delta\tilde{u}_\varepsilon \sim \varepsilon \quad \text{in } B_1^+(u) \cup B_1^-(u).$$

On $F(u)$,

$$\left|\nabla u^+\right| = \alpha\,|e_n + \varepsilon\nabla\tilde{u}_\varepsilon(x)| \sim \alpha\left(1 + \varepsilon\,(\tilde{u}_\varepsilon)_{x_n} + \varepsilon^2\,|\nabla\tilde{u}_\varepsilon|^2\right)$$

and

$$\begin{aligned} G(|\nabla u^-|, x) = G(|\beta e_n + \varepsilon\beta\nabla\tilde{u}_\varepsilon|, x) &\sim G\left(\beta\left(1 + \varepsilon\,(\tilde{u}_\varepsilon)_{x_n} + \varepsilon^2\,|\nabla\tilde{u}_\varepsilon|^2\right), x\right) \\ &\sim G_0(\beta) + \varepsilon G_0'(\beta)\left(\beta\,(\tilde{u}_\varepsilon)_{x_n} + \varepsilon\beta\,|\nabla\tilde{u}_\varepsilon|^2\right) + \varepsilon^2. \end{aligned}$$

As before, letting $\varepsilon \to 0$, we get formally for "the limit" $\tilde{u} = \tilde{u}_0$ the following problem:

$$\Delta\tilde{u} = 0, \ \text{in } B_{1/2}^+ \cup B_{1/2}^- \tag{17}$$

and ($\alpha = G_0(\beta)$) the transmission condition (linearization of the free boundary condition)

$$\alpha\,(\tilde{u}_{x_n})^+ - \beta G_0'(\beta)\,(\tilde{u}_{x_n})^- = 0 \text{ on} B_{1/2} \cap \{x_n = 0\} \tag{18}$$

where $(\tilde{u}_{x_n})^+$ and $(\tilde{u}_{x_n})^-$ denote the $e_n$-derivatives of $\tilde{u}$ restricted to $\{x_n > 0\}$ and $\{x_n < 0\}$, respectively.

Thus, at least formally, we have found an asymptotic problem for the limits of the renormalizations $\tilde{u}_\varepsilon$. The crucial information we were mentioning before is contained in the following regularity result. Consider the transmission problem, ($\tilde{\alpha} \neq 0$)

$$\begin{cases} \Delta\tilde{u} = 0 & \text{in } B_1 \cap \{x_n \neq 0\}, \\ \tilde{\alpha}^2(\tilde{u}_n)^+ - \tilde{\beta}^2(\tilde{u}_n)^- = 0 & \text{on } B_1 \cap \{x_n = 0\}. \end{cases} \tag{19}$$

**Theorem 4.1** *Let $\tilde{u}$ be a viscosity solution to (19) in $B_1$ such that $\|\tilde{u}\|_\infty \leq 1$. Then $\tilde{u} \in C^\infty\left(\bar{B}_1^{\pm}\right)$ and in particular, there exists a universal constant $\bar{C}$ such that*

$$|\tilde{u}(x) - \tilde{u}(0) - (\nabla_{x'}\tilde{u}(0)\cdot x' + \tilde{p}x_n^+ - \tilde{q}x_n^-)| \leq \bar{C}r^2, \quad \text{in } B_r \tag{20}$$

*for all $r \leq 1/2$ and with $\tilde{\alpha}^2\tilde{p} - \tilde{\beta}^2\tilde{q} = 0$.*

The question is now how to transfer the estimate (20) to $\tilde{u}_\varepsilon$ and then read it in terms of flatness for $u$ through the formulas (16).

The right way to obtain the proof of Lemma 4.1 is to proceed by contradiction.

Fix $r \leq r_0$, to be chosen suitably. Assume that for a sequence $\varepsilon_k \to 0$ there is a sequence $u_k$ of solutions of our free boundary problem in $B_1$, with right hand side $f_k$ such that $\|f_k\|_{L^\infty(B_1)} \leq \varepsilon_k^2 \min\{\alpha_k, \beta_k\}$,

$$\|G_k(\eta,\cdot) - G_k(\eta, 0)\|_\infty \leq \varepsilon_k^2, \quad \forall 0 \leq \eta \leq L, \tag{21}$$

and

$$U_{\beta_k}(x_n - \varepsilon_k) \leq u_k(x) \leq U_{\beta_k}(x_n + \varepsilon_k) \quad \text{in } B_1, \; 0 \in F(u_k), \tag{22}$$

with $0 \leq \beta_k \leq L$, $\alpha_k = G_k(\beta_k, 0)$, but the conclusion of Lemma 4.1 does not hold for every $k \geq 1$.

Construct the corresponding sequence of renormalized functions

$$\tilde{u}_k(x) = \begin{cases} \frac{u_k(x) - \alpha_k x_n}{\alpha_k \varepsilon_k}, & x \in B_1^+(u_k) \cup F(u_k) \\ \\ \frac{u_k(x) - \beta_k x_n}{\beta_k \varepsilon_k}, & x \in B_1^-(u_k). \end{cases}$$

Up to a subsequence, $G_k(\cdot, 0)$ converges, locally uniformly, to some $C^1$-function $\tilde{G}_0$, while $\beta_k \to \tilde{\beta}$ so that $\alpha_k \to \tilde{\alpha} = \tilde{G}_0(\tilde{\beta})$. At this point we need compactness to show that the graphs of $\tilde{u}_k$ converge in the Hausdorff distance to a Hölder continuous $\tilde{u}$ in $B_{1/2}$. The compactness is provided by the Harnack inequality stated in Theorem 4.2 below and its corollary, as we shall see later, and is inspired by the work of Savin [20].

It turns out that the limit function $\tilde{u}$ satisfies the linearized problem (17) and (18) in the viscosity sense. Hence, from (20), having $\tilde{u}(0) = 0$,

$$|\tilde{u}\,(x) - (x' \cdot \nu' + \tilde{p}x_n^+ - \tilde{q}x_n^-)| \leq Cr^2, \quad x \in B_r, \tag{23}$$

for all $r \leq 1/4$ (say), with

$$\tilde{\alpha}^2\tilde{p} - \tilde{\beta}^2\tilde{q} = 0, \quad |\nu'| = |\nabla_{x'}\tilde{u}(0)| \leq C.$$

Since $\tilde{u}_k$ converges uniformly to $\tilde{u}$ in $B_{1/2}$, (23) transfers to $\tilde{u}_k$ :

$$|\tilde{u}_k\,(x) - (x' \cdot \nu' + \tilde{p}x_n^+ - \tilde{q}x_n^-)| \leq C'r^2, \quad x \in B_r. \tag{24}$$

Set

$$\beta_k' = \beta_k\,(1 + \varepsilon_k\tilde{q})\,, \quad \nu_k = \frac{1}{\sqrt{1 + \varepsilon_k^2|\nu'|^2}}\left(e_n + \varepsilon_k\left(\nu', 0\right)\right).$$

Then,

$$\begin{aligned}\alpha_k' &= G_k(\beta_k(1 + \varepsilon_k q), 0) = G_k(\beta_k, 0) + \beta_k G_k'(\beta_k, 0)\varepsilon_k q + O(\varepsilon_k^2)\\ &= \alpha_k(1 + \beta_k\frac{G_k'(\beta_k, 0)}{\alpha_k}q\varepsilon_k) + O(\varepsilon_k^2) = \alpha_k(1 + \varepsilon_k p) + O(\varepsilon_k^2)\end{aligned}$$

since from the identity $\tilde{\alpha}p - \tilde{\beta}\tilde{G}_0'(\tilde{\beta})q = 0$ we derive that

$$\beta_k\frac{G_k'(\beta_k, 0)}{\alpha_k}q = p + O(\varepsilon_k).$$

Moreover

$$\nu_k = e_n + \varepsilon_k(\nu', 0) + \varepsilon_k^2\tau, \quad |\tau| \leq C.$$

With these choices we can show that (for $k$ large and $r \leq r_0$)

$$\tilde{U}_{\beta_k'}(x \cdot \nu_k - \varepsilon_k\frac{r}{2}) \leq \tilde{u}_k(x) \leq \tilde{U}_{\beta_k'}(x \cdot \nu_k + \varepsilon_k\frac{r}{2}), \quad \text{in } B_r$$

where again we are using the notation:

$$\tilde{U}_{\beta'_k}(x) = \begin{cases} \frac{U_{\beta'_k}(x)-\alpha_k x_n}{\alpha_k \varepsilon_k}, & x \in B_1^+(U_{\beta'_k}) \cup F(U_{\beta'_k}) \\ \frac{U_{\beta'_k}(x)-\beta_k x_n}{\beta_k \varepsilon_k}, & x \in B_1^-(U_{\beta'_k}). \end{cases}$$

This will clearly imply that

$$U_{\beta'_k}(x \cdot \nu_k - \varepsilon_k \frac{r}{2}) \leq u_k(x) \leq U_{\beta'_k}(x \cdot \nu_k + \varepsilon_k \frac{r}{2}), \quad \text{in } B_r$$

leading to a contradiction.

In view of (24) we need to show that in $B_r$

$$\tilde{U}_{\beta'_k}(x \cdot \nu_k - \varepsilon_k \frac{r}{2}) \leq (x' \cdot \nu' + \tilde{p} x_n^+ - \tilde{q} x_n^-) - Cr^2$$

and

$$\tilde{U}_{\beta'_k}(x \cdot \nu_k + \varepsilon_k \frac{r}{2}) \geq (x' \cdot \nu' + \tilde{p} x_n^+ - \tilde{q} x_n^-) + Cr^2.$$

This can be shown after some elementary calculations as long as $r \leq r_0$, $r_0$ universal, and $\varepsilon \leq \varepsilon_0(r)$.

## 4.3 Compactness

We are left with compactness. The Harnack inequality takes the following form.

**Theorem 4.2** *Let $u$ be a solution of our f.b.p. in $B_1$ with Lipschitz constant $L$. There exists a universal $\tilde{\varepsilon} > 0$ such that, if $x_0 \in B_1$ and $u$ satisfies the following condition:*

$$U_\beta(x_n + a_0) \leq u(x) \leq U_\beta(x_n + b_0) \quad \text{in } B_r(x_0) \subset B_1 \tag{25}$$

*with*

$$\|f\|_{L^\infty(B_2)} \leq \varepsilon^2 \min\{\alpha, \beta\}, \quad 0 < \beta \leq L, \tag{26}$$

$$\|G(\eta, x) - G_0(\eta)\|_{L^\infty(B_1)} \leq \varepsilon^2, \quad \forall 0 \leq \eta \leq L, \tag{27}$$

*and*

$$0 < b_0 - a_0 \leq \varepsilon r$$

*for some* $0 < \varepsilon \leq \tilde{\varepsilon}$, *then*

$$U_\beta\left(x_n + a_1\right) \leq u\left(x\right) \leq U_\beta\left(x_n + b_1\right) \quad in \quad B_{r/20}\left(x_0\right)$$

*with*

$$a_0 \leq a_1 \leq b_1 \leq b_0 \ and \ b_1 - a_1 \leq \left(1 - c\right) \varepsilon r$$

*and* $0 < c < 1$ *universal.*

If $u$ satisfies (25) with, say $r = 1$, then we can apply Harnack inequality repeatedly and obtain

$$U_\beta(x_n + a_m) \leq u(x) \leq U_\beta(x_n + b_m) \quad \text{in } B_{20^{-m}}(x_0),$$

with

$$b_m - a_m \leq (1 - c)^m \varepsilon$$

for all $m$'s such that

$$(1 - c)^m 20^m \varepsilon \leq \bar{\varepsilon}.$$

This implies that for all such $m$'s, the oscillation of the renormalized functions $\tilde{u}_k$ in $B_r(x_0), r = 20^{-m}$, is less than $(1-c)^m = 20^{-\gamma m} = r^\gamma$. Thus, the following corollary holds.

**Corollary 4.1** *Let u satisfies at some point* $x_0 \in B_2$

$$U_\beta(x_n + a_0) \leq u(x) \leq U_\beta(x_n + b_0) \quad in \ B_1(x_0) \subset B_2, \tag{28}$$

*for some* $0 < \beta \leq L$, *with*

$$b_0 - a_0 \leq \varepsilon,$$

*and let (26)–(27) hold, for* $\varepsilon \leq \bar{\varepsilon}$, $\bar{\varepsilon}$ *universal. Then in* $B_1(x_0)$, $(\alpha = G_0(\beta))$

$$\tilde{u}_\varepsilon(x) = \begin{cases} \frac{u(x)-\alpha x_n}{\alpha\varepsilon}, & in \ B_2^+(u) \cup F(u) \\ \frac{u(x)-\beta x_n}{\beta\varepsilon}, & in \ B_2^-(u) \end{cases}$$

*has a Hölder modulus of continuity at* $x_0$, *outside the ball of radius* $\varepsilon/\bar{\varepsilon}$, *i.e for all* $x \in B_1(x_0)$, *with* $|x - x_0| \geq \varepsilon/\bar{\varepsilon}$

$$|\tilde{u}_\varepsilon(x) - \tilde{u}_\varepsilon(x_0)| \leq C|x - x_0|^\gamma.$$

Since in the proof of Lemma 4.1,

$$-1 \le \tilde{u}_k(x) \le 1, \quad \text{for } x \in B_1$$

we can implement the corollary above and use Ascoli-Arzela theorem to obtain that as $\varepsilon_k \to 0$ the graphs of the $\tilde{u}_k$ converge (up to a subsequence) in the Hausdorff distance to the graph of a Hölder continuous function $\tilde{u}$ over $B_{1/2}$.

Thus the improvement of flatness process in the nondegenerate case can be concluded.

# 5 The Degenerate Case

## 5.1 *Improvement of Flatness*

In this case, the negative part of $u$ is negligible and the positive part is close to a one-plane solution (i.e. $\beta = 0$). Precisely, assume that for some $\varepsilon > 0$, small, we have

$$U_0(x_n - \varepsilon) \le u^+(x) \le U_0(x_n + \varepsilon), \quad \text{in } B_1. \tag{29}$$

Again one would like to get in a smaller ball an improvement of (29). This time the key lemma is the following.

**Lemma 5.1** *Let the solution $u$ satisfies (29) with*

$$\|f\|_{L^\infty(B_1)} \le \varepsilon^4,$$
$$\|G(\eta,\cdot) - G_0(\eta)\|_{L^\infty} \le \varepsilon^2, \quad 0 \le \eta \le C\varepsilon^2,$$

*and*

$$\|u^-\|_{L^\infty(B_1)} \le \varepsilon^2. \tag{30}$$

*There exists a universal $r_1$, such that if $0 < r \le r_1$ and $0 < \varepsilon \le \varepsilon_1$ for some $\varepsilon_1$ depending on $r$, then*

$$U_0(x \cdot \nu_1 - r\frac{\varepsilon}{2}) \le u^+(x) \le U_0(x \cdot \nu_1 + r\frac{\varepsilon}{2}) \quad \text{in } B_r, \tag{31}$$

*with $|\nu_1| = 1$, $|\nu_1 - e_n| \le C\varepsilon$ for a universal constant $C$.*

The proof follows the same pattern of the nondegenerate case. For simplicity, we outline it in the model case $G(\eta, x) = \sqrt{1 + \eta^2}$.

Fix $r \leq r_1$, to be chosen suitably. By contradiction assume that, for some sequences $\varepsilon_k \to 0$ and $u_k$, solutions of our f.b.p. in $B_1$ with r.h.s. $f_k$ such that $\|f_k\|_{L^\infty(B_1)} \leq \varepsilon_k^4$ and

$$\|u_k^-\|_{L^\infty(B_1)} \leq \varepsilon_k^2,$$

$$U_0(x_n - \varepsilon_k) \leq u_k(x) \leq U_0(x_n + \varepsilon_k) \quad \text{in } B_1, 0 \in F(u_k)$$

the conclusion of the lemma does not hold.

Then one proves via a Harnack inequality (see below), that the sequence of normalized functions

$$\tilde{u}_k(x) = \frac{u_k(x) - x_n}{\varepsilon_k} \qquad x \in B_1^+(u_k) \cup F(u_k)$$

converges to a limit function $\tilde{u}$, Hölder continuous in $B_{1/2}$.

The limit function $\tilde{u}$ is a viscosity solution of the linearized problem

$$\begin{cases} \Delta\tilde{u} = 0, \text{ in } B_{1/2} \cap \{x_n > 0\}, \\ \tilde{u}_n = 0, \quad \text{on } B_{1/2} \cap \{x_n = 0\}. \end{cases}$$

The regularity of $\tilde{u}$ is not a problem and the contradiction argument proceeds as before with obvious changes.

The Harnack inequality takes the following form.

**Theorem 5.1** *Let $u$ be a solution of our f.b.p. in $B_1$. There exists a universal $\tilde{\varepsilon} > 0$ such that, if $x_0 \in B_1$ and $u$ satisfies the following condition*

$$(x_n + a_0)^+ \leq u^+(x) \leq (x_n + b_0)^+, \quad \text{in } B_r(x_0) \subset B_1 \tag{32}$$

*with*

$$\|f\|_{L^\infty(B_2)} \leq \varepsilon^4, \quad \|u^-\|_{L^\infty(B_2)} \leq \varepsilon^2$$

*and*

$$0 < b_0 - a_0 \leq \varepsilon r$$

*for some $0 < \varepsilon \leq \tilde{\varepsilon}$, then*

$$(x_n + a_1)^+ \leq u^+(x) \leq (x_n + b_1)^+ \quad \text{in } B_{r/20}(x_0)$$

*with*

$$a_0 \leq a_1 \leq b_1 \leq b_0 \text{ and } b_1 - a_1 \leq (1 - c)\,\varepsilon r$$

*and $0 < c < 1$ universal.*

Lemma 5.1 provides the first step in the flatness improvement. Notice that this improvement is obtained through the closeness of the positive phase to a *one plane solution*, as long as inequality (30) holds. This inequality expresses in another quantitative way the degeneracy of the negative phase and should be kept valid at each step of the iteration of Lemma 5.1. However, it could happen that this is not the case and in some step of the iteration, say at the level $\varepsilon_k$ of flatness, the norm $\|u^-\|_{L^\infty(B_1)}$ becomes of order $\varepsilon_k^2$. When this occurs, a suitable rescaling restores a nondegenerate situation. This give rise in the final iteration to the dychotomy we have mentioned in Sect. 2.

The situation is precisely described in the following lemma.

**Lemma 5.2** *Let u be a solution in $B_1$ satisfying*

$$U_0(x_n - \varepsilon) \leq u^+(x) \leq U_0(x_n + \varepsilon) \quad \text{in } B_1 \tag{33}$$

*with*

$$\|f\|_{L^\infty(B_1)} \leq \varepsilon^4,$$

*and for $\tilde{C}$ universal,*

$$\|u^-\|_{L^\infty(B_2)} \leq \tilde{C}\varepsilon^2, \quad \|u^-\|_{L^\infty(B_1)} > \varepsilon^2. \tag{34}$$

*There exists (universal) $\varepsilon_1$ such that, if $0 < \varepsilon \leq \varepsilon_1$, the rescaling*

$$u_\varepsilon(x) = \varepsilon^{-1/2} u\left(\varepsilon^{1/2} x\right)$$

*satisfies, in $B_{2/3}$ :*

$$U_{\beta'}(x_n - C'\varepsilon^{1/2}) \leq u_\varepsilon(x) \leq U_{\beta'}(x_n + C'\varepsilon^{1/2})$$

*with $\beta' \sim \varepsilon^2$ and $C'$ depending on $\tilde{C}$.*

Let us see how the dychotomy arises. To prove Lemma 2.3 **in the degenerate case**, $\beta < \tilde{\varepsilon}$, choose $\bar{r} \leq \min\{r_0, r_1, 1/16\}$ and $\tilde{\varepsilon} \leq \min\{\varepsilon_0(\bar{r}), \varepsilon_1(\bar{r})/2, 1/(2\tilde{C})\}$. In view of our choice of $\tilde{\varepsilon}$, we obtain that $u$ satisfies the relation

$$U_0(x_n - \tilde{\varepsilon}) \leq u^+(x) \leq U_0(x_n + \tilde{\varepsilon}) \quad \text{in } B_1.$$

Since

$$\left\|u - U_\beta\right\|_{L^\infty(B_1)} \leq \bar{\eta} = \tilde{\varepsilon}^3$$

we infer

$$\|u^-\|_{L^\infty(B_1)} \le \beta + \tilde{\varepsilon}^3 \le 2\tilde{\varepsilon}.$$

Call $\varepsilon' = \sqrt{2\tilde{\varepsilon}}$. Then

$$U_0(x_n - \varepsilon') \le u^+(x) \le U_0(x_n + \varepsilon') \quad \text{in } B_1$$

and

$$\|f\|_{L^\infty(B_1)} \le \left(\varepsilon'\right)^4, \|u^-\|_{L^\infty(B_1)} \le \left(\varepsilon'\right)^2.$$

From Lemma 5.1, we get

$$U_0(x \cdot \nu_1 - \bar{r}\frac{\varepsilon'}{2}) \le u^+(x) \le U_0(x \cdot \nu_1 + \bar{r}\frac{\varepsilon'}{2}) \quad \text{in } B_{\bar{r}}$$

with $|\nu_1| = 1, |\nu_1 - e_n| \le C\varepsilon'$ for a universal constant $C$.

We now rescale considering a blow up sequence

$$u_k(x) = \frac{u(\rho_k x)}{\rho_k} \qquad \rho_k = \bar{r}^k, \quad x \in B_1 \tag{35}$$

and set $\varepsilon_k = 2^{-k}\varepsilon'$

$$f_k(x) = \rho_k f(\rho_k x) \quad x \in B_1.$$

Note that

$$\|f_k\|_{L^\infty(B_1)} \le \rho_k \left(\varepsilon'\right)^4 \le \frac{1}{16}\left(\varepsilon'\right)^4 = \varepsilon_k^4.$$

We can iterate Lemma 5.1 and obtain

$$U_0(x \cdot \nu_k - \varepsilon_k) \le u_k^+(x) \le U_0(x \cdot \nu_k + \varepsilon_k), \quad \text{in } B_1$$

with $|\nu_k - \nu_{k-1}| \le C\varepsilon_{k-1}$, **as long as**

$$\|u_k^-\|_{L^\infty(B_1)} \le \varepsilon_k^2.$$

Let $k^* > 1$ be the first integer for which this fails:

$$\|u_{k^*}^-\|_{L^\infty(B_1)} > \varepsilon_{k^*}^2$$

and

$$\|u^-_{k^*-1}\|_{L^\infty(B_1)} \le \varepsilon^2_{k^*-1}.$$

We also have

$$U_0(x \cdot \nu_{k^*-1} - \varepsilon_{k^*-1}) \le u^+_{k^*-1}(x) \le U_0(x \cdot \nu_{k^*-1} + \varepsilon_{k^*-1}), \quad \text{in } B_1.$$

By usual comparison arguments we can write

$$u^+_{k^*-1}(x) \le C\,|x_n - \varepsilon_{k^*-1}|\,\varepsilon^2_{k^*-1} \quad \text{in} \quad B_{19/20}$$

for $C$ universal. Rescaling, we have

$$\|u^-_{k^*}\|_{L^\infty(B_1)} \le C_1 \varepsilon^2_{k^*}$$

where $C_1$ universal ($C_1$ depends on $\bar{r}$). Then $u_{k^*}$ satisfies the assumptions of Lemma 5.2 and therefore the rescaling

$$v\,(x) = \varepsilon^{-1/2}_{k^*} u_{k^*}(\varepsilon^{1/2}_{k^*} x)$$

satisfies in $B_{2/3}$

$$U_{\beta'}(x \cdot \nu_{k^*} - C'\varepsilon^{1/2}_{k^*}) \le v(x) \le U_{\beta'}(x \cdot \nu_{k^*} + C'\varepsilon^{1/2}_{k^*})$$

with $\beta' \sim \varepsilon^2_{k^*}$. Call $\hat{\varepsilon} = C'\varepsilon^{1/2}_{k^*}$. Then $v$ is a solution of our f.b.p. in $B_{2/3}$ with r.h.s.

$$g\,(x) = \varepsilon^{1/2}_{k^*} f_{k^*}(\varepsilon^{1/2}_{k^*} x)$$

satisfying the flatness assumption

$$U_{\beta'}(x \cdot \nu_{k^*} - \hat{\varepsilon}) \le v(x) \le U_{\beta'}(x \cdot \nu_{k^*} + \hat{\varepsilon}).$$

Since $\beta' \sim \varepsilon^2_{k^*}$, we have

$$\|g\|_{L^\infty(B_1)} \le \varepsilon^{1/2}_{k^*} \varepsilon^4_{k^*} \le \hat{\varepsilon}^2 \beta'$$

as long as $\hat{\varepsilon} \le \min\left\{\varepsilon_0\,(\bar{r})\,, \frac{1}{2\bar{C}}\right\}$, which is true if $C'\,(2\tilde{\varepsilon})^{1/4} \le \min\left\{\varepsilon_0\,(\bar{r})\,, \frac{1}{2\bar{C}}\right\}$ or

$$\tilde{\varepsilon} \le \frac{1}{2C'^4} \min\left\{\varepsilon_0\,(\bar{r})\,, \frac{1}{2\bar{C}}\right\}^4.$$

With these choices, $v$ satisfies the assumptions of the nondegenerate case and we can proceed accordingly.

This concludes the proof of the main lemma.

## 6 Remarks and Further Developments

Theorem 1.2 holds for more general operators (see [11, 12]). For instance when $\mathscr{L}$ is a uniformly elliptic operator in nondivergence form with Hölder continuous coefficients,

$$\mathscr{L}u = \mathrm{Tr}\left(A(x)\,D^2u\right) + b(x)\cdot\nabla u$$

or a fully nonlinear operator

$$\mathscr{L}u = \mathscr{F}\left(D^2u\right),\quad (\mathscr{F}(O) = 0)$$

where $D^2u$ is the Hessian matrix of $u$.

Notably, in the fully nonlinear case we do not need to assume for $\mathscr{F}$ neither concavity nor homogeneity of degree one. In this case in Theorem 4.1, $u \in C^{\infty}(\overline{B}_1^{\pm})$ has to be replaced by $u \in C^{1,\alpha}(\overline{B}_1^{\pm})$ and, in formula (20), $r^2$ is replaced by $r^{1+\alpha}$.

In general, we need to assume Lipschitz regularity of our solution. Indeed, in this generality, the existence of Lipschitz viscosity solutions with proper measure theoretical properties of the free boundary is an open problem and it will be object of future investigations.

However, if $\mathscr{L}$ is linear and can be written in divergence form an estimate like in Theorem 1.1 is available (see [19]) and one can reproduce the proof of Theorem 4.5 in [6], to recover the Lipschitz continuity of a viscosity solution. Observe that then $f = f(x, u, \nabla u)$ is allowed, with $f(x, \cdot, \cdot)$ locally bounded.

Theorem 1.3 continues to hold when $\mathscr{L}$ is linear or if $\mathscr{F}$ (positively) homogeneous of degree one (or when $\mathscr{F}_r(M)$ has a limit $\mathscr{F}^*(M)$, as $r \to 0$, which is homogeneous of degree one).

With the two Theorems 1.2 and 1.3 the regularity theory for two phase problems has reached a reasonably satisfactory level. However many questions remain open, object of future investigations.

The first one is to provide an existence results for viscosity solutions satisfying a Dirichlet boundary condition, extending for instance the results in the homogeneous case in [4].

Another question is the $C^{\infty}$-smoothness (resp. analyticity) of the free boundary in presence of $C^{\infty}$ (resp. analytic) coefficients and data.

We shall deal with these two questions in forthcoming papers.

Also of great importance, we believe, is to have information on the Hausdorff measure or dimension of the *singular* (*nonflat*) points of the free boundary. For instance, in 3 dimensions, the free boundary for local energy minimizer in the variational problem

$$\int_{\Omega} \left\{ |\nabla u|^2 + \chi_{\{u>0\}} \right\} \to \min$$

is a smooth surface (see [7]). In dimension $n = 7$, De Silva and Jerison in [9] provided an example of a minimizer with singular free boundary. The conjecture is that energy minimizing free boundaries should be smooth for $n < 7$.

Nothing is known in the nonhomogeneous case.

**Acknowledgements** Daniela De Silva and Fausto Ferrari are supported by the ERC starting grant project 2011 EPSILON (Elliptic PDEs and Symmetry of Interfaces and Layers for Odd Nonlinearities). Daniela De Silva is supported by NSF grant, DMS-1301535. Fausto Ferrari is supported by Miur Grant (Prin): Equazioni di diffusione in ambiti sub-riemanniani e problemi geometrici associati. Sandro Salsa is supported by Miur Grant, Geometric Properties of Nonlinear Diffusion Problems.

## References

1. Alt, H.W., Caffarelli L.A., Friedman A.: Variational problems with two phases and their free boundaries. T.A.M.S. **282**, 431–461 (1984)
2. Argiolas, R., Ferrari F.: Flat free boundaries regularity in two-phase problems for a class of fully nonlinear elliptic operators with variable coefficients. Interfaces Free Bound. **11**, 177–199 (2009)
3. Caffarelli, L.A.: A Harnack inequality approach to the regularity of free boundaries. Part I: Lipschitz free boundaries are $C^{1,\alpha}$. Rev. Mat. Iberoam. **3**, 139–162 (1987)
4. Caffarelli, L.A.: A Harnack inequality approach to the regularity of free boundaries. Part III: Existence theory, compactness and dependence on $x$. Ann. Scuola. Norm. Ser. IV **XV**, 583–602 (1988)
5. Caffarelli, L.A.: A Harnack inequality approach to the regularity of free boundaries. Part II: Flat free boundaries are Lipschitz. Commun. Pure Appl. Math. **42**, 55–78 (1989)
6. Caffarelli, L.A., Jerison, D., Kenig, C.E.: Some new monotonicity theorems with applications to free boundary problems. Ann. Math. **155**, 369–404 (2002)
7. Caffarelli, L.A., Jerison, D., Kenig, C.E.: Global energy minimizers for free boundary problems and full regularity in three dimensions. Contemp. Math. **350**, 83–97 (2004)
8. De Silva, D.: Free boundary regularity for a problem with right hand side. Interfaces Free Bound. **13**, 223–238 (2011)
9. De Silva, D., Jerison, D.: A singular energy minimizing free boundary. J. Reine Angew. Math. **635**, 1–21 (2009)
10. De Silva, D., Roquejoffre, J.: Regularity in a one-phase free boundary problem for the fractional Laplacian. Ann. Inst. H. Poincare (C) Non Linear Anal. **29**, 335–367 (2012)
11. De Silva, D., Ferrari, F., Salsa, S.: Two-phase problems with distributed source: regularity of the free boundary. Anal. PDE **7**(2), 267–310 (2014)
12. De Silva, D., Ferrari, F., Salsa, S.: Free boundary regularity for fully nonlinear non-homogeneous two-phase problems. J. Math. Pures Appl. (9) **103**(3), 658–694 (2015)
13. Feldman, M.: Regularity for nonisotropic two-phase problems with Lipshitz free boundaries. Differ. Integr. Equ. **10**, 1171–1179 (1997)

14. Feldman, M.: Regularity of Lipschitz free boundaries in two-phase problems for fully nonlinear elliptic equations. Indiana Univ. Math. J. **50**, 1171–1200 (2001)
15. Ferrari, F.: Two-phase problems for a class of fully nonlinear elliptic operators, Lipschitz free boundaries are $C^{1,\gamma}$. Am. J. Math. **128**, 541–571 (2006)
16. Ferrari, F., Salsa, S.: Regularity of the free boundary in two-phase problems for elliptic operators. Adv. Math. **214**, 288–322 (2007)
17. Friedman, A., Liu, Y.: A free boundary problem arising in magnetohydrodynamic system. Ann. Scuola Norm. Sup. Pisa Cl. Sci. **22**, 375–448 (1994)
18. Lederman, C., Wolanski, N.: A two phase elliptic singular perturbation problem with a forcing term. J. Math. Pures Appl. **86**, 552–589 (2006)
19. Matevosyan, N., Petrosyan, A.: Almost monotonicity formulas for elliptic and parabolic operators with variable coefficients. Comm. Pure Appl. Math **44**, 271–311 (2011)
20. Savin, O.: Small perturbation solutions for elliptic equations. Commun. Partial Differ. Equ. **32**, 557–578 (2007)
21. Varvaruca, E.: On the existence of extreme waves and the Stokes conjecture with vorticity. J. Differ. Equ. **246**, 4043–4076 (2009)
22. Varvaruca, E., Weiss, G.S: A geometric approach to generalized Stokes conjectures. Acta Math. **206**, 363–403 (2011)

# The Role of Fundamental Solution in Potential and Regularity Theory for Subelliptic PDE

**Andrea Bonfiglioli, Giovanna Citti, Giovanni Cupini, Maria Manfredini, Annamaria Montanari, Daniele Morbidelli, Andrea Pascucci, Sergio Polidoro, and Francesco Uguzzoni**

*Tu se' lo mio maestro e 'l mio autore;*
*tu se' solo colui da cu' io tolsi*
*lo bello stilo che m'ha fatto onore*[1]
Dante Alighieri

**Abstract** In this survey we consider a general Hörmander type operator, represented as a sum of squares of vector fields plus a drift and we outline the central role of the fundamental solution in developing Potential and Regularity Theory for solutions of related PDEs. After recalling the Gaussian behavior at infinity of the kernel, we show some mean value formula on the level set of the fundamental solution, which allow to obtain a comprehensive parallel of the classical Potential Theory. Then we show that a precise knowledge of the fundamental solution leads to global regularity results: estimates at the boundary or on the whole space. Finally in the problem of regularity of non linear differential equations we need an ad hoc modification of the parametrix method, based on the properties of the fundamental solution of an approximating problem.

[1] *You are my master, and indeed my author;*
*It is from you alone that I have taken*
*The exact style for which I have been honoured.*
Dante Alighieri, *The Divine Comedy*, translated by C.H. Sisson, Oxford University Press, New York, 2008.

A. Bonfiglioli • G. Citti • G. Cupini • M. Manfredini • A. Montanari • D. Morbidelli (✉) •
A. Pascucci • F. Uguzzoni
Dipartimento di Matematica, Università di Bologna, Bologna, Italy
e-mail: andrea.bonfiglioli6@unibo.it; giovanna.citti@unibo.it; giovanni.cupini@unibo.it; maria.manfredini@unibo.it; annamaria.montanari@unibo.it; daniele.morbidelli@unibo.it; andrea.pascucci@unibo.it; francesco.uguzzoni@unibo.it

S. Polidoro
Dipartimento di Matematica, Università di Bologna, Bologna, Italy

Dipartimento FIM, Università di Modena e Reggio Emilia, Modena, Italy
e-mail: sergio.polidoro@unimore.it



G. Citti et al. (eds.), *Geometric Methods in PDE's*, Springer INdAM Series 13,
DOI 10.1007/978-3-319-02666-4_18

**Keywords** Potential theory • Subelliptic PDEs • Hörmander operators • Poincaré inequality

**Mathematical Subject Classification:** 35A17, 35A08, 35R03, 35K70, 35H20, 35B65, 53C17

# 1 Introduction

In this paper we consider a general operator of the form

$$L_A = \sum_{i,j=1}^{m} a_{ij}\,(t,x)\,X_iX_j - X_0 \tag{1}$$

where

$$X_i = \sum_{j=1}^{N} \sigma_{ij}\partial_{x_j}, \quad X_0 = \partial_t + \sum_{j=1}^{N} \sigma_{0j}\partial_{x_j},$$

and the coefficients $\sigma_{ij}$ only depend on the spatial variables $x \in \mathbb{R}^N$. We also require that $X_0, X_1, X_2, \ldots, X_m$ is a system of real smooth vector fields defined in some domain $D \subset [0,T[\times\mathbb{R}^N$ satisfying the Hörmander's rank condition at any point:

$$\operatorname{rank}(\operatorname{Lie}(X_0,\ldots,X_m)(t,x)) = N+1, \quad \forall\ (t,x) \in D.$$

The matrix $A = \left\{a_{ij}\,(t,x)\right\}_{i,j=1}^{m}$ is real symmetric and uniformly positive definite, that is

$$\lambda^{-1}\,|\xi|^2 \leq \textstyle\sum_{i,j=1}^{m} a_{ij}\,(t,x)\,\xi_i\xi_j \leq \lambda\,|\xi|^2 \tag{2}$$

for some $\lambda > 0$ and for every $\xi \in \mathbb{R}^N$ and every $(t,x) \in D$. We will assign degree 1 to the vector fields $(X_i)_{i=1,\ldots,m}$, (denoted $d(X_i) = 1$), while $d(X_0) = 2$. We will denote $d((t,x),(\tau,\xi))$ the Carnot-Carathéodory distance generated in $D$ by the vector fields $X_0, X_1, \ldots, X_m$ with their degrees. Precisely for every pair of points $(t,x)$ and $(\tau,\xi)$, we define

$$\begin{aligned} d((t,x),(\tau,\xi)) = \inf\Big\{r > 0 \ \Big|\ & \text{there is a Lipschitz path } \gamma \text{ such that} \\ & \gamma(0) = (t,x),\ \gamma(1) = (\tau,\xi),\ \text{and, for a.e. } s,\ \gamma'(s) = \sum_{i=0}^{m} \beta_i(s)X_i(\gamma(s)) \\ & \text{with } |\beta_i(s)| \leq r \text{ for } i = 1,\ldots,m, \text{ and } |\beta_0(s)| \leq r^2\Big\}. \end{aligned} \tag{3}$$

The Carnot-Carathéodory metric generated by the vector fields $X_0, X_1, \ldots, X_m$ plays a crucial role in the regularity theory for subelliptic degenerate operators.

After the celebrated Hörmander's paper [65], where the explicit fundamental solution of a Kolmogorov-type operator was computed, Folland [55], Rothschild and Stein [97], Jerison and Sánchez-Calle [66] proved existence and asymptotic behavior of the fundamental solution, under the assumption that $A$ is the identity. Almost at the same time, Franchi and Lanconelli [58] studied regularity of sum of squares of diagonal vector fields and established a Poincaré type inequality. The equivalence of several distances was proved by Nagel et al. in [88]. After that, in the last 20 years we witnessed an extraordinary development of the theory of subelliptic operators. We refer the reader to the book [23] and to the introduction of each section of this paper, for more historical remarks and references.

A significant contribution to the development of Potential Theory of subelliptic PDEs is due to Ermanno Lanconelli. His personal and original approach is based on a far-reaching use of the fundamental solution in order to prove, in this setting, a complete parallel of the classical Potential and Regularity Theory.

In this paper we take this perspective, and we describe from a unitary point of view a number of results obtained by the authors in collaboration with him. In Sect. 2 we will recall Gaussian estimates of the fundamental solution of large classes of operators of the type (1). In particular for the heat equation we discuss the results of Bonfiglioli et al. [21, 22], Bramanti et al. in [28], and for the Kolmogorov operator we quote the results of Polidoro [94], Lanconelli and Polidoro [76], Lanconelli and Pascucci in [74]. In Sect. 3 we describe the quasi-exponential mappings, introduced in Lanconelli and Morbidelli [73], which are a tool to obtain a Poincaré inequality. Level sets of the fundamental solution are special families of balls, on which mean value formulas have been proved by Citti et al. (see [42]), Lanconelli and Pascucci (see [75]), which lead to another proof of the Poincaré inequality. Using the mean value formulas, characterizations of subharmonicity were obtained by Bonfiglioli and Lanconelli [13, 15, 17]. The optimality of these sets have been investigated by Lanconelli [72], Abbondanza and Bonfiglioli [1], Kogoj et al. [69] and Kogoj and Tralli [68]. The properties of the fundamental solution immediately imply internal regularity of solutions. Here we are also interested in global regularity of solutions which will be presented in Sect. 4. Precisely we will recall Schauder regularity up to the boundary, by Manfredini in [79], and the estimates on the whole of space by Bramanti et al. [29, 31]. Finally in Sect. 5 we conclude our survey with a discussion on regularity of solutions of non linear-equations with nonlinearity in the vector fields, and in particular of the Levi equation. For the case of $\mathbb{C}^2$ see [40, 44]. For the Levi equation in $\mathbb{C}^{n+1}$ with $n > 1$, we refer to the regularity results in [49, 81, 83], the counterexamples by Gutierrez et al. [62], and the symmetry results by Martino and Montanari [80].

## 1.1 Applications to Complex Analysis, Finance and Vision

Equation (1) is a natural generalization of the classical equation which models particle interactions in phase spaces. In this case the drift term expresses the coupling position-velocity:

$$X_0 = \sum_{j=1}^{m} p_j \partial_{q_j} + \partial_t$$

and the matrix $(a_{ij})$ is the identity in the space of velocities:

$$L = \frac{1}{2}\sum_{j=1}^{m} \partial_{p_j}^2 - \sum_{j=1}^{m} p_j \partial_{q_j} - \partial_t, \qquad (t, q, p) \in \mathbb{R} \times \mathbb{R}^m \times \mathbb{R}^m. \tag{4}$$

Kolmogorov constructed already in 1934 an explicit fundamental solution of (4) (see (13) below) which is a $C^\infty$ function outside the diagonal [70].

As it is well known, in this problem the propagation is expressed as a $2m$-dimensional stochastic process $Y = (P, Q)$, and the fundamental solution of (4) describes its transition density. As a result of the random collision, the propagation of the $P$-variables is driven by a $m$-dimensional standard Brownian motion $W$, while $Q$ variables are related to the $P$ by a natural differential equality. Then the propagation is formalized as solution of the Langevin's equation

$$\begin{cases} dP(t) = dW(t), \\ dQ(t) = -P(t)dt. \end{cases} \tag{5}$$

In the deterministic expression, the differential relation between the variables is coded as a 1-form. Clearly the fundamental solution of the more general Eq. (1) has an analogous probabilistic meaning.

These models, introduced at microscopical level for the description of kinetic theory of gases (see [36]), can be applied at meso-scopical level in biological models, where the atoms are replaced by cells. Indeed simple cells of the cortex are able to detect not only the intensity of the visual input, but also secondary variables, typically gradient of perceived images or velocities of objects. The differential relation between these variables allow to identify the cortical space as a phase space, and to describe propagation of the visual signal with instruments similar to the ones recalled above. Consequently propagation of the signal have been modeled with a Kolmogorov-Fokker-Planck equation by Mumford [87], Williams and Jacobs [107], August and Zucker [3], models with non linear differential equations are due to [41].

Also in financial mathematics, stochastic models involving linear and non linear Kolmogorov type equations are relevant because they appear when considering path-dependent contingent claims (see, for instance, [90]). More precisely, let us assume that the price $S$ of an asset is defined as in the Black-Scholes framework

[10]: $S_t = \exp\left(\left(r - \frac{\sigma^2}{2}\right)t + \sigma W_t\right)$ where $r$ and $\sigma$ denote the constant interest rate and volatility respectively. Then the price $u = u(t, S_t, Y_t)$ of a contingent claim which depends on $Y_t = \int_0^t \log S_s ds$, solves a Kolmogorov type equation (see, for instance, [4]). Other examples of path-dependent models arising in finance can be found in [56, 63].

As a generalization of the phase space, we can consider a general CR structure or a real hyper-surface in $\mathbb{C}^n$: in this case the analogous of the coupling position-velocity is realized by the quasicomplex structure. The basis of the complex tangent bundle is a lower dimensional distribution, described by a family of vector fields. In particular, curvature equations are naturally expressed in terms of vector fields and provide examples non linear Hörmander type PDE (see [39, 84]).

## 2 Fundamental Solutions of Linear Operators

The first aspect of the problem we want to face is the existence and Gaussian estimates and of a fundamental solution for the operator of (1) with Hölder continuous coefficients. The first existence results for operators of Hörmander type operators, refer to sum of squares of vector fields, plus a drift term.

$$L_I = \sum_{i=1}^{m} X_i^2 - X_0. \tag{6}$$

In this case the matrix $(a_{ij})$ in (1) is the identity. In particular Hörmander pointed out in the introduction of his celebrated paper on hypoelliptic second order differential equations [65] that the Kolmogorov method can also be applied to a class of operators which generalize Eq. (4), but fall in the general framework (1). Uniform but not Gaussian estimates, for families of Hörmander operators of this type, were proved by Rothschild and Stein [97]. Gaussian but not uniform estimates were proved by Jerison and Sánchez-Calle [66], via Gevrey regularity methods, Varopoulos et al. [105], via semi-group theory, and by Kusuoka and Stroock [71], via probabilistic techniques.

The results we plan to present here refer to non divergence form operators, with $C^\alpha$ coefficients, and the main results regarding the heat equation are due to Bonfiglioli et al. [21, 22], Bramanti et al. in [28] while for the Kolmogorov operator we quote the results of Polidoro [94], Lanconelli and Polidoro [76], Lanconelli and Pascucci in [74].

The contribution of these papers are twofold: from one side they establish uniform Gaussian bounds for the fundamental solution of a model operator of the form

$$L_w = \sum_{i,j=1}^{m} a_{ij}(w) X_{i,w} X_{j,w} - X_{0,w} \tag{7}$$

where $(a_{ij}(w))$ are constant coefficients while the family $(X_{i,w})$ is can be the given operators or a nilpotent and stratified approximation. This goal can be reached either with probabilistic instruments or with an analytic approach:

They apply the Levi's parametrix method to prove the results for operators with Hölder continuous coefficients $a_{ij}$. The method is based on the approximation of the fundamental solution $\Gamma_A(z;\zeta)$ of the given operator by the fundamental solution $\Gamma_w(z;\zeta)$ of a model operator belonging of the previous studied class and obtained by evaluating the coefficient at a point $w$ and approximating the vector fields in a neighborhood of each point $w$.

We present here the application of the method in two particularly significant cases of Eq. (1): the Kolmogorov equation, which will be studied with stochastic instruments and the heat equation, which will be studied with deterministic ones.

## 2.1 Kolmogorov Type Operators

We will call Kolmogorov type operators an operator of the form

$$L_A u(t,x) := \sum_{i,j=1}^{m} a_{ij}(t,x)\partial_{x_i x_j} u(t,x) + \sum_{i,j=1}^{N} b_{ij} x_j \partial_{x_i} u(t,x) - \partial_t u(t,x), \tag{8}$$

where $a_{ij}$ satisfy condition (2). This operator clearly falls in the general framework of Eq. (1), by choosing $X_0 = \partial_t - \sum_{i,j=1}^{N} b_{ij} x_i \partial_{x_j}$, $X_j = \partial_j$. In order to study its fundamental solution, we will preliminary study a model operator

$$Ku(t,x) := \frac{1}{2}\sum_{j=1}^{m} \partial_{x_j}^2 u(t,x) + \sum_{i,j=1}^{N} b_{ij} x_j \partial_{x_i} u(t,x) - \partial_t u(t,x). \tag{9}$$

The linear stochastic differential equation in $\mathbb{R}^N$ associated to $K$ is the following:

$$dZ_t = BZ_t dt + \sigma dW_t, \qquad Z_s = z, \tag{10}$$

where $W$ is a standard $m$-dimensional Brownian motion, $B$ is a $N \times N$ constant matrix and $\sigma$ is the $N \times m$ constant matrix

$$\sigma = \begin{pmatrix} I_m \\ 0 \end{pmatrix} \tag{11}$$

where $I_m$ denotes the identity matrix in $\mathbb{R}^m$. Then the solution of (10) is a Gaussian process with mean vector

$$E[Z_t] = e^{(t-s)B} z,$$

and covariance matrix $\mathscr{C}_0(t-s)$ where

$$\mathscr{C}_0(t) = \int_0^t e^{(t-\rho)B}\sigma\sigma^* e^{(t-\rho)B^*}\, d\rho, \qquad t \geq 0.$$

Since $\sigma$ has dimension $N \times m$, the matrix $\mathscr{C}_0(t)$ is generally only positive *semi-definite* in $\mathbb{R}^N$, that is $Z_t$ possibly has *degenerate* multi-normal distribution. We recall the well-known Kalman condition from control theory provides an operative criterion for the positivity of $\mathscr{C}_0(t)$: the matrix $\mathscr{C}_0(t)$ is positive definite for $t > 0$ if and only if

$$\operatorname{rank}\left[\sigma,\, B\sigma,\, B^2\sigma,\, \ldots\,,\, B^{N-1}\sigma\right] = N. \tag{12}$$

Then (12) ensures that $Z_t$ has a Gaussian transition density

$$G(s,y;t,x) = \frac{1}{\sqrt{(2\pi)^N \det \mathscr{C}_0(t-s)}} \exp\left(-\frac{1}{2}\langle \mathscr{C}_0^{-1}(t-s)(x - e^{(t-s)B}y), x - e^{(t-s)B}y\rangle\right). \tag{13}$$

Furthermore $G$ is the fundamental solution of the Kolmogorov PDE associated to (10).

The fundamental solution under special assumptions has been found by Kolmogorov and Hörmander [65], but a systematic study of the operator (9) has been done by Lanconelli and Polidoro in [76]. In particular they recognized that the hypoellipticity is equivalent to the following explicit expression of $B$, with respect to a suitable basis of $\mathbb{R}^N$: $B = (b_{ij})_{i,j=1,\ldots,N}$ writes in the form

$$B = \begin{pmatrix} * & * & \cdots & * & * \\ B_1 & * & \cdots & * & * \\ 0 & B_2 & \cdots & * & * \\ \vdots & \vdots & \ddots & \vdots & \vdots \\ 0 & 0 & \cdots & B_r & * \end{pmatrix}, \tag{14}$$

where each $B_j$ is a $p_j \times p_{j-1}$ matrix with rank $p_j$, with

$$p_0 = m \geq p_1 \geq \cdots \geq p_r \geq 1, \quad \sum_{j=0}^{r} p_j = N, \tag{15}$$

and the $*$-blocks are arbitrary. Let us explicitly recall that the stratification condition implies in a standard way that in canonical coordinates there is a dilation and a translation naturally associated to the vector fields.

Using the existence of the fundamental solution for the constant coefficient operator, from the parametrix method it follows:

**Theorem 2.1** *Assume that* $\left(a_{ij}(t,x)\right)_{i,j=1,\dots,m}$ *is symmetric with Hölder continuous entries and satisfies* (2) *for some positive constant* $\lambda$. *Then the operator L defined in (8) has a fundamental solution* $\Gamma$. *Moreover, for any* $T > 0$ *there exist some positive constants* $c^-, c^+, \lambda^-, \lambda^+$ *such that*

$$c^-\Gamma^-(t,x;\tau,\xi) \le \Gamma(t,x;\tau,\xi) \le c^+\Gamma^+(t,x;\tau,\xi),$$

$$\left|\partial_{x_j}\Gamma(t,x;\tau,\xi)\right| \le \frac{c^+}{\sqrt{\tau - t}}\Gamma^+(t,x;\tau,\xi),$$

*for any* $(t,x), (\tau,\xi)$ *with* $0 < \tau - t < T$. *Here* $\Gamma^\pm$ *is the fundamental solution of L in (9) with constant coefficients* $a_{ij}^+ = \lambda^+\delta_{ij}$, $a_{ij}^- = \lambda^-\delta_{ij}$.

We outline the proof of Theorem 2.1 given in Polidoro [95] and Di Francesco and Pascucci [51].

*Sketch of the Proof.* For fixed $w \in \mathbb{R}^{1+N}$, we denote by $\Gamma_w(z;\zeta)$ the fundamental solution of the model operator $L_w$, with constant coefficients evaluated at the point $w$

$$L_w u := \sum_{i,j=1}^{m} a_{ij}(w)X_iX_ju - X_0.$$

Then we call parametrix the function

$$Z(z;\zeta) = \Gamma_\zeta(z;\zeta). \tag{16}$$

We remark that $Z$ is a good approximation of $\Gamma$ near $\zeta$ and the expression of $Z$ can be estimated explicitly. Then we suppose that the fundamental solution takes the form:

$$\Gamma(z;\zeta) = Z(z;\zeta) + \int_0^t \int_{\mathbb{R}^N} Z(z;w)G(w;\zeta)dw. \tag{17}$$

In order to find the unknown function $G$, we impose that $\Gamma$ is the solution to the equation $L\Gamma(\cdot;\zeta) = 0$ in $]0,+\infty[\times\mathbb{R}^N$: we wish to point out one more time, to make this totally transparent, that the operator $L$ acts on the variable $z$ while the point $\zeta$ is fixed. Then formally we obtain

$$\begin{aligned} 0 = L\Gamma(z;\zeta) &= LZ(z;\zeta) + L \iint_{]0,T[\times\mathbb{R}^N} Z(z;w)G(w;\zeta)dw \\ &= LZ(z;\zeta) + \iint_{]0,T[\times\mathbb{R}^N} LZ(z;w)G(w;\zeta)dw - G(z;\zeta), \end{aligned}$$

hence

$$G(z;\zeta) = LZ(z;\zeta) + \iint\limits_{]0,T[\times\mathbb{R}^N} LZ(z;w)G(w;\zeta)dw. \tag{18}$$

Therefore $G$ is a solution of an integral equation equivalent to a fixed-point problem that can be solved by the method of successive approximations:

$$G(z;\zeta) = \sum_{k=1}^{+\infty}(LZ)_k(z;\zeta), \tag{19}$$

where

$$(LZ)_1(z;\zeta) = LZ(z;\zeta),$$
$$(LZ)_{k+1}(z;\zeta) = \iint\limits_{]0,T[\times\mathbb{R}^N} LZ(z;w)(LZ)_k(w;\zeta)dw, \qquad k \in \mathbb{N}.$$

It is possible to prove that there exists $k_0 \in \mathbb{N}$ such that, for all $T > 0$ and $\zeta = (0, y) \in \mathbb{R}^{1+N}$, the function $(LZ)_k(\cdot;\zeta)$ is continuous and bounded for any $k \geq k_0$. Moreover the series

$$\sum_{k=k_0}^{+\infty}(LZ)_k(\cdot;\zeta)$$

converges uniformly on the strip $]0, T[\times\mathbb{R}^N$. Furthermore, the function $G(\cdot,\zeta)$ defined by (19) is a solution to the integral equation (18) in $]0, T[\times\mathbb{R}^N$ and $\Gamma$ in (17) is a fundamental solution to $L$.

*Remark 2.1* The method also gives some pointwise estimates of the fundamental solution and its derivatives. We refer to Corielli et al. [48] where the accuracy of the parametrix method is studied to obtain numerical approximations for financial problems.

*Remark 2.2* There exists a positive constant $M$ and, for every $T > 0$, there exists $c = c(T) > 0$ such that

$$\frac{e^{-Md((t,x),(\tau,y))^2/(t-\tau)}}{c\left|B\left((t,x), \sqrt{t-\tau}\right)\right|} \leq \Gamma_A\left((t,x),(\tau,y)\right) \leq \frac{ce^{-d((t,x),(\tau,y))^2/M(t-\tau)}}{\left|B\left((t,x), \sqrt{t-\tau}\right)\right|} \tag{20}$$

for any $(t, x), (\tau, y)$ with $0 < \tau - t < T$, where $d$ is the distance defined by the vector fields. Gaussian estimates for a general equation like (1) have been obtained by Boscain and Polidoro [26] and Cinti and Polidoro [37].

## 2.2 Gaussian Estimates for the Fundamental Solution of Heat Operators

An other particularly notable class of operators of type (1) is given by the heat operators

$$L_A = \sum_{i,j=1}^{m} a_{ij}(t,x) X_i X_j - \partial_t. \tag{21}$$

For sum of squares of vector fields operators of the kind (1) with left invariant homogeneous vector fields on Lie groups, Gaussian bounds have been proved by Varopoulos (see [105] and references therein). In absence of a group structure, Gaussian bounds have been proved, on a compact manifold and for finite time, by Jerison and Sánchez-Calle [66], with an analytic approach and, on the whole $\mathbb{R}^{N+1}$, by Kusuoka and Stroock, (see [71] and references therein), using the Malliavin stochastic calculus.

In a long series of papers Bonfiglioli et al. [20–22], Bramanti et al. [28] and Capogna et al. [35], proved new Gaussian bounds for the operator $L_A$ with Hölder continuous coefficients. In the first papers the vector fields were assumed to belong to a Carnot group. Then, in [28] the results are presented in the full generality of $C^\infty$ vector fields satisfying the Hörmander condition. In this last case, the operator $L_A$ is initially assumed defined only on a cylinder $\mathbb{R} \times \Omega$ for some bounded $\Omega$, but, in order to obtain asymptotic estimates, it is extended to the whole space $\mathbb{R}^{N+1}$, in such a way that, outside a compact spatial set, it coincides with the classical heat operator. Henceforth all our statements will be referred to this extended operator.

**Theorem 2.2 (Gaussian Bounds)** *There exists a positive constant M and, for every $T > 0$, there exists a positive constant $c = c(T)$ such that, for $0 < t - \tau \le T$, $x, \xi \in \mathbb{R}^N$, the following estimates hold*

$$\frac{e^{-Md(x,\xi)^2/(t-\tau)}}{c|B(x,\sqrt{t-\tau})|} \le \Gamma_A(t,x;\tau,\xi) \le \frac{ce^{-d(x,\xi)^2/M(t-\tau)}}{|B(x,\sqrt{t-\tau})|}$$

$$|X_i\Gamma_A(t,\cdot;\tau,\xi)(x)| \le \frac{ce^{-d(x,\xi)^2/M(t-\tau)}}{(t-\tau)^{1/2}|B(x,\sqrt{t-\tau})|}$$

$$\left|X_iX_j\Gamma_A(t,\cdot;\tau,\xi)(x)\right| + |\partial_t\Gamma_A(\cdot,x;\tau,\xi)(t)| \le \frac{ce^{-d(x,\xi)^2/M(t-\tau)}}{(t-\tau)|B(x,\sqrt{t-\tau})|}$$

*where $|B(x,r)|$ denotes the Lebesgue measure of the purely spatial d-Carnot-Carathéodory ball in $\mathbb{R}^N$.*

We explicitly note that this estimate is analogous to the estimate (20) for the Komogorov equation, but here the distance in $[0,T] \times \mathbb{R}^N$ splits in the sum of a purely spatial one and a purely temporal one. Hence in this case

$$\frac{d_2\left((t,x),(\tau,\xi)\right)^2}{M\left(t-\tau\right)} \leq \frac{d\left(x,\xi\right)^2}{M\left(t-\tau\right)} + C$$

allowing to discard the temporal part of the distance in the estimate.

As a main step in the proof of these bounds, they first consider constant coefficients operators: the point here is to handle carefully the dependence on the matrix $A$ and obtain *uniform* estimates, in the ellipticity class of the matrix $A$. To prove these uniform bounds, in [21] the authors exploited direct methods and the previous results in [18, 19]. While in [28] the authors have followed as close as possible the techniques of Jerison and Sánchez-Calle [66], the main new difficulties being the following: first, they have to take into account the dependence on the matrix $A$, getting estimates depending on $A$ only through the number $\lambda$; second, the estimates have to be global in space, while in [66] they work on a compact manifold; third, they need estimates on the difference of the fundamental solutions of two operators which have no analogue in [66]. The procedure is technically involved, it makes use of the uniform estimates [21] on groups, and a crucial role is played by the Rothschild-Stein lifting theorem [97].

Once obtained the uniform estimates for the model operator with constant coefficients, one can apply the Levi parametrix method and establish existence and Gaussian bounds for the fundamental solution of the operators with variable Hölder continuous coefficients $a_{ij}$.

**Theorem 2.3 (Existence of a Fundamental Solution)** *Under the above assumptions, there exists a global fundamental solution $\Gamma_A(t,x;\tau,\xi)$ for $L_A$ in $\mathbb{R}^{N+1}$, with the properties listed below.*

*(i) $\Gamma_A$ is a continuous function away from the diagonal of $\mathbb{R}^{N+1} \times \mathbb{R}^{N+1}$; $\Gamma_A(t,x;\tau,\xi) = 0$ for $t \leq \tau$. Moreover, for every fixed $\zeta \in \mathbb{R}^{N+1}$, $\Gamma_A(\cdot;\zeta) \in C^{2,\alpha}_{loc}(\mathbb{R}^{N+1} \setminus \{\zeta\})$, and we have*

$$L_A\left(\Gamma_A(\cdot;\zeta)\right) = 0 \quad \text{in } \mathbb{R}^{N+1} \setminus \{\zeta\}.$$

*(ii) For every $\psi \in C_0^\infty(\mathbb{R}^{N+1})$, the function $w(z) = \int_{\mathbb{R}^{N+1}} \Gamma_A(z;\zeta)\,\psi(\zeta)\,d\zeta$ belongs to the class $C^{2,\alpha}_{loc}(\mathbb{R}^{N+1})$, and we have*

$$L_A w = -\psi \quad \text{in } \mathbb{R}^{N+1}.$$

*(iii) Let $\mu \geq 0$ and $T_2 > T_1$ be such that $(T_2 - T_1)\mu$ is small enough. Then, for every $f \in C^\beta([T_1,T_2] \times \mathbb{R}^N)$ (where $0 < \beta \leq \alpha$) and $g \in C\left(\mathbb{R}^N\right)$ satisfying the growth condition $|f(x,t)|, |g(x)| \leq c\,\exp(\mu\, d(x,0)^2)$ for some constant $c > 0$, the function*

$$u(x,t) = \int_{\mathbb{R}^N} \Gamma_A(t,x;T_1,\xi)\,g(\xi)\,d\xi + \int_{[T_1,t]\times\mathbb{R}^N} \Gamma_A(t,x;\tau,\xi) f(\tau,\xi)\,d\tau d\xi,$$

*$x \in \mathbb{R}^N$, $t \in (T_1, T_2]$, belongs to the class $C^{2,\beta}_{loc}((T_1, T_2) \times \mathbb{R}^N) \cap C([T_1, T_2] \times \mathbb{R}^N)$. Moreover, u is a solution to the following Cauchy problem*

$$L_A u = -f \ \text{in} \ (T_1, T_2) \times \mathbb{R}^N, \quad u(\cdot, T_1) = g \ \text{in} \ \mathbb{R}^N.$$

The proof follows the same ideas of the analogous presented in the previous section. The parametrix function is built starting from the fundamental solution $\Gamma_w$ of the constant coefficient operator

$$L_w = \sum_{i,j=1}^{m} a_{ij}(w)\,(t,x)\,X_i X_j - \partial_t.$$

## 2.3 Fundamental Solution of More General Operators

Operators in the form of sum of squares of Hörmander vector fields with drift

$$L = \sum_{j=1}^{m} X_j^2 - X_0 \tag{22}$$

write in the form (1) as $A$ is the $m \times m$ identity matrix. Kogoj and Lanconelli consider this kind of operators in [67], under the assumption that every pair of points $(t, x)$ and $(\tau, \xi)$ with $t < \tau$ can be joined by a Lipschitz path $\gamma$ which solves almost everywhere the non-autonomous ODE

$$\gamma'(s) = \sum_{i=1}^{m} \beta_i(s) X_i(\gamma(s)) + \beta_0(s) X_0(\gamma(s)) \tag{23}$$

with $\beta_0(s) \geq 0$ for almost every $s$. In the article [67], Kogoj and Lanconelli give a list of examples of operators satisfying (23), that include, among other examples, Kolmogorov operators, as well as heat operators with smooth coefficients. For this family of operators, they prove the existence of a fundamental solution $\Gamma(t, x, s, y)$, which is strictly positive in the set $\{(x, t) \in \mathbb{R}^{N+1} \mid t > s\}$, and Gaussian upper bounds for $\Gamma$. They also prove mean value formulas and Harnack inequalities for the positive solutions of $Lu = 0$.

Based on the Harnack inequality proved in [67], and on the translation invariance, Pascucci and Polidoro prove in [91] sharp lower bounds for the fundamental solution of operators satisfying (23). More recently, the method used in [91] has been extended in [38] to the study of Hormander operators that do not satisfy (23). For instance, the operator $L = \partial^2_{x_1} + x_1^2 \partial^2_{x_2} + \partial_t$ is considered in two space variables. The fundamental solution $\Gamma = \Gamma(t, x, \tau, \xi)$ is supported in the set $\{(t, x) \in \mathbb{R}^3 | t > \tau, x_2 > \xi_2\}$, then no Gaussian estimates can be proved for this example. On the other hand, upper and lower bound have been proved by combining PDE methods and Malliavin calculus.

## 3 Balls, Mean Value Formulas and Potential Theory

An important aspect in the study of the geometric analysis associated with many of the PDEs discussed so far is the investigation of the underlying geometric properties naturally associated to these PDEs. Starting from the celebrated papers of Bony [25] and of Nagel et al. [88], it became clear that the properties of the exponential maps associated with the smooth vector fields play a crucial role in understanding the equivalence of the distances of the spaces. This notion has been weakened by Lanconelli and Morbidelli [73] to the notion of quasi exponential for Lipschitz continuous vector fields. Then they proved a Poincaré inequality under a ball box type assumption.

A complementary point of view, largely adopted by Lanconelli, is to choose the level sets of the fundamental solution as privileged class of balls for the operator. The main advantage of this perspective is that the level sets of the fundamental solution reflect the main properties of the operator, and in particular they give information on the directions of propagation, allowing to express in a natural and intrinsic way the Poincaré inequality and the Potential Theory results, properties which are classically expressed on the balls of the metric.

The first results extending the mean-value formulas from the classical Laplace setting to the parabolic one are due to Pini [92], Watson [106], Fabes and Garofalo [54], Lanconelli and Garofalo [60, 61]. In the sub-Riemannian setting, a mean value theorem for sums of squares of vector fields has been proved by Hoh and Jacob [64], Citti et al. [42], while the formula for general Kolmogorov type operators of type (1) is due to Lanconelli and Pascucci [75]. It has been proved in [23, 33, 59] that there is a strict relation between the existence of representation formulas and the Poincaré inequality, which indeed are equivalent in some special cases.

The use of asymptotic average operators in the characterization of classical subharmonic functions has a long history, starting with the papers [9] of Blaschke, [93] of Privaloff, [6] of Beckenbach and Radó, up to the recent monograph [2] of Armitage and Gardiner. This direction of research has been deeply developed in the framework of Carnot groups by Bonfiglioli et al. in the monograph [23], and then by Bonfiglioli and Lanconelli, who obtained new results concerning with: Harnack and Liouville type theorems [11]; characterizations of subharmonicity [13] (see also the very recent paper [17]); average formulas and representation theorems [17]; formulas of Poisson and Jensen type; maximum principles for open unbounded sets [12]; the Dirichlet problem with $L^p$ boundary data and the Hardy spaces associated with them [14]; the Eikonal equation and Bôcher-type theorems for the removal of singularities [15]; convexity properties of the mean-value formulas with respect to the radius [24]; Gauss-Koebe and Montel type normality results [5].

Finally we quote some results of Lanconelli [72], Abbondanza and Bonfiglioli [1], Kogoj et al. [69], Kogoj and Tralli [68], who characterized the set on which a mean value formula can be proved as the level sets of the fundamental solution.

## 3.1 Almost Exponential Maps and Poincaré Inequality

The most classical result on exponential mappings and properties of control balls is due to Nagel et al. [88].

An abstract version of these notions was provided in the paper [73] for a family $X_1, \dots, X_m$ of Lipschitz continuous vector fields in $\mathbb{R}^N$. Indeed the authors introduced the notion of *controllable almost exponential map* and they showed that, if a suitable *ball-box* inclusion holds, then one can get a proof of a Poincaré-type inequality for the family $X_1, \dots, X_m$. Next we will describe such result.

**Definition 3.1** Let $\Omega \subset \mathbb{R}^N$ be an open set and let $Q$ be an open neighborhood of the origin in $\mathbb{R}^N$. We say that a $C^1$ map $E : \Omega \times Q \to \mathbb{R}^N$ is an *almost exponential map* if:

(i) The map $Q \ni h \mapsto E(x,h)$ is one-to-one for each $x \in \Omega$.
(ii) There is $C_0 > 1$ such that

$$0 < C_0^{-1} \left| \det \frac{\partial E}{\partial h}(x,0) \right| \le \left| \det \frac{\partial E}{\partial h}(x,h) \right| \le C_0 \left| \det \frac{\partial E}{\partial h}(x,0) \right| \quad \text{for all } h \in Q.$$

An almost exponential map is *controllable* if there are a *hitting time* $T > 0$ and a *control function* $\gamma : \Omega \times Q \times [0,T] \to \mathbb{R}^N$ such that:

(iii) For each $(x,h) \in \Omega \times Q$, the path $t \mapsto \gamma(x,h,t)$ is subunit and it satisfies $\gamma(x,h,0) = x$ and $\gamma(x,h,T(h)) = E(x,h)$ for some $T(h) \le T$.
(iv) For each $h \in Q$ and $t \in [0,T(h)]$, the map $x \mapsto \gamma(x,h,t)$ is one-to-one, of class $C^1$ and it satisfies for some $C_0 > 1$

$$\left| \det \frac{\partial \gamma}{\partial x}(x,h,t) \right| \ge C_0^{-1} \quad \text{for all } x \in \Omega, h \in Q \text{ and } t \in [0,T(h)].$$

Let us recall also the *local doubling condition* for the Lebesgue measure of control balls: for any compact $K$ there is $C_D$ and $r_0 > 0$ such that

$$|B(x_0,2r)| \le C_D |B(x_0,r)| \quad \text{for all } x_0 \in K \text{ and } r \le r_0.$$

Now we are ready to give a condition which ensures the Poincaré inequality.

**Theorem 3.1 ([73], Theorem 2.1)** *Let $X_1, \dots, X_m$ be a family of locally Lipschitz-continuous vector fields in $\mathbb{R}^N$. Assume that the Lebesgue measure of Carnot–Carathéodory balls is locally doubling. Let $K \subset \mathbb{R}^N$ be a compact set and let*

*$B = B(x_0, r)$ be a ball such that $x_0 \in K$ and $r \leq r_0$. Assume that for a suitable $C_0 > 0$ there are open sets $Q \supset \{0\}$, $\Omega \subset B$ and an almost exponential map $E : \Omega \times Q \to \mathbb{R}^N$ such that*

*(1) $|\Omega| > C_0^{-1}|B|$.*
*(2) The map E is controllable with a control $\gamma$ having hitting time $T \leq C_0 r$.*
*(3) we have the inclusion $B \subset E(x, Q)$ for each $x \in \Omega$.*

*Then, there is a constant $C_1$ depending on $C_0$ and $C_D$ such that*

$$\int_B |u(x) - u_B| dx \leq C_1 r \int_{C_1 B} |Xu(x)| dx \quad \text{for all } u \in C^1(C_1 B). \tag{24}$$

*Remark 3.1* Here $u_B$ denote the standard average on the ball $B$:

$$u_B = \frac{1}{|B|} \int_B u(y) dy.$$

In the next section we will give a different definition of mean, to be used when the vector fields are associated to an operator.

For the proof we refer to the original paper [73]. Here we note that the method has been tested successfully in the case of Hörmander vector fields, with regular and non regular coefficients and on a class of vector fields introduced by Franchi and Lanconelli [57, 58] which have the form

$$X_1 = \partial_{x_1},\ X_2 = \lambda_2(x_1)\partial_{x_2}, \quad \ldots \quad , X_n = \lambda_n(x_1, \ldots, x_{n-1})\partial_{x_n},$$

where the functions $\lambda_j$ satisfy suitable assumptions. We refer to the discussion in [73, Sect. 3] for the proof that these vector fields fit in the framework of controllable almost exponential maps. Further results with minimal assumptions on the coefficients are due to [85], and with a slightly different technique to [30].

## 3.2 Mean Value Formulas on Level Sets and Poincaré Inequality

We will present here *mean value formulas*, which have been constructed for different operators: in the subelliptic setting: by Citti et al. for sum of squares [42], Lanconelli and Pascucci for Kolmogorov-type operators [75]. The first corollary will be a Poincaré formula, to be compared with the one established in the previous section.

Let us consider a particular operator of type (1)

$$L_A := \sum_{i,j=1}^{N} a_{ij} X_i X_j + X_0 \tag{25}$$

for a constant coefficient matrix $(a_{ij})$ satisfying (2). We will denote

$$\Omega_r(x) := \{y \in \mathbb{R}^N : \Gamma(x,y) > 1/r\},$$

so that $\partial\Omega_r(x)$ will be the level set of $\Gamma$.

Let $\Omega \subseteq \mathbb{R}^N$ be an open set and suppose $u$ is u.s.c. on $\Omega$. For every fixed $\alpha > 0$, and every $x \in \mathbb{R}^N$ and $r > 0$ such that $\overline{\Omega_r(x)} \subset \Omega$, we define the Surface Mean $m_r$ and the Solid Mean $M_r$ for a function $u$:

$$m_r(u)(x) = \int_{\partial\Omega_r(x)} u(y) \frac{a_{ij}(y)X_i\Gamma_x(y)X_j\Gamma_x(y)}{|\nabla_E\Gamma_x(y)|} \, d\sigma(y), \tag{26}$$

$$M_r^\alpha(u)(x) = \frac{\alpha+1}{r^{\alpha+1}} \int_0^r \rho^\alpha \, m_\rho(u)(x) \, d\rho, \tag{27}$$

where $\nabla_E$ denotes the Euclidean gradient, and $\Gamma_x(y) = \Gamma(x,y)$. Here $\sigma$ denotes the Hausdorff $(N-1)$-dimensional measure in $\mathbb{R}^N$. We also denote

$$I_r(x) = \frac{\alpha+1}{r^{\alpha+1}} \int_0^r \rho^\alpha \Big( \int_{\Omega_\rho(x)} a_{ij}X_i\Gamma(x,y)X_ju(y)\, dy \Big) d\rho. \tag{28}$$

The following theorem, for the special case $X_0 = 0$ has been proved in [42], A general formula has been established by Lanconelli and Pascucci [75] for Kolmogorov equations, which reduces to the following one, when $divX_0 = 0$.

**Theorem 3.2** *Then, for every function $u$ of class $C^2$ on an open set containing $\overline{\Omega_r(x)}$, we have the following mean value formulas:*

$$u(x) = m_r(u)(x) - \int_{\Omega_r(x)} a_{ij}X_i\Gamma(x,y)X_ju(y)\, dy, \quad u(x) = M_r^\alpha(u)(x) - I_r(x). \tag{29}$$

In [45] the authors remarked that the Poincaré inequality can be obtained by means of the mean value formula for a very special class of vector fields, with minimal regularity of the coefficients in the same spirit of Bramanti et al. [30] and Montanari and Morbidelli [85]. Precisely when $X_0 = 0$, and there exists a continuous function $\varphi$ such that

$$X_i = \partial_{x_i} - x_{i+n}\partial_{x_{2n}}, \; X_n = \partial_{x_n} + 2\varphi(x)\partial_{x_{2n}}, X_{i+n} = \partial_{x_{i+n}} + x_i\partial_{x_{2n}}, \tag{30}$$

$i = 1, \dots, n-1$. These vector fields satisfy the Hörmander condition, and $\varphi$ is continuous, so that there is a CC distance associated to these vector fields.

**Theorem 3.3** *Let $\Omega$ be an open set. Assume that the functions $\varphi$ and $u$ are Lipschitz continuous defined on $\Omega$ with respect to the CC distance associated to these vector fields. For every compact set $K \subset \Omega$ there exist positive constants $C_1, C_2$*

*with $C_2 > 1$ (depending continuously on the Lipschitz constant of $\varphi$) such that if $\overline{\Omega_{C_2 r}(\bar{x})} \subset K$, we have*

$$\int_{\Omega_r(\bar{x})} |u(x) - u_{\Omega_r(\bar{x})}| dx \leq C_1 r \int_{\Omega_{C_2 r}(\bar{x})} |\nabla u|.$$

## 3.3 L-Subharmonicity and Average Operators

As we shall see soon, mean value formulas naturally allow the characterizations of the $\mathscr{L}$-subharmonic functions, and the derivation of an in-depth *Potential Theory* for $\mathscr{L}$.

Let

$$\mathscr{L} := \sum_{i,j=1}^{N} \partial_{x_i}(a_{i,j}(x)\, \partial_{x_j}) = \mathrm{div}(A(x)\, \nabla) \tag{31}$$

be a linear second order PDO in $\mathbb{R}^N$, in divergence form, with $C^\infty$ coefficients and such that the matrix $A(x) := (a_{i,j}(x))_{i,j \leq N}$ is *symmetric* and *nonnegative definite* at any point $x = (x_1, \dots, x_N) \in \mathbb{R}^N$. The operator $\mathscr{L}$ is (possibly) degenerate elliptic. However, we assume that $\mathscr{L}$ is *not totally degenerate* at every point. Precisely, we assume that the following condition holds: there exists $i \in \{1, \dots, N\}$ such that $a_{i,i} > 0$ on $\mathbb{R}^N$. This condition, together with $A(x) \geq 0$, implies the well-known Picone's Maximum Principle for $\mathscr{L}$.

A function $h$ will be said *$\mathscr{L}$-harmonic* in an open set $\Omega \subseteq \mathbb{R}^N$ if $h \in C^2(\Omega, \mathbb{R})$ and $\mathscr{L}h = 0$ in $\Omega$. An upper semicontinuous function (u.s.c. function, for short) $u : \Omega \to [-\infty, \infty)$ will be called *$\mathscr{L}$-subharmonic* in $\Omega$ if:

1. The set $\Omega(u) := \{x \in \Omega \mid u(x) > -\infty\}$ contains at least one point of every (connected) component of $\Omega$, and
2. For every bounded open set $V \subset \overline{V} \subset \Omega$ and for every $\mathscr{L}$-harmonic function $h \in C^2(V, \mathbb{R}) \cap C(\overline{V}, \mathbb{R})$ such that $u \leq h$ on $\partial V$, one has $u \leq h$ in $V$.

We shall denote by $\mathscr{S}(\Omega)$ the family of the $\mathscr{L}$-subharmonic functions in $\Omega$.

It is well known that the subharmonic functions play crucial roles in Potential Theory of linear second order PDEs (just think about Perron's method for the Dirichlet problem) as well as in studying the notion of convexity in Euclidean and non-Euclidean settings.

Our main assumption on $\mathscr{L}$ is that it is *$C^\infty$-hypoelliptic* in every open subset of $\mathbb{R}^N$. We further assume that, in the spirit of the rest of the present paper, $\mathscr{L}$ admits a *nonnegative global fundamental solution*

$$\mathbb{R}^N \times \mathbb{R}^N \setminus \{x = y\} \ni (x, y) \mapsto \Gamma(x, y) \in \mathbb{R},$$

with pole at any point of the diagonal $\{x = y\}$ of $\mathbb{R}^N$ and vanishing at infinity.

We are then able to define suitable mean value operators on the *level sets* of $\Gamma$.

We explicitly remark that study of the integral operators related to general PDOs considered in this paper is complicated by the presence of non-trivial kernels. For instance, when $\mathscr{L}$ in (31) is a sub-Laplacian on a stratified Lie group $\mathbb{G}$, the kernels appearing in the relevant mean-integrals cannot be identically 1, *unless* $\mathbb{G}$ *is the usual Euclidean group* $(\mathbb{R}^N, +)$, as it is proved in [15].

**Definition 3.2 (Mean-Integral Operators)** Let $x \in \mathbb{R}^N$ and let us consider the functions, defined for $y \neq x$,

$$\Gamma_x(y) := \Gamma(x,y), \qquad \mathscr{K}_x(y) := \frac{\langle A(y)\nabla\Gamma_x(y), \nabla\Gamma_x(y)\rangle}{|\nabla\Gamma_x(y)|}.$$

We will call *surface mean integral operator* and *solid mean integral operator*, the two mean operators $m_r$ and $M_r$ defined in (26) and (27), respectively. Furthermore, for every $x \in \mathbb{R}^N$ and every $r > 0$, we set

$$q_r(x) = \int_{\Omega_r(x)} \left(\Gamma_x(y) - \frac{1}{r}\right) \mathrm{d}y, \qquad Q_r(x) = \frac{\alpha+1}{r^{\alpha+1}} \int_0^r \rho^\alpha\, q_\rho(x)\, \mathrm{d}\rho,$$

$$\omega_r(x) = \frac{1}{\alpha\, r^{\alpha+1}} \int_{\Omega_r(x)} \left(r^\alpha - \Gamma_x^{-\alpha}(y)\right) \mathrm{d}y.$$

Remarkable mean-value formulas generalizing the classical Gauss-Green formulas for Laplace's operator and the ones in Theorem 3.2 hold true also in this more general setting:

**Theorem 3.4 (Mean-Value Formulas for $\mathscr{L}$)** *Let $m_r, M_r^\alpha$ be the average operators in Definition 3.2. Let also $x \in \mathbb{R}^N$ and $r > 0$.*

*Then, for every function $u$ of class $C^2$ on an open set containing $\overline{\Omega_r(x)}$, we have the following $\mathscr{L}$-representation formulas:*

$$u(x) = m_r(u)(x) - \int_{\Omega_r(x)} \left(\Gamma(x,y) - \frac{1}{r}\right) \mathscr{L}u(y)\, \mathrm{d}y, \tag{32}$$

$$u(x) = M_r^\alpha(u)(x) - \frac{\alpha+1}{r^{\alpha+1}} \int_0^r \rho^\alpha \left( \int_{\Omega_\rho(x)} \left(\Gamma(x,y) - \frac{1}{\rho}\right) \mathscr{L}u(y)\, \mathrm{d}y \right) \mathrm{d}\rho. \tag{33}$$

*We shall refer to* (32) *as the* Surface Mean-Value Formula for $\mathscr{L}$, *whereas* (33) *will be called the* Solid Mean-Value Formula for $\mathscr{L}$.

Before stating our main theorem, we need two definitions. With the same notations as in the previous paragraph, an u.s.c. function $u$ defined on an open subset $\Omega$ of $\mathbb{R}^N$ will be called *m-continuous in* $\Omega$ if

$$\lim_{r\to 0} m_r(u)(x) = u(x), \quad \text{for every } x \in \Omega.$$

Analogously, $u$ is said to be $M^{\alpha}$*-continuous in* $\Omega$ if $\lim_{r\to 0} M_r^{\alpha}(u)(x) = u(x)$, for every $x \in \Omega$.

Finally, let $I \subseteq \mathbb{R}$ be an interval and suppose that $\varphi : I \to \mathbb{R}$ is a strictly monotone continuous function. We say that $f : I \to \mathbb{R}$ is $\varphi$-convex if

$$f(r) \leq \frac{\varphi(r_2) - \varphi(r)}{\varphi(r_2) - \varphi(r_1)} f(r_1) + \frac{\varphi(r) - \varphi(r_1)}{\varphi(r_2) - \varphi(r_1)} f(r_2),$$

for every $r_1, r, r_2 \in I$ such that $r_1 < r < r_2$.

We are ready to present our main result (see [17, 24]). This generalizes previous results in [13]; in the case of sub-Laplacians on Carnot groups, the paramount role of mean value operators is shown in [1, 5, 11, 12, 14–16]; see also the comprehensive monograph [23].

**Theorem 3.5 (Characterizations of Subharmonicity)** *Suppose* $\mathscr{L}$ *satisfies the above axioms. Let* $\Omega$ *be an open subset of* $\mathbb{R}^N$ *and let* $u : \Omega \to [-\infty, \infty)$ *be an u.s.c. function such that* $\Omega(u) = \{x : u(x) > -\infty\}$ *contains at least one point of every component of* $\Omega$.

*Let* $q_r, Q_r, \omega_r$ *be as in Definition 3.2. Let also* $R(x) := \sup\{r > 0 : \Omega_r(x) \subseteq \Omega\}$. *Then, the following conditions are equivalent:*

1. $u \in \mathscr{S}(\Omega)$ *with respect to* $\mathscr{L}$.
2. $u(x) \leq m_r(u)(x)$, *for every* $x \in \Omega$ *and* $r \in (0, R(x))$.
3. $u(x) \leq M_r^{\alpha}(u)(x)$, *for every* $x \in \Omega$ *and* $r \in (0, R(x))$.
4. *It holds that*

$$\limsup_{r\to 0} \frac{m_r(u)(x) - u(x)}{q_r(x)} \geq 0, \quad \textit{for every } x \in \Omega(u).$$

5. *It holds that*

$$\limsup_{r\to 0} \frac{M_r^{\alpha}(u)(x) - u(x)}{Q_r(x)} \geq 0, \quad \textit{for every } x \in \Omega(u).$$

6. $u$ *is m-continuous in* $\Omega$, *and* $r \mapsto m_r(u)(x)$ *is monotone increasing on* $(0, R(x))$, *for every* $x \in \Omega$.
7. $u$ *is* $M^{\alpha}$*-continuous in* $\Omega$, *and* $r \mapsto M_r^{\alpha}(u)(x)$ *is monotone increasing on* $(0, R(x))$, *for every* $x \in \Omega$.
8. $u$ *is m-continuous in* $\Omega$, *and*

$$M_r^{\alpha}(u)(x) \leq m_r(u)(x),$$

*for every* $x \in \Omega$ *and every* $r \in (0, R(x))$.

9. $u$ *is m-continuous in* $\Omega$, *and*

$$\liminf_{r\to 0} \frac{m_r(u)(x) - M_r^{\alpha}(u)(x)}{\omega_r(x)} \geq 0, \quad \textit{for every } x \in \Omega(u).$$

10. *$u$ is $M^\alpha$-continuous in $\Omega$, $u \in L^1_{\mathrm{loc}}(\Omega)$ and $\mathscr{L}u \geq 0$ in the weak sense of distributions.*
11. *$u$ is m-continuous and the map $r \mapsto m_r(u)(x)$ is $\frac{1}{r}$-convex on $(0, R(x))$, for every $x \in \Omega$ (or, equivalently, for every $x \in \Omega(u)$).*
12. *$u$ is $M^\alpha$-continuous and, for every $x \in \Omega$ (or, equivalently, for every $x \in \Omega(u)$), the map $r \mapsto M_r^\alpha(u)(x)$ is $\frac{1}{r^{\alpha+1}}$-convex on $(0, R(x))$, for some (or for every) $\alpha > 0$.*

*Furthermore, if $u \in \mathscr{S}(\mathbb{R}^N)$ we have the following results:*

13. *The functions $x \mapsto m_r(u)(x)$, $M_r^\alpha(u)(x)$ are $\mathscr{L}$-subharmonic in $\mathbb{R}^N$, finite valued and continuous.*
14. *Let $\mu_u$ be the $\mathscr{L}$-Riesz measure of $u$; the maps $r \mapsto m_r(u)(x)$ and $r \mapsto M_r^\alpha(u)(x)$ can be prolonged with continuity up to $r = 0$ if and only if $x \in \Omega(u)$.*

    *Furthermore, for every $x \in \Omega$ and $r \in (0, R(x))$, one has the following representation formulas (of Poisson-Jensen type):*

$$u(x) = m_r(u)(x) - \int_0^r \frac{\mu_u(\Omega_\rho(x))}{\rho^2}\,\mathrm{d}\rho$$
$$= m_r(u)(x) - \int_{\Omega_r(x)} \Big(\Gamma(x,y) - \frac{1}{r}\Big)\,\mathrm{d}\mu_u(y),$$
$$u(x) = M_r^\alpha(u)(x) - \int_0^r \frac{\alpha+1}{\rho^{\alpha+2}}\left(\int_{\Omega_\rho(x)} \Big(f_\alpha(\rho) - f_\alpha\Big(\frac{1}{\Gamma(x,y)}\Big)\Big)\,\mathrm{d}\mu_u(y)\right)\mathrm{d}\rho$$
$$= M_r^\alpha(u)(x) - \frac{\alpha+1}{r^{\alpha+1}}\int_0^r \rho^\alpha\left(\int_{\Omega_\rho(x)} \Big(\Gamma(x,y) - \frac{1}{\rho}\Big)\,\mathrm{d}\mu_u(y)\right)\mathrm{d}\rho.$$

    *When $x \notin \Omega(u)$, all the sides of these formulas are $-\infty$, and this happens if and only if $\mu_u(\{x\}) > 0$.*

The equivalences (1)–(9) do *not* require the hypoellipticity of $\mathscr{L}$, which is only used in (10)–(14) (requiring Riesz-type representation results).

We observe that Theorem 3.5 provides new insight on the Potential Theory for operators in divergence form, which are *not necessarily* in the form of Hörmander sums of squares, nor left invariant on some Lie group (see [17, 24]).

Finally we mention some results of Lanconelli [72], Abbondanza and Bonfiglioli [1], Kogoj et al. [69], Kogoj and Tralli [68]: in these papers, it is proved, for several classes of PDOs, that the sets on which a mean value formula can be obtained are precisely the level sets of the fundamental solution. For instance the *inverse mean value theorem* for $\mathscr{L}$ states the following: let $\mathscr{K}_0(y)$ be as in Definition 3.2 and let us set $\mathrm{d}\nu(y) := \mathscr{K}_0(y)\,\mathrm{d}y$; let $D$ be a bounded open neighborhood of 0 such that

$$u(0) = \frac{1}{\nu(D)} \int_D u(y)\,\mathrm{d}\nu(y), \tag{34}$$

for every $u$ which is $\mathscr{L}$-harmonic and $\nu$-integrable on $D$. Then, necessarily, $D = \Omega_r(0)$ for some $r > 0$. More precisely, it suffices to suppose that (34) holds for the family of the $\mathscr{L}$-harmonic functions on $D$ of the form $D \ni y \mapsto \Gamma(y,x)$, for $x \notin D$.

# 4 Global Regularity Results

Interior Schauder and $L^p$ estimates can be obtained as a direct consequence of existence of the fundamental solution. A much more delicate problem is the problem of global regularity results, namely regularity on the whole space, or regularity at the boundary.

We provide here a couple of results, obtained using potential theory and existence of the fundamental solution, proved in the previous section.

## 4.1 A First Regularity Result at the Boundary

An essential play in study of the existence and regularity theory of the equation $Lu = f$ where $L$ is the operator in (8) is the derivation of the Schauder estimates in terms of weighted interior norms. Such apriori estimates allow to extends the results of potential theory to the class of $L$ having Hölder continuous coefficients and to establish the solvability of the Dirichlet problem in the generalized sense. For continuous boundary values and a suitably wide class of bounded open set the proof of solvability of the Dirichlet problem can be achieved entirely with interior estimates.

Interior Schauder's estimates for the Kolmogorov operator (6) are proved in Shatyro [99], for the operator (8) in the homogeneous case by Manfredini in [79] and in the non homogeneous case in Di Francesco and Polidoro [52]. In Lunardi [78] global estimates with respect to the spatial variable are proved for operator (8) with constant coefficients $a_{ij}$.

We denote by $C_d^\alpha(\Omega)$ the space of the Hölder continuous function whose norms $|\cdot|_{\alpha,d;\Omega}$ are weighted by the distance to the boundary of the bounded open set $\Omega$. Schauder's type estimate can be proved using classical arguments, based on a representation formula for the second derivatives of smooth functions in terms of the fundamental solution of the operator and on its bounds in Theorem 2.1.

**Theorem 4.1 (Schauder Interior Estimates)** *Let $\Omega$ be a bounded open set, $f \in C_d^\alpha(\Omega)$, and let $u$ be a bounded function belonging to $C_{loc}^{2+\alpha}(\Omega)$ such that $Lu = f$ in $\Omega$. Then $u \in C_d^{2+\alpha}(\Omega)$ and there exists a positive constant $c$, independent of $u$, such that*

$$|u|_{2+\alpha,d;\Omega} \leq c\,(\sup_{\Omega}|u| + |d^2 f|_{\alpha,d;\Omega}).$$

Here $|d^2 f|_{\alpha,d;\Omega}$ denotes the following norm :

$$|d^2 f|_{\alpha,d;\Omega} = \sup_{z\in\Omega} d_z^2 |f(z)| + \sup_{z,\zeta\in\Omega, z\neq\zeta} d_{z,\zeta}^{2+\alpha} \frac{|f(z)-f(\zeta)|}{d(z,\zeta)^\alpha}$$

where $d_z = \inf_{w\in\Omega} d(z,w)$ and $d_{z,\zeta} = \min\{d_z, d_\zeta\}$.

And

$$\begin{aligned}|u|_{2+\alpha,d;\Omega} &= \sup_{z\in\Omega}|u(z)| + \sum_{i=1}^{m} \sup_{z\in\Omega} d_z |\partial_{x_i} u(z)| + \sup_{z,\zeta\in\Omega, z\neq\zeta} d_{z,\zeta}^{2+\alpha} \frac{|u(z)-u(\zeta)|}{d(z,\zeta)^\alpha} \\ &+ \sum_{i=1}^{m} \sup_{z,\zeta\in\Omega, z\neq\zeta} d_{z,\zeta}^{2+\alpha} \frac{|\partial_{x_i} u(z) - \partial_{x_i} u(\zeta)|}{d(z,\zeta)^\alpha} + |d^2 X_0 u|_{\alpha,d;\Omega} \\ &+ \sum_{i,j=1}^{m} |d^2 \partial^2_{x_i x_j} u|_{\alpha,d;\Omega}.\end{aligned} \tag{35}$$

Using Schauder apriori estimates we can extend potential theory to the operator $L$ with Hölder continuous coefficients. In fact, $L$ endows $\mathbb{R}^{N+1}$ with a structure of $\beta$-harmonic space (according to the classical definition in [47]). Precisely, if $U$ is a bounded subset of $\mathbb{R}^{N+1}$ the space $(U, H^L)$ of $L$-harmonic $C^{2+\alpha}_{loc}(U)$ functions satisfies the axiom of positivity and separation, the Doob convergence property and finally the property of resolutivity. In particular the last axiom requires that there exists a basis (for the Euclidean topology) of bounded open set $V$ called $H^L$-regular set such that the Dirichlet problem

$$\begin{cases} Lu = 0 & \text{in } V, \\ u = \varphi & \text{in } \partial V, \ \varphi \in C(\partial V) \end{cases} \tag{36}$$

is univocally solvable. We cannot expect that the parabolic cylinders are $H^L$-regular set. A geometric condition on $\partial V$ ensuring the solvability of (36) is a generalization of the Poincaré exterior ball condition. Precisely, we assume that for every $z_0 \in \partial V$ there exists a $L$-non-characteristic outer normal $\nu \in \mathbb{R}^{N+1}$ such that $B_{eucl}(z_0 + \nu, |\nu|) \subset \mathbb{R}^{N+1} \setminus V$ and

$$\sum_{i,j=1}^{m} a_{ij} z_0 \langle \nu, X_i \rangle \langle \nu, X_j \rangle > 0.$$

The construction of a basis of $H^L$-regular sets is proved using an argument due a Bony [25] and the method of continuity.

The general potential theory ensures the existence of a generalized solution in the sense of Perron-Wiener-Brelot-Bauer of the Dirichlet problem in an arbitrary bounded open set $\Omega$. This solution assumes the boundary data at every $L$-regular point. A point $z_0$ is $L$-regular if there exists a local barrier at $z_0$.

**Theorem 4.2** *Manfredini [79] (Existence of a Generalized Solution). Let $\Omega$ be a bounded open set, $f \in C^{\alpha}(\Omega)$ and $\varphi \in C(\partial\Omega)$. Then, there exists a solution $u \in C^{2+\alpha}_{loc}(\Omega)$ of $Lu = f$ in $\Omega$ such that $\lim_{z\to z_0} u(z) = \varphi z_0$ for every $L$-regular point $z_0 \in \partial\Omega$.*

Geometric properties of the boundary determine the continuous assumption of boundary values. In the paper [79] the author introduce an exterior cone type condition which extends the classical Zaremba criteria for the regularity of the boundary points and a boundary condition for the Kolmogorov operator in $\mathbb{R}^3$ proved in [89]. Besides, a geometric condition ensures the regularity for the $L$-characteristic boundary point, when the Fichera function $X_0 \nu z_0$ is positive definite.

Related results on the regularity of boundary points for the Dirichlet problem are also proved in [77, 102–104].

## 4.2 A Global Regularity Result in $L^p$ Spaces

We conclude this section with $L^p$-regularity results on the whole space for degenerate Ornstein-Uhlenbeck operators obtained by Bramanti et al. in [29] (constant coefficients) and [31] (variable coefficients).

The class of operators considered in [29] is

$$\mathscr{A} = \sum_{i,j=1}^{p_0} a_{ij} X_i X_j + X_0,$$

where $1 \le p_0 \le N,\ X_i = \partial_{x_i}$ and $X_0 = \sum_{i,j=1}^{N} b_{ij} x_i \partial_{x_j}$. Here $A = (a_{ij})_{1\le i,j\le p_0}$ and $B^t = (b_{ij})_{1\le i,j\le N}$ are constant coefficient matrices. Moreover, $B^t$ has the structure described in (14) and $A$ satisfies the ellipticity assumption (2), with $p_0$ in place of $m$.

The evolution operator corresponding to $\mathscr{A}$, that is

$$L = \mathscr{A} - \partial_t,$$

is a Kolmogorov-Fokker-Planck ultraparabolic operator, studied in Sect. 2.1. For this operator, $L^p$ global estimates on the strip $S = \mathbb{R}^N \times [-1, 1]$ have been proved in [29].

**Theorem 4.3** *For every $p \in (1, \infty)$ there exists a constant $c > 0$ such that*

$$\left\| \partial^2_{x_i x_j} u \right\|_{L^p(S)} \le c \, \|Lu\|_{L^p(S)} \ \ \textit{for } i,j = 1, 2, \ldots, p_0, \tag{37}$$

*for every* $u \in C_0^\infty(S)$. *The constant c depends on* $p, N, p_0$, *the matrix B and the number* $\lambda$ *in (2).*

As a by-product of the above result, global $L^p(\mathbb{R}^N)$ estimates are deduced for the operator $\mathscr{A}$.

**Theorem 4.4** *For every* $p \in (1,\infty)$ *there exists a constant* $c > 0$, *such that for every* $u \in C_0^\infty(\mathbb{R}^N)$ *one has:*

$$\left\| \partial^2_{x_i x_j} u \right\|_{L^p(\mathbb{R}^N)} \leq c \left\{ \|\mathscr{A} u\|_{L^p(\mathbb{R}^N)} + \|u\|_{L^p(\mathbb{R}^N)} \right\} \text{ for } i,j = 1,2,\dots,p_0. \tag{38}$$

We point out that the estimates of [97] for these type of operators only allow to get local estimates in $L^p$, while the results presented here are global.

The same authors prove in [31] similar estimates in the case of variable coefficients $a_{ij}$, entries of the matrix $A$. Precisely, if $a_{ij}$ are uniformly continuous and bounded functions in $\mathbb{R}^N$, estimates analogous to (37) (with $S = \mathbb{R}^N \times [-T,T]$, for some $T > 0$) and (38) still hold true. The proofs of the results in [31] rely on a freezing argument, that allows to exploit results and techniques contained in [29], and useful estimates proved in [52, 76].

Let us describe now the general strategy of the proof of Theorem 4.3, as well as the main difficulties.

Since $B$ has the structure described in (14), with the $*$-blocks possibly not null, the operator $L$ is left invariant with respect to a suitable Lie group of translations, but, in general, is not homogeneous. A basic idea is that of linking the properties of $L$ to those of another operator of the same kind, which not only is left translation invariant, but is also homogeneous of degree 2 with respect to a family of dilations. Such an operator $L_0$ always exists under our assumptions by Lanconelli and Polidoro [76], and has been called "the principal part" of $L$. Note that the operator $L_0$ fits the assumptions of Folland's theory [55]. The authors exploit the fact that, by results proved in [52], the operator $L$ has a fundamental solution $\Gamma$ with some good properties. First of all, $\Gamma$ is translation invariant and has a fast decay at infinity, in space; this allows to reduce the desired $L^p$ estimates to estimates of a singular integral operator whose kernel vanishes far off the pole. Second, this singular kernel, which has the form $\eta \cdot \partial^2_{x_i x_j}\Gamma$ where $\eta$ is a radial cutoff function, satisfies "standard estimates" (in the language of singular integrals theory) with respect to a suitable "local quasisymmetric quasidistance" $d$, which is a key geometrical object in the paper under consideration. Namely,

$$d(z,\zeta) = \left\|\zeta^{-1} \circ z\right\| \tag{39}$$

where $\zeta^{-1} \circ z$ is the Lie group operation related to the operator $L$, while $\|\cdot\|$ is a homogeneous norm related to the principal part operator $L_0$ (recall that $L$ does not have an associated family of dilations, and therefore does not have a natural homogeneous norm). This "hybrid" quasidistance does not fulfill enough good properties in order to apply the standard theory of "singular integrals in spaces of homogeneous type" (in the sense of Coifman and Weiss [46]). Hence, an ad hoc theory of singular integrals in nonhomogeneous spaces (see [27]) and a nontrivial covering argument are applied to get the desired $L^p$ bound.

## 5 Non Linear Curvature Equations

We conclude this review studying non linear PDE's. The standard prototype of non linear equations have always been minimal surfaces and curvature equations. Also in the setting of CR manifolds and subriemannian spaces, curvature equations can be chosen as the prototype of non linear equations. These equations describe the curvature or the evolution of a graph, with respect to vector fields, (or a metric) dependent on the graph itself. This is why curvature equation in this setting can be expressed in the form (1), where the coefficients $\sigma_{ij} = \sigma_{ij}(u, X_i u)$ of the vector fields depend on the solution or its intrinsic derivatives. Equations of this type naturally arise while studying curvature equations in CR manifolds, called Levi Equation [39, 84], Monge-Ampère equation [96, 98] or minimal graphs in the Heisenberg group (see for instance [34, 50, 86]), as well as in mathematical finance [43, 53].

Here we will focus in particular on Levi equations, for which much of the technique has been developed.

### *5.1 Regularity Results for the Levi Equations*

The Levi curvatures of a graph is the formal complex analogous of the curvature operator in $\mathbb{R}^N$. Namely it is the determinant of Levi form of a real hypersurface in $\mathbb{C}^{n+1}$ (or elementary symmetric functions of it). We can always assume that the surface $M$ is the graph of a $C^2$ function $u : \Omega \to \mathbb{R}$, where $\Omega \subseteq \mathbb{R}^{2n+1}$ is open. We identify $\mathbb{R}^{2n+1} \times \mathbb{R}$ with $\mathbb{C}^{n+1}$, and denote

$$z = (x, y) = (z_1, \dots, z_n, z_{n+1}), \; z_j = x_{2j-1} + ix_{2j}, 1 \le j \le n, \;\; z_{n+1} = x_{2n+1} + iy.$$

We let

$$\gamma(u) = \{(x, y) \in \Omega \times \mathbb{R} \, : \, y = u(x)\} \equiv \text{ graph of } u.$$

Calling $f(x, y) = y - u(x)$, the Levi form associated with $f$ at the point $p = (x, u(x))$ is the following Hermitian form:

$$L_p(u, \zeta) = \sum_{j,k=1}^{n+1} f_{j,\bar{k}}(p)\zeta_j \bar{\zeta}_k, \qquad \zeta \in T_p^{\mathbb{C}}(\gamma(u)),$$

where $T_p^{\mathbb{C}}\gamma(u)$ denotes the complex tangent space to the graph of $u$. If we denote by $h$ the complexified second fundamental form, it turns our that $h_p(\zeta,\bar{\zeta}) = \frac{1}{|\partial f(p)|}L_p(u,\zeta)$ for all $\zeta \in T_p^{\mathbb{C}}(\gamma(u))$. Let $\lambda_1(p),\ldots,\lambda_n(p)$ be the eigenvalues of $h$. For $1 \le m \le n$, $\sigma^{(m)}$ denotes the $m$-th elementary symmetric function and

$$K_p^{(m)}(\partial D) := \frac{1}{\binom{n}{m}}\sigma^{(m)}(\lambda_1,\ldots,\lambda_n),$$

we define the *m-th Levi curvature operator* as

$$\mathscr{L}^{(m)}(u)(x) := K_p^{(m)}(\gamma(u)), \quad x \in \Omega,$$

(see the papers by Bedford and Gaveau [7], by Tomassini [101], and by Lanconelli and Montanari [84]).

The Levi form has been introduced by E.E. Levi and used by Oka, Bremmerman and Norgouet in order to characterize domains of holomorphy. The first existence results were obtained in the Levi flat case, i.e. null Levi form, by Bedford and Gaveau [7] and by Bedford and Klingenberg [8]. They used a purely geometric approach, which does not work in the non Levi flat case. Slodkowski and Tomassini in [100] introduced a PDE's approach in studying boundary value problems for the prescribed Levi curvature equation with curvature different from zero at any point and proved $L^\infty$ a-priori bound for the gradient. However the degeneracy of the equation did not allow the mentioned authors to obtain internal regularity with standard instruments. Almost 10 years later the work by Slodkowski and Tomassini, in [39] G. Citti recasted the problem in dimension $n = 1$ in the set of sum of squares of vector fields. Precisely, choosing the coefficients of the vector fields $X_i = \sigma_{ij}\partial_{x_j}$ of (1) as $\sigma_{ij} = \delta_{ij}$ for $i = 1, 2$,

$$\sigma_{13}(Du) = \frac{u_{x_2} - u_{x_1}u_{x_3}}{1+u_{x_3}^2}, \qquad \sigma_{23}(Du) = -\frac{u_{x_1} + u_{x_2}u_{x_3}}{1+u_{x_3}^2}$$

and the Levi Curvature operator for $n = 1$ can be expressed as

$$\mathscr{L}^{(1)}u = (X_1^2 u + X_2^2 u)\,(1 + u_{x_3}^2) \text{ and } [X_1\,,\,X_2] = -\frac{\mathscr{L}^{(1)}u}{1+u_{x_3}^2}\,\partial_{x_3}.$$

This representation tells us that, while prescribing the curvature, we can control the rank of the Lie algebra generated by the vector fields, allowing to apply to the equation the theory of subriemannian operators. E. Lanconelli and A. Montanari studied the problem in full generality (see [84]) proving that $\mathscr{L}^{(m)}$ can be written as follows:

$$\mathscr{L}^{(m)}(u)(x) = \sum_{j,k=1}^{2n} a_{j,k}Z_jZ_k(u), \quad u \in C^2(\Omega,\mathbb{R}), \quad \Omega \subset \mathbb{R}^{2n+1}$$

where

- $Z_j = \partial_{x_j} + a_j \partial_{x_{2n+1}}, \quad a_j = a_j(Du), \ j = 1, \cdots, 2n.$
- $(a_{j,k})_{j,k=1,\dots,2n}$ is symmetric and $a_{j,k} = a_{j,k}(Du, D^2 u)$.

Then, if $j \neq k$, we have : $Z_{j,k} := [Z_j, Z_k] = q_{j,k}\, \partial_{x_{2n+1}}$. When computed on $m$-strictly pseudoconvex functions, i.e., on functions satisfying $\mathscr{L}^{(k)}(u)(x) > 0$ for every $x \in \Omega$ and $1 \leq k \leq m$, the operator $\mathscr{L}^{(m)}$ displays a *subelliptic property*. Precisely:

- For every $x \in \Omega, \ q_{j,k}(x) \neq 0$ for suitable $j, k$.
- The matrix $(a_{j,k}(x))_{j,k=1,\dots,2n}$ is strictly positive definite at any point $x \in \Omega$.

Therefore, if $u$ is $m$-strictly pseudoconvex, $\mathscr{L}^{(m)}$ is elliptic only along the $2n$ linearly independent directions $Z_j = \partial_{x_j} + a_j \partial_{x_{2n+1}} \equiv e_j + a_j e_{n+1}, j = 1, \dots, 2n$, and the *missing ellipticity direction* $e_{2n+1}$ is recovered by commutation. This commutation property can be restated as follows:

$$\dim(\mathrm{span}\{Z_j(x), Z_{j,k}(x) \ : \ j,k = 1, \dots, 2n\}) = 2n + 1, \quad \text{for every } x \in \Omega.$$

We would also like to stress that $\mathscr{L}^{(m)}$ is a PDO in $\mathbb{R}^{2n+1}$, which is fully nonlinear if $n > 1$.

From the subelliptic properties of $\mathscr{L}^{(m)}$ several crucial results follow. Here we only mention a Strong Comparison Principle and a regularity result.

**Theorem 5.1 (Strong Comparison Principle)** *Let $u, v : \Omega \to \mathbb{R}$, where $\Omega \subseteq \mathbb{R}^{2n+1}$ is open and connected. Assume $u$ and $v$ strictly $m$-pseudoconvex and*

*(i) $u \leq v$ in $\Omega$, $u(x_0) = v(x_0)$ at $x_0 \in \Omega$.*
*(ii) $\mathscr{L}^{(m)}(u) \geq \mathscr{L}^{(m)}(v)$ in $\Omega$.*

*Then $u = v$ in $\Omega$ (see [39] for $n = 1$, [84] for the general case).*

**Theorem 5.2 (Smoothness of Classical Solutions)** *Let $u \in C^{2,\alpha}(\Omega)$ be a strictly $m$-pseudoconvex solution to the $K$-prescribed Levi curvature equation*

$$\mathscr{L}^{(m)}(u) = K(\cdot, u) \quad in \ \Omega.$$

*If $K$ is strictly positive and $C^\infty$ in its domain, then $u \in C^\infty(\Omega)$ (see [40] for $n = 1$, [83] for $1 \leq m$ and [81] for the general case $1 \leq m \leq n$).*

*Strategy of the Proof for n=1.* In the low dimensional case, the proof of regularity is based on an ad hoc approximation method, similar to the parametrix method (see Sect. 2). The difficult here is the fact that the approximation has to be applied to the vector fields, not to the metric of the space. Following [97], the approximating vector fields $X_{i,w}$ of $X_i$ are obtained via Taylor approximation. The additional difficulty here is due to the fact that a function differentiable in the direction of the vectors $X_i$ will not necessarily be differentiable in the direction $X_{i,w}$. We explicitly note that not even the more recent results of Bramanti et al. [32] could allow to obtain the result.

On the contrary a completely new approach to singular integrals has been introduced in order to deal with these non linear vector fields.

*Strategy of the Proof in Higher Dimension.* Since the prescribed Levi curvature equations present formal similarities with the real and complex Monge-Ampère equations, which are elliptic PDE's if evaluated on strictly convex and plurisubharmonic functions, respectively, we would like to briefly recall how the smoothness follows from the classical Schauder theory for the real Monge-Ampère equation. The real Monge-Ampère equation in a domain $\Omega \subset \mathbb{R}^n$ is of the form $\det(D^2u) = f(x, u, Du)$. If $u \in C^{2,\alpha}(\Omega)$ is a strictly convex solution to this equation, then the linearized operator $L$ (at u) is elliptic with $C^\alpha$ coefficients, and $Du$ satisfies a linear uniformly elliptic equation of the type $L(Du) = F \in C^\alpha(\Omega)$. By the classical Schauder theory, $Du \in C^{2,\alpha}(\Omega)$. Repeating this argument one proves $u \in C^\infty(\Omega)$. In our case it is not possible to argue in the same way, because the Levi curvature equations are not elliptic at any point, also when restricted to the class of strictly pseudoconvex functions. However, in [82] Montanari proved interior Schauder-type estimates for solutions of $Hv = f$ with $H$ a linear second order subelliptic operator of the type $H = \sum_{m,j=1}^n h_{mj} Z_m Z_j$ with Hölder continuous coefficients and with $Z_j$ first order partial differential operator with $C^{1,\alpha}$ coefficients. This result is obtained by a non standard freezing method and on the lifting argument by Rothshild and Stein. The study of the operator $H$ is reduced to the analysis of a family of left invariant operators on a free nilpotent Lie group, whose fundamental solutions are used a parametrix of the operator $H$, and provides an explicit representation formula for solutions of the linear equation $Hv = f$. Once this is established, the strategy to handle the prescribed $m$-th Levi curvature equation in higher dimension is to apply the a priori estimates in [82] to first order Euclidean difference quotients of a strictly $m$-pseudoconvex solution $u$, in order to prove that the function $Du$ has Hölder continuous second order horizontal derivatives. The smoothness result is then obtain by a bootstrap argument.

## 5.2 A Negative Regularity Result

We want to stress that, in dimension $n > 1$, the classical $C^{2,\alpha}$ solvability of the Dirichlet problem for the $K$-prescribed Levi curvature equations is still a widely open problem. Even though it is possible to give a definition of Lipschitz continuous viscosity solutions (we refer [83]), these solutions are not expected to be smooth if the data are smooth. Indeed, very recently, Gutierrez et al. [62] proved the following negative regularity result. To state the theorem, we need some more notation. With $B_r$ we denote the Euclidean ball in $\mathbb{R}^{2n+1}$ centered at the origin and with radius $r$, $K$ denote a function of class $C^\infty$ defined on the ball $(B_1 \times \mathbb{R})$, strictly positive and such that $s \mapsto K(\cdot, s)$ is increasing. Then, we have the following result.

**Theorem 5.3 (Gutierrez et al. [62])** *There exist $r \in (0,1)$ and a pseudoconvex function $u \in Lip(\bar{B}_r)$ solving*

$$\mathscr{L}^{(n)}(u) = K(x,u) \quad in \quad B_r,$$

*in the weak viscosity sense and such that*

- $u \notin C^1(B_r)$ *if* $n = 2$.
- $u \notin C^{1,\beta}$ *for any* $\beta > 1 - \frac{2}{n}$ *when* $n > 2$.

This equation is the motivation for a number of interesting problems: symmetry problems and isoperimetric integral inequalities [80] of surfaces with prescribed Levi curvature, and regularity results of radially symmetric solutions.

## References


1. Abbondanza, B., Bonfiglioli, A.: On the Dirichlet problem and the inverse mean value theorem for a class of divergence form operators. J. Lond. Math. Soc. **87**, 321–346 (2013)
2. Armitage, D.H., Gardiner, S.J.: Classical Potential Theory. Springer Monographs in Mathematics. Springer, London (2001)
3. August, J., Zucker, S.: Sketches with curvature: the curve indicator random field and markov processes. IEEE Trans. Pattern Anal. Mach. Intell. **25**(4), 387–400 (2003)
4. Barucci, E., Polidoro, S., Vespri, V.: Some results on partial differential equations and Asian options. Math. Models Methods Appl. Sci. **11**,475–497 (2001)
5. Battaglia, E., Bonfiglioli, A.: Normal families of functions for subelliptic operators and the theorems of Montel and Koebe. J. Math. Anal. Appl. **409**, 1–12 (2014)
6. Beckenbach, E.F., Radó, T.: Subharmonic functions and minimal surfaces. Trans. Am. Math. Soc. **35**, 648–661 (1933)
7. Bedford, E., Gaveau, B.: Hypersurfaces with bounded Levi form. Indiana Univ. J. **27**(5), 867–873 (1978)
8. Bedford, E., Klingenberg, W.: On the envelope of holomorphy of a 2-sphere in $\mathbb{C}^2$. J. Am. Math. Soc. **4**(3), 623–646 (1991)
9. Blaschke, W.: Ein Mittelwertsatz und eine kennzeichnende Eigenschaft des logarithmischen Potentials. Leipz. Ber. **68**, 3–7 (1916)
10. Black, F., Scholes, M.: The pricing of options and corporate liabilities. J. Polit. Econ. **81**, 637–654 (1973)
11. Bonfiglioli, A., Lanconelli, E.: Liouville-type theorems for real sub-Laplacians. Manuscripta Math. **105**, 111–124 (2001)
12. Bonfiglioli, A., Lanconelli, E.: Maximum principle on unbounded domains for sub-Laplacians: a potential theory approach. Proc. Am. Math. Soc. **130**, 2295–2304 (2002)
13. Bonfiglioli, A., Lanconelli, E.: Subharmonic functions on Carnot groups. Math. Ann. **325**, 97–122 (2003)
14. Bonfiglioli, A., Lanconelli, E.: Dirichlet problem with $L^p$-boundary data in contractible domains of Carnot groups. Ann. Sc. Norm. Super. Pisa Cl. Sci. **5**, 579–610 (2006)
15. Bonfiglioli, A., Lanconelli, E.: Gauge functions, Eikonal equations and Bôcher's theorem on stratified Lie groups. Calc. Var. Partial Differ. Equ. **30**, 277–291 (2007)
16. Bonfiglioli, A., Lanconelli, E.: A new characterization of convexity in free Carnot groups. Proc. Am. Math. Soc. **140**, 3263–3273 (2012)

17. Bonfiglioli, A., Lanconelli, E.: Subharmonic functions in sub-Riemannian settings. J. Eur. Math. Soc. **15**, 387–441 (2013)
18. Bonfiglioli, A., Uguzzoni, F.: Families of diffeomorphic sub-Laplacians and free Carnot groups. Forum Math. **16**, 403–415 (2004)
19. Bonfiglioli, A., Uguzzoni, F.: A note on lifting of Carnot groups. Rev. Mat. Iberoam. **21**(3), 1013–1035 (2005)
20. Bonfiglioli, A., Uguzzoni, F.: Harnack inequality for non-divergence form operators on stratified groups. Trans. Am. Math. Soc. **359**(6), 2463–2481 (2007)
21. Bonfiglioli, A., Lanconelli, E., Uguzzoni, F.: Uniform Gaussian estimates of the fundamental solutions for heat operators on Carnot groups. Adv. Differ. Equ. **7**, 1153–1192 (2002)
22. Bonfiglioli, A., Lanconelli, E., Uguzzoni, F.: Fundamental solutions for non-divergence form operators on stratified groups. Trans. Am. Math. Soc. **356**(7), 2709–2737 (2004)
23. Bonfiglioli, A., Lanconelli, E., Uguzzoni, F.: Stratified Lie Groups and Potential Theory for Their Sub-Laplacians. Springer Monographs in Mathematics, vol. 26. New York, Springer (2007)
24. Bonfiglioli, A., Lanconelli, E., Tommasoli, A.: Convexity of average operators for subsolutions to subelliptic equations. Anal. PDE **7**(2), 345–373 (2014). doi:10.2140/apde.2014.7.345
25. Bony, J.M.: Principe du maximum, inégalité de Harnack et unicité du problème de Cauchy pour les opérateurs elliptiques dégénérés. Ann. Inst. Fourier **19**, 277–304 (1969)
26. Boscain, U., Polidoro, S.: Gaussian estimates for hypoelliptic operators via optimal control. Rend. Lincei Mat. Appl. **18**, 343–349 (2007)
27. Bramanti, M.: Singular integrals in nonhomogeneous spaces: $L^2$ and $L^p$ continuity from Hölder estimates. Rev. Mat. Iberoam. **26**(1), 347–366 (2010)
28. Bramanti, M., Brandolini, L., Lanconelli, E., Uguzzoni, F.: Non-divergence equations structured on Hörmander vector fields: heat kernels and Harnack inequalities. Mem. Am. Math. Soc. **204**(961), vi+123 pp. (2010)
29. Bramanti, M., Cupini, G., Lanconelli, E., Priola, E.: Global $L^p$ estimates for degenerate Ornstein-Uhlenbeck operators. Math. Z. **266**(4), 789–816 (2010)
30. Bramanti, M., Brandolini, L., Pedroni, M.: Basic properties of nonsmooth Hörmander's vector fields and Poincaré's inequality. Forum Math. **25**(4), 703–769 (2013)
31. Bramanti, M., Cupini, G., Lanconelli, E.: Global $L^p$ estimates for degenerate Ornstein-Uhlenbeck operators with variable coefficients Math. Nach. **286**, 1087–1101 (2013)
32. Bramanti, M., Brandolini, L., Manfredini, M., Pedroni, M.: Fundamental solutions and local solvability for nonsmooth Hïmander's operators. Mem. Amer. Math Soc.
33. Capogna, L., Danielli, D., Garofalo, N.: Subelliptic mollifiers and a basic pointwise estimate of Poincaré type. **226**(1), 147–154 (1997)
34. Capogna, L., Citti, G., Manfredini, M.: Regularity of non-characteristic minimal graphs in the Heisenberg group H1. Indiana Univ. Math. J. **58**(5), 2115–2160 (2009)
35. Capogna, L., Citti, G., Manfredini, M.: Uniform Gaussian bounds for sub elliptic heat kernels and an application to the total variation flow of graphs over Carnot groups. Anal. Geom. Metric Spaces **1**, 233–234 (2013)
36. Chapman, S., Cowling, T.G.: The Mathematical Theory of Nonuniform Gases, 3rd edn. Cambridge University Press, Cambridge (1990)
37. Cinti, C., Polidoro, S.: Pointwise local estimates and Gaussian upper bounds for a class of uniformly subelliptic ultraparabolic operators. J. Math. Anal. Appl. **338**, 946–969 (2008)
38. Cinti, C., Menozzi, S., Polidoro, S.: Two-sided bounds for degenerate processes with densities supported in subsets of $R^N$. Potential Anal. **42**, 39–98 (2015)
39. Citti, G.: A comparison theorem for the Levi equation. Rend. Mat. Accad. Lincei **4**, 207–212 (1993)
40. Citti, G.: $C^\infty$ regularity of solutions of the Levi equation. Ann. Inst. H. Poincaré Anal. Non Linéaire **15**, 517–534 (1998)
41. Citti, G., Sarti, A.: A cortical based model of perceptual completion in the roto-translation space. J. Math. Imaging Vis. **24**(3), 307–326 (2006)

42. Citti, G., Garofalo, N., Lanconelli, E.: Harnack's inequality for sum of squares of vector fields plus a potential. Am. J. Math. **115**, 699–734 (1993)
43. Citti, G., Pascucci, A., Polidoro, S.: On the regularity of solutions to a nonlinear ultraparabolic equation arising in mathematical finance. Differ. Integral Equ. **14**(6), 701–738 (2001)
44. Citti, G., Lanconelli, E., Montanari, A.: Smoothness of Lipschitz continuous graphs with non vanishing Levi curvature. Acta Math. **188**(1), 87–128 (2002)
45. Citti, G., Manfredini, M., Serra Cassano, F., Pinamonti, A.: Poincaré-type inequality for intrinsic Lipschitz continuous vector fields in the Heisenberg group. J. Math. Pures Appl.
46. Coifman, R., Weiss, G.: Analyse Harmonique Non-Commutative sur Certains Espaces Homogenes. Lecture Notes in Mathematics, vol. 242. Springer, Berlin (1971)
47. Constantinescu, C., Cornea, A.: Potential Theory on Harmonic Spaces. Springer, Berlin (1972)
48. Corielli, F., Foschi, P., Pascucci, A.: Parametrix approximation of diffusion transition densities. SIAM J. Financ. Math. **1**, 833–867 (2010)
49. Da Lio, F., Montanari, A.: Existence and uniqueness of Lipschitz continuous graphs with prescribed Levi curvature. Ann. Inst. H. Poincaré Anal. Non Linéaire **23**(1), 1–28 (2006)
50. Danielli, D., Garofalo, N., Nhieu, D.M., Pauls, S.D.: Instability of graphical strips and a positive answer to the Bernstein problem in the Heisenberg group H1. J. Differ. Geom. **81**, 251–295 (2009)
51. Di Francesco, M., Pascucci, A.: On a class of degenerate parabolic equations of Kolmogorov type. AMRX Appl. Math. Res. Express **3**, 77–116 (2005)
52. Di Francesco, M., Polidoro, S.: Schauder estimates, Harnack inequality and Gaussian lower bound for Kolmogorov type operators in non-divergence form. Adv. Differ. Equ. **11**(11), 1261–1320 (2006)
53. Escobedo, M., Vazquez, J.L., Zuazua, E.: Entropy solutions for diusion-convection equations with partial diffusivity. Trans. Am. Math. Soc., **343**, 829–842 (1994)
54. Fabes, E.B., Garofalo, N.: Mean value properties of solutions to parabolic equations with variable coefficients. J. Math. Anal. Appl. **121**(2), 305–316 (1987)
55. Folland, G.B.: Subelliptic estimates and function spaces on nilpotent Lie groups. Ark. Mat. **13**, 161–207 (1975)
56. Foschi, P., Pascucci, A.: Calibration of a path-dependent volatility model: empirical tests. Comput. Stat. Data Anal. **53**, 2219–2235 (2009)
57. Franchi, B., Lanconelli, E.: Une métrique associée à une classe d'opérateurs elliptiques dégénérés. Rend. Sem. Mat. Univ. Politec. Torino, Special Issue, 105–114 (1984). Conference on Linear Partial and Pseudodifferential Operators, Torino (1982)
58. Franchi, B., Lanconelli, E.: Hölder regularity theorem for a class of linear nonuniformly elliptic operators with measurable coefficients. Ann. Scuola Norm. Sup. Pisa Cl. Sci. (4) **10**(4), 523–541 (1983)
59. Franchi, B., Lu, G., Wheeden, R.L.: A relationship between Poincaré type inequalities and representation formulas in spaces of homogeneous type. Int. Math. Res. Not. **1**, 1–14 (1996)
60. Garofalo, N., Lanconelli, E.: Wiener's criterion for parabolic equations with variable coefficients and its consequences. Trans. Am. Math. Soc. **308**, 811–836 (1988)
61. Garofalo, N., Lanconelli, E.: Asymptotic behavior of fundamental solutions and potential theory of parabolic operator with variable coefficients. Math. Ann. **283**, 211–239 (1989)
62. Gutierrez, C., Lanconelli, E., Montanari, A.: Nonsmooth Hypersurfaces with Smooth Levi Curvature. Nonlinear Anal. Theory Methods Appl. **76**, 115–121 (2013)
63. Hobson, D.G., Rogers, L.C.G.: Complete models with stochastic volatility. Math. Financ. **8**, 27–48 (1998)
64. Hoh, W., Jacob, N.: Remarks on mean value properties of solutions of second order differential operators. Expo. Math. **9**, 367–377 (1991)
65. Hörmander, L.: Hypoelliptic second order differential equations. Acta Math. **119**, 147–171 (1967)
66. Jerison, D.S., Sánchez-Calle, A.: Estimates for the heat kernel for a sum of squares of vector fields. Indiana Univ. Math. J. **35**(4), 835–854 (1986)

67. Kogoj, A., Lanconelli, E.: An invariant Harnack inequality for a class of hypoelliptic ultraparabolic equations. Mediterr. J. Math. **1**, 51–80 (2004)
68. Kogoj, A., Tralli, G.: Blaschke, Privaloff, Reade and Saks Theorems for diffusion equations on Lie groups. Potential Anal. **38**(4), 1103–1122 (2013)
69. Kogoj, A., Lanconelli, E., Tralli, G.: An inverse mean value property for evolution equations. Adv. Differ. Equ. **19**(7–8), 783–804 (2014)
70. Kolmogorov, A.: Zufllige Bewegungen. (Zur Theorie der Brownschen Bewegung.). Ann. Math. II. **35**, 116–117 (1934)
71. Kusuoka, S., Stroock, D.: Long time estimates for the heat kernel associated with a uniformly subelliptic symmetric second order operator. Ann. Math. (2) **127**(1), 165–189 (1988)
72. Lanconelli, E.: Potato Kugel for Sub-Laplacians. Isr. J. Math. **194**, 277–283 (2013)
73. Lanconelli, E., Morbidelli, D.: On the Poincaré inequality for vector fields. Ark. Mat. **38**(2), 327–342 (2000)
74. Lanconelli, E., Pascucci, A.: On the fundamental solution for hypoelliptic second order partial differential equations with non-negative characteristic form. Ricerche Mat. **48**(1), 81–106, 0035–5038 (1999)
75. Lanconelli, E., Pascucci, A.: Superparabolic functions related to second order hypoelliptic operators. Potential Anal. **11**, 303–323 (1999)
76. Lanconelli, E., Polidoro, S.: On a class of hypoelliptic evolution operators. Rend. Semin. Mat. Univ. Politecnico Torino **52**, 29–63 (1994)
77. Lanconelli, E., Uguzzoni, F.: Potential analysis for a class of diffusion equations: a Gaussian bounds approach. J. Differ. Equ. **248**, 2329–2367 (2010)
78. Lunardi, A.: Schauder estimates for a class of parabolic operators with unbounded coefficients in $R^n$. Ann. Scuola Norm. Sup. Pisa (4) **24**(1), 133–164 (1997)
79. Manfredini, M.: The Dirichlet problem for a class of ultraparabolic equations. Adv. Differ. Equ. **2**, 831–866 (1997)
80. Martino, V., Montanari, A.: Integral formulas for a class of curvature PDE's and applications to isoperimetric inequalities and to symmetry problems. Forum Math. **22**, 253–265 (2010)
81. Montanari, A.: On the regularity of solutions of the prescribed Levi curvature equation in several complex variables. Nonlinear elliptic and parabolic equations and systems (Pisa, 2002). Commun. Appl. Nonlinear Anal. **10**(2), 63–71 (2003)
82. Montanari, A.: Hölder a priori estimates for second order tangential operators on CR manifolds. Ann. Scuola Norm. Sup. Pisa Cl. Sci (5) **II**, 345–378 (2003)
83. Montanari, A., Lascialfari, F.: The Levi Monge-Ampère equation: smooth regularity of strictly Levi convex solutions. J. Geom. Anal. **14**(2), 331–353 (2004)
84. Montanari, A., Lanconelli, E.: Pseudoconvex fully nonlinear partial differential operators: strong comparison theorems. J. Differ. Equ. **202**(2), 306–331 (2004)
85. Montanari, A., Morbidelli, D.: Step-$s$ involutive families of vector fields, their orbits and the Poincaré inequality. J. Math. Pures Appl. (9) **99**(4), 375–394 (2013)
86. Monti, R., Serra Cassano, F., Vittone, D.: A negative answer to the Bernstein problem for intrinsic graphs in the Heisenberg group. Boll. Unione Mat. Ital. **9**, 709–727 (2009)
87. Mumford, D.: Elastica and computer vision. In: Algebraic Geometry and Its Applications (West Lafayette, IN, 1990), pp. 491–506. Springer, New York (1994)
88. Nagel, A., Stein, E.M., Wainger, S.: Balls and metrics defined by vector fields I: basic properties. Acta Math. **155**, 130–147 (1985)
89. Negrini, P., Scornazzani, V.: Superharmonic functions and regularity of boundary points for a class of elliptic parabolic partial differential operators. Boll. Unione Mat. Ital. (6) **3**, 85–107 (1984)
90. Pascucci, A.: PDE and Martingale Methods in Option Pricing. In: Bocconi Springer Series, vol. 2. Springer, Milan/Bocconi University Press, Milan (2011)
91. Pascucci, A., Polidoro, S.: Harnack inequalities and Gaussian estimates for a class of hypoelliptic operators. Trans. Am. Math. Soc. **358**(11), 4873–4893 (2006)
92. Pini, B.: Sulle equazioni a derivate parziali lineari del secondo ordine in due variabili di tipo parabolico. Ann. Mat. Pura Appl. **32**, 179–204 (1951)

93. Privaloff, I.: Sur le fonctions harmoniques. Rec. Math. Moscou (Mat. Sbornik) **32**, 464–471 (1925)
94. Polidoro, S.: A global lower bound for the fundamental solution of Kolmogorov-Fokker-Planck equations. Arch. Ration. Mech. Anal. **137**, 321–340 (1997)
95. Polidoro, S.: On a class of ultraparabolic operators of Kolmogorov-Fokker-Planck type. Matematiche (Catania) **49**, 53–105 (1994)
96. Rios, C., Sawyer, E.T., Wheeden, R.L.: Regularity of subelliptic Monge-Ampre equations. Adv. Math. **217**(3), 967–1026 (2008)
97. Rothschild, L., Stein, E.M.: Hypoelliptic differential operators and nilpotent groups. Acta Math. **137**, 247–320 (1976)
98. Sawyer, E.T., Wheeden, R.L.: Regularity of degenerate Monge-Ampère and prescribed Gaussian curvature equations in two dimensions. Potential Anal. **24**(3), 267–301 (2006)
99. Shatyro, Y.I.: Smoothness of solutions of certain singular second order equations. Mat. Zametki **10**, 101–111 (1971); English Translation Math. Notes **10**, 484–489 (1971)
100. Slodkowski, Z., Tomassini, G.: Weak solutions for the Levi equation and envelope of holomorphy. J. Funct. Anal. **101**, 392–407 (1991)
101. Tomassini, G.: Geometric properties of solutions of the Levi equation. Ann. Mat. Pura Appl. **152**(4), 331–344 (1988)
102. Tralli, G., Uguzzoni, F.: Wiener criterion for X-elliptic operators. Arxiv 1408.6789
103. Uguzzoni, F.: Cone criterion for non-divergence equations modeled on Hörmander vector fields. In: Subelliptic PDE's and Applications to Geometry and Finance. Lect. Notes Semin. Interdiscip. Mat, vol. 6, pp. 227–241. Semin. Interdiscip. Mat. (S.I.M.), Potenza (2007)
104. Uguzzoni, F.: Estimates of the Green function for X-elliptic operators. Math. Ann. **361**(1–2), 169–190 (2014)
Math. Ann. (2014)
105. Varopoulos, N.Th., Saloff-Coste, L., Coulhon, T.: Analysis and geometry on groups. In: Cambridge Tracts in Mathematics, vol. 100. Cambridge University Press, Cambridge (1992)
106. Watson, N.A.: A theory of subtemperatures in several variables. Proc. Lond. Math. Soc. **26**, 385–417 (1973)
107. Williams, L., Jacobs, D.: Stochastic completion fields: a neural model of illusory contour shape and salience. In: Proceedings of ICCV, pp. 408–415 (1995)

MIX
Papier aus verantwortungsvollen Quellen
Paper from responsible sources
FSC® C105338

If you have any concerns about our products,
you can contact us on
**ProductSafety@springernature.com**

In case Publisher is established outside the EU,
the EU authorized representative is:
**Springer Nature Customer Service Center GmbH**
**Europaplatz 3, 69115 Heidelberg, Germany**

Printed by Libri Plureos GmbH
in Hamburg, Germany